MW00718649

WATER AND FLORIDA CITRUS:

Use, Regulation, Irrigation, Systems, and Management

Edited by Brian J. Boman

University of Florida
Institute of Food and Agricultural Sciences
July 2002

SP 281 Water and Florida Citrus: Use, Regulation, Irrigation, Systems, and Management
Edited by Brian Boman

© **Copyright July 2002 by the University of Florida, Institute of Food and Agricultural Sciences**
Printed in the United States of America. All rights reserved. Parts of this publication may be reproduced for educational purposes only. Please provide credit to the "University of Florida, Institute of Food and Agricultural Sciences," accompanied by the published date.

ISBN 0-916287-38-6

Produced by IFAS Communication Services
Editors: Ray Ford, Jr., Jennifer Stimson, Chana J. Bird, and Andee Cohen
Graphic Designers: Scott Burton, Joni Craig, Stephen Moree, and Tracy D. Zwillinger
Index by Linda Herr Hallinger

Use pesticides and water additives safely. Read and follow directions on the manufacturer's label. The use of trade names in this publication is solely for the purpose of providing specific information. It is not a guarantee or warranty of the products, and does not signify that they are approved to the exclusion of others of suitable composition.

IFAS-Extension Bookstore
Building 440, Mowry Road
PO Box 110011
Gainesville, FL 32611
352-392-1764 • 352-392-2628 (fax)
800-225-1764   VISA/MasterCard orders only
http://IFASbooks.ufl.edu

Cooperative Extension Service University of Florida, Institute of Food and Agricultural Sciences, Christine Taylor Waddill, Director, in cooperation with the United States Department of Agriculture, publishes this information to further the purpose of the May 8 and June 30, 1914 Acts of Congress; and is authorized to provide research, educational information and other services only to individuals and institutions that function without regard to race, color, age, sex, disability or national origin. Information about alternate format is available from IFAS Communication Services, University of Florida, P.O. Box 11810, Gainesville, FL 32611-0810. This information was published July 2002, as SP 281, Florida Cooperative Extension Service.

Dedication
This publication is dedicated to the memory of Dr. Allen Smajstrla, whose devotion to teaching, research, and extension laid the foundation for microirrigation of citrus in Florida. Dr. Smajstrla served on the faculty of the Agricultural and Biological Engineering Department at the University of Florida from 1977 until his passing in 1999. Allen was the leader of the Department's statewide soil and water engineering extension efforts, and provided leadership both within Florida and also on the national level. In 1987, Dr. Smajstrla was awarded the Presidential Gold Medal Award by the Florida State Horticultural Society as a recognition of the work on citrus irrigation published in the Citrus Section of the Society over the previous six years. Even though we recognize Dr. Smajstrla's many contributions to teaching, research, and extension in Florida, we remember Allen most as a friend with a wonderful sense of humor who touched everyone with whom he associated.

Contributing Authors

Howard Beck	University of Florida, Agricultural and Biological Engineering Dept.
Brian Boman	University of Florida, Indian River Research and Education Center
Scott Hall	University of Florida, Levin College of Law, Gainesville
Dorata Haman	University of Florida, Agricultural and Biological Engineering
Jack Hebb	University of Florida, Multi-County Citrus Agent, Ft. Pierce
Nigel Morris	AMS Engineering and Environmental, Inc., Punta Gorda, FL
Tom Obreza	University of Florida, Southwest FL Research and Education Center
Michael Olexa	University of Florida, Food Resources & Economics Dept., Gainesville
Esa Ontermaa	Agricultural Resource Associates, Inc., Alva, FL
Larry Parsons	University of Florida, Citrus Research and Education Center
Mark Ritenour	University of Florida, Indian River Research and Education Center
Steve Smith	Aqua Engineering, Inc. Fort Collins, CO
Ed Stover	University of Florida, Indian River Research and Education Center
Dave Tucker	University of Florida, Citrus Research and Education Center
Bill Tullos	Rain Bird Sales, Inc.
Vernon Vandiver, Jr.	University of Florida, Ft. Lauderdale Research and Education Center

Mark Wade   University of Florida, Indian River Research and Education Center
Chris Wilson   University of Florida, Indian River Research and Education Center
Wilfred Wardowski   University of Florida, Citrus Research and Education Center

Acknowledgments

Ann Workman and Marion Parsons are thanked for their work on preparation of the manuscript. Eva Squires, Ray Ford, Jr., Jennifer Stimson, and Chana J. Bird provided editorial review and guided the manuscript through the publication process. Special thanks to Joni Craig, Stephen Moree, and Tracy D. Zwillinger for the wonderful job they did in laying out the manuscript and getting it to print stage. The following individuals are gratefully acknowledged for the time and effort they spent reviewing the document and providing suggestions for improvement. Their input led to changes that improved the overall quality of the document and made it more user-friendly.

Ilan Bar	Jim Brigham	Ron Cohen
Steve Futch	Dorota Haman	Jeff Kreiger
Yuncong Li	Esa Ontermaa	Charlie Shinn
Harold Thompson	Adair Wheaton	Jesse Wilson

Disclaimer

The mention of a specific product or company is for information purposes only and does not constitute an endorsement or criticism of that product or company by the authors or the University of Florida, Institute of Food and Agricultural Sciences.

Funding from the following agencies to print this document is gratefully acknowledged:

South Florida Water Management District

Saint Johns River Water Management District

Southwest Florida Water Management District

Florida Department of Agriculture and Consumer Services

Table of Contents

Chapter 1. History of Florida Citrus Irrigation

by Dave Tucker, Brian Boman, and Larry Parsons

Introduction

Citrus originates from the wet tropics of Southeast Asia, but it has adapted to a range of conditions, ranging from the humid subtropics to desert environments. From the Orient, citrus moved westward—first to India, then the Mediterranean, and finally across the Atlantic. Citrus was brought to the Americas by Columbus in November 1493 when he brought citrus seeds that were planted on the island of Haiti. The first seeds planted on the mainland of the Americas were brought by the expedition of Juan de Grijalva when he landed in Central America in July 1518.

The exact date of the introduction of citrus trees into Florida is not known, but from a statement made by Pedro Menéndez, dated April 2, 1579, it appears that citrus fruits were grown in abundance around St. Augustine at that time. Early settlers in Florida some two centuries later found wild citrus trees scattered over the state. One of the oldest cultivated groves planted in Florida is thought to have been the Don Phillipe grove in Pinellas County. It was planted sometime between 1809 and 1820.

Citrus production in Florida reached an all-time high of more than 5 million boxes when the Great Freeze of 1894-95 almost wiped out the citrus industry. It was not until 1909-10 that this level was reached again (Fig. 1-1). Production increased at a steady pace, reaching nearly 154 million boxes during the 1961-62 season. Freezes dropped the total production to about 90 million boxes in the 1963-64 season. Production increased dramatically during the late 1960s and through the 1970s. Prior to the 1981 freeze, production had reached 283 million boxes/yr (1979-80). Even with the severe freezes in the 1980s, newer planting at higher densities in South Florida resulted in production reaching over 300 million boxes in the 1997-98 season.

Citrus acreage peaked at over 941,000 acres in 1970 (Fig. 1-2). Since that time the effects of freezes, tree decline, and urbanization have resulted in a significant loss of acreage. By the time

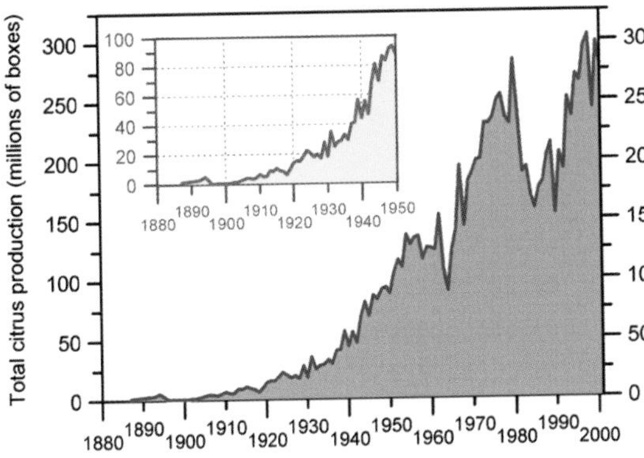

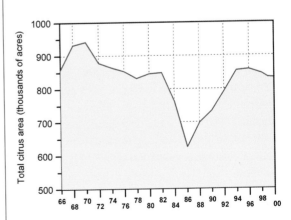

Figure 1-1. Citrus production (all varieties) from the 1888/89 season to the 2000/01 season.

Figure 1-2. Total Florida citrus acreage for 1966 through 2000.

1

of the 1985 freeze, the total citrus acreage had dropped to less than 625,000 acres. Since then, the acreage has climbed up to nearly 850,000 acres. However, due to higher density plantings and younger, more productive trees, overall production in the late 1990s reached all-time highs.

Irrigation

The Florida citrus belt is located in a subtropical climate zone that normally receives 48-59 inches of rainfall annually. This rainfall amount is adequate to meet the requirements of citrus production on the state's predominantly sandy, low water-holding capacity Ridge soils, and poorly drained, high water table flatwoods soils of coastal and southern Florida. However, its annual distribution does not usually satisfy peak seasonal demands during the critical fruit set and early fruit development phase corresponding to the stress-prone dry season of March through early June. During this period, good irrigation management is particularly critical to reduce stress and the associated young fruit drop and yield loss.

Irrigation is an important cultural factor in Florida's subtropical climate to maintain yields and fruit quality that would otherwise be reduced during short-term dry periods. Studies have shown that money spent for improved weed and pest control, pruning, and fertilization do not result in higher yields if irrigation practices are inferior. It is only when such improvements are combined with good irrigation management practices that the benefits can be fully realized.

History of Florida Citrus Irrigation

Early citrus production in Florida prior to the 1940s was accomplished essentially without supplemental watering, other than during tree establishment. Irrigation was not an established component of citrus production, but rather a hit-and-miss operation. Trees were watered to keep them out of a severe wilt and to keep production at acceptable levels. In their favor, early Ridge groves were often planted at wide spacings on rough lemon rootstock, under which conditions trees developed extensive root systems that foraged widely through the soil profile. However, without irrigation, acceptable levels of production are not possible under today's management systems using cultivars on deep and shallow soils, at closer tree spacings, and with more precise inputs of nutrients and other agrichemicals.

Historically, citrus has been irrigated using water wagons, surface flood, subirrigation, sprinkler, and microirrigation systems (Figs. 1-3 to 1-11). Sprinkler systems used to irrigate the deep sandy soils of Florida's Central Ridge included portable galvanized or aluminum perforated pipe. Later, permanent overhead sprinkler systems and portable guns and traveling guns became popular. The perforated pipe systems became less popular with time due to their high labor requirements, and only a few, if any, remain in operation today. Most of the Ridge citrus was planted on rolling hills; thus, flood irrigation was not practical.

On the inherently poorly drained flatwoods areas where citrus can only be grown with artificial drainage, field ditches or combinations of ditches and soil bedding were used to obtain the required surface and subsurface drainage. Typically, these areas have highly permeable surface soils, and surface flood systems require that the water table be raised on top of slowly permeable subsurface soil layers or on existing high water tables. Ditches are required to maintain or change water table levels in the tree root zone. With seepage (flood) irrigation, water is maintained at or just below the tree root zone to permit water to move into this area by capillary action. For flood irrigation, bedded groves are irrigated by filling field ditches to near the tops of beds, and the water table is maintained for a period of up to three days so the water can thoroughly infiltrate the root zone. Ditches are then drained by pumping excess water into adjacent groves via the system of canals.

In crown flood irrigation, enough water is pumped into the grove to fill the water furrows and allow water to overflow the tops of the beds. Instead of relying on capillary action and horizontal seepage, crown flood irrigation allows the water to infiltrate and move downward into the soil profile.

During the late 1940s and early 1950s, portable galvanized and aluminum perforated pipe systems were first utilized. Shortly before, and following the great freeze of 1962, perforated pipe systems gave way to mobile high-volume guns called "travelling guns" on the Ridge and also the permanent overhead sprinkler systems. These latter systems were widely installed throughout the Ridge growing area, particularly in the southern Ridge where growers were less impacted by the freeze and had capital to invest in such systems. These overhead systems were widely used for cold protection during the 1962 freeze (without research-based information), mostly with disastrous consequences. Due to the advective nature of the freeze and lack of enough volume, severe limb breakage and tree collapse occurred. Overhead irrigation during the freeze often resulted in such a large ice buildup on tree limbs that they collapsed from the sheer weight of the ice.

Evaporative cooling enhanced by the wind lowered the wetted tree temperature below air temperature and increased damage. However, as the central and northern Ridge areas recovered, these systems remained popular and continued to be installed on new acreage. As large new citrus developments moved to the Indian River and Southwest Florida flatwoods areas, new installations generally did not utilize permanent overhead systems due to the poor water quality.

Prior to the 1960s, when substantial commercial acreage already existed on the East Coast Indian River area, flood systems were fed from artesian wells located on relatively high ground so that water could be well distributed. When land was not properly leveled and the water volume was low, it usually took too long for water to be distributed through groves, and problems occurred due to root damage. In these cases, water pumped into the water furrows was allowed to seep into the beds during irrigation events (seepage irrigation).

Properly engineered seepage systems worked well and remain popular today.

Large new developments occurred through the 1960s, particularly in Southwest Florida. Crown flood and seepage continued to be widely used, with seepage being more popular in the southwest area. Volume guns drawing water out of the system of ditches and canals were also widely used through the 1960s and 1970s, but were slowly replaced with microirrigation during the 1980s. The extended drought period during the early 1970s precipitated the passage of legislation that included the Florida Water Resources Act, which modified common law water use doctrines and created five water management districts with broad statutory powers. Another extended dry period in the early 1980s prompted further concerns over water allocation among agriculture, industry, and the urban sector. This resulted in further regulations governing land development, water retention, and other environmental issues.

The early 1970s saw the introduction of the first subcanopy low-volume drip (trickle) and later microsprinkler systems, first evaluated in other countries, notably Israel and South Africa. As efficiency of irrigation became a concern, overhead sprinkler systems came under scrutiny, and fewer installations were made in new plantings throughout the state. New groves were developed with low-volume systems, and some existing acreage under permanent overhead and volume guns were converted to microsprinklers.

During the 1980s, another factor hastening conversion to microsprinkler systems was the discovery of their effectiveness for cold protection in both young plantings and in older groves. Design and layout of such microsprinkler systems therefore took into account both irrigation and cold protection efficiency. This was hastened by the fact that traditional heating systems were phased out due to fuel costs and the advent of environmental regulations targeting air pollution.

Today most of the citrus on the Ridge area is under microsprinkler irrigation with a much reduced acreage of permanent overhead systems. In the flatwoods production areas, drip and microsprinkler systems have replaced volume guns, and in large part also seepage and flood irrigation. Although most of the latter systems are still effectively used, new or renovated acreage is developed with microirrigation systems. The increasing popularity of microirrigation systems over the past twenty years has been due to:

• the technology's many production benefits;

• periodic severe drought events with water shortages and restrictions, particularly in water use caution areas;

• negative impact of saline water on tree vigor and productivity;

• public perception of inefficiency and waste portrayed by highly visible overhead systems;

• frost protection;

• fertigation.

In recent years, it has become more difficult to obtain water use permits for groves using less efficient irrigation methods in some of the water management districts. While permits are still held for some flood and high-volume sprinkler systems, the trend is irreversibly toward microirrigation. However, the relatively high maintenance costs (due to clogging, repair, and replacement) of microirrigation systems leave some growers with fond memories of earlier systems.

In the selection of irrigation management systems throughout the industry's history, many factors played a role, including emerging technologies, economics, labor availability, soil types and terrain, freeze potential, available water supplies and salinity, emerging regulations, and associated political and environmental considerations. The above factors will, to a greater or lesser extent, continue to play roles in determining future water policy and water management strategies in Florida citrus production.

Figure 1-3. Portable perforated pipe irrigation.

Figure 1-4. Portable volume gun irrigation.

Figure 1-5. Overhead irrigation on mature trees.

Figure 1-6. Overhead irrigation risers on young trees.

Figure 1-7. 1/4-circle microsprinklers on young trees.

Figure 1-8. Microsprinkler stake assembly and lateral tubing on young tree.

Figure 1-9. Double driplines on young citrus trees.

Figure 1-10. Flood- or seepage-irrigated flatwoods grove.

Figure 1-11. Water furrow flooded during irrigation.

Chapter 2. Overview of Grove Design and Development

by Brian Boman, Nigel Morris, and Mark Wade

There are many factors that are involved in water management of citrus in Florida. However, since no two groves are the same, the optimum water management strategy for each grove has to consider the specific conditions of that particular grove. Not only do the physical grove conditions differ from site to site, but the experiences and management philosophies of each manager differ. As a result, there are no hard and fast rules that delineate good and poor water management for Florida citrus. All factors need to be evaluated together in order to accurately assess potential water management strategies for a particular site. These factors include the following:

1. Soil survey
2. Boundary survey
3. Topographic survey
   a. Ground
   b. Aerial photographs
4. Wetland delineations
5. Irrigation source and consumptive use permit
6. Surface Water Management Plan
   a. On-site ditches and canals
   b. Above- or below-ground reservoirs
   c. Drainage pump station
   d. Off-site roadway or drainage improvements
   e. Off-site drainage through or around site
7. Construction plans and specifications
   a. Clearing
   b. Grading and bedding
   c. Drainage, including ditches, reservoirs and dikes, culverts, discharge structures, emergency overflow structures, and pump stations
8. Permits
   a. County or local municipality
   b. Local water control district
   c. Regional water management district
      (1) Surface water management/natural resources
      (2) Consumptive use of irrigation water
      (3) District right-of-way permit
   d. Florida Department of Environmental Protection (FDEP)
   e. U.S. Corps of Engineers
   f. Florida Department of Transportation (FDOT) for drainage or access rights-of-way

This chapter briefly outlines some of the important factors involved in selecting a grove site, things to consider for water management district permits, and various aspects of irrigation and drainage systems. Later chapters will cover these topics in greater detail.

Soil Considerations

The rolling sand hills in Central Florida (the Ridge) are largely well drained. These light sandy soils are usually underlain by a sandy clay layer. Clay or organic matter in the soil helps to retain moisture and nutrients for use by the trees. The citrus growing areas in the Indian River area and in Southwest Florida are generally poorly drained flatwoods soil. These are usually white sands underlain by an organic layer or hardpan that inhibits internal drainage. The water table in these soils is normally at or near the soil surface and must be lowered artificially to facilitate citrus growth.

The poorly drained coastal areas are underlain with shell, marl, or limestone. The oolitic limestone soils of extreme South Florida are predominately found in parts of Dade, Broward, and Monroe counties. Both soil types tend to be quite alkaline and may need special consideration in the cultural program. Soil types, drainage, texture, and the amount of organic matter in flatwoods soils can vary markedly, even within a given area. The importance of a careful survey prior to planting cannot be overemphasized as this may help to prevent mistakes that the grower will have to contend with for the life of the grove.

Location

There are many other factors that govern the suitability of a site for citrus. Access to the property, land value, taxes, and water may be as important as soil type or freeze potential. Access to any property is quite important, especially in selecting a grove site. Heavy equipment will need to be moved in and out of the grove periodically and roads need to be in good condition to accommodate this traffic. Distance from a processing or packing facility should also be considered as hauling costs will only go up in the future.

Land value and taxes should also be considered when selecting grove sites. Some land is simply too valuable to grow citrus on and good grove land is also desirable real estate, especially if the area is near a population center. Land purchased at a high price is also likely to carry high property taxes. Property tax may vary from area to area and should be thoroughly checked out. Special exemptions for agriculture may not always be in effect.

Land Preparation

On well-drained sandy soils of the Central Florida Ridge (soil series such as Astatula, Archbold, Tavares, Candler, etc.), planting on unaltered ground is adequate. In areas of native vegetation, the main requirement for land preparation is to clear, pile, and burn the existing vegetation. Most areas can be cleared with front-end loaders fitted with root rakes. Some may require heavier equipment, such as bulldozers or backhoes for pine stump removal. Further physical alteration of the land surface is generally unnecessary. Soil amendments are usually required and should be guided by the results of soil testing. On previously cleared land, the chief requirement for land preparation would be discing.

In poorly drained sandy soils of the South Florida Flatwoods (soil series such as Basinger, Immokalee, Myakka, Pineda, Riviera, Smyrna, Wabasso, Winder, etc.), control of the naturally high water table and rapid removal of excess surface water from rainfall are essential in addition to land clearing. Trees (long leaf pine, slash pine, and oaks), including stumps and shrubs (palmetto, wax myrtle, etc.), can be cleared in a single operation with heavy machinery, then piled and burned. Alternatively, pines can be sold as timber, after which stumps must be removed with a backhoe. Native flatwoods soils will usually require adjustment of pH via the application of lime or dolomite. All soil amendments should be based on the results of soil tests.

Drainage

On the Ridge, drainage is not usually required. In the Flatwoods, the naturally high water table must be controlled and provision must be made for the rapid removal of excess surface water from rainfall. Control of the water table is achieved via the construction of evenly spaced ditches in addition to beds. The use of drain tiles may be employed, particularly in problem areas resulting from the general nonuniformity of soils. However, some growers routinely install drain tile to control the water table in addition to beds, swales, and ditches.

Beds and Water Furrows

Rows are typically oriented north-south and consist of 2 1/2- to 3 1/2-foot high beds constructed with vee-ploughs and/or motorgraders between water furrows that are generally 48 to 55 feet apart. Water furrows are cut 2 to 3 feet deep and the soil is mounded between them to provide a 2 1/2- to 3 1/2-foot bed height from the bottom of the water furrow to the crown of the bed. Beds of these dimensions are the most common and they accommodate two rows of trees 24 to 27.5 feet apart. Single-row beds are used by some growers, but they are becoming increasingly rare. Wider multiple (4 to 6) row beds are sometimes seen, particularly in areas where shallow fractured limestone is encountered. Prior to bedding it is sometimes advantageous to laser-level the land to facilitate rapid surface water removal by the swales.

Lateral Ditches

Lateral drainage ditches should be cut at right angles to the beds and water furrows, and spaced no further apart than 1320 feet. Topsoil spoil from the ditches can be used to provide fill for low areas in the adjoining fields. Subsoil spoil provides a grove road base on either side of the lateral ditch. Swales drain into the ditches via 6 to 8 inch flexible polyethylene or rigid pipe that can be installed either before or after swale construction. A laser level is sometimes employed in this operation, but it is not essential. The pipe is installed in the bottom of the water furrow and sloped to discharge approximately 1 foot above the bottom of the ditch. Ditch size will vary depending upon the area served and other water management district criteria. In general, lateral ditches should have a minimum top width of 14 to 15 feet, bottom width of 4 feet, 2:1 side slopes, and a depth of at least 5 feet.

Collector Ditches

Drainage water from several lateral ditches runs into collector ditches and is conveyed off-site. Gravity drainage is preferred if topographic relief allows. However, discharge pumps are required where there is insufficient relief. Size of the collector ditches and any related pumping facilities is dependent on several factors, such as size of the area being served, soils, bed and water furrow design, and slope of ditches. The surface water drainage system should be designed to remove at least 4 inches per day from the grove.

Tile Drainage

Drain tiles may be installed for additional control of the water table. Perforated, 4-inch diameter, flexible polyethylene pipe covered with a nylon fabric sock installed down the center of every other bed normally provides effective control. The pipe should be installed on a slope corresponding to the flow of the swales at depths averaging 3 to 4 feet or less, depending upon the location of spodic or clay horizons. Generally speaking, drain tile should not be installed below the depth of the hardpan horizons.

Perimeter Grading

In order to intercept and control the off-site water table and off-site surface flows, it is necessary to construct a perimeter ditch and dike. The dike is located external to the ditch. Frequently the ditches can serve as collector ditches. The actual size of the ditches depends on anticipated flow rates. High water tables or natural drainage from adjacent undeveloped properties may result in subsurface flow towards a grove. Pumps may be required in the perimeter ditches to intercept this seepage water in order to maintain satisfactory water table depths in the developed grove.

Resource Protection and Permitting Considerations

Listed Species

The Central Florida Ridge and other well-drained sandy soils in Florida support a unique flora and fauna with a high degree of endemism (species found only in these habitats). Due to the initial, relatively sparse distribution of the habitat types associated with these soils and the rapid conversion of them to citriculture, housing developments, golf courses, and urban encroachment, many of the species unique to them have been listed as threatened or endangered by federal, state, and local agencies. Consequently, those areas remaining in their natural vegetative state have many development limitations associated with the protection of listed species. Therefore, when selecting land for new citrus plantings, it is often beneficial to select land that has already been cleared for other purposes.

In the Flatwoods, loss of existing, native, upland habitats is of concern in certain areas but generally land clearing regulations are far less restrictive than on the Ridge. In this regard (i.e., land clearing), many County Comprehensive Plans contain exemptions for bona fide agricultural operations. When considering site selection for new groves in flatwoods areas, it is still advantageous to procure already cleared land. The best lands for citrus development generally are former row-crop farming areas where land will have been cleared, leveled, and have received soil amendments. Before starting site clearing, check with state and local agencies to ensure that the project is consistent with all regulatory criteria.

Nutrient Leaching

Due to the highly permeable nature of Ridge-type soils, the potential for leaching of nutrients and agrichemicals to aquifers is high. Recently, elevated levels of nitrates have been discovered in potable well water on the Ridge. Conscientious use of Best Management Practices (BMPs) in the application of all agrichemicals should be given paramount importance in both new and existing citrus operations in these areas.

Water Use

Water use is regulated in Florida by five water management districts (WMDs). The districts issue water use permits (WUPs) to allocate water for reasonable beneficial uses. Citrus production is considered to be such a use. However, water is generally allocated from lower-quality sources for agricultural uses. Consumptive use permits are granted for fixed periods of time. Duration may not exceed 20 years. However, permits are generally issued for shorter durations because the districts have lacked the information needed to commit the resource for longer periods of time. Permits are revocable only for material false statements, willful violation of permit conditions, and nonuse of the water supply. Thus, except during times of water shortage or emergency, permittees have full rights to the water specified in their permit.

Reasonable beneficial use is defined as the use of water in such quantity as is necessary for economic and efficient utilization for a purpose and in a manner that is both reasonable and consistent with the public interest. Applicants must present evidence of the quantity of water requested, the need, purpose, and value of the use, and the suitability of the use to the source. The method and efficiency of use, water conservation measures, and the practicality of reuse or the use of lower-quality water are also considerations.

Applicants must show the proposed use will not interfere with any presently existing legal use of water. The extent and amount of harm caused, whether that harm extends to other lands, and the practicality of mitigating harm by adjusting the quantity or method of use, also are considered. Any adverse actions from using the water such as water quality degradation or increased flood damage are also factors considered in the permit. In the case of groundwater withdrawals, extensive modeling (and sometimes testing) of the aquifers concerned is necessary to achieve this requirement.

The permitted amounts must also be consistent with the public interest. The public interest can be defined further through rule making by the Florida Department of Environmental Protection (FDEP) and the water management districts. Generally, consistency with the public interest is determined on a case-by-case basis in the permitting process.

Surface Water Management

Because of the more drastic changes necessary in terms of water table control and drainage that are associated with the successful production of citrus on the Flatwoods, the chief resource protection concerns center around effects on wetlands and water quality. In addition, changes to surface water discharge rates must be addressed to meet criteria adopted by Florida's WMDs. A surface water management system for citrus production in the Flatwoods should be designed to remove at least 4 inches in 24 hours. However, most native Flatwoods areas drain at rates of 1 inch in 24 hours or less.

Increased Runoff Rates

Properly designed surface water management systems can minimize storm water runoff rates. Runoff rates are reduced by designing surface water detention areas that are interspersed between the grove area and the ultimate off-site discharge points. Typically, these are diked-off areas that receive inflow from the grove area either via gravity or pumped discharge. Outflow from the detention areas (often called reservoirs) passes through discharge structures that are designed to restrict the flow rate to predevelopment peak rates. Water levels thus build up in detention areas for a short period of time following major rainfall events. A civil or agricultural engineer registered in the state of

Florida must be retained to design a system in order to obtain a permit from the appropriate WMD or other regulatory agency.

Water Quality

Pollutant loadings in the form of sediments and dissolved agrichemicals, including pesticides, can and should be minimized via the use of BMPs. In order to obtain an Environmental Resource Permit from the appropriate WMD, a grove developer must either agree to provide ongoing monitoring of discharge water quality or retain a professional engineer to design a system to meet the criteria for water quality treatment. In most cases this is achieved via detention areas. The detention areas provide attenuation of peak runoff rates and allow the drainage water to be released at lower rates over a longer period. Studies have also shown that the detention areas are effective in providing water quality treatment and improved quality of runoff water.

An alternative form of water quality treatment acceptable to the Southwest Florida Water Management District (SWFWMD) is the use of grassed filter strips. For smaller systems, less than 40 acres, the USDA Natural Resource Conservation Service (NRCS) engineers can provide designs to obtain a so-called permanent agricultural exemption from SWFWMD. For larger systems, a professional engineer must be retained.

Impacts to Wetlands and Diversion of Watersheds

The WMDs now have a fairly clear legislative mandate to protect wetland resources. In addition, the U.S. Army Corps of Engineers (COE) regulates direct impacts to wetlands via the discharge or deposition of fill. Both agencies are also tasked with assessing and regulating secondary and cumulative impacts. It is important, therefore, to try to eliminate impacts to wetlands from proposed designs. The issues can be fairly complex to the extent that it usually becomes cost effective to retain consultants in the field to prepare designs and permit applications to address them.

Grove design should start with a thorough investigation of predevelopment drainage patterns and accommodation of same in the design. That is, care must be taken to eliminate or minimize diversion of natural watersheds, particularly if they are feeding downstream, off-site wetland systems. Off-site, up-gradient flows can be taken through projects via natural flowways or constructed channels, or they can be taken around projects in constructed ditches provided they outflow at their predevelopment location. Most often, off-site flows must be kept separate from project flows because of intermingling results in having to provide water quality treatment for the entire volume.

Avoiding direct impacts to wetlands can be achieved by setting back development lines to provide an upland buffer. This also eliminates the need to obtain a permit from the COE. When setbacks are large enough, the indirect impact of water table drawdown can be minimized or eliminated. Another benefit is that the requirement for incorporation of upland buffers to accommodate wildlife in groves is realized. Some WMDs have developed technical procedures for determining the adequacy of development setbacks to eliminate the effect of drawdown on wetlands.

Incorporation of wetlands in detention areas provides a mechanism to counteract the effect of drawdown. However, when detention areas are serving the function of attenuation of peak runoff rate, water levels will rise within them. The maximum safe depth for small impoundments is about 4 feet. This, however, would be too great a depth to stack on natural wetland systems for prolonged periods without deleteriously impacting them. Therefore, it is necessary to analyze the results of an engineered storm routing and adjust the design, if necessary, to reduce the depth and/or duration of excess inundation to avoid negative impacts.

Directing project discharges to off-site wetlands can counteract the effect of drawdown from adjacent citrus development. When this is done, however, it is important to try to maintain the predevelopment delivery system. That is, if it was via sheet flow, then the project should discharge to a spreader swale to restore such flow prior to entering the wetlands. Maintaining hydraulic head within the water table of an adjacent wetland can be achieved by interposing a detention area between a citrus development and the wetlands.

Incorporating wetlands into natural flowways carrying off-site flows through a project can provide for a habitat mix of wetlands and uplands, and connectivity between off-site and on-site undeveloped areas. This also meets the goal of accommodating wildlife species that are sensitive to habitat fragmentation.

Tree and Row Layout

Tree planting patterns should be selected to provide spacing between trees for ample growth, maximum exposure to sunlight, and good air circulation, as well as to facilitate management, cultural practices, and harvesting operations. Trees are typically planted in a geometric design. The planting system to be used will be selected based on several considerations: slope, maximum use of the land area, final tree size, required spacing between trees, and type of fruit grown. In addition, the space needed for the passage of cultural and harvest equipment is a very important consideration. Generally, the triangular systems are better if the trees are to be allowed to attain their natural size and form (no hedging). However, if distance between rows is more important than distance between trees and if tree rows are to be hedged to control tree size, then the rectangular designs are recommended.

Generally, row orientation should be north-south to maximize exposure to sunlight. North-south orientation is more important for groves that are closely spaced within the rows and for hedgerows. When closely spaced trees are aligned in east-west rows, typically one side of the tree receives significantly more sunlight than the other.

Cold Protection

There are three basic considerations that determine how cold a grove site will be during a freeze when compared with other sites in the same general geographic area. The first consideration is topography. Cold air is heavier than warm air and drains downhill, collecting in low areas that are often known as pockets. These low areas should be avoided to minimize the cold hazard. The site under consideration would have an adequate outlet provided by adjacent low areas into which cold air can drain.

Land elevation and air drainage are the most important factors affecting the temperature in any one location. An elevation difference of as little as 4 feet can cause a temperature difference of as much as 5° F on clear nights with little wind movement. In selecting sites for the growing of citrus, consideration should be given to the microclimate of the various areas and types of locations. Weather records and experience attest to the fact that well-drained, high-ground slopes are the most desirable.

Low areas or pockets in the best and warmest citrus areas are often as cold or colder than areas north of the citrus area where it is not profitable to grow citrus. Moreover, citrus in cold pockets in warm areas is often more tender to cold than that grown farther north because the latter becomes more cold hardy by virtue of the generally cooler winter temperatures under which it is grown.

The second consideration is proximity to large bodies of water. Freezes come into Florida from the north or northwesterly direction. Therefore, groves situated near lakes or large rivers will be afforded some protection by warming of the cold air by the water. Groves located near the Atlantic Ocean or Gulf of Mexico will also be afforded some degree of cold protection by proximity.

The third consideration is latitude or how far south the grove is located in the Florida peninsula. As a generalization, the farther south the grove is located, the less likely it is to be damaged during freeze conditions in Florida. Arctic air masses that bring freezing temperatures to Florida are moderated somewhat as they traverse the length of the state. It is difficult to classify Florida into broad climatic zones based upon the anticipated or historical incidence of freezing temperatures. Each potential site should be appraised separately based upon all considerations.

Irrigation System

Research since the 1960s has consistently shown that irrigation can significantly increase production on most commercial cultivars. The research has shown that it is critical to maintain relatively high soil moisture levels from the spring flush until the young fruit is at least one inch in diameter. The soil moisture should be maintained so that less than 1/3 of the available water is depleted from the soil during this time. Field capacity is that amount of water remaining in the soil after the excess has drained away. During the remainder of the year, about 2/3 of the available water can be depleted from the soil without adversely affecting production. Irrigation can have an influence on other aspects of citrus production. Generally, irrigation can increase tree size, yield, fruit size, and juice content. Under certain conditions, irrigation can reduce fruit color, soluble solids, acid, and may affect the Brix:acid ratio. Excessive irrigation in the fall can reduce the fruit solids, affecting the net returns.

Methods of Irrigation

Various methods of irrigation have been used in Florida over the years. All have their advantages and disadvantages. However, in recent years, nearly all new groves have had microirrigation systems installed. Freeze protection, rising costs of high pressure systems, and water use limitations placed on growers by water management districts are the primary reasons for the switch to these types of systems.

Portable perforated pipe has almost become a thing of the past due to high labor costs. There are still a few systems still in operation, but these are limited to smaller acreages. The portable and self-propelled volume guns, while still in use in many groves, are losing favor due to the high energy costs. They are inefficient under high wind conditions and are not considered to be effective on bedded groves since much of the water runs off into the furrows and ditches, and is thereby lost to the tree's root zone. In addition, they provide little freeze protection.

Seepage and crown flood irrigation is practiced on Flatwoods areas in the Indian River area and in Southwest Florida. With a proper match of water control structures, ditches, soils, and topography, these types of irrigation can be quite efficient. With this type of system, irrigation and drainage are managed together. This system requires large, expensive pumping equipment capable of moving large volumes of water. Care must be exercised not to allow water to remain in the root zone for extended periods as damage to the tree root system may occur. Since large volumes of water are required, even though a high percentage of the water is returned to the shallow aquifer and retention ponds, the grower must be able to justify its use to the water management districts. These systems also offer good cold protection as growers can flood the furrows and perimeter ditches during freeze periods.

Permanent overhead sprinkler systems have declined in popularity over the last 20 years due to high fixed costs and inefficiency during periods of high winds. Pop-up tree sprinklers have also lost favor since they are susceptible to damage by grove equipment and impairment of riser movement by sand contamination. High-volume, under-tree sprinklers are still used on limited acreage in Florida. Low-hanging branches interfere with the water distribution pattern of these under-tree systems.

Microirrigation systems have the advantage of lower initial and operating costs as compared to other systems. They operate at lower pressures, reducing energy consumption, and are easily automated. Since lower volumes of water are required, these types of systems can utilize less productive water sources. If operated properly, these systems can minimize water loss due to evaporation and deep percolation below the root zone of the tree. The spray-jet systems offer some degree of cold protection during radiation type freezes. Maintenance of microirrigation systems is needed on a frequent basis. Clogging of emitters by algae, insects, ants, and precipitates can be a major problem with these systems.

Pump Units

The pumping unit must have enough capacity to irrigate all intended zones during the most extreme conditions (such as freeze events). Ideally, all zones should be of about the same size and have about the same pressure requirements because an irrigation pump operates most efficiently at a single flow rate and pressure. The system design should meet all local and state standards, and should be approved by a professional engineer.

Consideration should be given to the pressures required to operate the critical subunit, friction losses through the mainline (including all losses through valves, filters, meters, fittings, etc.), and elevation changes including pumping lift. For surface water supplies and water at pumping levels of less than 20 feet in wells, centrifugal pumps are the most economical option. For water at depths greater than 20 feet in wells, turbine pumps must be used. For large systems, deep-well turbines with power units on the surface are commonly used. For smaller units, submersible turbines are a less expensive option.

With submersible turbines, electric motors are directly connected to the pumps and lowered into the well. For automatic operation, turbine pumps have the advantage in that they do not require priming for the pump to operate (conventional centrifugal pumps require priming). For automatic operation, electric motors are recommended as power units for microirrigation systems. They have lower initial costs than internal combustion engines, especially for smaller sizes. There may be a demand charge on the electric bill for their use, especially for larger units. Most power companies now have off-peak rates for irrigation pumps. Some power companies have also eliminated standby or demand charges for off-peak users. Local power company policies will dictate actual costs.

Diesel power units are the most common type of internal combustion engines used for irrigation in Florida. They are more efficient than other types of internal combustion engines. Internal combustion engines are recommended when irrigation systems will be used for cold protection because of the possibility of electric power interruption and loss of pumping capability on cold nights.

Pump Station

The pumping station should be protected from the elements as rain can cause moisture in motor windings and direct sun can raise motor operating temperatures significantly. A sturdy shed with a concrete floor provides a clean, dry, well-ventilated environment free of weeds and debris, and is conducive to proper operation and good maintenance. Safety requirements should be considered in the design of the pump station. There may be high-voltage power lines, tanks of gasoline or diesel fuel, toxic agricultural chemicals, chlorine solutions, etc. present in or near the pumping site. There may be problems associated with water and fuel spills, leaks in pressurized pipes or discharges from pressure relief valves. The possibility of vandalism, theft, and entry by unauthorized persons should be considered. The mounting for pumps and drivers (motors or engines) should be a sturdy concrete base. For permanent diesel engine installations, construct a concrete mounting isolated from the concrete floor to minimize vibration.

Filters

Filters should always be used on microirrigation systems. Filters remove small particles that may clog the tiny orifices in emitters. The type of filtration system required depends on the type of emitter used and the source and quality of the irrigation water. Filters should be selected based on emitter manufacturer's recommendations. The mesh size selected should be small enough to remove all particles larger than 1/7 the size of the emitter orifice.

If organic matter is a problem when pumping from surface waters, media (sand) or self-cleaning screen filters should be used as the primary filter. A strainer should be used on the pump intake to exclude as much organic matter as possible. The intake should be positioned below the water surface to avoid floating debris, and above the bottom to avoid pumping sediment from the bottom of the pond or canal. Self-cleaning strainers for the pump inlet are available to prevent larger particles from entering the irrigation system.

When pumping from wells, screen filters alone are normally adequate unless large amounts of sand are being pumped. If large amounts of sand are being pumped, a vortex-type sand separator may be used, followed by a screen filter. Settling basins may also be used to remove large amounts of sand, but basins may cause problems if organic matter such as algae is present in the basins.

Valves

Valves are required to control the filling of irrigation systems at pump startup, to control flows to the desired subunits of a system, and to allow flushing of irrigation pipes. Only properly pressure-rated irrigation valves must be used to avoid failures due to system pressure and water hammer problems. Valve materials and components must be resistant to corrosion by the irrigation water and any chemicals injected during irrigation. Valves should also be sized to avoid excessive pressure losses. Typically, manual zone valves are the same size as the pipe (or possibly one size smaller if the valves are not operating in series). Installing smaller valves to save initial costs will result in higher operating costs for the life of the system due to friction losses. Automatically controlled irrigation systems will require the use of automatic valves. These may be controlled by electric solenoids or hydraulic pressures, depending upon the type of timer or controller used. Automatic valves usually require some pressure differential across the valve to operate in a timely manner. Automatic valves should be sized according to the manufactures' specifications (typically, one size smaller than the pipe).

Backflow Prevention

Florida law requires that a backflow prevention system be installed on most irrigation systems.

Backflow prevention systems are always required when chemicals are injected into an irrigation system. The minimum backflow prevention system required when fertilizers or chemicals are injected includes a check valve, low pressure drain, and vacuum breaker on the irrigation pipe to prevent water and chemicals from flowing back to the water source. Florida law also requires interlocked power supplies to prevent chemical injection unless the irrigation water is flowing, a check valve on the injection line to prevent water flow to the chemical supply tank, and a positive shutoff valve on the chemical tank to prevent accidental drainage from the tank.

When chemical toxicity Category I pesticides are injected into irrigation systems, a double check valve, low-pressure drain, and vacuum relief valve assembly are required. These pesticides are marked with the keywords "Danger" or "Poison" on the label. The Environmental Protection Agency (EPA) requires that all pesticide products be labeled to clearly state whether injection into irrigation systems is permitted. Pesticide labels must also list the backflow prevention equipment requirements and application instructions.

Chemical Injection Equipment

Pressurized irrigation systems are often used to apply chemicals, especially fertilizers. Growers can obtain yield increases and minimize leaching losses (and pollution) by injecting nutrients and other chemicals through the irrigation systems. Many growers currently inject fertilizer through sprinkler and microirrigation systems. Chemical injection equipment is required to add the correct amount and rate of chemical. Several types of chemical injectors are commercially available ranging from the low-cost venturi devices to high-cost positive displacement pumps. If a high degree of precision is required (such as for pesticides), more precise injection methods must be used. These include the high-precision but more costly positive displacement injection pumps such as diaphragm- and piston-type pumps. Micro-irrigation systems require high-precision chemical injection pumps to precisely control biocides and water amendments used to prevent emitter plugging.

Irrigation Controllers

Irrigation controllers are devices that automatically turn the irrigation system and associated equipment, such as chemical injection pumps, on and off. Controllers are not mandatory for system operation, but they are time- and labor-saving devices. They are especially economical and efficient for management of microirrigation systems on Florida's sandy soils because of the requirement for frequent irrigations. Controllers range in capabilities from simple timers that can turn a single valve on and off at preset times, to computers that allow remote control of pumps, valves, and control devices at several locations. Computer controllers can collect data from sensors, make calculations, and adjust water and chemical application schedules in response to plant needs and environmental conditions.

Emitters

The emitter selection process should consider uniformity as well as other factors such as cost, wind effects, system constraints, maintenance, tree density, and soil type so that the best emitter for a particular field condition is selected. The wetting pattern of emitters needs to be compatible with the soils and rooting pattern of the trees. Emitters with distinct spoke patterns may not be as efficient as more uniform patterns where shallow root systems and very sandy soils are encountered. Consideration also must be given to the water requirements of the mature tree. Higher density plantings with smaller trees will require less water per tree than more widely spaced trees. Larger wetting patterns may be more desirable for more widely spaced trees. In all cases, designers should ensure that tree water requirements can be met with reasonable run times, and not result in percolation of nutrients and water below the root zone.

Protection during severely cold freezes can be improved by using a higher application rate. For young trees, this can be done by changing from a 360° to a 180° or 90° spray pattern. While a high application rate may be beneficial for freeze protection, this is not necessarily desirable for normal irrigation during the rest of the year. A high application rate can also lead to leaching of nutrients and pesticides. Application rate can be increased by increasing emitter output and decreasing spray pattern size. With small diameter spray patterns and moist soil, irrigation duration of one hour or more can drive water below the main root zone. By adjusting spray diameter, irrigation duration, and emitter output, microirrigation systems can be managed to meet tree water needs while reducing overirrigation and chemical leaching.

The need for a high application rate for freeze protection and a lower application rate to reduce chemical leaching means that there may not be an ideal microsprinkler that provides adequate freeze protection, good root zone coverage, and reduced chemical leaching. Nevertheless, with knowledge of the factors involved, emitters can be chosen that will perform all these functions reasonably well.

Benefit Cost Analysis

The decision to plant citrus in many areas of Florida can be made even more complex due to issues related to water management. Irrigation systems can positively influence both fruit quantity and quality, but are not inexpensive undertakings. The decision to invest in a high performance irrigation system can be difficult to evaluate, but must be included as part of the overall citrus production strategy. Project evaluation provides a means of evaluating the positive benefits versus the costs of undertaking a given project. In other words, project evaluation determines empirically whether projects are worthwhile economically. One way to evaluate an irrigation project's value is to conduct a benefit cost analysis.

The objective of a benefit cost analysis (BCA) is to determine if the projected benefits from implementation of a proposed project exceed the projected costs within a stated time frame. While BCA is commonly used to evaluate publicly funded projects, it can also be used to evaluate private investments. The BCA calculation should be the final determinant as to whether a project should be undertaken.

Benefit Cost Analysis examines the net benefits of a project, recognizing that dollars earned and costs incurred in the future must be discounted to current values. BCA also allows for evaluation over any time frame within the life expectancy of the project, and is accomplished through calculation of the project's Net Present Value (Eq. 2-1). If net present value (NPV) is greater than zero, the project benefits exceed project costs and can justifiably be undertaken.

$$NPV_t = \sum_{t=0}^{T} \frac{F_t}{(1+r)^t} \qquad \text{Eq. 2-1}$$

where,

$F_t = B_t - C_t$ = net benefits accruing in year t
B_t = benefits accruing in year t
C_t = costs accruing in year t
T = last year the project influences productivity
r = discount rate (usually the interest rate)

Example:
An irrigation project is proposed with a 20-year life expectancy. The lending institution is concerned that the project is too expensive and will not pay for itself quickly enough. Conduct a BCA for the first five years given a discount rate of 6%. Benefits and costs are as follows:

Year	Costs	Benefits
1	$50,000	$0
2	$15,000	$8,000
3	$10,000	$10,000
4	$5,000	$15,000
5	$5,000	$18,000

In this example, the net present value is negative indicating that the irrigation project costs exceed the benefits during the five-year period of analysis. If the investment parameters require a positive NPV within the first five years, the BCA indicates that the irrigation project should not be done.

Using Eq. 2-1,

$$NPV = \sum_{t=1}^{5} \frac{F_t}{(1+r)^t} = \frac{B_t - C_t}{(1+r)^t}$$

$$= \frac{0 - 50,000}{(1+.06)^1} + \frac{8,000 - 15,000}{(1+.06)^2} + \frac{10,000 - 10,000}{(1+.06)^3}$$

$$+ \frac{15,000 - 5,000}{(1+.06)^4} + \frac{1,800 - 5,000}{(1+.06)^5}$$

$$= \frac{50,000}{1.06} + \frac{7,000}{1.124} + \frac{0}{1.191} + \frac{10,000}{1.262} + \frac{13,000}{1.338}$$

$$= -47,169.81 - 6,227.76 + 0 + 7,923.93 + 9,715.99$$

$$= -\$35,757.65$$

Chapter 3. Environmental Acts and Regulatory Agencies

by Michael Olexa, Brian Boman, and Scott Hall

Clean Water Act

The Clean Water Act is directed at maintaining and restoring the quality of navigable waters within the United States. Navigable waters include large to small bodies of water, as long as they have even a remote potential to affect interstate commerce or people involved in interstate commerce. Wetlands, which are generally defined as lands that are covered periodically with enough water to support vegetation especially suited to a wetlands environment, are also included. The Act's purpose is to limit the amount of pollutants that may be released into these waters, in an attempt to keep the waters safe for a variety of users as well as for fish and aquatic life.

The Act in general is enforced by the EPA, but dredge and fill permitting is enforced by the Army Corps of Engineers. The EPA has established national standards for the maximum amount of pollutants that may be released under its permits. States are authorized under the Act to establish their own standards for allowable levels of pollutants, as long as such standards are at least as stringent as those mandated by the EPA. Furthermore, the EPA may also delegate permitting authority to the state. Currently, Florida has been delegated enforcement of only portions of the Act.

The Act requires all operators of point sources of pollution to get permits. A point source can be any confined and measurable location from which a pollutant may be discharged. Point sources include pollutants that are discharged from ditches, rinsed pesticide containers, or any other source that releases pollutants into a specific area. Agricultural stormwater discharges and irrigation systems are considered nonpoint sources, and are not covered by the Act. National Pollutant Discharge Elimination System (NPDES) permits are the main avenue for the enforcement of this Act. These permits specify the amount and concentration of pollutants the holder is authorized to discharge, direct when compliance must be achieved, and specify the requirements for testing and reporting to the permitting authority. In Florida, NPDES permitting is conducted by the EPA as it has not been delegated to the state.

The NPDES permits impose two types of limitations on point-source polluters:

- Technology-Based Effluent Limitations set limits on the contents of the effluent based upon the available treatment technology.

- Water Quality-Based Effluent Limitations depend on the standards established for the quality of the water body (including groundwater bodies) into which the discharge takes place. These cases are viewed on a case-by-case basis.

The Clean Water Act requires separate permits for the discharge of dredge and fill material into navigable waters or wetlands. Dredge and fill permits are issued by the Corps of Engineers, but the EPA has a veto power over Corps-issued permits. The EPA may enforce permits issued by the Corps or those issued by the state.

Resource Conservation and Recovery Act

The Resource Conservation and Recovery Act (RCRA) is intended to be comprehensive authority for all aspects of managing hazardous wastes. Separate requirements apply for categories of generators, transporters, and facilities for treatment, storage, or disposal of hazardous waste. These requirements, which include permitting for many facilities and extensive record keeping for all phases of management, are intended to track the movement and handling of the waste until it reaches its final point of disposal. This is known as the cradle-to-grave approach of monitoring wastes. The Act also sets out standards for the disposal of solid wastes, which include wastes resulting from agricultural operations. However, irrigation return flow, or pollutants covered under the NPDES permits of the Clean Water Act as point sources, are not considered solid wastes, and are therefore not subject to RCRA's disposal standard.

The EPA is responsible for implementing and enforcing RCRA, establishing the criteria for classifying hazardous wastes, and listing the wastes for that the Act applies. The Act also permits states to enact their own hazardous waste programs that a state, instead of the EPA, may enforce. In order for a state to act in lieu of the EPA, the state acts must be at least as strict as the EPA regulations and must be approved by the EPA. Florida is authorized to administer all aspects of RCRA's base program, which includes all facets of the Act that were passed before 1986.

Permits are required under the RCRA for anyone who owns or operates a facility for the treatment, storage, or disposal of hazardous wastes. Generators and transporters may operate without an EPA permit, but must obtain an identification number and may be required to file reports with the EPA at regular intervals, depending upon the quantity and type of wastes they handle. Additionally, the EPA retains broad authority to require tests, inspections, or additional monitoring when it determines there is an enhanced danger to health or the environment from the facility. This is especially true with regard to permitted hazardous waste facilities. Conversely, no federal permits are required for solid waste disposal, although there are federal regulations on the subject.

Comprehensive Environmental Response, Compensation and Liability Act

The Comprehensive Environmental Response, Compensation and Liability Act (CERCLA or Superfund) was passed in 1980 and amended in 1986. It empowers and provides a trust fund for the EPA to investigate and clean up sites contaminated by hazardous substances. CERCLA also extends liability for site pollution to several tiers of

potential defendants at once, and is a potent measure for forcing responsible parties to contribute to the costs of cleanup.

The EPA has created a list of hazardous substances that are covered by CERCLA regulations. The Act includes all hazardous substances or hazardous pollutants that are identified by the federal RCRA, the Clean Air Act, or the Clean Water Act. The only express exclusions from CERCLA coverage are petroleum (although the EPA reserves the power to classify specific petroleum products as hazardous) and natural or synthetic gas.

The EPA is the chief enforcer of CERCLA. However, the EPA must consult with the relevant state and local officials before deciding upon remedies for pollution at federal facilities, especially where the facilities or the remedies chosen fall within the reach of state environmental law. The administrator of the Act has authority to begin investigations whenever there is reason to believe that a release has occurred or may occur. The EPA, or a state or local authority acting under agreement with the EPA, may require the person or entity under investigation to provide them with information about the nature and handling of all hazardous materials on the site, as well as information related to the subject's ability to pay for the cleanup. The Act also authorizes entry at reasonable times to any site dealing with hazardous materials and further authorizes the taking of samples from the site.

Under CERCLA, owners may be held liable for contaminated groves, even if they purchased land without the knowledge of previously buried hazardous waste. This has been a source of great concern to land buyers, banks, and others on the verge of acquiring land. CERCLA requires that the location of any site containing hazardous materials be reported to the EPA. Prompt notification is also required after any spill or release of contaminated materials into the environment. In many cases, liability can be placed on both the present and past owners irrespective of culpability. However, the security interest exemption protects lenders (such as banks) from liability when the lender does not participate in the management of the facility. Furthermore, persons who apply pesticides that are registered under FIFRA will not be subject to liability under CERCLA if the pesticides are applied according to the labeling instructions.

Potential land purchasers may want to use a method such as an environmental audit to lessen the fear associated with CERCLA liability. Environmental audits are an evaluation of the land's condition and an appraisal of the consequent likelihood of the lender becoming subject to some type of enforcement lien that might impair the lender's security. Such a lien might arise, for example, from the liability CERCLA imposes upon owners for hazardous substances buried on their land. CERCLA is threatening even to innocent buyers, since it applies even if the pollution was left by a previous owner and the buyer had no knowledge of it.

Safe Drinking Water Act

The Safe Drinking Water Act (SDWA) was passed in 1974, and has been amended several times to expand both its breadth and the EPA's power to enforce it. The Act's primary purpose, which is to stop organic chemicals from entering drinking water systems, is accomplished by establishing quality standards for drinking water, monitoring public water systems, and guarding against groundwater contamination from injection wells. For example, under the Act no hazardous waste may be disposed of by underground injection within 1/4 mile of underground drinking water. Injection wells usually imply very deep wells. However, the definition of injection under the Act may encompass several types of runoff, including irrigation return flow that enters the groundwater. Regulatory agencies, therefore, regulate the activity of the injection wells and not the wells themselves.

In Florida, the EPA has given up enforcement of the SDWA, and now serves only to supervise the state programs approved to take its place. The 1986 amendments to the Act, however, gave the EPA increased authority to step in and enforce the Act if the state takes no action within 30 days of receiving notice from the EPA that water quality standards of the Act have been violated. The states must also adopt all new and revised national regulations in order to retain primary enforcement powers.

Federal Insecticide, Fungicide, and Rodenticide Act

The Federal Insecticide, Fungicide, and Rodenticide Act (FIFRA) was originally passed in 1947 and significantly amended in the 1970s and in 1988 to regulate all phases of pesticide sale, use, handling, and disposal. A pesticide is broadly defined within the meaning of the Act as any substance used to regulate, prevent, repel, or destroy any pest or plant. Pests include insects, rodents, nematodes, fungus, weeds, terrestrial and aquatic plants, viruses, bacteria, and any other living organism that the EPA designates as a pest.

FIFRA is administered by the EPA, but the Act specifies that states are to have primary enforcement responsibility if they demonstrate to the EPA that they have adopted adequate regulations and enforcement mechanisms. Florida has entered into several cooperative agreements with the EPA, and now shoulders the responsibilities for testing and training permit applicants. In these areas, the EPA now has only a supervisory role. However, the registration of pesticides and the monitoring of pesticide producers is still regulated entirely by the EPA's central office. States may impose additional conditions on pesticide use where special problems related to their use are encountered.

There are two broad classes of pesticides: general-use and restricted-use. General-use pesticides may be applied by anyone and no permit is required, although the user must still comply with labeling requirements and other regulations. Restricted-use pesticides may be applied by three different categories of applicators:

• Private applicators who apply pesticides in producing an agricultural commodity on their own lands or on lands under their control.

Applicators must be aware of regulatory restrictions to avoid adverse effects on the environment. Additionally, private applicators must be licensed to use pesticides.

- Commercial applicators who apply pesticides to other people's lands in exchange for a fee. Commercial applicators are also required to be licensed.

- Experimental-use applicators, who are usually manufacturers or researchers, are required to have an experimental-use permit to test an unregistered pesticide in order to gather data to support its registration.

Furthermore, restricted-use applicators must keep records comparable to commercial applicators. Also, the state may require specific minimum training for all applicators. Each permitting category is subject to separate testing and certification procedures, and may be subject to different penalties for violations.

Toxic Substances Control Act

The Toxic Substances Control Act (TSCA) of 1976 governs the manufacturing, disposal, importing, distributing, and processing of all toxic chemicals. TSCA requires that all such chemicals be inspected and approved by the EPA before they enter the market. The EPA has the option to restrict the chemical, run toxological tests, and gather data during its preapproval determination of whether the chemical represents a threat to health or the environment.

TSCA does not cover pesticides (which are covered by FIFRA) and chemicals that are covered by the Federal, Food, Drug, and Cosmetic Act. All other chemicals, if they are not already on the EPA's approval list, are subject to review before they are released into the stream of commerce. The Act specifically dictates that the EPA create restrictions for PCBs (the only chemical identified by name) as part of the mandate of TSCA.

The EPA is the sole authority for enforcement of TSCA, although the Act specifically provides that the states are not prohibited from enacting their own legislation to regulate chemicals. The Act limits the states' power only to the extent that the states may not test new chemicals if the EPA is testing for the same purpose. Additionally, the states will be restricted in creating different requirements than the requirements furnished by the EPA concerning chemicals that have already been regulated by the EPA.

The Act also gives the EPA power to enforce its provisions by injunctions, restraint orders, forced inspections, and seizure of a product, as well as other means. In addition, courts have the power to restrain or force action on the part of producers in order to enforce TSCA. For obvious reasons, these powers are more broad with respect to chemicals that are identified as posing an imminent and unreasonable risk of serious or widespread harm. The EPA may also require cleanup of areas where violations resulted in environmental damage.

Endangered Species Act

The Endangered Species Act was passed in 1973 to protect fish, wildlife, and plants that are threatened with extinction, and also to protect the ecosystems that are determined critical to their survival. In many cases these ecosystems are wetlands, so that endangered species protection often equals wetland protection. The Act requires all federal agencies to consult with an appropriate federal department (i.e., Department of the Interior, USDA) to determine what effects its land use or other actions will have upon endangered species. The agencies are prohibited from taking any action that will threaten an endangered species unless the agency obtains a special waiver from a committee headed by the Secretary of the Interior.

Agency action might include the issuance of a permit to a private party, such as when the Corps of Engineers issues permits for dredge and fill activities. The Act also applies to private and state actions, although the prohibitions are slightly different. The ESA grants enforcement authority to the Secretary of the Interior, the Secretary of

Commerce, and the Secretary of Agriculture. However, the Secretary of Agriculture is limited to enforcement authority in cases regarding the importation or exportation of plants.

The ESA prohibits the taking of any endangered species. Taking under the ESA means killing, trapping, harassing, hunting, collecting, or harming the species in any way. The definition of taking has been extended by the federal courts to include the destruction of areas designated as critical habitats, where the destruction might reasonably be expected to result in a reduction in the number or distribution of an endangered species. Therefore, any adverse modification, as well as destruction of a critical habitat, is restricted under the ESA.

The agencies administering the Act are authorized to grant exceptions, usually by way of limited permits, to the prohibitions against takings. Permits may be granted for scientific purposes, and in cases where the applicant became economically dependent upon the species before it was placed on the endangered species list. These permits are usually limited to allow takings for only one year following the designation of the species as endangered.

State Water Quality Plan

The Florida Department of Environmental Protection (DEP) has been directed to prepare the state water use plan and to foster interagency agreements to achieve the State Water Quality Plan. The water use plan is developed by DEP through consultation with federal, state, and local agencies, and particularly the water management districts (WMDs). The plan includes all water in the state. Its main purpose is to recognize various interests competing for water use rights, and to allocate for these rights while retaining reasonable water quality and quantity control. Thus, the plan promotes the goals of environmental protection, proper drainage, flood control, and water storage. The DEP is responsible for the collection of scientific data and information regarding water resources, while permitting responsibilities are assigned to the WMDs. The DEP also retains direct regulatory power over:

1. point source discharges;

2. dredge and fill;

3. groundwater discharges;

4. solid and hazardous waste concerns;

5. leaking underground storage tanks.

State Agencies Administering Acts

Department of Environmental Protection (DEP)

The DEP was created by the Florida Environmental Reorganization Act passed in 1993. Its primary responsibility is to preserve the environmental integrity of Florida's air and water. Although this includes a large number of duties, only those duties relevant to the scope of this chapter are discussed. These include:

- the permitting of dredging and filling in waters of the state;

- administering the Water Resources Act of 1972;

- review of water control districts;

- regulation of air, water, and noise pollution;

- solid and hazardous waste management;

- public drinking water supplies;

- controlling noxious aquatic weeds;

- regulation of injection wells and wells related to oil exploration;

- the prevention or cleanup of pollutant spills or discharges into inland waters or lands of the state;

- the administration of such federal acts as the Clean Water Act and the Safe Drinking Water Act within the state of Florida.

The Department of Environmental Protection has specifically delegated to the water management districts (WMDs) power to administer and enforce the provisions of Chapter 373 of the Florida Statutes. The DEP has also given authority to the Department of Agriculture and Consumer Services (DACS). This agency regulates certain open burning activities through the Division of Forestry.

Fish and Wildlife Conservation Commission (FWCC)

The FWCC was created in 1968 for the management, protection, and conservation of wild animal life and fresh water aquatic life. Some of the agency's activities include habitat improvement, research, inspection of construction and development projects, development of public recreational areas, running a conservation information service program, and updating the threatened and endangered freshwater aquatic and land species lists. Although the FWCC has jurisdiction over any fresh running waters of the state, the DEP is the agency primarily responsible for enforcement of water regulations. Jurisdictional directives in this area are overlapping, as many agencies have been granted the authority to address this concern.

Department of Agriculture & Consumer Services (DACS)

DACS carries out functions related to farming practices and products. Responsibilities include the registration, labeling, and inspection of commercial fertilizers and pesticides; registration of pesticide applicators; and soil and water conservation. Like the other state agencies, DACS is divided into divisions, each with a separate concern. Among them are:

- Division of Agricultural and Environmental Services. This division regulates and licenses pest control operators and provides assistance to the Soil and Water Conservation Districts. Within this division is the Bureau of Pesticides, which registers pesticides and oversees pesticide programs that are related to the protection of ground water; farm workers, and endangered species. This division also includes the Bureau of Soil and Water Conservation.

- Standards Division. The Standards Division is responsible for petroleum and underground tank inspections. Its functions include testing fuel quality and assuring fuel dispenser, scale, and measuring device accuracy.

- Division of Forestry. This division manages forest resources. Within this division is the Bureau

of Planning Services, which includes the Watershed Protection Section (WPS). The WPS offers professional hydrology services to the public and other state agencies. It is also responsible for developing the silviculture elements of the State Water Quality Plan. Most importantly, the WPS monitors impacts of water regulation on forestry and could be most helpful in assisting the farmer in managing any forest resources that he owns.

- Division of Administration. The Division of Administration provides support services to all other divisions. This division includes the Office of Agricultural Law Enforcement, which provides for the inspection of agricultural products.

Litigation

One form of water quality regulation is through private law suits. These law suits are often based on the nuisance and negligence principles of law. Some federal laws have clauses that allow citizens to enforce the federal laws. It is important to understand how one may be held liable for actions by the public.

Nuisance, which is defined as using one's property in a way that causes harm to others, is a common basis for pollution litigation. Nuisances are categorized as either public or private, depending upon whether the nuisance affects the rights of the public or only the rights of an individual. Public nuisance actions may be brought by a public official on behalf of the public at large, and certain types of public nuisances may be criminal acts. In nuisance actions, the court will often balance the social value of the nuisance against the harm it causes. If the harm is slight and the social value is great, the suit will fail. But if the social value is small and the harm is great, the plaintiff may recover monetary damages and prohibit the defendant from continuing the activity.

The Florida Right-to-Farm Act restricts nuisance suits against farmers by providing that no farm, which was not a nuisance when it was established, will constitute a public or private nuisance after one year of operation. Change of ownership does not affect the provisions of the Act. This does not give the farmer a license to violate the principles of negligence or nuisance. Therefore, contaminating a water well or misapplying pesticides will still open the farmer up to a potential law suit. Additionally, the Act does not extend protection to unsanitary conditions or health hazards, or changes of use, either in type or intensity.

Negligence basically means causing harm to someone else by failing to do what a reasonable person would have done under the same circumstances. The harm may be economic, physical, or emotional. Anyone seeking to recover damages for someone else's negligence must prove four legal points: duty, breach of duty, causation of damage, and value of damage.

• Duty is an individual's responsibility to govern his or her own conduct so that others are not harmed.

• Breach of duty occurs where one of the parties does not fulfill his duty of care. That is, he does not act with the degree of caution or foresight that a reasonably prudent person would have used in the same situation.

• Causation is when the defendant's failure to use due care was the cause of the plaintiff's harm. Proving this step may be difficult if the damage is only indirectly related to the defendant's act, or if there are other possible causes for the harm.

• Damage. The plaintiff must prove that she/he suffered actual damage from the defendant's act. If no damage resulted, even where the defendant's conduct was admittedly negligent, the plaintiff has no claim for negligence.

Where a defendant's acts subject him/her to private suit and, at the same time, violate a statute, the court will usually regard the violation of the statute as enough evidence of wrongful conduct to find the defendant guilty of negligence in the private suit as well. This rule applies, however, only if two conditions exist. First, the damage complained of in the lawsuit is of the type the statute is intended to

prevent. Second, the plaintiff is a member of the class of persons the statute is intended to protect. It should be noted that compliance with all statutes does not guarantee immunity from negligence actions, as lawful behavior may still be negligent.

Strict liability means liability imposed without evidence of negligence. That is, the defendant may be found guilty upon a showing that his action resulted in harm, without consideration of whether or not he acted reasonably. Strict liability is usually imposed upon those who engage in abnormally dangerous or hazardous activities, like handling explosives, or other activities defined by statute, e.g., CERCLA.

Where two or more parties, acting independently, are the source of a plaintiff's injuries, the law (or the courts) may impose the principle of joint and several liability. This principle allows the plaintiff to recover from either defendant the full amount of damages, and the defendants are forced to apportion the cost among themselves.

Groundwater
State Provisions

Discharge of waste into state waters is prohibited unless permitted by a state agency. Because underground water is included in the definition of water, this also applies to groundwater. A discharge activity will not be permitted if contaminants reduce ground or surface water quality below the required DEP classification standard. A contaminant is any substance that is harmful to plant or animal life. However, applications of chemicals for agricultural purposes such as to control insects and aquatic weeds are exempt. Additionally, these exempt chemical applications must be approved for the particular use by the EPA or DACS; application must be made according to the label, and state standards as well as the Florida Pesticide Law must be followed.

Groundwater is classified into four categories, based first on whether the water is potable (drinkable) or nonpotable, and on the total of dissolved solids the water contains. Under the classification scheme, aquifers retain the highest protection and are known as G-1 waters. Class G-4 waters are nonpotable, located in confined aquifers only, and receive the least amount of protection. Additionally, unconfined groundwater always receives more protection, as it is susceptible to contamination from another aquifer.

Maximum contaminant standards specify the maximum amount of particular contaminants that will be tolerated in a particular class of water. Maximum contaminant levels (Primary Drinking Water Standards) are generally in accord with the EPA standards contained in the Federal Safe Drinking Water Act. Permits will not be issued when maximum contaminant levels are exceeded by a discharge activity. Secondary standards are also monitored in new facilities for compliance.

The Water Management Districts have the power to control consumptive uses of groundwater in areas of known groundwater contamination. In other words, through Florida Statutes Chapter 373, the WMDs can restrict consumptive use through permitting whenever contamination is found. Furthermore, the EPA regulates stormwater discharges associated with agriculture. The EPA exempts, via permit, agricultural stormwater discharges and agricultural return flows composed entirely of return flow from irrigated areas. However, the EPA does not exempt agricultural return flows that are not composed entirely of return flows from irrigated agriculture.

Federal Provisions

Federal regulation of groundwater consists of a variety of statutory directives administered by a host of administrative agencies. More than sixteen pieces of federal legislation have some effect on groundwater, or have the potential to affect activities and programs relevant to its use. The most important are:

• Clean Water Act. The most important piece of federal legislation is the Clean Water Act. The chief purpose of the Act is the elimination of point source pollution to surface water. Groundwater is directly implicated due to the natural linkage of surface and groundwater resources. When a party pollutes the surface water, the hydrologic water cycle makes it more likely than not that groundwater is simultaneously being contaminated.

• National Pollutant Discharge Elimination System (NPDES). NPDES places flow limitations on point sources (a recognizable origin of pollution, for example, a pipe, well, or leaking container) of water pollution. Currently, the EPA administers the program. If Florida were to adopt the NPDES program, the DEP would be the sole agency issuing the NPDES permits, except for stormwater discharge permitting, which is the responsibility of WMDs.

• Safe Drinking Water Act. This act establishes maximum contamination levels for drinking water and serves to protect sole source aquifers (aquifers whose main use is to provide drinking water). Perhaps most significantly, the Act monitors the underground injection of contaminants into groundwater used for the public drinking water supply. The Underground Injection Control (UIC) program is the vehicle used to protect underground sources of drinking water. The UIC program is delegated to Florida by the EPA.

Wetlands

Federal protection of wetlands stems from several sources. Under the Clean Water Act, the Corps of Engineers is authorized to issue permits for the discharge of dredge and fill material into waters of the United States. Coupled with this authority, and pursuant to the National Environmental Policies Act (NEPA), the Corps must complete an environmental assessment study before issuing a permit for work in sensitive areas, especially wetlands. If the Corps determines that the activity will have a significant impact on the wetlands, a more thorough environmental impact study will be required as well. As added checks on the Corps' authority to issue dredge and fill permits, the Corps is required to consult with other federal and state agencies whenever relevant, and, by virtue of the Clean

Water Act, the Administrator of the EPA retains the power to veto Corps-issued permits at the EPA's discretion.

On the state level, the DEP is primarily responsible for wetland protection. DEP has been given authority to protect environmentally sensitive wetland areas and designated areas of critical state concern (prominent examples include the Florida Keys, Big Cypress Area, Green Swamp Area, the Everglades Protection Area, and the Apalachicola Bay Area). In addition, the appropriate water management district should be consulted with regard to any work to be done near wetlands, whether individually owned or not. The districts have adopted management and storage of surface water rules that regulate activities in wetlands under DEP supervision. Also, the districts are authorized to establish specific permitting criteria for dredge and fill operations in connected and isolated wetlands.

For the purpose of Corp of Engineers permitting, wetlands include those areas with a prevalence of vegetation adapted to live in saturated soil conditions. With regard to the DEP's jurisdiction, detailed indexes have been compiled to determine which soil and plant characteristics identify wetlands. On request, the DEP or the appropriate water management district will issue declaratory statements for particular sites regarding whether or not its jurisdiction extends to that area. They also can issue formal or informal determinations as to the extent of wetlands. These declarations are valid for up to five years, as long as physical conditions on the property do not change. The Districts' definitions of wetlands depend on hydrologic, vegetative, and soil characteristics of an area.

Dredge and fill permits are required for any nonexempt activity. The first step in obtaining a permit is assuring DEP (or the WMD) that state water quality standards will not be violated by the proposed dredging and filling activity. Second, the project cannot be contrary to the public interest. Several factors must be considered in this determination, including the effect on public health, safety, welfare, and property rights, the effects on fish and wildlife (especially on threatened or endangered species), adverse effects on navigation or harmful erosion, a variety of other factors, like the effects on marine productivity, the temporary or permanent nature of the project, and effects on historical and archaeological resources. In the final analysis, factors that mitigate destruction of the wetlands are also considered where some damage to the environment is inevitable. Duration of permits is generally for five years and does not exceed ten years, unless the project cannot reasonably be completed within that time. In these exceptional circumstances, the DEP may issue 25-year permits.

Under federal regulations, normal farming, silviculture and ranching activities in wetlands are exempt from the fill permitting mandates of the Clean Water Act. This exemption does not apply to mechanized equipment used in land clearing, as this has been held to represent a point source of pollution. In Florida, generally speaking, agricultural activities and agricultural water management systems are exempt from wetland regulation, provided that the activities are consistent with agricultural activities. Agricultural activities are defined to include all necessary farming and forestry activities that are normal and customary for a particular area, provided such operations do not impede or divert the flow of surface waters. Agricultural water management includes farming or forestry water management systems and farm ponds, which are permitted pursuant to Chapter 373 or exempted from the permitting requirement of the surface water management statutes.

The Everglades Forever Act empowers the South Florida Water Management District to do the following:

- Adopt the Surface Water Improvement Management plan.

- Acquire land by eminent domain for treatment and storage of water prior to its release into the Everglades Protection Area.

- Create and administer a stormwater management system.

- Develop, implement, and enforce plans and programs along with DEP for the improvement and management of the Everglades Protection Area.

Solid Waste Management

Both Florida and the federal governments have enacted Resource Conservation and Recovery (RCRA) legislation to deal with the disposal of solid wastes. The state rules are administered by the DEP, and the federal rules by the EPA. Both RCRAs are aimed at protecting health and the environment and at recycling or reclaiming solid wastes to the greatest extent practicable. Because improperly handled solid waste represents such a grave threat to water quality, it also falls within the scope of other federal regulations, most notably the National Pollutant Discharge Elimination System (NPDES) of the Federal Clean Water Act. NPDES is an EPA- and state-enforced system requiring permits for all point source pollution discharges into navigable waters.

Aside from the areas of special state and federal interest, local authorities are basically free to set their own rules, within broad state guidelines, for the collection and disposal of solid wastes. Cities, counties, or municipalities should be consulted regarding problems involving the locations of disposal facilities, frequency of waste pickup, and other day-to-day waste disposal questions.

Solid waste has been defined as any garbage, refuse, sludge, or other discarded materials, including liquid, semisolid, or contained gaseous materials. Furthermore, waste from agricultural activities is expressly included. The only relevant materials that are expressly listed as NOT within the RCRA definition of solid waste are domestic sewage; waste from irrigation return flows; and point source discharges that are permitted for release under NPDES of the Clean Water Act. If none of these exceptions apply, and if the material involved may be said to be discarded or to have served its useful purpose, it is a solid waste.

Solid waste may be recycled, reused, discarded, reclaimed, or stored, depending on its nature.

Many solid wastes may also be hazardous wastes, and therefore subject to other, more restrictive regulations. Therefore, the section on hazardous waste must also be consulted when considering solid wastes.

NPDES coverage extends only to discharges into navigable waters, and sets different standards for the amount of waste that may be discharged based upon the type of industry involved, the toxicity of the waste, and the acceptable maximum amounts of harmful elements that can be released into the body of water that is being polluted. Facilities that are disposing of wastes in any way not in accordance with the federal RCRA guidelines or NPDES requirements for discharging water are deemed to be violators of the law.

The DEP has established detailed regulations controlling the application of sludge to land. Sludge is basically solid, liquid, or semisolid waste generated from wastewater treatment plants, air pollution control facilities, septic tanks, grease traps, portable toilets or related operations, or any other waste having similar characteristics. This system requires all producers to have their sludge tested so that it may be classified as one of the three grades of sludge recognized by DEP. (Grade I is the least hazardous, Grade III, the most hazardous.) No permit is required to apply sludge to the land owned by the sludge generator as long as the sludge is treated for bacteria and insect infestation and handled in accordance with DEP rules. No permit is required for normal farming operations, which include the application to farm land of Grade I domestic, Grade I composite sludge, domestic septate that has been properly treated to kill microorganisms, or food service sludge. The application of Grade II sludge is not a normal farming operation under the rules. Since Grade III sludge must generally be disposed of where it will have the least harmful effect on the environment, it is unsuitable for application to farm land. All uses that are not normal farming operations will require DEP permits. The DEP may require a general permit for application of less harmful sludge, but may require more restrictive permits (including solid waste disposal site permits) for the application or disposal of more harmful sludge.

Pollutant Storage Tank Systems

Statewide regulation of stationary storage tanks is primarily the domain of the DEP. The Pollutant Discharge Prevention and Control Act gives the DEP the power to regulate pollutant storage tanks. Subsequently, most of the rules covered in this section are those of the DEP. It is important to note, however, that the regulations allow individual county governments to promulgate their own regulations. These local regulations can be more stringent than those of the DEP. County authorities should be consulted even if the storage tank activity or condition is apparently within the statewide standards.

A tank will fall within the scope of the regulations if it holds a pollutant. A pollutant is defined as any type of oil or gasoline; any pesticide; or any ammonia or chlorine compound or derivative. The bulk of the regulations only apply to storage tank systems whose individual storage capacity is greater than 110 gallons. Tanks smaller than this need only comply with the general requirement that they do not discharge their contents into the environment, and that they conform to fire prevention standards.

Twenty-one types of storage tank systems are on a list that makes them exempt from this Act. For example, any agricultural storage tank system of 550 gallons or less capacity is exempt. For the most extensive and updated list, check with the DEP. The DEP makes important distinctions between new tanks and those that were already in operation before 1992. New tanks are subject to the strictest safety standards, but existing tanks must be brought into compliance with many of the same standards for overfill protection, monitoring systems, and tank linings within a given period. The time allowed for this retrofitting varies depending upon the year the tank was installed, but the DEP contemplates the complete retro-fitting of all existing tanks by 2009.

Aboveground systems are subject to less restrictive regulation, and are subject to this chapter if they have a storage capacity greater than 550 gallons. (An aboveground tank has no more than 10 percent of its volume buried, including integral piping.) These tanks, however, are subject to similar record keeping rules as buried tanks, and must be equipped with an impervious containment barrier to catch spills. Further, any part of the tank in contact with the ground must be protected against corrosion.

Operators of storage facilities are required to keep records for DEP inspection. These records must include the results of all tests and inspections, as well as maintenance and inventory notes on the tanks or their contents. The records must date back at least two years and must be made available to the DEP within five working days notice.

Pesticides

The use of agricultural pesticides falls within the purview of several federal statutes as well as Florida law. Under federal law, pesticides are primarily regulated by the Federal Insecticide, Fungicide and Rodenticide Act (FIFRA). Many other federal statutes and agencies come into play, though, in more particular situations. Florida has passed additional legislation designed specifically to cover the use of pesticides, with enforcement authority vested primarily in the Florida Department of Agriculture and Consumer Services (DACS). Many of these laws and regulations are similar to those enacted by the federal government.

Pesticide labels have become a central tool for enforcement of both FIFRA and state law. Under FIFRA, users must comply with all the application and precautionary instructions on the label. In this sense, the label is the law. Failure to comply with the label can result in strict penalties. It is important to note that the farmer can be penalized even if an employee of the farmer was responsible for the failure to follow the label instructions. The Florida statutes excuse pesticide users from following the instructions on the label only to the extent that:

- the pesticide may be applied in lesser dosages, concentrations, or frequencies than recommended by the label;

- the pesticide may be used against a pest not specified as a target pest by the label, so long as the application is to a crop, animal or site specified by the label and the label does not prohibit the application;

- applicators may employ any method of application not prohibited by the labeling;

- applicators may mix the pesticide with a fertilizer when not prohibited by the labeling.

Under the federal Clean Water Act (as outlined in the Solid Waste section), any point source pollution of a navigable waterway must have a NPDES permit from the EPA. While irrigation return flow is exempted from this requirement, other agricultural polluting activities are not. For instance, a ditch containing fertilizers or pesticides entering navigable waterways is a point source of pollution, and is subject to the permitting requirements of this act.

Any pollutant introduced into a source of drinking water, including wells and aquifers, is subject to regulation under the Federal Safe Drinking Water Act. It should be mentioned that this act has been interpreted to cover irrigation return flow, which seeps back into subsurface water as an underground injection if the flow carries any pollutants.

The FIFRA requirements for adherence to label instructions also extend to the storage and disposal of pesticides. Further, pesticides often fall within the hazardous waste classifications of RCRA, and are usually subject to solid waste disposal regulations. The RCRA or FIFRA sections of this publication should be reviewed before storing or disposing of any pesticides. The EPA provides guidelines for the storage and disposal of pesticides that, while not mandatory, are helpful in dealing with specifics of locating storage sites, protecting groundwater and other ways of avoiding environmental contamination.

Florida law requires that all irrigation systems that are used to administer crop management materials, such as pesticides or fertilizers, be equipped with antisyphon devices to prevent the backflow of pesticide- or fertilizer-contaminated water to the aquifer. These rules also apply to pesticides premixed with fertilizer, although such mixtures must be clearly labeled as such before they may be offered for sale.

Chapter 4. Water Management Districts

by Brian Boman and Larry Parsons

Water Management Districts

The Florida Water Resources Act of 1972 was enacted following a severe drought in 1971, as part of a package of major environmental and land use bills. (The South Florida and the Southwest Florida Water Management Districts had been created earlier to address flooding and water shortage problems.) The act provides for comprehensive water resource management and the division of the state into five water management districts (WMDs, Fig. 4-1). The boundaries of the WMDs were established by hydrologic surface water runoff boundaries. The WMDs that govern water in Florida's citrus belt include the Southwest Florida Water Management District (SWFWMD, Fig. 4-2), the St. Johns River Water Management District (SJRWMD, Fig. 4-3), and the South Florida Water Management District (SFWMD, Fig. 4-4). The WMDs play a very significant role in the regulation of water use and natural resource protection. The WMDs' responsibilities include:

- management of water and related land resources via promotion of conservation;

- proper utilization of surface and groundwater resources;

- regulation of dams, impoundments, reservoirs, and other structures to alter surface water movement;

- combating damage from floods, soil erosion, and excessive drainage;

- assisting local governments in developing comprehensive water management plans, particularly by providing data on water resources (to accomplish this, WMDs are authorized to perform various field investigations and to provide works for the beneficial storage of water);

- maintenance of navigable rivers and harbors and the promotion of the health, safety, and general welfare of the people of the state (directly attached to this general welfare consideration is the power to implement water shortage emergency plans);

- participation in flood control programs, and the reclamation, conservation, and protection of lands from water surplus or deficiencies;

- maintainance of district water management and use facilities, and determination of levels of water to be maintained in the district bodies of water (to do so, districts may establish minimum flows for their works and water courses, below which further withdrawals would be significantly harmful to the water resources or ecology; this also involves the establishment of minimum water levels for surface water and groundwater).

Each district is autonomous, has ad valorem taxing authority, and is run by a governing board consisting of nine members (except for SWFWMD, which has eleven). The members serve four-year terms and are appointed by the governor and

Figure 4-1. Location map of Florida's five water management districts.

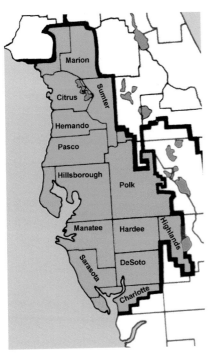

Figure 4-2. Southwest Florida Water Management District boundaries.

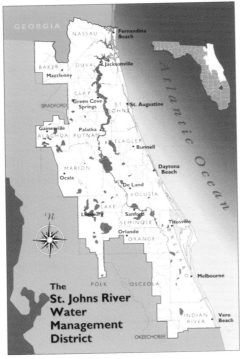

Figure 4-3. St. Johns River Water Management District boundaries.

confirmed by the state. The governing board sets the policies that will best accomplish the district's mission. Generally, an executive director is responsible for the operation of the district, including the implementation of policies and rules. Typically, each district is divided into departments that handle the various programs, including permitting.

WMD governing board powers include administering the permit programs of Chapter 373, Florida Statutes. Thus, WMDs are responsible for permitting wells, management and storage of surface waters, and issuing consumptive use permits (SFWMD and SJRWMD) or water use permits (SWFWMD). Each WMD has specific criteria detailing the types of activities that require permits, the contents of permit applications, the procedures that surround submission of an application, and areas specifically exempted from permitting requirements. The governor and cabinet, sitting as the Land and Water Adjudicatory Commission, have authority to review any order or rule of a WMD.

Permitting System

Each of the WMDs has implemented a water use permitting program. In the SFWMD and SJRWMD, the water permits are called Consumptive Use Permits (CUP) and in the SWFWMD, the permits are called Water Use Permits (WUP). However, not all users are required to apply for a permit. Uses for domestic consumption of water by individual users are exempt. In addition, the WMDs have varying thresholds of use, withdrawal capacity, or well size, for which permits may not be required. Other users may qualify for general permits or exemptions.

Applicants proposing new water uses are required to establish that the proposed use:

• is a reasonable beneficial use,

• will not interfere with a presently existing legal use of water,

• and is consistent with the public interest.

Water use permits are granted for fixed periods of time. Duration may not exceed twenty years, except that public facilities may be permitted for up to fifty years, if necessary, for bonding. In practice, permits have been issued for much shorter durations because the districts have lacked the information needed to commit the resources for longer periods of time. Temporary permits also are authorized. Permits are revocable only for material false statements, willful violation of permit conditions of the Act, and for nonuse of the water supply. Thus, except during times of water shortage or emergency, permittees have certainty of use. The districts generally allow free transfer of permits, provided the use and conditions of withdrawal remain the same.

Reasonable Beneficial Use

Reasonable beneficial use is defined as the use of water in such quantity as is necessary for economic and efficient utilization for a purpose and in a manner that is both reasonable and consistent with the public interest. It requires efficient economic use of water, a characteristic of beneficial use. In addition, a water user must consider the rights of the general public. Wasteful use of water is not permitted under the reasonable, beneficial-use standard. Reasonableness and consistency with the public interest also are required, both with regard to the purpose of the use and the manner in which it is carried out.

Harm To Other Users

Applicants must show that the proposed use will not interfere with any presently existing legal use of water. Existing legal users thus have been granted protected rights for the duration of their permits. Other users cannot gain rights conflicting with existing rights except through the renewal process, at which time a use that better serves the public interest may be permitted instead of the existing use. Often, however, harm to other water users may not become apparent until after the new use has been permitted and begun. In such a situation, state water policy provides for modification of the permit to curtail or abate the adverse impacts.

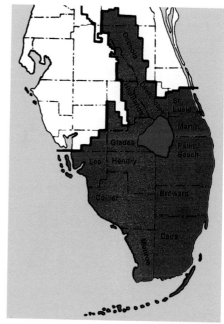

Figure 4-4. South Florida Water Management District boundaries.

Public Interest

A CUP/WUP must be consistent with the public interest. Public interest can be defined through rule making by the Florida Department of Environmental Protection (FDEP) and the WMDs. Some of the WMDs have listed criteria for determining the public interest. Generally, however, consistency with the public interest is determined on a case-by-case basis in the permitting process.

Planning

FDEP has been charged with the responsibility of developing an integrated, coordinated plan for the use and development of the waters of the state. This responsibility extends to studies of existing water resources, contemplated uses and needs for water, and other subjects such as drainage and floodplain zoning. The product of these efforts is the State Water Use Plan, which is to be combined with state water quality standards and classifications to form the Florida Water Plan. Other FDEP responsibilities include giving consideration to the requirements of public recreation, and to the protection and procreation of fish and wildlife. The WMDs are part of the process in developing these plans.

Additional planning requirements were imposed by the 1982 and 1997 Legislatures. Each WMD was required to develop a groundwater basin resource availability inventory, identifying and analyzing specified basic information regarding groundwater basins and associated recharge areas. Planning to assist rapidly urbanizing areas in meeting water supply needs was mandated, as was the development of water shortage plans. The 1997 Water Resource Development Act required that each District develop a water supply plan to provide water needs for a 1-in-10-year drought.

Environmental Considerations

The effects of consumptive uses on the environment are important considerations in planning, permitting, and preservation of natural resources, fish, and wildlife. The state water use plan may designate desirable uses that are to be given a preference in the granting of CUP/WUPs. Such uses might include preservation of the environment, protection of recharge areas, or recreation. Once established, such a preference must be recognized in permitting. Similarly, certain uses might be declared undesirable because of the nature of the activity or the amount of water required. In such cases, the governing board of a water management district is authorized, but not compelled, to deny a CUP/WUP.

The Water Resources Act mandates the establishment of minimum flows for surface bodies as well as minimum lake and groundwater levels. Commercial navigation, recreational boating, fishing, hunting, swimming, and functioning of the ecosystem are some of the nonconsumptive public purposes that are protected under the minimum flow and level concept in Florida. Minimum flows are to be established at the level at which further withdrawals would be significantly harmful to the water resources or ecology of the area. Minimum levels are the levels of surface or groundwater

below which further withdrawals would be significantly harmful to the water resources of the area.

Water Shortages and Emergencies

Most WMDs allocate water for CUP/WUPs based on the supply of water required to meet irrigation needs for a 2-in-10-year irrigation requirement. When the supply of water is reduced by drought or overuse, other measures must be implemented to protect the resource and users. The governing board of each water management district is required to adopt a water shortage plan, which includes a system of permit classifications according to source, method of withdrawal, and use. The governing board may implement the water shortage plan in the entire district or a portion of the district when insufficient water is available to meet the requirements of the regular permit system, or when conditions require temporary reduction in use to protect the area's water resources from serious harm. Restrictions may be imposed on all permits, or on specific classes of permits to protect or restore the water resources.

If the water shortage is an emergency and implementation of the water shortage plan would be inadequate to protect the public, reasonable uses, or the ecosystem, the executive director of the WMD may declare a water emergency. Unlike the water shortage plan, which applies to classes of permits, water emergency orders may be issued to specific users. An order may limit, rotate, apportion, or prohibit use of the water resources and must be complied with immediately.

Each of the water management districts has adopted water shortage plans. These plans classify uses by source within surface water and groundwater categories. Within each category, the source is classified further. For example, withdrawals from the Upper Hawthorne Aquifer are classified separately from withdrawals from the Sandstone Aquifer. Uses are classified within four major groups:

• domestic

• commercial (utility, mining)

- agricultural

- recreation

A water shortage may be declared for specific sources. Once declared, predetermined restrictions are imposed on each of the use classes. A level of restrictions is imposed depending on the severity of the shortage. Each WMD has its own set of restrictions and criteria. The plan also provides criteria and procedures for declaring a water shortage emergency. Although local authorities are encouraged to adopt and enforce local ordinances to implement the plan, the district is authorized to enforce the plan.

Water Management Permits for Citrus Operations

Water allocation and resource oversight are accomplished in the Florida's citrus producing regions through the regulation programs of the SFWMD in South Florida and the Kissimmee River Basin, SWFWMD in the West Central region, and SJRWMD in the Northeast region. The WMDs issue permits to manage the quantity of surface and ground water used, and to reduce pollution of water supplies. The WMDs also keep track of how much water is being used throughout the WMDs, and protect associated property and environmental resources from harmful activities. This permitting protects the water needs of the public, existing and future users, as well as the environment.

A permit represents a contract between the land owner or permittee and the WMD. Failing to obtain a permit, if one is required, or to comply with the conditions of a permit, can result in enforcement actions such as fines and other legal action. WMDs issues three main types of permits:

- water use

- well construction

- environmental resource or surface water

Within those main categories are two tiers: general and individual permits. General permits are issued for water quantities or surface water impacts below a specified level, and are approved by the staff of the appropriate WMD. Individual permits typically involve larger quantities or impacts, and require approval of the WMD governing board.

A consumptive water use allows a user to withdraw a specified amount of water, either from a groundwater well or from a surface water source. A Well Construction Permit (WCP) is required prior to the drilling or construction of a new well, and the repair or plugging of an existing well. WCPs ensure that wells are constructed by qualified contractors to meet safety, durability, and resource protection standards. An Environmental Resource Permit (ERP) must be obtained before beginning any construction activity that would affect wetlands, alter surface water flows, or contribute to water pollution. An ERP is needed to regulate activities such as dredging and filling in wetlands, construction of drainage facilities, stormwater containment and treatment, construction of dams or reservoirs, and other activities affecting state waters. The ERP combines wetland resources permitting and management and storage-of-surface-water permitting into a single permit, in an effort to streamline the permitting process.

Citrus operations will usually need well construction permits. An environmental resource permit might also be needed. A complete ERP application may be needed prior to obtaining a water use permit, and a CUP/WUP may be needed prior to obtaining a well construction permit. In general, the following may be required for citrus operations.

CUP/WUPs are needed to:

- irrigate crops with water withdrawn from either a well or surface water facility;

- provide water from a well or pond for frost protection;

- refill irrigation ponds or ditches with water from a well;

- furnish well water for product spraying or other support activities.

Well construction permits are required to:

• construct a new well, and

• repair or plug an existing well.

Environmental resource permits are needed to:

• develop a grove in an area that currently contains wetlands, or excavate or fill in natural wetlands;

• construct any facility that will alter surface water flows or runoff, intrude into a floodplain, or contribute to water pollution;

• construct a dam, canal, impoundment or reservoir.

Wells

FDEP has delegated most of its authority to regulate water wells to the individual WMDs. Therefore, the appropriate WMD should always be contacted before taking any action involving water wells. The rules are designed to safeguard both the quality of water extracted from the wells and the quality of the aquifer water, both of which could potentially be polluted by intruding wells.

All proposed well sites must be preapproved by the appropriate WMD. This guards against the possibility that a well will unknowingly be drilled into an area of existing groundwater contamination. The FDEP provides continually updated maps of contaminated sections of the aquifer to the WMDs and pertinent county health departments. The WMDs and county health departments also prescribe the minimum distances from the contaminated areas in which wells may be constructed.

Each WMD sets permit application fees depending primarily upon the nature of the activity and the size of the well. The fees vary with each WMD. Construction permits are valid for a period of one year. If the construction or repair cannot be finished within one year, the WMDs can either extend the limit or require a new permit. Most Districts require a CUP/WUP before a well construction permit will be approved.

CUP/WUP

The various WMDs have exclusive authority to issue Consumptive Use or Water Use Permits, since they are charged with maintaining the state's reserves of usable water at an acceptable level.

When a grower's required water usage reaches the WMDs' specified threshold levels, a CUP or WUP will be required. Threshold levels are determined by the individual WMDs, and it is important to consult the appropriate WMD regarding this threshold. All uses must be permitted. However, there are some uses that may be exempted, including:

• individual residential consumption,

• wells for testing or monitoring,

• private, shallow wells,

• certain heating and cooling systems,

• dewatering activities necessary for construction, if they are completed in less than six months.

Exemptions, however, do not absolve growers from complying with the intent of the Water Resources Act, and the use of the water supply must be reasonable. Reasonable use may include the purpose of use, suitability of the body of water and social values, degree of harm to the environment, and the practicality of avoiding the harm.

Permits are granted for fixed periods of time. Except for public facilities, they may not exceed twenty years and are usually granted for much shorter periods. When the nature of a proposed use is such that the permit application process may be lengthy, the district may issue a temporary permit. Transfer of permits between activities identical in nature at the same location is usually allowed and conditions of the permit usually remain the same. Any failure to continually observe the terms provided by a permit may result in its revocation.

Environmental Resource Permits

The governing boards of the WMDs and the FDEP are vested with the authority to require Environmental Resource Permits (ERPs). This authority is delegated almost entirely to the WMDs. Therefore, they should be consulted before any alteration of surface water is undertaken.

The rules governing surface water management include the construction, operation, or alteration of any stormwater management system, dam, impoundment, or reservoir. In effect, the rules apply to virtually every type of artificial or natural structure or construction that can be used to connect to, draw water from, drain water into, or be placed in or across surface water. They include all structures and constructions that can have an effect on surface water, including dredging, filling, and activities that create canals, ditches, culverts, impoundments, fill roads, buildings, and other impervious surfaces.

The statute contains an exception for agricultural operations. Citrus growers may alter any tract of land without an ERP permit as long as the practices are normal occupational activities whose sole or predominant purpose is not to obstruct or impound surface water. These activities can include site preparation, clearing, fencing, or contouring to prevent soil erosion, soil preparation, plowing, planting or harvesting. The exemption is qualified in that the impoundment or obstruction of surface waters may not be the chief purpose of the alteration. Construction or maintenance performed on dikes, dams, or levees in an agricultural closed system will be exempt from ERP permitting requirements. ("Closed system" means a self-contained irrigation system used in farming that does not discharge off-site.) Nonetheless, these works must still comply with generally accepted engineering standards and, where the engineering practice is regulated by the state, this might require proper certification of the project and strict adherence with the original plans. It is always wise to consult with the specific WMD when attempting to determine if a proposed activity is exempt.

Certain WMDs rely on the threshold concept to determine when a permit is required. For example, once a certain quantity of water is impounded or a certain size project is proposed, the WMD will require a permit unless the activity is somehow exempt. Permitting thresholds and exemptions adopted by rule vary among the WMDs. The applicant must show that the planned activity will not be harmful to the water resources, and that the activity will not be inconsistent with the objectives of the WMD by showing the project is not against the public interest.

If the planned project significantly degrades the water quality, the applicant can still obtain a permit by showing that the planned activity will be clearly in the public interest. In determining whether a planned activity is not contrary to public interest, there are some criteria that are evaluated:

- whether the activity will adversely affect the public health and safety of others,
- whether the activity will adversely affect fish and wildlife conservation,
- whether the activity will adversely affect navigation or the flow of water,
- whether the activity will adversely affect fishing or recreational values in the area of the activity,
- whether the activity is permanent or temporary in nature,
- whether the activity will adversely affect or will enhance significant historical and archaeological resources,
- whether the current condition and value of activities occurring in the area is affected by the planned activity.

If an applicant has trouble meeting the permit criteria, he can still get a permit through mitigation. Mitigation is the creation, maintenance, or restoring of a surface water area in exchange for the degradation of another area. If a project will degrade a surface water body, it may be still possible to get a permit, but that applicant will have to mitigate the damage by creating or restoring another area. The applicant's mitigation plan needs to approved by the WMD.

After permit approval, following certain best management practices is an excellent way to help ensure continuing compliance with government regulation. Some examples include establishing buffer strips and streamside management zones around a system; maintaining streams and culverts so as not to affect upstream or downstream culverts; and careful construction of access roads.

Permits and Regulation by Other Agencies

Water management WMD permits and regulation will not eliminate the need to obtain other required authorizations from federal, state and local agencies, or other special jurisdictions. The WMD staff are willing to discuss this issue with growers and provide information when possible. Landowners should be especially aware of the provisions of the federal Food Security Act. Unless federal permitting is approved prior to construction, activities that affect any wetlands or threatened/endangered wildlife may jeopardize their eligibility for USDA benefits, such as crop insurance, disaster payments, etc. Violations can cause ineligibility on the landowner's entire land holdings. Prior to any land use changes, landowners should contact the local USDA-NRCS office to help delineate the wetlands and provide further assistance as to federal natural resources regulations.

Chapter 5. Conservation Practices and BMPs

by Brian Boman, Chris Wilson, and Jack Hebb

Recent concerns with the health and sustainability of plant and animal species in and around citrus-producing areas have highlighted the need for all users to minimize the adverse environmental effects of their operations. Nutrient and agricultural chemical movement into the groundwater have been identified as the most important concern in Ridge citrus areas. The most pressing concerns identified in flatwoods areas are:

- excessive releases of fresh water during heavy rainfall periods;

- movement of sediment from groves and ditches into water bodies;

- transport of pesticides and heavy metals into public water bodies;

- nutrient movement into canals, lakes, and streams;

- the proliferation of aquatic plants in canals and waterways.

The subtropical climate of the Florida citrus industry requires intense agricultural management to ensure that the citrus produced has the quality to succeed in an ever increasing competitive global market. Since the citrus industry is the largest and most visible agricultural enterprise in the state, it has become the target for the media, general public, and governmental organizations that are concerned with the long-term health of the state's many diverse environments. One of the major concerns of these groups is the protection of the state's water resources for the benefit of future generations. In order to meet society's demands to reduce the off-site effects of citrus operations, Best Management Practices (BMPs) should be employed whenever possible. BMPs are production systems and management strategies that have been scientifically shown to minimize adverse impacts of agricultural

production. BMPs can also be defined as those on-farm operational procedures that are designed to achieve greatest agronomic efficiency in food and fiber production with minimum off-site effects, while simultaneously maintaining an economically viable farming operation.

Citrus growers have been among the first in the state to embrace many of the BMPs that help to minimize adverse impacts of fertilization and agricultural chemical applications. The intent of these BMPs is to ensure that water quantity and quality are preserved to the benefit of all interests. Not all BMPs are applicable to any one particular citrus operation. Therefore, each BMP should be considered in the context of the overall citrus operation prior to adoption, to ensure that it will achieve the proper objective.

Water Volume

The drainage infrastructure that has been developed to make agricultural land productive has significantly increased drainage frequency, discharge volumes, and the velocity of water discharged from structures within watersheds. Excess rainfall from high-intensity thunderstorms, tropical storms, and hurricanes must be drained off in order to protect agricultural and urban areas from flooding damage. The resulting discharges from these events can cause significant environmental impacts. More stable and lower flows help maintain water quality in the receiving waters and help to ensure that dissolved oxygen (DO) levels remain above the minimum standard required for a healthy aquatic system.

Water quality in receiving bodies is directly affected by nutrients, pesticides, and sediment in freshwater inflows. There is a correlation between runoff from the watershed and algal blooms in receiving bodies. Therefore, management strategies need to address both freshwater quantity and quality. All types of land uses within a watershed contribute surface water runoff and pollutant loads to water bodies. However, in many parts of the state, citrus is the largest and most visible land use type, and the activities associated with citrus production are an important consideration in the overall health of nearby water bodies.

All activities on grove lands potentially can affect the water resources of the watershed. Wherever feasible, citrus growers should consider implementing surface water management strategies that can provide additional storage and reduce the impacts associated with excessive freshwater discharges. These surface water management strategies can range from additional on-site canal storage to the construction of detention reservoirs for holding excess drainage water. It is important to conduct site-specific evaluations to determine if any additional storage can be provided on-site, and to plan long-term water management strategies that will minimize off-site discharges during periods with intense rainfall. Potential BMPs to consider for reducing off-site impacts of drainage water include:

- Water Table Management
  Water table control can be managed more efficiently by having sufficient hydraulic capacity in the ditch/canal system, using water control structures on culverts, laser land leveling where appropriate, constructing and maintaining a properly designed drainage system, and actively monitoring the water table.

- Scheduling Irrigation and Drainage
  Drainage and irrigation schedules should focus on optimal crop production by encouraging deep rooting, and by maintaining a water table that minimizes water quantity and quality impacts. During excessive rainfall periods, when drainage rates are insufficient to prevent upward fluctuations of the water table, root pruning can occur. Therefore, irrigation and drainage practices should be focused on maintaining a

well-defined root zone that can be managed during both drought and wet periods.

- Moderating Discharge Rate
  Adjust the rate of discharge proportionate to the rate of lateral movement of water through soils. This can lessen turbulence, reduce sediment movement, reduce erosion, and moderate the impacts on the receiving water body.

- Water Furrow Maintenance
  Maintain a consistent bottom slope on water furrows between beds to achieve more uniform drainage. Avoid rutting and sloughing of water furrow areas.

- Monitoring Soil Moisture
  Use tensiometers and water table observation wells for irrigation and drainage management to avoid excess soil moisture depletion and minimize water volume requirements during irrigation cycles.

- Drainage Management Plan
  Implement and maintain a written drainage management plan that provides specific responses to various types and levels of rainfall. The goal of the plan should be a reduction in volume of off-site discharge while maintaining a healthy environment for citrus growth. The plan should include target water table levels and pump or drainage structure operating procedures that will be used for typical and extreme rainfall events. Consideration should be given to the use of existing canals and ditches for temporary water storage. USDA-NRCS Conservation Plans can be used to help develop drainage management plans.

- Drainage Rate and Volume
  Drainage rate and volume following excess rainfall events should be consistent with maintaining an adequately drained root zone while minimizing off-site impacts. When the water table approaches the target level, off-site discharges should be moderated. Depending on the grove design, this may require reducing pump rpm, and adjusting the discharge structure or employing

pulse drainage (discharging for short periods of time and then allowing for recharge in the ditches). If adequate drainage in a portion of a grove results in the water table dropping below target levels in another area, ditch cleaning, drainage system redesign, or auxiliary pumps may be needed to achieve more uniform drainage.

- Discharge Structures
  Structures and/or pumps that regulate off-site discharge of water should be adequately designed, constructed, and maintained so that target water table levels within the grove can be achieved. If safety or operational concerns prevent structures from being adjusted to regulate discharges during stormwater off-site drainage events, they should be rehabilitated or replaced (i.e., modifying riser-board structures to allow easier water level control).

- Detention
  Where possible, on-site detention should be provided to reduce both the rate and volume of off-site discharges following heavy rains. Detention areas allow all or a portion of the drainage water to be temporarily stored on-site. The excess water can then be gradually released.

Sediment Transport
Suspended solids or sediments are recognized forms of water pollution, and often result in loss of ditch or canal capacity. Sediment losses may also be associated with reductions in water clarity, which may lead to a reduction in dissolved oxygen levels due to decreased light penetration and photosynthesis. Other water quality parameters such as nutrients, pesticides, and metals are also associated with sediments. These solids originate from four primary sources:

- soil particles eroded into ditches

- soil particles eroded from ditches

- plant material washed into the ditches

- plant and biological material growing within the ditches and canals

In addition to potential downstream water quality impacts, the buildup of silts and sediments in the grove- or farm-level, primary, and secondary drainage canals reduces the ditch and canal cross-sectional area. This reduction in cross-sectional area results in higher water velocities, as compared to an unfilled ditch or canal. This higher water velocity (compared to unfilled ditches or canals) may induce greater amounts of erosion of fine and coarse particles from ditch and canal banks. The presence of shoals and sandbars is a good indicator of soil losses. Field erosion also results in site degradation, resulting in increased costs for ditch cleaning and reshaping of beds and furrows. In order to minimize effects of sediment transport in surface water, efforts should focus on keeping soils in the fields and along canal and ditch banks.

Minimizing downstream transport of sediments from groves and canal/ditch banks requires an integrated approach of managing erosion at the grove level, the secondary canal system level, and the primary canal system level. It should be noted that maximum sediment losses from groves are expected during construction of new groves or renovation of older ones. Losses from mature, well-managed groves will be much lower. The selection and implementation of the following sediment-related BMPs must be based upon site-specific circumstances and management styles.

- Riser-Board Water Control Structures
  Place and maintain culverts with risers and boards on laterals and/or field ditch connections.

- Sediment Settling Basins in all Ditches
  Create and maintain localized settling basins (sumps) to trap sediments at field ditch connections to lateral canals, at lateral and collector ditch connections, and prior to water discharge points from the grove.

- Ditch Construction
  Construct ditches and canals with side slopes consistent with soil types. Refer to NRCS engineering tables and local county soil surveys for information on preferred side slopes for specific soil types.

- Stabilize Bare Soils
  Stabilize bare soils and canal/ditch banks by encouraging coverage by vegetation. Vegetation types selected should be adapted to grove conditions and should provide maximum stabilization by roots and foliage. Vegetative buffer strips can also serve to reduce the erosion of soil particles.

- Ditch Bank Vegetation Maintenance
  Broadleaf weed control using herbicides or maintenance mowing of slopes and ditch banks increases grass cover and decreases the proliferation of shade-producing shrubs and weeds, thus reducing erosion from wind and rainfall.

- Aquatic Plant Management
  When removing vegetation from ditch bottoms, avoid disrupting grassed slopes.

- Ditch Cleaning Program
  Develop and implement a systematic management plan for removing sediments from canals and farm ditches on a regular basis. Removal of excess sediment increases the canal cross-sectional area and reduces water velocities (compared to the same water volume in filled-in systems), thus reducing the potential for bank scouring.

- Ditch Bank Contours
  Slope ditch bank top edges or berms to divert water away from the drainage ditch. This practice will minimize overland flow of stormwater directly down the banks.

- Protect Ditch Banks
  Protect canal and ditch banks from erosion in areas subject to high water velocities. This may be accomplished by using rip-rap, concrete, headwalls, or other materials that buffer against turbulence associated with high flow velocity.

- Vegetative Stabilization (Water Furrows)
  Plant vegetation or maintain desirable vegetation within all water furrows to prevent erosion and trap sediments that may result from stormwater runoff or irrigation drainage.

- Herbicide Applications (Water Furrows)
  Restrict the area of tree-row applied herbicides to within the canopy dripline. This will maximize the width of grassed water furrow slopes. Grassed water furrows serve as filters, preventing sediment movement from the fields. For new plantings, minimize the width of tree-row-applied herbicides.

- Water Furrow Maintenance
  Use water furrow drain tiles with managed vegetation in furrows to reduce surface water transfer velocity from the furrows to drainage ditches and canals.

- Settling Basins (Sumps)
  Dig out and maintain small settling basins in front of the drainage inlets within water furrows.

- Water Furrow Drain Tiles
  Use PVC pipe or flexible pipe to connect all water furrows or field ditches to lateral ditches. Extend the pipe on the downstream side away from the ditch bank to prevent bank scouring.

- Take Precautions During Construction
  Special precautionary measures should be taken to reduce sediment transport during construction and other land-altering events.

- Sediment Traps Upstream of Pump Intake
  Prevent discharges of sediments by locating sediment settling basins upstream of the discharge pump intake. The settling basin should be located far enough upstream so that it is removed from the influence of the pump intake.

Pesticides

Water quality monitoring programs conducted by the Florida Department of Environmental Protection and the various WMDs have detected a variety of pesticides within receiving water bodies in citrus-producing areas. The term "pesticides" is inclusive of herbicides, fungicides, and insecticides. Pesticide detections in water are of concern because of their potential toxicity to nontarget plant, invertebrate, fish, and wildlife species within surrounding waters. As with all surface runoff generated in watersheds, all land uses may potentially contribute negative water quality constituents such as pesticides. The following are potential BMPs for controlling pesticide contamination of water bodies. The selection and implementation of particular BMPs must be based upon site-specific circumstances and crop management styles.

- Reduce Spray Drift
Reduce the potential for drift through appropriate selection of nozzles, spray pressure, and application methods/techniques for the formulation applied. Make applications when the wind is blowing away from any highly sensitive nontarget areas and the wind velocity ranges between 3 to 10 mph. Extremely low winds should be avoided because they indicate inversion conditions; winds greater than 10 mph should be avoided because relatively large droplets can be transported long distances in the air.

- Timing of Application
Time pesticide applications in relation to current soil moisture, anticipated weather conditions, and irrigation schedule to achieve greatest efficiency and reduce potential for off-site transport. Do not apply pesticides when rain and high-velocity winds are expected.

- Turn Sprayer Nozzles Off at Row Ends
Turn sprayer nozzles off at the trunk of the last tree within rows. The end trees can be sprayed by making a final pass around the outside perimeter with only the inside nozzles turned on (wrapping). This practice prevents overspray of nontarget areas such as surface water within drainage ditches and canals. Some pesticide labels may be more restrictive. Always consult the label before applying a pesticide.

- Equipment Calibration and Maintenance
Proper calibration and maintenance of pesticide application equipment is essential for minimizing the potential for misapplication of agricultural chemicals.

- Training
Proper training of field operators responsible for handling, loading, and operating spray machinery can minimize the potential for misapplication of agricultural chemicals. It is essential that information learned at continuing education classes be transferred to individuals of the application staff. Special efforts should be taken to ensure that non-English speaking field personnel understand proper handling, loading, and operating techniques.

- Integrated Pest Management (IPM)
Adopt an Integrated Pest Management (IPM) program. IPM is an integrated system using a combination of two or more mechanical, cultural, biological, and chemical systems. This approach provides better and more economical control of most pests.

- Pesticide Spill Management
Establish a plan for action in the event that chemical spills occur. Potential for off-site movement of spilled agrochemicals in water is reduced the sooner the spill is controlled, contained, and cleaned up.

- Precision Application
Use precision applications of reduced amounts of material to smaller trees in order to minimize application of pesticides to nontarget areas and result in more efficient utilization of applied materials. The method of pesticide application, such as ground or aerial spraying, wicking, granules, etc., is important since the degree of drift and volatilization can vary considerably.

- Maintain Soil pH
Where feasible, soil pH should be maintained at a minimum of 6.5 to ensure that any excess copper in the soil remains immobile.

- Read and Understand the Label
Read and understand the pesticide label. The label is the law. Pay special attention to the "Environmental Hazards" section of the label.

- Pesticide Application Equipment Washwater
Wash-water from pesticide application equipment must be managed properly since it may contain pesticide residues.

- Prevent Backflow to Water Sources
  Protect your water source by keeping the water pipe or hose well above the level of the pesticide mixture. This prevents contamination of the hose and keeps pesticides from back-siphoning into the water source. If you are pumping water directly from the source into a tank, use a check valve, antisiphoning device or backflow preventer to prevent back-siphoning if the pump fails.

- Mixing and Loading Activities
  (Permanent Location)
  When available, utilize permanent mix-load stations to reduce pesticide spillage. Spills can result in expensive hazardous waste cleanup. Locate pesticide loading stations away from groundwater wells and areas where runoff may carry spilled pesticides into surface water bodies. If such areas cannot be avoided, protect wells by properly casing and capping them and use berms to keep spills out of surface waters.

- Mixing and Loading Activities
  (Nonpermanent Location)
  When permanent facilities are not available or feasible, utilize portable mix-load stations to reduce pesticide spillage. Pesticide loading areas should be conducted at random locations in the field with the aid of nurse tanks. Spills can result in expensive hazardous waste clean-up. Locate pesticide loading stations away from groundwater wells and areas where runoff may carry spilled pesticides into surface water bodies. If such areas cannot be avoided, protect wells by properly casing and capping them and use berms to keep spills out of surface waters.

- Pesticide Container Management
  Develop and implement procedures to appropriately rinse and dispose or recycle agricultural chemical containers.

- Pesticide Selection
  Select target-specific active ingredients that complement natural systems in epidemiological cycles and modes of action (i.e., insect growth regulators, botanicals, and biologicals). Be aware of pesticide-leaching and/or runoff potential with each soil type within the grove. Select pesticides that have the least potential for impacting water quality negatively.

- Pesticide Record Keeping
  The Florida pesticide law requires certified applicators to keep records of all restricted-use pesticides (RUP) applied. The federal worker protection standard (WPS) requires employers to inform employees of all pesticides applied. Maintain accurate pesticide records to meet legal responsibilities and document production methods.

- Pesticide Storage
  Store pesticides in a secure structure.

- Excess Pesticide Mixture
  Mix only the amount of pesticides needed during an application period. However, it is not always possible to avoid generating excess. The appropriate practices to be followed depend on the type of pesticide waste. If there is excess pesticide material, use it in accordance with the labeled instructions.

- Excess Formulation (Raw Product)
  When possible, return excess formulated materials to the dealer where they were purchased. In most cases, the excess material must be in an unopened, original container. Contact local dealers for their requirements.

Nutrients

If not handled properly, fertilizers can be a significant source of water pollution. Nitrogen and phosphorus are of particular concern within most of the state's water bodies. These nutrients originate from a variety of land uses, including agricultural, urban, suburban, and natural areas. Excess nutrients stimulate algal blooms and growth of noxious plants in receiving water bodies. This stimulation of growth may result in reduced dissolved oxygen concentrations due to plant respiration and decomposition. Lower dissolved oxygen concentrations stress desirable game fish, while promoting less desirable fish that are more tolerant.

Nitrate-nitrogen is a special health concern according to the Environmental Protection Agency. Excessive levels of nitrate in drinking water can cause methemoglobinemia (blue baby syndrome) in infants. Case studies show that the likelihood of this condition increases rapidly when water contains nitrate-nitrogen above 20 parts per million. Because of the extensive interconnection of Florida's aquifers and surface waters, Florida requires that all potable ground waters meet drinking water standards. For nitrate-nitrogen, federal and state regulations set this standard at 10 parts per million. Extremely shallow wells (less than 50 feet), and old wells that may have faulty casings, are at the highest risk for nitrate contamination.

Good fertilizer management for every agricultural commodity is essential for profitable production and environmental protection. The key to success is knowing crop needs and matching fertilization programs to those needs. Rules to follow are simply to apply as little fertilizer as necessary to meet optimum production requirements, to seek professional advice in determining appropriate recommendations, and to practice other conservation measures that reduce nutrient losses.

The following sections list potential BMPs for controlling nutrient contamination of water bodies. The selection and implementation of particular BMPs must be based upon site-specific circumstances and management styles. In many cases, implementation of specific BMPs will be based on best professional judgment (BPJ).

- Education
  Proper training of the field operators responsible for handling, loading, and operating fertilizer spreading equipment, and for correct maintenance of field equipment can help achieve desired placement of fertilizers, avoid waste, and prevent contamination of open waters. Reinforce training with checklists of critical operating points before application of materials. Confirm that each newly assigned employee is adequately informed about machine operation, rates of

discharge, and the intended nutrient placement zone for "feeding the tree."

- Nutrient Management
  Develop a nutrient management plan based upon soil, water, plant and organic material sample analyses, and expected crop yields. USDA-NRCS routinely develops nutrient management plans, and requires them for practices that receive cost share benefits.

- Employ Tissue and Soil Analyses
  Fertilizer applications based on these tests will help avoid overfertilization and subsequent losses of nutrients in runoff water. Application of mobile elements such as N (nitrogen) and K (potassium) should be made on the basis of leaf tissue analysis and production levels instead of routine applications. Applications of less mobile elements such as Ca (calcium), Mg (magnesium), and P (phosphorus) should be based on soil testing and leaf analysis instead of regular applications of specific amounts. The comparison of both types of testing will give production standards for applications that are based on plant need and response, rather than routine applications of standard amounts. Proper fertilization results in high yields and minimal environmental effects.

- Use Appropriate Application Equipment
  Operate machinery as designed in order to achieve precise and desired placement of nutrient materials at specified rates consistent with the form and source of nutrient materials.

- Equipment Calibration and Maintenance
  Proper calibration and maintenance of fertilizer application equipment is essential to avoid misapplication of nutrients.

- Apply Materials to Target Sites
  Place nutrients within the root zone of individual trees or dripline bands along hedgerows of trees. Avoid placement in areas prone to off-site transport of nutrients, especially water furrows.

- **Avoid High Risk Applications**
  Do not apply materials under "high risk" situations, such as before forecasted rainfall. Avoid applications of nutrients during intense rainfall, on bare soils with extreme erosion potential, or when water tables are near the surface.

- **Fertilizer Storage**
  Take precautions when storing fertilizer to prevent contamination of nearby ground and surface water. Always store fertilizer in an area that is protected from rainfall.

- **Spilled Fertilizers**
  Immediately remove any fertilizer materials spilled on ground surfaces and apply at recommended rates to crops. When possible, place a tarp over ground surfaces where fertilizer transfer operations are conducted. Spilled materials should be transferred to application equipment.

- **Use Caution When Loading Near Ditches, Canals, and Wells**
  Minimize the potential for spilled materials to pollute surface waters. When possible, locate mixing and loading activities away from groundwater wells, ditches, canals, and other areas where runoff may carry spilled fertilizer into surface water bodies. If such areas cannot be avoided, protect wells by properly casing and capping them and use berms to keep spills out of surface waters. Recover and apply spilled materials to the intended zone of application.

- **Alternate Loading Operation Sites**
  Use multiple fertilizer loading and transfer sites to prevent concentration of nutrients in a single area.

- **Use Backflow Prevention Devices**
  Use backflow prevention devices on irrigation and spray tank filling systems to preclude entry of nutrients into surface waters. Never leave a filling tank unattended.

- **Split Applications Throughout Season**
  Dividing the annual fertilizer requirement into three or more applications can minimize leaching and runoff losses during the summer rainy season, and help maintain the supply of nutrients over the long growing season of Florida. Frequent fertigations can be an efficient method of application for nitrogen (N) and potassium (K), while minimizing the potential for leaching of nutrients during excessive rainfall events. The trade-off between costs and fertilizer-use efficiency/resource protection must be considered.

- **Erosion Control**
  Erosion-control practices should be considered to minimize soil loss and runoff that can carry dissolved and attached nutrients on soil particles to surface waters. Vegetative filter strips are effective in reducing the levels of suspended solids and nutrients.

- **Irrigation Management**
  Irrigation should be limited to wetting only the root zone where possible. Excessive irrigation can transport nutrients below the root zone through leaching. Proper scheduling and uniform water distribution are necessary to assure control.

- **Incorporate Organic Materials**
  Increase the use of organic matter (i.e., mulch) when possible to retain nutrients, increase soil moisture retention, improve soil structure, and improve biological ecosystems.

- **Well Protection**
  Prevent groundwater contamination by decommissioning wells that are not in use.

- **Use Appropriate Sources and Formulations**
  Reduce the potential for nutrient leaching and off-site movement by choosing appropriate sources and formulations of fertilizer based on nutritional needs, season (rainy or dry), and anticipated weather conditions to achieve greatest efficiency and reduce potential for off-site transport. Utilize controlled-release and slow-release formulations when feasible.

Aquatic Plants

Where there is water, there are weeds. Aquatic plants maintain a balance of nature, offering food, protection, oxygen, and shelter to aquatic species. In addition, aquatic plants may be beneficial in removing nutrients and pesticides from surface water. However, maintaining a balance in the aquatic system while sustaining crop success and avoiding loss of income can be a challenge.

Overabundant aquatic weed growth can clog or restrict drainage following heavy rains, resulting in severe root pruning with increased disease incidence and subsequent fruit drop. The discharge of floating aquatic vegetation from citrus grove drainage ditches results in additional ditch and canal maintenance costs. Another serious problem associated with excess aquatic weed growth is the amount of nutrient-rich organic sediments transported into the receiving water bodies during storm events.

As aquatic weeds decompose, nutrients are released into the water. The organic particles that result from decomposition are lightweight, and are readily transported in drainage discharge water. As these particles are discharged, they remain suspended and contribute to the "ooze" that accumulates in canals and water bodies. These particles are easily distributed in the water column by wind or wave action. The organic particles contribute to the turbidity of the water and block light from penetrating, often resulting in a reduction in sea grasses and other species of aquatic life in the affected areas. The following are potential BMPs that should help minimize the discharge of aquatic weeds off-site from citrus groves.

I. Physical Control
A. Utilize barriers, traps, screen devices and debris baffles to exclude and/or remove aquatic weeds. Avoid structural designs or placement that may severely limit discharges during periods with high flow rates.

   1. Debris baffles on outfall structures reduce off-site discharges of floating aquatic vegetation into the primary canal system. Accumulated debris should be removed periodically to ensure free flow through the structure. Debris baffles should be used in conjunction with other aquatic weed control strategies.

   2. Ribbon barriers installed upstream of outfall structures reduce off-site discharges of floating aquatic vegetation into the primary canal system. Ribbon barriers are most effective when attached to the bank and allowed to move vertically according to canal stage levels. Ribbon barriers should be utilized in conjunction with chemical or biological aquatic weed control programs.

   3. Hyacinth barriers installed upstream of outfall structures reduce discharges of floating aquatic vegetation into the primary canal system. Hyacinth barriers are not suitable for all sites, and should only be installed in ditches or canals with low flow potentials. Hyacinth barriers should be installed in combination with an aquatic weed removal program.

B. Use mechanical harvesters for removal of floating weed species. The objective of mechanical weed removal is to physically remove aquatic weed debris from grove canals and ditches. Removal is accomplished by different types of equipment such as cranes, track hoes, and backhoes, depending on the location and the situation. Weed debris should be removed and placed on a nearby service area, so that drainage will be away from the canal or ditch. Oftentimes, significant amounts of sediment are removed from ditch bottoms during the process. Weeds chopped along canal banks with a disc or mower can wash back into the waterway, thus recreating the weed problem. Weeds chopped along ditch banks also can wash down the ditch bank slope with rainfall, enter the drain pipes, and reenter the waterway.

II. Biological Control
A. Introduce biological control agents with

appropriate constraints and management programs to permanently or recurrently sustain the life cycles of control agents.

1. Triploid grass carp feed upon aquatic vegetation. Triploid grass carp (*Clenopharyngodon idella*) are exotic, hybrid fish (with 3 chromosomes, one sterile by nature) that cannot reproduce. Their introduction into water bodies requires permitting from the Florida Fish and Wildlife Conservation Commission in Tallahassee. Usually, the permitting requires a fish barrier retention structure on outfall structures to contain the grass carp and reduce the chances of introduction to main canals, streams, or lakes.

2. All grass carp used in Florida under this permit system must be certified as triploid by the Florida Fish and Wildlife Conservation Commission or the U.S. Fish and Wildlife Service. Instructions on purchasing these fish and the certification process are included with the Commission's permit, which must be received prior to possession of any grass carp. An application for certification and possession of grass carp for control of aquatic weeds may be obtained by contacting:

   Florida Fish and Wildlife Conservation Commission
   620 South Meridian Street
   Tallahassee, FL 32301
   Telephone: (850) 488-4066
   or (850) 488-4069

B. Augment existing biological control agents to achieve a desired level of control on target weed species.

1. Some exotic plant species have been controlled by introduction of biological control agents. The alligator weed flea beetle (*Agasicles hygrophila*) was introduced into the United States from South America in 1964. This beetle has done a remarkable job of reducing the problems with alligator weed. In fact, alligator weed is not considered a major aquatic weed in most areas of the state. However, most exotic aquatic plant species have not been adequately managed by biological controls alone.

2. Various biological control agents have been tested on water-hyacinths throughout the years. Of these predator introductions, the most effective have been two types of water hyacinth weevils (*Neochetina eichorniae* and *Neochetina bruchi*) and the water-hyacinth mite (*Orthagalumna terebrantis*). In addition, the fungus *Cercospona rodmanic* has been imported and found to have some effect on the water-hyacinth. When infected water-hyacinths are introduced into areas with healthy hyacinths, some control results as the disease spreads and infects healthy plants.

III. Chemical Control

Utilize selective herbicides for suppression or elimination of weeds that impair the desired utility of canals and waterways. Carefully select herbicide materials with consideration for target species, mode of action, nontarget toxicity, and residual behavior.

A. Apply herbicides registered for aquatic applications with consideration for chemical drift, target species habitat (submersed, emergent, or floating), and movement of herbicide materials into nontarget areas.

B. Use herbicides in combination with other types of control.

IV. Post-Pump Settling Basin (PPSB)

Post-Pump Settling Basins (PPSB) are aquatic detention areas designed to receive and discharge excess surface water from the grove. Water is pumped from on-site farm ditches into the PPSB. The PPSB is generally located adjacent to a major canal and is typically connected by a riser culvert to the canal. PPSBs should be managed in coordination with other physical, chemical, or biological aquatic weed control programs.

Storm Events

Hurricanes, tropical storms, winds, and flooding can wreak havoc with agricultural operations. Severe weather events can cause both dollar loss and environmental pollution with respect to agricultural chemicals. Fertilizers, pesticides, solvents, fuels, etc., can be physically lost, contaminated, or contaminate the surrounding environment and environments downstream of chemical storage and use areas.

Be aware of weather predictions on the morning, noon, and evening newscasts when there are potential storms. If storms are imminent, do not delay preparations. Conduct an inventory of the pesticides and other chemicals you have on hand. Such an inventory will be useful for insurance purposes, or in the event of necessary pesticide or chemical cleanups. Include product and active ingredient names, and container sizes. Receipts for the purchase of these materials are useful for this, or in some cases may suffice themselves. Do the inventory before you take other measures. Put the inventory in a safe location.

You should know exactly what kind of insurance coverage you have. Find out if the policy covers your chemical inventory or the damage it could cause. Do not make applications of agricultural chemicals until the potential of impending severe weather is resolved. Delay purchase or delivery of additional chemicals to your operation until after the impending storm risk has passed. If you have

any such deliveries scheduled, cancel delivery until after the storm has passed.

Secure all of your chemicals. This includes fertilizers, pesticides, solvents, fuels, etc. Close and secure container lids, move containers and application equipment to the most secure location. Raise chemicals from the floor or cover materials that could be damaged by water. Do what you can to protect product labels and labeling. Doors, windows and other points of access to storage locations should be secured and locked. If you are going to board up windows on your house, do the same for pesticide and other chemical storage areas. Don't leave chemicals in vehicles, or in application equipment.

As you prepare for a storm and hurry to put many things into secure locations, be sure all of these items are compatible. Read the storage and spill containment sections of your MSDSs. Round up your pesticide and other chemical MSDSs and put these in a secure location, and if you have not done so, provide local emergency first responders with a copy of these, along with a copy of your chemical inventory. Secure your personal protective equipment. You may need it as part of your own cleanup operations after the storm.

Be sure that your buildings will stay where they are as much as possible. Make sure the roofs are tied into the building. Tie down small storage buildings and storage tanks. If you leave your location during a severe weather event, be sure that buildings that store pesticides and other chemicals are well signed. Have on hand all emergency phone numbers that you may need. Consult your chemical dealer and insurance agent for additional suggestions.

Chapter 6. Environmental Concerns - Pesticides

by Chris Wilson, Brian Boman, and Jack Hebb

Introduction

Water is one of the most valuable natural resources in Florida. Of all the freshwater used in Florida, approximately two-thirds is returned to surface streams or to aquifers. With the increased competition for water due to development, water quality is becoming one of the most important issues associated with water use. The term "water quality" refers to the content and/or magnitude of a broad array of specific parameters that affect the usefulness of the water by human interests, or by fish and wildlife. Water quality parameters of concern may include turbidity, nutrients, inorganic and organic pesticides, or organic materials that contribute excess biological oxygen demand in the water body. The use and return of irrigation water back to the environment through drainage potentially contributes a variety of agrichemicals to the water, depending on management practices and the physical environment. Irrigation managers should especially be familiar with the environmental concerns associated with their use and return of water to the environment. When considering the issues of water quality as related to irrigation and pesticides, one should consider:

- how much of the applied pesticide moves from the application site in runoff;

- how the receiving water quality will be affected, considering all later uses of the water;

- what the fate of the pesticide is in the environment;

- how changes in management practices can reduce losses.

The presence of pesticides in ground and surface water has become an issue in a number of agricultural counties in Florida. Their presence may exceed limits set by the Florida Department of Environmental Protection (FDEP). In addition, the presence of pesticides in ground and surface waters unnecessarily exposes humans and other organisms to hazards that they normally would not be exposed to. Many pesticide molecules interact with nontarget as well as target organisms, and many parent compounds may readily break down into more toxic metabolites in the environment. In the environment, pesticide fate depends on three basic factors:

- its occurrence within the environmental compartments, i.e., air, soil, water, and biota;

- its transport between the compartments, i.e., leaching, diffusion, volatilization, runoff;

- its chemical transformations within the compartments by processes such as metabolism, photodecomposition, and hydrolysis.

Partitioning into the different environmental compartments is due to a combination of physical and chemical properties of the pesticide and of the surrounding environment. Properties such as vapor pressure, water solubility, polarity, photodecomposition, and adsorption to organic matter are key chemical properties. The following sections describe many of the factors influencing pesticide fate in the environment, ecological effects of pesticides, and current water quality criteria (relative to pesticides) for the state of Florida.

Factors and Processes That Affect Water Contamination by Pesticides

The fate of chemicals in terrestrial and aquatic environments may differ markedly. Possible fates of pesticides include sorption to clays and/or organic matter, chemical decomposition, microbial decomposition, uptake by plants and animals, photolytic breakdown, complexation with minerals, and volitilization into the atmosphere. All of these processes are dynamic and interrelated. No one

51

process usually accounts for the full fate of a pesticide. Rather, many or all of the processes collectively contribute to the ultimate fate of the compound. The following sections describe many of the factors and processes influencing the fate of pesticides in the environment.

Soil Properties

Soil properties are important factors determining the potential for certain pesticides to leach into groundwater. Leaching is the vertical movement of water and solutes through the soil profile (as opposed to horizontal movement over the soil surface). Florida's unique hydrogeologic features (i.e., a thin surface soil layer, high water table, and porous limestone in many areas) make the soil vulnerable to downward movement of water-soluble nutrients and pesticides. Soil properties such as permeability, soil structure, and organic matter content all influence the degree to which a soil is vulnerable to leaching.

Permeability is a measure of how fast water can move through the soil. It is affected by the texture and structure of the soil. Soils with coarse, sandy textures are generally more permeable. Soils with higher permeability have greater potential for groundwater contamination than less permeable soils. Soil moisture affects how fast water will travel through the soil. If soils are already wet or saturated before rainfall or irrigation, excess moisture will run off. Soil moisture may also influence pesticide breakdown.

Soil structure describes how the soil is aggregated. Uncompacted soils allow more water to pass through the profile. Soils with a loose structure, or soils with hollow channels such as dried root channels or animal or worm tunnels, will also allow for increased flow of water through the soil profile. Soils that permit rapid flow of water through the soil profile present a higher potential for groundwater contamination than less permeable soils. Sandy soils with their coarse textures and low water-holding capacity will allow for greater infiltration than finer, heavier clay soils. Intense or sustained periods of rainfall may result in large amounts of water passing through the soil profile, especially where little runoff occurs. Depth to groundwater is another factor affecting the potential for pesticides to reach groundwater. If the top of the water table is close to the ground surface, pesticides have less distance to travel to reach groundwater.

Organic matter is the single most important factor affecting adsorption of pesticides in soils. Organic matter content of soils may be increased by the addition of manure and incorporation of crop residues. Many pesticides are adsorbed (bound) by soil organic matter, which reduces their rate of downward movement. Soils high in organic matter tend to hold more water, which may make less water available for leaching.

Many soils commonly used for citrus production in the Ridge area are particularly subject to leaching, and are referred to as vulnerable soils. These soils are well drained with low organic matter content, and provide ideal conditions for leaching of many soil-applied agrichemicals. The soils in Table 6-1 are categorized by the Florida Department of Agriculture and Consumer Services as highly permeable, well drained, and therefore at risk in terms of groundwater contamination through agrichemical leaching. Although such soils are dominant throughout Ridge citrus production areas, many do occur scattered throughout other Florida citrus-producing areas (Fig. 6-1). Leaching is typically of most concern in Florida since most of the soils are sandy and permeable.

Properties of the Agrichemical

The fate of pesticides in soils and water is not determined by a single property of the pesticide, but by a combination of properties. Information on chemical properties of pesticides is available from the manufacturer. It is not available on the pesticide label, although several pesticides (aldicarb, bromacil, and simazine) have groundwater advisory statements on their labels because of their tendency to leach. Two of the most important properties that determine whether a pesticide will be

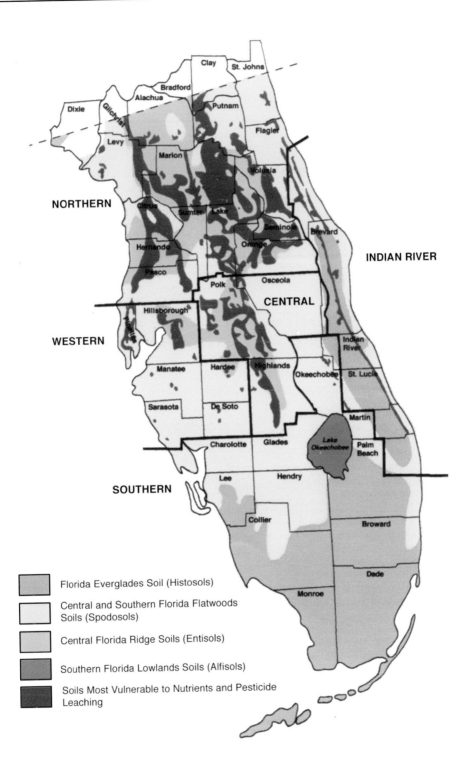

Figure 6-1. Approximate location of major soil types and vulnerable soils in citrus-producing regions.

Table 6-1. Soils classified as vulnerable to leaching.

Adamsville	Bahiahonda	Candler	Florahome
Lake	Orlando	Satellite	Tavares
Archbold	Broward	Cocoa	Fort Meade
Lakewood	Palm Beach	St. Augustine	Orsino
Astatula	Canaveral	Dade	Gainesville
Neilhurst	Paola	St. Lucie	

present in water or sediments are the water solubility and sorption coefficient (K_{oc}). Persistence of the compound is another important property influencing the likelihood of contamination over long periods of time. These properties are discussed below.

Solubility and Sorption

One of the most important properties influencing pesticide movement with water is its solubility. Solubility is the tendency of a chemical to dissolve in a solvent. It is another property that affects the behavior of a pesticide in the soil. As water from rainfall or irrigation percolates through soil, it carries water-soluble chemicals with it. The higher the water solubility value, the more soluble the chemical. For instance, a pesticide with a water-solubility value of 10,000 ppm at 80° F is much more soluble than a chemical having a solubility of 10 ppm at 80° F. The former is also more likely to leach. Solubility values for pesticides can be obtained from the Material Data Safety Sheets (MSDS), which are available from pesticide dealers. The loss of pesticides with low water solubilities is normally a result of sediment movement off-site.

Sorption is the tendency for a compound to adhere to water-insoluble matrices. It is inversely related to the water solubility. A useful index for quantifying pesticide sorption to soils is the partition coefficient (K_{oc}). The K_{oc} value is defined as the ratio of pesticide concentration in the sorbed state (bound to soil particles) to the concentration in the solution phase (dissolved in the soil water). Thus, for a given amount of pesticide applied, the smaller the K_{oc} value, the greater the concentration of pesticide in solution. Pesticides with small K_{oc} values are more likely to be leached, compared to those with large K_{oc} values. The degree to which a pesticide sorbs varies with environmental conditions, such as moisture content and temperature, and with chemical properties of the compound. In addition, pesticides sorbed to insoluble matrices may also be detached from soil particles through a process called desorption. Desorption is likely to occur as more rain or irrigation water enters the soil. This process is usually much slower than the adsorption process, and is minimal for herbicides such as diquat.

Values of partition coefficients ($K_{oc's}$) are independent of soil type and are characteristic of each pesticide. The partition coefficient is determined by a pesticide's chemical properties such as solubility and melting point. The partition coefficient makes it possible to estimate a particular pesticide's relative chance of being lost via runoff or leaching in a specific soil. These calculations are based on the sorption index of a pesticide on a particular soil (K_p), the organic matter percentage in the soil (OM), and the partition coefficient of the pesticide normalized to organic carbon (K_{oc}). Five important aspects of pesticide sorption by soils should be recognized:

1. For pesticides not adsorbed by soils, K_{oc} has a low value. Such pesticides will leach in a manner similar to inorganic ions (such as nitrate) that are not adsorbed by most soils.

2. For a given pesticide, sorption is greater in soils containing more organic matter (OM). Thus, pesticide leaching in soils with higher OM is expected to be slower compared to soils low in OM.

3. In most soils, OM decreases with depth. This is particularly so in the sandy soils of the Ridge. In these soils, pesticide sorption decreases with increased soil depth. Consequently, leaching is likely to be more rapid in subsoils. In contrast, many of the flatwoods soils are Spodosols. These soils have organic layers (spodic horizons) that are generally within 2-4 feet of the surface. These soils have a higher capacity for sorption in the spodic layer than in the soil above.

4. For a given soil, pesticides with smaller K_{oc} values are sorbed to a lesser extent, and are therefore more likely to be leached when compared to pesticides having larger K_{oc} values.

5. Pesticide sorption to sediments within the bottoms of drainage ways and ponds may differ markedly from that in soils. This is primarily due to increased amounts of amorphous material and organic matter in the sediments.

Persistence

Persistence describes how long a pesticide remains unchanged in the environment. A pesticide that is persistent will maintain its structure and presence in the environment for a long period of time. Persistence varies greatly among different pesticides. Under terrestrial and aquatic conditions, pesticides are degraded at different rates by many processes. Degradation rates are influenced by soil and environmental factors such as temperature, water content, soil pH, and organic carbon (C) content, as well as by microbial activity and solar radiation. Degradation rates ultimately affect the amount of pesticide available for runoff or leaching in the environment.

A common measurement of persistence is the half-life ($T_{1/2}$) of the pesticide. The half-life is a measure of the amount of time it takes for one-half the original amount of a pesticide to be deactivated. Half-life is sometimes defined as the time required for half the amount of applied pesticide to be completely degraded and released as carbon dioxide. Usually, the half-life of a pesticide measured by the latter basis is longer than that based on deactivation only. This is especially true if metabolites that are more resistant to degradation accumulate in the soil. $T_{1/2}$ values in subsoils and in groundwater are usually much larger. Thus, as pesticides are leached to lower depths, their persistency increases. Values for pesticide degradation ($T_{1/2}$) in subsoils and groundwater are scarce. In some cases, sediments play an important role in protecting pesticides from degradation through irreversible adsorption, and in others they may promote chemical degradation by adsorption-catalysis.

Agrichemical Availability

Applied pesticides must be "available" in order to be washed from a field or application site in surface runoff water. The "available" fraction of water-soluble pesticides is simply that fraction of the applied material that is easily dissolved in runoff water or that is easily dissolved and carried off with particulates. Water solubility and herbicide formulations are important factors affecting this availability. Pesticides must be either near the soil surface or on exposed plant surfaces where they can be dislodged by rainfall, irrigation, or surface water flow. Such exposed deposits are also susceptible to degradation by other processes such as volatilization, photolysis, and leaching. The carrier particle type for a given pesticide formulation may also affect the availability of pesticides for losses in runoff water. Pesticides bound to lightweight dust particles are more readily available to move in runoff water. Use of spray additives such as stickers can reduce the availability of the pesticides for wash-off from foliage and fruits.

Irrigation/Rainfall Timing

Nonpoint pesticide pollution from agricultural fields occurs primarily during "runoff events," or periods when rainfall rates exceed the soil-water-infiltration capacity, resulting in over-the-land-surface flow of water. Water infiltration into the soil profile is indirectly proportional to soil water content; that is, infiltration into the soil profile decreases as water content increases, and vice versa. Pesticide losses in runoff water are usually greatest when runoff occurs shortly after pesticide applications were made. Avoid applying pesticides prior to predicted heavy rains or irrigation.

Site Characteristics

Factors such as slope and the presence of vegetative ground covers affect both runoff velocity and water infiltration. These factors affect the amount of ground-applied pesticides that may move from the application site in runoff water. Water infiltration into the soil profile will also affect the degree to which surface runoff occurs. This is dependent on antecedent conditions, or the amount of water that resides in the soil profile prior to the rainfall or irrigation event. Runoff increases as infiltration decreases.

Plant Uptake

Pesticides may enter into a plant by many pathways, including root uptake from the surrounding medium, vegetative uptake of vapor from the surrounding air, and uptake by external contamination of the stems and foliage. Root uptake of pesticides involves movement from the soil solution into the root with the transpiration stream. This is a common uptake route for many soil-deposited pesticides. Once in the roots, the compound may be distributed throughout the plant, depending on its chemical properties. The pesticide may be broken down inside of the plant; be sequestered in vacuoles, membranes, or cell walls; or it may be re-emitted back into the environment through plant secretions or upon plant death and dessication. The physical and chemical properties of the pesticide and environmental conditions discussed in this section influence plant exposure and pesticide uptake pathways.

Ingestion by Organisms

Some pesticides may directly adsorb onto animals in the environment, or to their food sources. Upon ingestion, the pesticide may be broken down inside the organism, it may be excreted, it may kill the organism, or it may be stored in fatty tissues. Common organisms where this might be an issue include birds, worms, fish, mammals, and aquatic and terrestrial invertebrates.

Photodecomposition

Photodecomposition reactions account for the degradation of many pesticides in both aquatic and terrestrial environments. In order for these reactions to occur, the pesticide molecules must absorb light energy. Water turbidity, or the presence of suspended sediments and foreign particles, directly affects light penetration into the surface water, and indirectly affects light absorption by the pesticide molecules. Substances absorbing ultraviolet light in the spectral region of sunlight (wavelengths >290 nm) may either:

• undergo direct destruction due to photolysis, or

• undergo indirect destruction as a result of photolysis, which is caused by other constituents in the water that absorbed light energy and either transmitted it to the pesticide (sensitization) or lead to the formation of reactive species that enter into chemical reactions.

Microbial Degradation

Under moist and warm soil conditions, microbes may use certain pesticide molecules as a food source and transform them into harmless molecules of carbon dioxide and water. Breakdown processes occur primarily in the root zone. Breakdown is considerably slower in deeper soils and sediments. Some pesticides form intermediate substances during the breakdown process, which can be more toxic than the original compound.

Within drainage ditches, canals, and retention ponds, the uppermost sediment layer is the most biologically active area within the sediment. This stratum may contain 10 to 100 times more microbial populations than the strata 1 m below the sediment surface. As a result of microbial respiration in stagnant areas, O_2 concentrations decrease. In response to the formation of this anaerobic zone, the biochemical composition of the system changes, resulting in an accumulation of organic matter providing additional adsorbent and an energy source for pesticide degradation at the sediment surface. Accumulated Fe^{2+} and Mn^{2+} in anaerobic environments may also complex with and stabilize negatively charged organic components, resulting in the formation and deposition of amorphous minerals at the sediment surface. Complex formation between acidic pesticides and Fe^{2+} and Mn^{2+}, or returning amorphous colloids, may stabilize and protect pesticides against destruction, thus increasing the persistence of the chemical in the aquatic system.

Pond/Lake Characteristics

Retention/detention pond and lake characteristics such as size, thermal stratification, and lake age have important effects on mixing, chemical regime, and sediment distribution within the water body. Adsorption of pesticides to soil/sediment particles is not a uniform process. In addition, the distribution of the soil/chemical complex within the pond can not be considered uniform. These complexes will segregate on a particle size basis, with the largest-sized particles remaining close to the runoff inflow point, and the smaller-sized particles moving progressively farther from the inflow point based on decreasing density. Clay distribution is usually more uniform in small ponds that are subjected to large inflows of runoff and no thermal stratification. These clay particles eventually settle to the bottom and distribute vertically according to size. In larger ponds, thermal stratification may result in an increasing density gradient with depth. As a result, incoming sediments/complexes tend to move vertically into levels of equal density before moving laterally. This movement results in more

localized deposition of clays and less dilution of the runoff due to the lack of normal mixing. A common result of thermal stratification is the formation of anoxic conditions at the bottom layer of the lake. These conditions form as a result of microbial oxygen consumption in the uppermost layer of the lake sediment. Oxygen diffusion from the upper lake layers is slowed due to the density gradient formed, and is usually not sufficient to meet the oxygen demands of the microbial populations.

Water Chemistry

Water pH can affect the adsorption of pesticides to sediments. Lake water sometimes possesses higher pH values than the incoming runoff water. As a result, weakly basic pesticides such as the s-triazine, 2, 4-D, and 2, 4, 5-T tend to desorb from sediment/soil complexes. Protonation of the basic triazine molecules and adsorption of the organic cation to exchange sites on clays and acidic functional groups on organic colloids are likely responsible for this pH-dependent adsorption. Values of pH approaching the pK of specific compounds often result in maximum adsorption to organic matter of many compounds such as the s-triazine herbicides. Changes in pH may also result in increased desorption of pesticides from sediments, and in some cases pH has no effect on adsorption/desorption phenomena. Water salinity may also influence the ionization of some acidic herbicides. In addition, water pH may also accelerate the chemical degradation of the pesticide. Many of the organophosphate and carbamate pesticides readily degrade with alkaline pHs.

Volatilization

Volatility describes how quickly a compound will evaporate when it is in contact with air. Highly volatile chemicals are easily lost to the atmosphere. Some pesticides, such as fumigants, must be volatile in order to move and provide uniform distribution through the soil profile. Highly volatile pesticides are not as likely to pose a significant threat to water quality as those that are less volatile.

Estimating Pesticide Losses

Qualitative predictions of the likelihood for a pesticide to contaminate surface water or groundwater are possible using K_{oc} and $T_{1/2}$ values as indices of pesticide sorption and persistence (Table 6-2). Strongly sorbed and persistent pesticides (i.e., compounds with large K_{oc} and large $T_{1/2}$) are likely to remain near the ground surface, thus increasing the chances of being carried to canals or lakes via runoff. In contrast, weakly sorbed but persistent pesticides (i.e., compounds with small K_{oc} and large $T_{1/2}$) may be readily leached through the soil, and are more likely to contaminate groundwater. For nonpersistent pesticides with small $T_{1/2}$, the possibility of surface water or groundwater contamination depends primarily on whether heavy rains (or irrigation) occur soon after pesticide application. Without water to move them downward, pesticides with short half-lives remain within the biologically active crop root zone and may be degraded readily. In terms of water quality, pesticides with intermediate K_{oc} values and short $T_{1/2}$ values may be considered low risk because they are not readily leached and are degraded fairly rapidly (Table 6-3).

Ecological Effects of Pesticide Exposures

An ecosystem is the combination of the physical environment and all of the organisms living within the area. The most notable pesticide causing impacts to ecosystems was DDT. This insecticide was associated with declines in eagle, hawk, and other top predator populations due to eggshell thinning and other effects. Pesticides may affect ecosystems at many levels, ranging from energetics and nutrient cycling to community structure and ecosystem functioning. Two examples of these effects include alterations in plant and algal production, and increases in nutrient turnover and decreased nutrient cycling. Both of these effects are expected with exposure of nontarget plant communities to herbicides.

Alteration of plant communities may render the habitat useless for other animal species dependent on the plants for food or shelter. Effects at the community level usually result from disruption of normal ecosystem structure and interdependencies. Community structure refers to the physical environment normally created by plants and geological features, as well as the relationships between populations of biota. Examples of adverse effects at the community level include promotion of undesirable species (weeds, trash-fish) due to the elimination of species essential for the functioning of the community; reduction in the number of species present in the community; and disruption of food webs by breaking dietary linkages between species.

Effects at the population level are concerned primarily with effects on population size, extinction

Table 6-2. Potential pesticide contamination on ground- and surface waters based on persistence and sorption.

Persistence	Sorption	Potential contamination	
		Groundwater	Surface Water
Nonpersistent	Low-moderate	Low	Low
Nonpersistent	Moderate-high	Low	Moderate
Moderately persistent	Moderate-high	Moderate	Moderate
Moderately persistent	Low-moderate	High	High
Persistent	Moderate-high	Moderate	High
Moderately persistent and persistent	Low-high	Site-specific conditions determine groundwater or surface water impacts.	

Table 6-3. Sorption coefficients and half-life data for many pesticides labeled for use in Florida citrus production (adapted from Rao and Hornsby, 1999).

Common Name	Sorption Coefficient K_{oc} (ml/g)	Half-Life $(T_{1/2})$ Days	Common Name	Sorption Coefficient K_{oc} (ml/g)	Half-Life $(T_{1/2})$ Days
1,3-dichloropropene	32	10	fosetyl-aluminum	0	0.1
2,4-D acid	20	10	glyphosate	24,000	47
abamectin	5000	7-60	malathion	1800	1
acephate	2	3	metalaxyl	50	70
aldicarb	30	30	metham sodium salt	10	7
azinphos-methyl	1000	10	methidathion	400	7
benomyl	1900	67	methomyl	72	30
bromacil	32	60	methyl bromide	22	55
carbaryl	300	10	methyl isothiocyanate	6	7
chloropicrin	62	1	NAA ethyl ester	300	10
chlorpyrifos	6070	30	NAA sodium salt	20	10
diazinon	1000	40	napropamide	700	70
dichlobenil	400	60	norflurazon	700	30
dicofol	5000	45	oryzalin	600	20
dimethoate	20	7	oxamyl	25	4
diuron	480	90	oxydemeton-methyl	10	10
EPTC	200	6	oxyfluorfen	100,000	35
ethephon	100,000	10	oxythioquinox	2300	30
ethion	10,000	82	paraquat dichloride salt	1,000,000	1,000
ethoprop	70	25	pendimethalin	5000	90
fenoxycarb	1000	1	petroleum oils	1000	10
fenamiphos	100	50	propargite	4000	56
fenbutatin oxide	2300	90	sethoxydim	100	5
ferbam	300	17	simazine	130	60
fluazifop-p-butyl	3000	15	trifluralin	8000	60
formetanate hydrochloride salt	10,000,000	100			

of species, and selection of more tolerant individuals within a particular species. Common methods for evaluating effects at the community and population level include laboratory model ecosystem studies, intermediate-sized model systems, and full field trials.

Information on the adverse effects at the individual organism level is most common. These data typically describe the effects, or dose-response relationship of pesticides, on individual organisms believed to be important within the ecosystem of concern. Under ideal conditions, dose-response studies would be conducted on all species within the ecosystem. However, this is not possible because of the great expenses associated with these tests.

To combat the tremendous monetary costs and time restraints, sentinel organisms are usually adopted for testing. These organisms are believed to be more sensitive to the pollutant of choice. The idea is that by protecting the most sensitive species within an ecosystem, all others will also be protected. Typical sentinel species include, but are not limited to, trout, bass, bluegill, water fleas, scuds, oysters, quail, and earthworms. Typical ecotox studies measure the acute and chronic effects of the pesticide. Acute effects are usually due to exposure to a large amount of the pesticide over a short period of time, whereas chronic effects are due to long-term exposures to typically low concentrations of the pesticide. These studies are typically conducted under controlled conditions through a combination of laboratory toxicity tests and small-scale field trials using artificial ponds or streams. Effective concentrations are then compared to concentrations that the organism is expected to be exposed to in the environment in order to quantify ecological risks.

While these types of studies are practical and easy to interpret from a toxicity standpoint, they do not predict recovery of the observed population to the toxic insult. Recovery cannot be determined from the toxicity data alone, but from a variety of complex, often confounding factors. The long-term

viability of a population of organisms may be reduced by relatively small declines in population size of some species, while a marked reduction in population size may have little impact on the longer-term average density of other species. These processes either contribute to the natural selection of species that are more tolerant to the pesticides, or to extinction of intolerant species. Furthermore, standard toxicity tests do not consider the interactions of species with one another and with their environment.

Keeping the limitations of ecological toxicity data in mind, the following section describes the ecological toxicity data that is known for many pesticides labeled for use in and around citrus production areas. Irrigation managers should be familiar with pesticides currently used in their operations, and should adjust irrigation strategies to minimize pesticide exposures to nontarget species where possible.

References
All of the pesticides described are currently labeled for use in and around citrus production. However, information was not available for all of the pesticides. The majority of descriptions listed were taken from *Pesticide Profiles*, edited by Michael Kamrin and published by CRC Press. Another source of information is *Pesticide Properties in the Environment*, written by Arthur G. Hornsby, Albert E. Herner, and R. Don Wauchope, and published by Springer-Verlag. These resources should be consulted when more detailed information is desired.

Ecotoxicity
Where two or more pesticides within the same pesticide class are used, a brief description of the mode of action and ecosystem components at risk for the general class is included, as well as a description of ecological toxicity data for each compound. The following sections list the categories for the ecotoxicity summaries.

Organophosphate Insecticides
Organophosphates are ecologically important

Table 6-4. Summary of categories used to qualitatively describe pesticide ecotoxicity.

Toxicity Category	Birds acute oral LD$_{50}$* (ppm)	Birds dietary LC$_{50}$** (ppm)	Fish water LC$_{50}$** (mg/L)
Very highly toxic	<10	<50	<0.1
Highly toxic	10-50	50-500	0.1-1
Moderately toxic	>50-500	>500-1000	>1-10
Slightly toxic	>500-2000	>1000-5000	>10-100
Practically nontoxic	>2000	>5000	>100

*LD$_{50}$ is the dose that kills 50% of the test organisms.

**LC$_{50}$ is the concentration at which 50% of the organisms die.

because of their acute toxicity to mammals and insects. In both cases, the mode of action is the same—inhibition of the enzyme acetylcholinesterase (AChE). This enzyme breaks down a substance (acetylcholine (ACh)) that is normally produced in nerve cells in response to a stimulus. ACh allows the transfer of nerve impulses from one nerve cell to receptor cells, such as a muscle or another nerve cell. The nerve impulse continues until the ACh is broken down by AChE. Organophosphate pesticides inactivate AChE, thus allowing continuous transmission of nerve impulses, resulting in a variety of symptoms such as paralysis of muscles or weakness. Death usually results from the inability to breathe caused by paralysis of the diaphram muscles. A summary of the ecotox data for specific organophosphates labeled for use in citrus follows.

Azinphos-methyl
Azinphos-methyl is considered slightly to moderately toxic to birds. This compound is moderately to very highly toxic to freshwater fish, and is highly toxic to aquatic invertebrates, shellfish, frogs, and toads. Several studies have also indicated that azinphos-methyl causes adverse effects in wildlife. Wild mammals and aquatic organisms appear to be more vulnerable than birds to the

acute hazards created by this material. Azinphos-methyl is toxic to honeybees and other beneficial insects, so care should be taken to minimize their exposures.

Chlorpyrifos
Chlorpyrifos is moderately to highly toxic to birds, and is very highly toxic to freshwater fish, aquatic invertebrates, and estuarine and marine organisms. Chlorpyrifos toxicity to fish may be related to water temperature. Chlorpyrifos exposures for long periods of time have been associated with decreased survival and growth of first generation offspring, as well as increased incidence of deformities. This compound accumulates in the tissues of aquatic organisms. Chlorpyrifos may present a hazard to sea-bottom dwellers due to its high acute toxicity and its persistence in sediments. Smaller organisms generally appear to be more sensitive than larger ones. Aquatic and general agricultural uses of chlorpyrifos pose a serious hazard to wildlife and honeybees.

Diazinon
Diazinon is very toxic to birds. The most common route of exposure is ingestion of granules on the ground surface. Bird poisonings are common following broadcast applications of this compound.

Birds appear to be more susceptible to this compound than other wildlife. Diazinon is highly toxic to fish. Fish susceptibility may be lesser in hard water. Saltwater fish may be more suceptible to adverse effects than freshwater fish. This compound does not tend to bioconcentrate in fish. Diazinon is highly toxic to bees, beneficial insects, and many other nontarget insects.

Dimethoate

Dimethoate is moderately to highly toxic to birds. It is moderately toxic to fish and is more toxic to aquatic invertebrate species such as stoneflies and scuds. This compound is very highly toxic to bees and other nontarget insects.

Ethion

Ethion toxicity to birds ranges from highly toxic to practically nontoxic, depending on the species. It is highly toxic to songbirds, is moderately toxic to medium-sized birds such as quail, and practically nontoxic to larger upland game birds and waterfowl. This compound is persistent in the environment and may accumulate in plant and animal tissues. Ethion is highly toxic to invertebrates, freshwater and marine fish; it accumulates in fish tissues. This compound is practically nontoxic to honeybees.

Fenamiphos

Fenamiphos is highly toxic to birds and moderately to highly toxic to aquatic species. This compound is not expected to bioaccumulate in aquatic organisms. It is practically nontoxic to honeybees.

Malathion

Malathion is moderately toxic to birds, and has a wide range of toxicities in fish. It is highly toxic to aquatic invertebrates, and to the aquatic stages of amphibians. This compound may bioconcentrate in aquatic species.

Methidathion

Methidathion is highly toxic to birds. It is very highly toxic to aquatic fish and invertebrates, but does not tend to bioaccumulate. This compound is slightly toxic to bees.

N-Methyl Carbamate Insecticides

The carbamate insecticides are ecologically important because of their acute toxicity to biota. This class of pesticides shares a similar mode of action with the organophosphates. In this case, the carbamate insecticide also inactivates the enzyme acetylcholinesterase. This enzyme is needed for breaking down acetylcholine that is produced by nerve cells in response to a stimulus. Acetylcholine functions similarly to a light switch. When it is present, nerve impulses are transmitted; when it is not present, nerve impulse transmission ceases. Thus, acetylcholine accumulates, resulting in continous nerve transmission. This class of compounds differs from the organophosphates in that the inhibition of acetylcholinesterase is more reversible. A summary of the ecotox for specific carbamates labeled for use in citrus follows.

Aldicarb

Aldicarb is highly toxic to birds. The unincorporated granules are attractive as a food source, leading to ingestion of the compound. Another common route of exposure is through ingestion of contaminated earthworms. This compound is moderately toxic to fish, and it does not tend to accumulate in aquatic organisms. Aldicarb is not toxic to bees.

Carbaryl

Carbaryl is considered to be practically nontoxic to wild birds. It is moderately toxic to aquatic organisms. This compound may accumulate in aquatic species such as catfish, crawfish, snails, algae, and duckweed under conditions below a pH of 7. However, bioaccumulation should not occur to a great extent in alkaline waters. Carbaryl is lethal to bees, beneficial insects, and many other nontarget insects.

Fenoxycarb

Fenoxycarb is practically nontoxic to birds. It is moderately to highly toxic to fish. It is also highly toxic to aquatic invertebrates, and may affect growth and reproduction following chronic exposures, depending on species. This compound is practically nontoxic to bees.

Methomyl

Methomyl is highly toxic to birds, is moderately to highly toxic to fish, and is highly toxic to aquatic invertebrates. This compound is not likely to bioconcentrate in aquatic systems. It is highly toxic to bees.

Oxamyl

Oxamyl is highly toxic to birds, and is moderately to slightly toxic to fish and aquatic invertebrates. It is highly toxic to bees.

Phenoxy Acid Herbicides

Phenoxy acid herbicides are used for controlling a variety of broadleaf weeds. These compounds vary in their degree of selectivity for certain broadleaf species. Being herbicides, ecosystem components at risk from nontarget exposure are associated with broadleaf vegetation. Phenoxy herbicides such as 2,4-D are taken up by the roots or absorbed through the leaves of the plant. They are then distributed throughout the plant. These compounds accumulate in the active growth regions at the tips of stems and roots where they disrupt normal cell growth. Both the phenoxy and benzoic acid herbicides mimic the plant growth hormone auxin, stimulating the growth of old cells and the rapid expansion of new cells. This combination of rapid cell enlargement without normal cell division crushes the plant's water and nutrient transport system in the active growing regions.

2,4-D

2,4-D toxicity ranges from slightly toxic to waterfowl to slightly-moderately toxic to birds. Toxicity to aquatic organisms varies with the formulation.

Fluazifop-p-butyl

Fluazifop-p-butyl is practically nontoxic to bird species. This compound may be highly to moderately toxic to fish. It is only slightly toxic to other aquatic species such as invertebrates. The compound is of low toxicity to bees.

Other Pesticides Labeled for Use In and Around Citrus

This section describes ecotox data for other pesticides labeled for use in and around citrus. "Around" citrus includes pesticides labeled for use in controlling ditch bank and aquatic weeds in and around on-farm drainage ditches.

Abamectin

Abamectin is considered to be practically nontoxic to birds. This compound is highly toxic to fish and very highly toxic to aquatic invertebrates. Abamectin does not tend to accumulate or persist in fish. It is highly toxic to bees.

Bacillus thuringiensis

B.t. is not toxic to birds, and is practically nontoxic to fish. Exposures to this organism may adversely affect shrimp and mussels. B.t. is not toxic to most beneficial or predatory insects and honey bees. However, very high concentrations of the spores may reduce the longevity of adult honey bees, and may kill some nontarget species.

Benomyl

Benomyl is considered to be moderately toxic to birds. It is highly to very highly toxic to fish. The primary breakdown product of this fungicide is as toxic to fish as the parent compound. Toxicity to aquatic invertebrate species varies. This compound is very toxic to earthworms at low concentrations over a long period of time. It is relatively nontoxic to bees.

Bromacil

Bromacil is considered practically nontoxic to birds and nontoxic to aquatic invertebrates. It is slightly toxic to practically nontoxic to fish.

Copper Sulfate

Copper sulfate is practically nontoxic to birds, but more so to other animals. It is highly toxic to fish, especially in soft or acidic waters. Fish toxicity generally decreases as water hardness increases. Young fish are more susceptible to the toxic effects of copper sulfate than fish eggs. Copper sulfate is toxic to aquatic invertebrates, bees, and animal life within the soil. Extensive use of copper fungicides has often resulted in elimination of most animal life (including earthworms) in the soil.

Dicofol
Dicofol is manufactured from DDT. The modern manufacturing process is able to produce technical grade dicofol containing less than 0.1% DDT. Dicofol is slightly toxic to birds. Some researchers have noted eggshell thinning and reduced offspring survival in mallard ducks, American kestrals, ring doves, and screech owls. Dicofol is highly toxic to fish, aquatic insects, and algae. This compound is not toxic to bees.

Diuron
Diuron is slightly toxic to birds. This compound is moderately toxic to fish and highly toxic to invertebrates. It is nontoxic to bees.

EPTC
EPTC is slightly toxic to relatively nontoxic to birds. It is slightly toxic to fish and aquatic organisms. This compound does not significantly accumulate in aquatic species, and is practically nontoxic to bees.

Glyphosate
Glyphosate is considered to be slightly toxic to birds. It is slightly toxic to aquatic invertebrates and is practically nontoxic to fish.

Metalaxyl
Metalaxyl is practically nontoxic to birds and to freshwater fish. It is slightly toxic to aquatic invertebrates. This compound is not toxic to bees.

Napropamide
Napropamide is practically nontoxic to game birds. The compound is slightly to moderately toxic to freshwater fish species, and appears to be slightly toxic to aquatic invertebrates. It is slightly toxic to fiddler crabs and moderately toxic to pink shrimp and eastern oysters. It is not toxic to bees.

Oryzalin
Oryzalin is slightly toxic to practically nontoxic to birds. It is highly toxic to fish, and is not toxic to bees.

Oxyfluorfen
Oxyfluorfen is considered to be practically nontoxic to birds. It is highly toxic to aquatic invertebrates, freshwater clams, oysters, aquatic plants, and fish. Oxyfluorfen has a low to moderate potential for bioaccumulation in aquatic species. It is not toxic to honeybees.

Paraquat
Paraquat is moderately toxic to birds, and is slightly to moderately toxic to many species of aquatic life, including rainbow trout, bluegill, and channel catfish. This compound may bioaccumulate in aquatic weeds. Paraquat inhibits photosynthesis of some algae in stream waters at high levels (sublethal concentrations for weeds). It is not toxic to honeybees.

Pendimethalin
Pendimethalin is slightly toxic to birds. It is highly toxic to fish and aquatic invertebrates. This compound has a moderate potential to accumulate in aquatic organisms. Pendimethalin is not toxic to bees.

Sethoxydim
Sethoxydim is considered to be practically nontoxic to birds, and is moderately to slightly toxic to aquatic species. It has low toxicity to wildlife, and is not toxic to bees.

Simazine
Simazine is practically nontoxic to birds and is slightly to practically nontoxic to aquatic fish and invertebrate species.

Triclopyr
Triclopyr is slightly to practically nontoxic to birds. The parent compound and amine salt are practically nontoxic to fish. The compound is practically nontoxic to the aquatic invertebrates. This compound has very little potential to accumulate in aquatic organisms. It is not toxic to bees.

Water Quality Criteria for Florida
The Florida Department of Environmental Protection (FDEP) is responsible for maintaining and protecting the quality of water within the state. To

perform this function, all waters within the state fall into one of five surface water classifications with specific criteria applicable to each class of water (Table 6-5).

Classification of a water body according to a particular use or uses does not preclude use of the water for other purposes. Water quality classifications are arranged in order of the degree of protection required. Class I water generally has the most stringent water quality criteria and Class V the least. However, Class I, II, and III surface waters share water quality criteria established to protect recreation and the propagation and maintenance of a healthy, well-balanced population of fish and wildlife. Criteria developed for each classification are designed to maintain the minimum conditions necessary to assure the suitability of water for its designated use.

Even though surface waters are classified as Class III, all secondary and tertiary canals wholly within agricultural areas are classified as Class IV and are not individually listed as exceptions to Class III. According to the FDEP, "secondary and tertiary canals" refers to any wholly artificial canal or ditch which is behind a control structure and which is part of a water control system that is connected to the works of a water management district.

Agricultural areas are recognized as lands used solely for the production of food and fiber, and that are zoned for agricultural use where county zoning is in effect. Agricultural areas exclude lands which are platted and subdivided or in a transitional phase to residential use. While water quality criteria are generally less stringent in Class IV waters, problems may arise when these agricultural waters empty into Class I, II, or III water bodies. Effective irrigation management tailored for environmental and grove conditions can aid in preventing this from happening.

Water quality criteria have been established for each water classification to protect the most beneficial uses of the waters. While some criteria are intended to protect aquatic life, others are designed to protect human health. The listed criteria are

more protective of both aquatic life and human health, and are located in rules 62-302.500 and 62-302.530 of the Florida Administrative Code. Water quality standards also include general provisions for pollutants that are not specifically listed. These laws state that at a minimum, all surface waters of the State must be free from domestic, industrial, agricultural, or other man-induced components of discharges that, alone or in combination with other substances, may:

- settle to form putrescent deposits or otherwise create a nuisance;

- float as debris, scum, oil, or other matter in such amounts as to form nuisances;

- produce color, odor, taste, turbidity, or other conditions in such a degree as to create a nuisance;

- be acutely toxic;

- be present in concentrations which are carcinogenic, mutagenic, or teratogenic to human beings or to significant, locally occurring wildlife or aquatic species, unless specific standards are established for such components in Rules 62-302.500(2) or 62-302.530;

- pose a serious danger to the public health, safety, or welfare.

Irrigation managers should be cognizant of these criteria with regard to water discharged from their groves. A violation of any surface water quality criterion as set forth by the State constitutes pollution in the eyes of the FDEP. For certain pollutants, numeric criteria have been established to protect human health from an unacceptable risk of additional cancer caused by the consumption of water or aquatic organisms. These numeric criteria are based on annual average flow conditions. In some cases, numeric criteria are not listed. In those cases, each of the minimum criteria mentioned above must be addressed.

The absence of water quality constituents, such as many of the pesticides used in citrus production, from the listed water quality criteria does not mean

Table 6-5. Classes of Florida water.

Class	Use
Class I	potable water supplies
Class II	shellfish propagation or harvesting
Class III	recreation, propagation and maintenance of a healthy, well-balanced population of fish and wildlife
Class IV	agricultural water supplies
Class V	navigation, utility, and industrial use

that they are of no concern. In fact, they may be of greater concern in environmentally sensitive areas. The general guidelines that the FDEP and EPA use for determining acute and chronic toxicity are discussed below. These are the guidelines used for determining acute and chronic toxicity in cases where no numerical criteria are listed. "Acute Toxicity" refers to the presence of one or more substances, characteristics, or components of substances in amounts that:

a. are greater than one-third of the amount lethal to 50% of the test organisms in 96 hours (96 hr LC_{50}) where the 96 hr LC_{50} is the lowest value which has been determined for a species significant to the indigenous aquatic community; or

b. may reasonably be expected, based upon evaluation by generally accepted scientific methods, to produce effects equal to those of the concentration of the substance specified in (a) above.

"Chronic Toxicity" refers to the presence of one or more substances, or characteristics or components of substances in amounts that:

a. are greater than one-twentieth of the amount lethal to 50% of the test organisms in 96 hrs (96 hr LC_{50}), where the 96 hr LC_{50} is the lowest value which has been determined for a species significant to the indigenous aquatic community; or

b. may reasonably be expected, based upon evaluation by generally accepted scientific methods, to produce effects equal to those of the concentration of the substance specified in (a) above.

General criteria for certain metals often used as, or present in, pesticides and fertilizers (copper, nickel, selenium, and zinc) have been derived based on the acid-soluble fraction. However, the criteria should be based on the total recoverable fraction, unless it is demonstrated to the DEP in a permit application that an alternative fraction is more appropriate for a particular site. Consideration of an alternative fraction should take into account analytical interferences, bioavailability, toxicity, chemical transformations, and any other factors reasonably related to the toxic effects of the metal within the aquatic environment.

Summary

Irrigation management practices affect many soil-related factors that influence the potential for losses of pesticides by leaching or surface runoff. These pesticides may significantly impact nontarget organisms that are exposed to them. Irrigation managers should become familiar with the pesticides used in their operations, and try to avoid conditions that favor leaching or the generation of surface runoff water.

Chapter 7. Water Resources of Florida

by Brian Boman and Larry Parsons

Florida has abundant water resources. The state receives an average of 175 billion gallons per day (150 bgd in rainfall and 25 bgd inflow from Georgia and Alabama). Florida's aquifers contain more than a quadrillion gallons of water, which is 30,000 times the average daily discharge of Florida's thirteen largest rivers. But withdrawal of water for consumptive use continues to rise. Moreover, there is tremendous variability in the source of supply (rainfall). Most of the potential supply must be left in the hydrologic system for nonconsumptive uses such as navigation, recreation, and aesthetics, or to provide a habitat for fish and wildlife, or because it cannot be economically used. In addition, water is not evenly distributed in the state. The densely populated coastal areas of the peninsula have much less available water than interior, northern regions. The potential for conflict over the allocation of water is growing in Florida.

Water is one of Florida's most valuable resources. Freshwater supplies come from extensive subsurface beds of porous rock (aquifers) and from freshwater lakes, streams and reservoirs. More than 50% of the total freshwater used in Florida comes from groundwater, and more than 90% of the public rely on groundwater supplies for their drinking water. Of all the freshwater withdrawn in Florida, only about one-third is consumed by evaporation, transpiration, or production processes. The remaining two-thirds are returned to the environment, either to surface streams or to aquifers.

Groundwater and the Hydrologic Cycle

The continuous circulation of water from land and sea to the atmosphere and back again is called the hydrologic cycle (Fig. 7-1). Inflow to the hydrologic system is primarily rainfall. Outflow occurs as runoff, evaporation, transpiration by plants, and outflow from groundwater into wells, rivers, springs or oceans. Components of the hydrologic cycle include the following:

Evaporation:
The change of water from a liquid to a gaseous state is called evaporation. Evaporation occurs from raindrops, free water surfaces such as seas and lakes, water settled on vegetation and soil, and from human activities. Evaporation rate is affected by solar radiation, air temperature, vapor pressure, wind, and atmospheric pressure. During evaporation, moisture is moved into the atmosphere as water vapor.

Condensation:
The change of water from a vapor to a liquid state is termed condensation. Water vapor condenses onto small airborne particles to form dew, fog, and clouds. Condensation occurs when the saturation point of air is exceeded by cooling or by increasing the amount of vapor in the air.

Precipitation:
The fall of water particles from the atmosphere to the ground is precipitation. It occurs in two forms. In the coalescence subprocess, a larger lead drop that reaches a critical size in the air attracts a number of smaller drops to create precipitation. In the ice-crystal subprocess, ice develops under freezing temperatures in clouds and attracts water droplets that evaporate and condense on the crystals. Precipitation may fall onto both bodies of water and land.

Interception:
The interruption of the movement of water on the land surface is called interception. It takes place by vegetation or storage in land formations. Rainwater is stored on the surface of leaves and other organic materials up to their maximum storage capacity, above which the excess water falls to the ground.

Infiltration:

The movement of water from the air into the soil at the interface of the atmosphere and the earth's surface is called infiltration. The amount of water transfer depends on the texture and structure of the soil, the soil moisture content, and the atmospheric concentration of water.

Percolation:

Percolation is the movement of water through the soil by gravity and capillary forces. Water in the zone of aeration is called vadose water. Water in the zone of saturation is called groundwater. The two zones are separated by the water table. Percolation contributes to both underground water storage and to water movement.

Transpiration:

The transfer of water from plants to the atmosphere as vapor through leaf openings is called transpiration. The amount of transpiration depends on the plant species and the amount of light exposure. Transpiration increases movement of water to the atmosphere but reduces evaporation from the soil.

Runoff:

The flow of water from drainage basins to surface streams is runoff. It occurs in three main forms.

Surface runoff takes place on the land surface through channels. Subsurface runoff occurs with water that infiltrates and then moves laterally into the drainage system. Groundwater runoff results from water that percolates deeper into the soil and moves with groundwater gradients.

Storage:

Water is naturally stored in the atmosphere, on the surface of the earth, and in the ground. Water movement though surface and ground storage depends on the geologic features of the storage locations.

Vadose Zone:

Between the land surface and the water table is the unsaturated (vadose) zone where both water and air occur in the soil pores. Water in the unsaturated zone is either taken up by plants, evaporated, or drained by gravity into the saturated zone. In the flatwoods soils of Florida, the unsaturated zone is typically only the top 10 to 40 inches of the soil profile in the dry season, and it may be nonexistent in the wet season when the water table is at or above the ground surface. In the sandy soils of the Central Florida Ridge, however, the vadose zone can extend a depth of 100 feet or more.

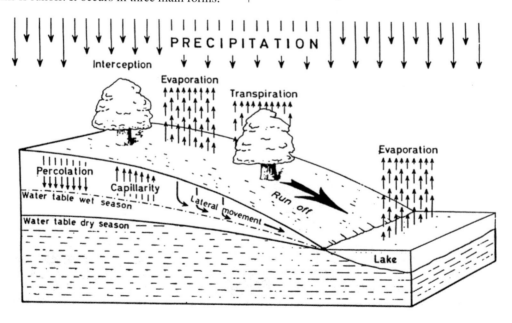

Figure 7-1. Typical water cycle under Central Florida conditions (Fitzpatrick, 1974).

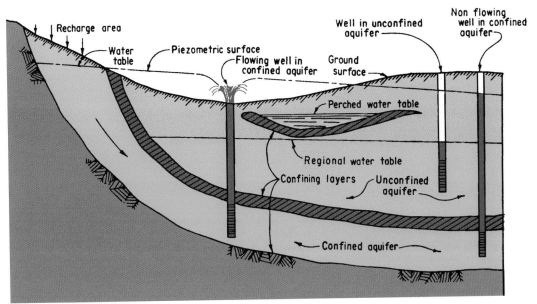

Figure 7-2. Typical hydrogeology of the surficial water table aquifer and the confirmed Floridan Aquifier system (with artesan wells) in South Florida.

Water Table:

In the saturated groundwater zone (water table), all pores and crevices are filled with water, and all of the air has been forced out. Water seeping into this zone is called recharge. Groundwater can occur either as an unconfined (phreatic) aquifer or as a confined (artesian) aquifer. In an unconfined aquifer, the water table forms the upper boundary of the aquifer, and the water level in a well will rest at this level. Water infiltrating from the surface has the potential to move rapidly into an unconfined aquifer; thus, there is potential for contamination from surface activities. In an unconfined aquifer, groundwater moves by gravity from areas of high water table elevation to areas of lower water table elevation.

Confined aquifers (Fig. 7-2) are overlain by an impermeable, or semipermeable confining layer, and are typically under pressure. Therefore, the level to which water will rise in a tightly cased well is above the top of its upper confining layer (artesian well). If the water level rises above the land surface, it is called a flowing artesian well.

Water in confined aquifers moves from areas of high potentiometric head (as measured by the level to which water will rise in a tightly cased well) to areas of low potentiometric head. Confined aquifers are less susceptible to contamination from local surface activities because infiltrating water typically moves very slowly through the confining layer. However, the confining layers may be fractured and missing in many places. Thus, contaminated water may move horizontally on top of the confining layer for some distance before recharging the confined aquifer through a breach in the confining layer.

Major Florida Aquifers

The most important aquifers that yield large quantities of water to wells, streams, lakes, and springs in Florida are shown in Fig. 7-3. The primary source of groundwater for most of the state is the Floridan Aquifer, which is one of the most prolific aquifers in the United States. However, in many areas Floridan Aquifer wells have high salinity levels, making them undesirable for citrus irrigation. In many areas, the Floridan Aquifer is confined by low permeability sediments of the Hawthorne formation (Fig. 7-4). The Hawthorne formation is absent in the north central part of the state along the

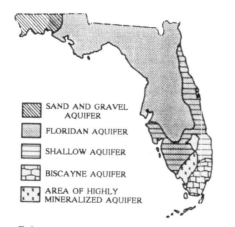

Figure 7-3. Major groundwater aquifers of Florida.

Ocala Uplift. In this area the aquifer is unconfined, and thus receives recharge from water infiltrating from the surface.

The origin of subsurface flow to the Floridan Aquifer in northern Florida is from Alabama and Georgia. In peninsular Florida, subsurface Floridan flow

originates in the Central Uplands of the state. In many coastal areas, the potentiometric surface is above the land surface; thus, artesian flow occurs in wells or along geologic openings (springs).

The unconfined Biscayne aquifer underlies an area of about 3000 square miles in Dade, Broward, and Palm Beach Counties. This aquifer is 100 to 400 feet thick near the coast, but thins to a thickness of only a few feet further inland. Water in the Biscayne aquifer is derived chiefly from local rainfall. However, during dry periods recharge can come from canals linked to Lake Okeechobee. A shallow, unconfined aquifer is present over much of the state, but in most areas it is not an important source of groundwater because a better supply is available from other aquifers. However, where water requirements are small, this aquifer is tapped by small diameter wells. In South Florida the shallow aquifer is a major source of groundwater in Martin, Palm Beach, Hendry, Lee, Collier, Indian

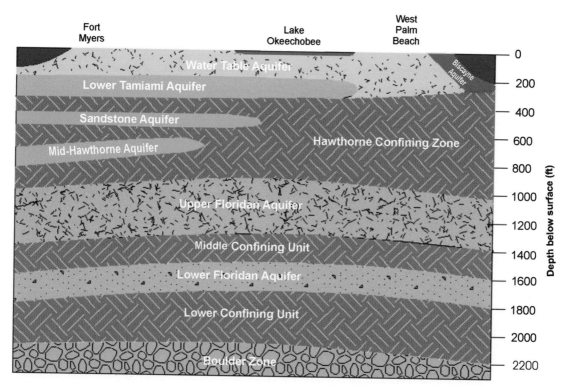

Figure 7-4. Generalized cross section of hydrogeology in South Florida.

River, St. Lucie, Glades and Charlotte Counties. The water in this shallow aquifer is derived primarily from local rainfall.

Sources of Groundwater Contamination

Florida's unique hydrogeologic features of a thin soil layer, high water table, porous limestone, and large amounts of rainfall, coupled with its rapid population growth, result in a groundwater resource extremely vulnerable to contamination. Nonpoint sources, which have potential for contributing to groundwater contamination, include coastal saltwater bodies, urban storm water, agricultural practices, and mining.

Florida's situation as a peninsula between two bodies of saltwater creates the potential for saltwater intrusion into the fresh groundwater supply. Saltwater is more dense than freshwater and thus exerts a constant pressure to flow into the freshwater aquifers. As long as freshwater levels in the aquifer are above sea level, the freshwater pressure limits the inland movement of the salt. Overpumping of coastal wells, however, can increase saltwater intrusion. If water is pumped out faster than the aquifer is replenished, the pressure of the freshwater is decreased. This causes the level at which the saltwater and freshwater meet to rise in the aquifer, degrading the freshwater quality. The problem of saltwater intrusion is aggravated by periods of drought during which there is not enough rainfall to replenish the freshwater aquifers.

Figure 7-5 shows areas of the Floridan Aquifer that contain chloride concentrations greater than 250 mg/l. In South Florida, where the Floridan Aquifer is artesian and underlies the Biscayne and shallow aquifers, its saline water may recharge the overlying freshwater aquifers, increasing their salt content. This type of recharge may occur naturally by upward seepage through the confining layer or it may be increased by flowing artesian wells.

Reclaimed Water Use on Citrus

The main studies using reclaimed municipal wastewater to irrigate young and mature orange and grapefruit trees in Florida have been conducted at the Water Conserv II project near Orlando and at a citrus grove adjacent to an Indian River County wastewater treatment plant near Vero Beach. Other smaller reclaimed water facilities are located in Pasco, Polk, Manatee, Sarasota, and Okeechobee Counties and involve primarily individual growers and local treatment facilities.

The Water Conserv II project provides citrus growers with a long-term source of reclaimed water (a twenty-year contract) to be used for irrigation of young and mature citrus trees. Water is provided free, although growers must absorb costs associated with connection to the system and purchase of water meters. Growers have the option of refusing water four weeks per year, but only two weeks consecutively. Water that is not applied to groves during high rainfall periods is diverted to rapid infiltration basins (RIBs), where it percolates through the sand and eventually reaches the aquifer. Water quality is regulated very strictly. The water may not have detectable levels of fecal coliforms or viruses and must have <30 mg/L biochemical oxygen demand (BOD) and 25 mg/L total suspended solids (TSS).

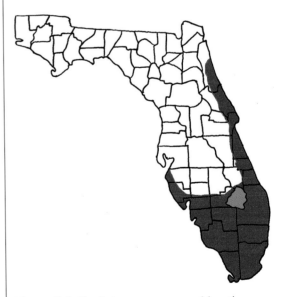

Figure 7-5. Shaded areas are general locations where the Floridan Aquifer contains more than 250 ppm Cl.

In the Water Conserv II study, trees receiving reclaimed wastewater had similar or greater yields and improved tree vigor compared to trees receiving well water. High application rates of reclaimed water (100 inches per year) decreased acid, and soluble solids (Brix and solids per box). By reducing water stress, high application rates of reclaimed water promoted greater canopy growth and yield. In contrast, tree vigor and fruit quality were not different for mature grapefruit trees growing in the flatwoods area near Vero Beach, which received reclaimed water or canal water. Yields, however, were higher for the reclaimed water treatments in one season, but the effect was variable. No adverse effects of applying high levels of reclaimed water were noted at either site.

The composition of reclaimed municipal wastewater at the Water Conserv II and Indian River County sites varied through the season. Water from both sites was low in heavy metals, reflecting the urban nature and lack of heavy industry in these areas. The water at both sites was quite similar in most characteristics, with the exception of significantly higher sodium, chloride, magnesium, and boron levels at the Vero Beach facility. After twelve years at Conserv II and three years at Vero Beach, no adverse effects on tree growth and development or yields have been observed related to elevated levels of these elements.

The reclaimed water provided some nutritional value from constituents that had not been removed in the treatment process. These nutrients are present at low levels. However, when reclaimed water is applied at high rates (1-1.5 inches per week) in order to dispose of as much water as possible per unit of land area, the amount of the nutrients adds up. Therefore, rates of nitrogen, phosphorus, and potassium fertilizers needed to be applied may sometimes be reduced. For example, about 65% of the nitrogen needed to produce a fresh grapefruit crop in the Indian River area was applied with the reclaimed water at the Vero Beach site. Considerable amounts of phosphorus and potassium were also provided by the reclaimed water at both sites. In fact, high levels of potassium

are a concern, because they may adversely affect fruit quality by increasing peel thickness to unacceptable levels.

Water Conserv II Site

At the Water Conserv II site, data have been collected from several scions and rootstocks. All groves were fertilized based on current recommendations and irrigated using microsprinklers or under canopy, high-volume sprinklers. Nine sites that received only well water and fertilizer served as controls. Data have been collected since 1987.

Tree canopy appearance and leaf color were, in general, better for trees receiving reclaimed water than for control untreated trees receiving well water. Trees receiving reclaimed water were deeper green in leaf color and in general more densely foliated. Fruit from trees receiving reclaimed water were larger, and in two years had slightly higher juice content than fruit receiving well water. However, TSS, acid, and lb-solids were reduced due to the higher levels of water applied in the reclaimed water treatments. These trends varied slightly among years because of differences in rainfall and soil water content, but were typical for citrus fruit from trees that received high levels of irrigation.

Soil water content was greater for the reclaimed water compared with the control at most times, but no evidence of water damage was observed in the well-drained sandy soils of the Ridge. Weed growth was greater in the reclaimed water treatments compared to the control. Leaf nitrogen, potassium, calcium, and magnesium levels were also similar. Leaf phosphorus and especially sodium levels were higher in the reclaimed water treatments, but sodium levels were below the toxic range. This reclaimed water can provide some nutrition, but it cannot provide sufficient nitrogen.

Indian River County Site

The Indian River County experimental site was much smaller than the Water Conserv II site, and compared three application rates of reclaimed water (0.9, 1.2, and 1.5 in./week) with a control (canal water) that received water based on soil water deficit. Microsprinkler irrigation was used to apply the water. Control trees were irrigated at one-third soil water depletion from January to June and at two-thirds depletion the rest of the year.

Tree vigor, monitored subjectively, and trunk circumference were unaffected by the treatments in the first year. However, there was a substantial numerical increase in yields for all reclaimed water treatments over the controls in the same year. The most surprising result from the Vero Beach study was the lack of effect of large quantities of water on grapefruit quality. At no time did juice content, TSS, TA, or TSS:TA differ among treatments. Fruit size, however, was smaller for controls compared to reclaimed water treatments.

Preliminary data suggest that applications of high rates of high-quality reclaimed water can result in some reductions in fertilizer use and increased yields. Irrigation of mature 'Redblush' grapefruit trees using reclaimed water on the spodic soils in Indian River County did not cause an accumulation of Sodium (Na), Boron (B), or heavy metals. Although a hardpan was present at about two feet, the upper sandy layer had a very high percolation rate and drained rapidly. In areas of the grove where drainage was impeded due to weed buildup or clogging of drainage pipes, trees became stunted and unproductive due to flooding damage. It is important that the groves to which high-rate applications are made should be designed without low spots to facilitate maximum drainage.

Horizontal Wells
In recent years, horizontal wells (Fig. 7-6) have become an economical water supply source alternative for many shallow water table areas in Florida. Horizontal wells typically utilize 6- to 10-inch diameter perforated drain (recovery) pipe installed 15 to 20 feet below the ground surface. Due to the high cost of installation, horizontal wells have traditionally been reserved for high-value projects where wells were installed to drop and/or maintain a lowered water table for construction purposes. Recently developed installation machinery has made the cost of horizontal wells competitive with other water supply sources.

Horizontal wells have been installed as primary water supplies for golf courses, citrus groves, ornamental nurseries, commercial landscapes, and vegetable farms. Traditional water supplies for these operations have come from deep aquifers. Concerns about slow recharge and aquifer depletion in the deep aquifers have made permitting additional users of this water difficult. In many areas, water withdrawn from deep wells in the Floridan Aquifer is highly mineralized and poorly suited for irrigation purposes. In contrast, the surficial aquifers generally have much less mineralization and have rapid recharge rates. The lower aquifer transmissivity and large drawdown caused by high-volume conventional vertical wells made tapping this resource unfeasible in many areas.

Horizontal Well Installation
The current technology allows extraction from groundwater aquifers at depths up to about 25 feet. The installation machine digs the trench, lays the recovery pipe, places drain envelope media (if required), and backfills in one operation (Fig. 7-7).

Typical installations use 400 to 800 feet of 8-inch diameter perforated recovery pipe (heavy-duty polyethylene tubing), normally installed with a double polyester filter (Fig. 7-8). A key to the installation is the attachment of a pumping riser that extends from the recovery pipe to the surface. This pumping riser is normally either 8- or 10-inch PVC pipe, depending on well capacity. The end of the well opposite the pumping header is brought to the

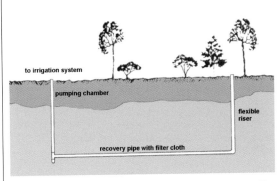

to irrigation system

pumping chamber

flexible riser

recovery pipe with filter cloth

Figure 7-6. Schematic diagram of horizontal well installation.

Figure 7-7. Horizontal well installation machine.

Figure 7-8. Horizontal well installation.

surface with nonperforated tubing. This cleanout end is normally capped, but it can also be used for pumping.

The actual well configuration is adaptable to the site conditions. Wells are installed at depths that are deemed most suitable from preliminary test and observation wells, which measure capacity and drawdown. Wells may be installed as "L" shaped, with the recovery tubing extending one direction from the pumping header, or as a "T" shape, with the recovery tubing extending in opposite directions from the header. Multiple wells have also been manifolded together to increase capacity or

minimize drawdown. Wells normally use either submersible pumps (located in the PVC pumping riser) or aboveground centrifugal pumps, depending on pumping needs and user preferences. Well capacities vary widely, depending on user needs, recovery tubing length, pump horsepower, and the nature of the aquifer. Studies have shown that the extent of the water table depression and rate of recovery following pumping are dependent upon the rate at which the well is pumped. Sustained flow rates of over 1800 gpm have been achieved in some areas with single wells.

Chapter 8. Well Design and Construction

by Brian Boman

Irrigation wells must be capable of producing adequate water during peak seasonal use and under drought conditions. Without a reliable, efficient and economical supply of water, the entire irrigation system, regardless of the most sophisticated surface equipment design, becomes nearly useless. The well is the "heart" of irrigation systems with groundwater supplies; it must be properly designed and compatible with the pump and distribution system to ensure long life, efficiency, and economic operation.

Water well construction (Fig. 8-1) in the state of Florida is regulated by statute and various agency rules enforced by the Department of Environmental Protection (FDEP), principally through delegation to the five water management districts (WMDs).

The potential groundwater sources of irrigation water in Florida include the surficial, intermediate, and Floridan Aquifer systems. The choice of aquifer is often dictated by location. It also depends on the quantity and quality of water desired. The surficial and intermediate systems generally produce limited quantities of water. For this reason, the Floridan Aquifer is the primary groundwater source for agriculture throughout the state. Exceptions occur in coastal and extreme South Florida locations where the Floridan Aquifer deteriorates in quality. In these areas, surficial and intermediate (lower and upper Hawthorne) aquifers are used along with surface water.

A well consists of many or all of the following key parts: casing, grout, screen, open bore hole, and a well head configuration. The standards enforced by the WMDs address each of these parts with alternatives given for variation in geology in various areas. The casing (or well pipe) is a very critical element in well construction. Casing may be metallic (black iron or galvanized steel) or nonmetallic

Figure 8-1. Typical well construction with hollow stem rotary drilling rig.

(PVC or ABS plastic). It must be adequately seated in a consolidated formation (limestone, sandstone, etc.) or attached to a screen suitably designed and situated in unconsolidated materials (shell, sand, gravel, etc.). The purpose of casing is to seal off materials that may enter a pumping system from strata other than the aquifer selected and prevent mixing between aquifers. To prevent contamination from surface flow into the well, the casing must be extended above surface flood water levels and the top portion must be grouted with cement grout or an approved alternate.

Rotary constructed wells are sealed with grout as a protective seal against contamination, surface entrance of pollutants, and as a seal for the casing seated into a consolidated formation. In areas where the beds of consolidated material that the casing is seated into are friable (crumble easily), the grout acts to prevent deterioration of the casing seat when turbulence develops during pumping. Poor grouting may create problems later as pump impellers and other mechanical parts are scoured by small particles moving around the casing seat. Poor grouting may also create voids where eventual corrosion of the casing wall allows unconsolidated matter to enter the well.

The cavernous nature of Florida's limestone formations produces abundant quantities of water from open bore holes (generally 4 to 12 inches in diameter) constructed into the limestone. The depth of each bore hole will depend on the percent of highly transmissive rock encountered and the needs of the irrigation system. In some areas of Florida, especially near the coast, the depth of bore holes may be limited due to increases of salinity with depth. If well yield is too low, additional properly spaced wells may be required. Back-plugging of some irrigation wells has been successful as a remedy against upconing of deep saline waters. Surficial aquifer wells usually do not produce adequate quantities of water for large operations. However, they may be very satisfactory for small systems. The local water management district should be contacted to apply for a consumptive use permit and obtain information on well specifications. In most cases, the district will be able to provide geologic data sufficient to help you secure the best well, and hopefully the least cost construction alternative. The contractor should be licensed and have irrigation experience. The irrigation system should be designed around the well capacity in order to minimize well and pumping costs.

Well Design and Construction Definitions

Drawdown
The amount the water level is lowered below the static level in a well when pumping is in progress is called drawdown. Drawdown is the difference, measured in feet of water, between the static water level and the pumping water level. This term represents the hydraulic head, in feet of water, that is needed to cause water to flow through the aquifer toward and into the well at the rate that water is being removed from the well.

Gravel-Packed Well
Gravel-packed wells have a borehole through the water-bearing formation that is larger in diameter than the well screen. The zone immediately surrounding the well screen is made more permeable than the aquifer by filling the space between the face of the borehole and well screen with graded sand or gravel that is coarser than the formation.

Naturally Developed Well
A well in which the well screen is placed directly in contact with the water-bearing sand and gravel is a naturally developed well. The width of the openings in the screen is selected so that fine sand in the aquifer immediately surrounding the screen can be removed by pumping during development to create a highly permeable zone consisting of the coarser formation particles.

Specific Capacity
The yield of the well per unit of drawdown, usually expressed as gallons per minute (gpm) per foot of drawdown, is called specific capacity. It is obtained by dividing the pumping rate by the drawdown, each measured at some specific time after pumping began. For example, if the pumping rate is 1500 gpm and the drawdown is 20 feet, the specific capacity of the well is 75 gpm per foot of drawdown.

Well Capacity or Yield
The volume of water per unit of time discharged from a well is its capacity. Well capacity is usually measured as the pumping rate in gallons per minute (gpm) or cubic feet per second (cfs).

Static Water Level
This is the level at which water stands in a well when no water is being removed from the well either by pumping or natural flow. It is generally expressed as the distance from the ground surface (or from a measuring point near the ground surface) to the water level in the well. The level to which the water level rises in a well that taps an artesian

aquifer is also referred to as the piezometric level. An imaginary surface representing the artesian pressure or hydraulic head throughout all or part of an artesian aquifer is called the piezometric surface. The piezometric surface is the real water surface or the water table in a water table aquifer.

Well Development

Well development is the process of using a variety of mechanical and chemical methods to correct damage to the formation that occurs during the drilling operation, and to remove the finer material from the aquifer adjacent to the screen or gravel pack. The process cleans the openings or enlarges passages in the water-bearing sand and gravel so that water can enter the well more freely.

Well Efficiency

The ratio of the actual specific capacity of a well at the design yield to the maximum specific capacity possible calculated from formation hydraulic characteristics and well geometry is the well efficiency. This is the same as the ratio of the theoretical drawdown that would be required to produce the design yield from a 100% efficient well to the actual drawdown measured in the pumped well when producing at the design yield. Efficiency is usually expressed as a percent. The difference (drawdown increase) between the theoretical drawdown and actual drawdown represents the head loss required to force water through the well screen. Obviously, this head loss should be a minimum. Well efficiency should not be confused with pump efficiency. Pump efficiency is a characteristic of the pump only and is completely independent of well efficiency. For example, the pump efficiency may be 75% while the well efficiency of a poorly designed and constructed well may be 45%.

Well Design

Preliminary Investigation

The preliminary investigation is the foundation upon which a well design depends. An examination of records from existing wells in the area should be made to determine yield, depth, and characteristics of the aquifers presently being used. Consultation should be made with the U.S.

Geological Survey (USGS) or any other agency that may have geologic information about the area in question. Progressive, reputable local well drillers are also an important source of useful information.

If sufficient records are not available, test holes should be drilled to allow selection of the site with the best water production potential, and to help formulate the production well design for the selected site. The information gained from test holes usually justifies the investment. In drilling test holes, samples of the aquifer should be collected so that sieve analyses and permeability tests can be made. From the completed test hole, the well designer should determine aquifer thickness, aquifer depth, and static water level of the aquifer, and estimate the yield and specific capacity of a full-sized production well. A water sample should be collected and analyzed to determine the corrosion and/or incrustation characteristics of the water.

Design Procedure

After the preliminary investigation and site selection is complete, a well design can be selected that best utilizes the hydrogeological conditions present at the site. The cased portion of the well should be designed first, and then the intake portion of the well.

The cased portion of the well consists of the well casing that serves as a housing for the pump and as a vertical conduit through that water flows upward from the intake portion of the well to the level where it enters the pump (Fig. 8-2). The casing must be large enough to accommodate the pump. In selecting the size of the casing, the controlling factor is usually the size of pump that is required for the design yield. Table 8-1 shows recommended casing sizes for various ranges of well yields (pumping rate).

The well should be of sufficient diameter to allow the ascending water to move at a velocity of 5.0 feet per second or less up the well casing. Data from the preliminary investigation and chemical analysis of water samples should be reviewed to determine if the water tends to be corrosive or

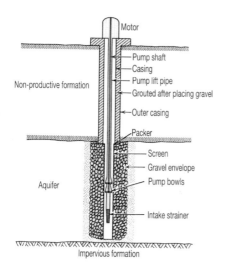

Figure 8-2. Components of gravel-packed well.

encrusting. When necessary, extra heavy steel casing should be installed. In severe corrosive cases, stainless steel, PVC, or fiberglass casing should be used.

The capacity of individual wells is highly variable from location to location. Approximate expected capacities for various size wells are given in Table 8-1, but the values are only average estimates. Much depends on the water level, yield characteristics of water-bearing formation and the pressure to be developed, but the table serves as a general guideline. In any case, the overall installation must be carefully evaluated. For instance, although 1000 gpm may be obtained from a 10-inch pump with

reasonably good efficiency, the life cycle cost of a 12-inch pump installation may be less, even including the higher cost of the larger well.

Well Screen

The screening device should be a commercially manufactured, quality well screen. To accomplish its intended purposes, the well screen must be of efficient design. A well screen is adequate only when it is capable of letting sand-free water flow into the well in ample quantity and with minimum hydraulic head loss. A properly designed well screen should have close spacing of slot openings to provide uniform open area distribution, maximum open area per foot of length, V-shaped slot openings that widen inwardly, corrosion resistance, and ample strength to resist external forces to which the screen may be subjected during and after installation. Screens with tapered slots provide hydraulic efficiency and offer self-cleaning properties. Sand grains smaller than the screen opening are easily brought into the well in the development process, while large grains are retained outside.

Screen length is an important design consideration because a screen that is too short seriously affects the efficiency of the well, whereas a well screen that is too long causes problems such as cascading water, entrained air, and accelerated corrosion and/or incrustation. The optimum length of well screen is chosen with relation to the thickness of the aquifer, available drawdown and stratification of the aquifer. In an artesian aquifer, the lower 70% to 80% of the thickness of the water-bearing sand

Table 8-1. Recommended well diameter for various yields in Central Florida Ridge areas (ID = inside diameter and OD = outside diameter).

Anticipated well yield (gpm)	Nominal size of pump bowls (inches)	Optimum size of well casing (inches)	Smallest size of well casing (inches)
200-600	6	8 ID	6 ID
600-1000	8	10 ID	8 ID
1000-1900	10	12 OD	10 ID
1900-2800	12	16 OD	14 OD

should be screened, assuming the pumping level is not expected to be below the top of the aquifer. It is generally not necessary to screen the entire thickness of artesian aquifers. About 90% of the maximum specific capacity can be obtained by screening only 75% of an artesian aquifer. An exception to this rule should be made when the aquifer is highly stratified and interbedded with low permeability layers. In this case all of the aquifer may need to be screened.

Optimum design practice dictates that the maximum available drawdown in an artesian well should be the distance from the static water level to the top of the aquifer. If it is necessary to lower the pumping level below the top of the aquifer to obtain greater yield, the screen length should be shortened and the screen should be set in the bottom of the aquifer. All attempts should be made to design and construct the well so that the pumping level stays above the top of the uppermost well screen.

For water table wells, selection of screen length is something of a compromise between two factors. High specific capacity is obtained by using as long a screen as possible. On the other hand, more available drawdown results from using as short a screen as possible. These two conflicting aims are satisfied, in part, by using an efficient well screen.

Available drawdown in a water table well is the distance between the static water level and the top of the screen. In a water table aquifer, screening the bottom 1/3 to 2/3 of the aquifer normally provides the optimum design. If the lower 1/3 of a water table aquifer is screened and the pumping level is lowered to the top of the screen, the well

will theoretically produce 88% of maximum yield. If the drawdown were increased to 95% of the maximum possible (almost to the bottom of the screen), the well yield would theoretically increase to 99% of its maximum capacity. The result will be only an 11% increase in yield for an additional 32% more drawdown.

Gravel Pack
Gravel-packed wells are particularly well suited to some geologic environments, but gravel packing is not a cure-all for every sand condition. Gravel pack construction is recommended:

• if the aquifer consists of fine sand,

• in thick artesian aquifers,

• in loosely cemented sandstone formations,

• in extensively stratified formations consisting of alternating layers of fine and coarse sediments or thin silt and clay layers.

Gravel packing makes the zone immediately surrounding the well screen more permeable by removing the formation materials and replacing it with artificially graded coarser materials (Fig. 8-3). The size of this artificially graded gravel should be chosen so that it retains essentially all of the formation particles. The well screen slot opening size is then selected to retain the gravel pack.

Gravel pack design includes specification of gradation, thickness, and quality of the gravel pack material. Samples collected during the test hole drilling representing that part of the borehole that may possibly be screened should be examined. Plain casing should be set in intervals with unfavorable strata of the aquifer, such as the finest sands. This

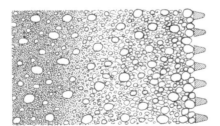 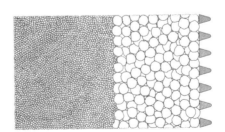

Figure 8-3. Differences in gravel pack for naturally developed well (left) and artificial gravel pack (right).

means it may be necessary to space plain casing between screen sections that are positioned in the best strata of the aquifer. One advantage of placing blank casing opposite strata composed of the finest sands and low permeability intervals is that a coarser gravel pack can be utilized. The coarser pack will allow the coarser strata of the water-bearing formation to yield maximum water. It is likely that little potential yield is lost by setting nonslotted casing opposite the finest sands and other low permeability strata because these unfavorable layers produce little water.

A sieve analysis should be prepared that represents the gradation of samples from the strata comprising the portion of the aquifer where the screen will be set (Fig. 8-4). The sieve analysis representing the finest stratum of those tested should be selected, and the grading of the gravel pack should be designed on the basis of these data. It is best to design as uniform a pack as possible because a more uniform gravel pack has significantly greater permeability and is easier to install without segregation. The gravel pack material should consist of clean and well-rounded grains that are smooth. These characteristics increase the permeability and porosity of the gravel pack. In addition, the particles should consist of siliceous (quartz) rather than calcareous material (calcareous material should be limited to <5%).

To ensure that an envelope of gravel will surround the entire screen, a thickness of 3 to 8 inches is recommended for field installation. This thickness will successfully retain formation particles regardless of how high the water velocity tends to carry the particles through the gravel pack. When more than 8 inches of gravel pack is provided, development of the aquifer is hampered. A thicker envelope does not significantly increase the yield of the well, and does little to control sand pumping because the controlling factor is the ratio of the grain size of the pack material to the formation material.

To ensure that the envelope of gravel completely surrounds the entire screen, centering guides should be used to center the screen in the borehole.

The pack material should be placed continuously, but slowly, to avoid bridging and sorting of the particles. If the screen is not centered in the borehole and is in direct contact with the formation material (no gravel pack between the well screen and formation), sand pumping will result.

Slot Openings

The retention of the water-bearing formation is accomplished by the proper selection of the gravel pack while the well screen retains the gravel pack particles. In a gravel-packed well, the size of the screen slot opening is selected to retain 90% or more of the gravel pack material. For the sand sieve analysis in Fig. 8-4, the proper size screen in a gravel-packed well would have a slot opening of 0.015 inch to retain 90% of the material in the water-bearing strata.

For naturally developed wells, the size or sizes of well screen slot openings will depend on the gradation of the sand, and are selected from a sieve analysis study of samples representing the water-bearing formation. A sand analysis curve, such as shown in Fig. 8-4, is plotted for each sand sample. The size of the screen opening is selected such that the screen will retain from 40% to 50% of the sand

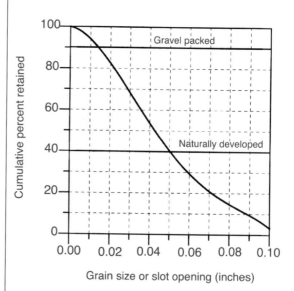

Figure 8-4. Example sieve analysis for materials from water-bearing strata.

(50% to 60% of the formation sand particles will pass through the openings in the screen during development). If the formation is nonhomogeneous, it may be necessary to select various sizes of slot openings for different sections of the well screen. The use of a multiple-slot screen to custom fit the gradation of each stratum will assist in attaining the highest specific capacity possible, and will greatly reduce the possibility of pumping sand with the water.

The 40% size is usually chosen when the ground-water is not particularly corrosive and when there is little doubt as to the reliability of the formation samples. For example, for the sieve analysis in Fig. 8-4, a slot size of 0.050 inch would provide 40% retention of the materials in the water-bearing strata. The 50% size is chosen if the water is corrosive or if the reliability of the sample is in question. If the water is corrosive, enlargement of the openings of only a few thousandths of an inch due to corrosion could cause the well to pump sand. If the water is encrusting, a 30% retained size may be selected. When this larger slot opening is selected, longer well life can be expected before plugging reduces the well yield.

Large slot size also makes it possible to develop a larger area of the formation surrounding the screen. This generally increases the specific capacity of the well by making the well more efficient.

Screen Diameter

One important consideration that must be kept in mind when selecting the screen diameter is that the diameter can be varied without greatly affecting the well yield. Doubling the diameter of the well screen can be expected to increase the well yield by only about 10%. Screen diameter can be varied after the length of the screen and size of the screen openings have been selected. Screen diameter is selected to provide enough total area of screen openings so that the average entrance velocity of the water through the slot openings does not exceed the design standard of 0.1 foot per second. A quality well screen with maximum open area offers a decided cost advantage when different types of

screening devices are compared at this entrance velocity.

The entrance velocity is calculated by dividing the expected or desired yield of the well by the total area of openings in the screen. If the velocity is greater than 0.1 foot per second, the diameter should be increased. If the calculated entrance velocity is less than 0.1 foot per second, the screen diameter may be reduced somewhat. However, the screen diameter should not be reduced to the point that the velocity of vertical water flow to the pump exceeds 5.0 feet per second. Laboratory tests and field experience show that if the screen entrance velocity is equal to or less than 0.1 foot per second, the friction loss through the screen openings is negligible, resulting in a higher well efficiency.

Open Area

The percentage of open area of the screen should be equal to or greater than the porosity of the sand and gravel water-bearing formation and the artificial gravel pack supported by the screen. Where the irrigation well screening device provides only 2% to 5% open area as in perforated pipe, flow restrictions are unavoidable. This is one of the most common reasons for low efficiencies of irrigation wells. Suppose that the water-bearing sand has 30% porosity (voids), and the screening device installed has only 5% open area. With such a small open area, there will be constriction of flow. As a result, there will be additional drawdown caused by increased head loss as water moves toward and into the well.

Adequate open area should be provided by the well screen to allow the desired or design yield to enter the well at an entrance velocity through the screen openings of 0.1 foot per second. This hydraulic characteristic of the screen is known as transmitting capacity. If the amount of open area of a screen is known and the recommended entrance velocity of 0.1 foot per second is used, the transmitting capacity of that screen can be readily calculated. For example, a 16-inch diameter quality well screen of continuous slot construction with 175 square inches (1.22 square feet) of open area per lineal foot of screen can transmit 55 gpm per foot of screen body at an entrance velocity of 0.1 foot per second. The calculated transmitting capacity of a screen is a hydraulic characteristic of that screen and not a measure of the yielding capability of the water-bearing formation in which the screen is installed.

Screen Material

The well screen should be fabricated of materials that are as corrosion resistant as necessary, as determined from analysis of data collected during the preliminary investigation. If the screen corrodes, sand and/or gravel will enter the well. Thus, the screen must be replaced or a new well must be drilled.

Corrosion of screens can occur from bimetallic corrosion if two different metals have been used in the fabrication, and therefore this type of bimetallic screen should always be avoided. Water with high total dissolved solids accelerates this type of corrosion because the water is a more effective electrolyte. Corrosion can also occur from dissolved gases in the water such as oxygen, carbon dioxide and hydrogen sulfide.

Irrigation wells are also commonly troubled with plugging by the deposits of incrustation. Such deposits plug the screen openings and the formation and/or gravel pack immediately surrounding the well screen. When incrustation is a problem, acid treatments can be used. Therefore, corrosion-resistant material should always be employed to resist the attack of strong acids introduced into the well screen during treatment.

Corrosion and incrustation can occur simultaneously in some groundwater environments. The products of corrosion can relocate themselves on the screen and form incrustation that plugs the screen openings much like waters that are naturally incrusting. Removal of those deposits often requires strong acids.

Choice of the well screen material sometimes is based on strength requirements regarding column load and collapse pressure. When a long screen supports a considerable weight of pipe, it functions as a slender column. The pressure of the formation and materials caving into the well pipe can squeeze the screen. Therefore, it must have good collapse resistance. It is impossible to accurately determine or calculate earth pressures with depth but generally greater strength is needed at greater depths.

Well screens can be constructed of materials that are especially adapted to resist corrosive attack of aggressive waters and acids. Stainless steel offers the maximum in corrosion resistance for most fresh groundwater environments, and also provides good strength. Galvanized steel is suitable for many irrigation wells where the water environment is not severe. It provides strength comparable to stainless steel. PVC well screens are resistant to corrosion and are often used in shallow wells. However, only limited open area can be provided and still maintain strength requirements. Therefore, nonmetallic well screens are not usually adequate for deep irrigation wells.

Well Development

Well development includes those steps in completion of a well that aim to clean, open, and enlarge passages in the formation near the borehole so that water can enter the well more freely. Three benefits of development are to:

• correct damage or clogging of the water-bearing formation that occurred as a result of the drilling operation;

• increase the porosity and permeability of the natural formation in the vicinity of the well;

• stabilize the sand formation around the screen or artificial gravel pack so that the well will yield sand-free water.

All these benefits can be obtained for wells in unconsolidated aquifers if the wells are properly screened and development procedures are properly applied.

The key to successful development is to cause vigorous reversals of water flow through the screen openings that will rearrange the formation particles. Provided that adequate energy is applied to the formation, this action breaks down bridging of groups of fine particles. Better results can be obtained if development begins slowly and increases in vigor with time and when the well is pumped during the development procedure. When the development method makes simultaneous pumping impractical, the well should at least be pumped occasionally. At times, it may be wise to incorporate chemical development with the mechanical methods. This is particularly true when silt and clay plugging of the formation is suspected.

No one particular development procedure is the best method for all geologic formations or types of well construction. Some methods are more adaptable to the particular type of drilling equipment used to construct the well, but other factors such as availability of water, air compressor, or pump may also dictate which development procedure is the most practical to use. The selection of the best method should be made on evaluation of the hydrogeologic conditions at the well site and past experience with similar irrigation wells in the same geologic formation. Once a method has been selected, the designer should specify the details of the procedure.

Well development methods are always needed and are generally economical regardless of the type of drilling methods used to construct the well. Proper development will improve almost any well. Surging with compressed air conveniently allows pumping from the well while development is in progress. Mechanical surging by operating a plunger in the casing like a piston in a cylinder is particularly adaptable when cable tool drilling equipment is used. Mechanical surging with the use of a bailer is adaptable to both cable tool and rotary drilling. Starting and stopping of a pump to produce a backlashing action is often called "rawhiding" the well. This is the simplest method of development, but it is usually the least effective since the surging effect that is created is usually not vigorous enough to obtain maximum results. High-velocity horizontal jetting with water is the most effective method of well development in most cases, and is especially useful for development of gravel-packed wells.

Test Pumping

Following completion of development, the well should be test pumped. The well should be pumped for at least twelve hours at a constant pumping rate, during which time drawdown measurements are taken within the pumped well and any nearby observation wells. The primary objectives of the pumping test are to obtain information about the performance and efficiency of the well, and to collect data which are used to select the permanent pumping equipment to ensure maximum pump efficiency. The information is used to evaluate the success of the design and development procedures and provides the basis to make other performance judgements and evaluations. In some cases, this information indicates that further development is necessary.

Well testing will also allow collection of data from which the hydraulic characteristics of the aquifer can be evaluated. Measurements of water table recovery at the end of a pumping test (rate at which the water table rises after pumping stops) can be beneficial in evaluating performance of irrigation wells. This data can also be used to make calculations of the aquifer hydraulic characteristics.

Drawdown

Drawdown is the distance that the water level in a well drops after pumping begins (Fig. 8-5). Drawdown will always occur because the water in the well is not replaced instantly when it is removed by the pump. Also, drawdown creates the gradient in water levels that causes flow to the well.

The amount of drawdown depends on well size and efficiency, aquifer properties and pumping. For the same well size and pumping rate, drawdown is greater in a sand or gravel aquifer as compared to the Florida limerock aquifer. Typical well drawdowns are in the range of 10 to 20 feet in the Floridan Aquifer, although drawdown can be significantly greater in specific wells. Accurate

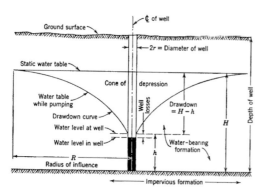

Figure 8-5. Drawdown characteristics for a well in an unconfined aquifer.

measurements of drawdown are important so that the pump can be properly positioned in a well.

Well Problems

Sand Pumping

Sand pumping causes excessive abrasion of the pump bowls and impellers, distribution pipe, emitters, and other irrigation system components. It reduces the useful life of the entire system and significantly increases maintenance costs. In addition, unnecessary costs result when sand must be periodically cleaned from the laterals, pipelines, and in some cases even from the well and pump. Sand pumping can create large underground cavities that collapse and cause land subsidence in the vicinity of the well bore. This eventually can cause total collapse of the well casing and the screen.

It is best to control sand at the well screen or gravel pack. A 2000 gpm irrigation well operated 800 hours per year that produces 20 ppm of sand will remove about 7 tons (or approximately 4 1/2 cubic yards) of fine sand from the water-bearing strata. In many cases, proper well design and development can prevent sand from being pumped.

In naturally developed wells, sand pumping is most often caused by using openings in the screen device that are too large to retain the proper amount of the water-bearing formation. An improper relationship between the grain size of the

gravel pack material and the size of the aquifer sand grains is the major cause of sand pumping from gravel-packed wells. Oftentimes, the slot opening in the well screens for gravel-packed wells is chosen first. A gravel pack size is then selected such that it will not pass through the openings in the screening device. This method gives no consideration to the size of the sand and gravel particles of the water-bearing formation. As a result, many times the gravel pack is too large to properly retain the water-bearing sand and the well pumps sand. A sieve analysis of formation samples is the basis for proper gravel pack and slot opening selection. The design procedure should proceed from the aquifer to the screening device rather than from the screening device to the formation.

In most cases, sand control cannot be achieved by merely installing a thick gravel pack. A thick gravel envelope does not significantly increase the yield of the well. Thick gravel packs don't reduce the possibility of sand pumping because the controlling factor is the ratio of the grain size of the gravel pack material to the aquifer material.

Well Inefficiency

Wells with poor efficiency result in high pumping costs due to excessive drawdown. Irrigation well efficiency should be at least 80%. Often the added operating cost in one pumping season can offset the slightly greater initial cost required for the design and construction of an efficient irrigation well. The cost resulting from inefficient operation for electric motors can be calculated by:

$$\text{Cost} = \frac{Q \times H_d \times 0.746 \times C_p}{3960 \times E_p \times E_m} \qquad \textbf{Eq. 8-1}$$

where,

Q = Pumping rate (gpm)
H_d = Additional head required as a result of increased drawdown (feet)
C_p = Cost per kWh (dollars)
E_p = Pump and drive efficiency (decimal)
E_m = Motor efficiency (decimal)

Example:
Assume a well is 90% efficient when producing 2000 gpm with 20 feet of drawdown. The specific capacity is 2000 gpm/20 ft = 100 gpm per foot of drawdown. If the well was only 45% efficient, the specific capacity would be only 50 gpm per foot and the drawdown required to produce the 2000 gpm would then be 40 feet rather than 20 feet.

Example:
For a 2000 gpm system operating 800 hours per year, calculate the cost of operation with the 45% efficient well compared to a 90% efficient well if the motor efficiency is 90%, the pump efficiency is 85%, and the power cost averages 10¢ per kWh.

$Q = 2000$ gpm
$H_d = 20$ feet
$C_p = \$\,0.10$ per kWh
$E_p = 85\%$
$E_m = 90\%$

$$\text{Cost} = \frac{2000 \times 20 \times 0.746 \times 0.10}{3960 \times 0.85 \times 0.80}$$

$$= \$1.10 \text{ per hour}$$

$1.10 per hour x 800 hour = $880 per year

Many times well inefficiency is related to the improper selection of the gravel pack particle size and the size of slot openings in the well screen. If the ratio of the gravel pack particle size to the formation particle size is too large, it allows migration of formation particles into the gravel pack. As this migration proceeds for a period of time, successively smaller and smaller particles become lodged in the gravel pack. Eventually its permeability is drastically reduced. As a result the movement of water to the well is impeded.

Low efficiency is also related to skimping on construction costs with cheap, makeshift screens including various types of perforated, punched, sawed or cut casing or pipe. These types of screens generally have limitations such as low open area percentage, poor distribution of slot openings, and slots that are inaccurate and vary in size. Usually these lower cost screens do not have openings small enough to control fine sand or retain finely graded gravel packs.

Hand-perforated or torch-slotted casing normally provides less than 3% open areas. In addition, the shape of the openings in punched or slotted pipe is such that the openings lend themselves to rapid plugging by sand particles. Slotted pipe screens typically have less than 5% of the total surface area as openings or passageways for water to enter the well. With these types of screens, the flow of water from the aquifer is restricted, since porosity of the formation is generally greater than the amount of open area provided by the well screen. Therefore, additional drawdown is required to force the water into the well, which makes the well inefficient.

A properly designed well screen allows both radial and horizontal flow to the entire well screen. Perforated pipe contains only a small number of holes, resulting in only a small percentage of the water approaching the well having direct access. As a result, there is excessive convergence of flow near the individual slot openings, which is a common cause of excessive drawdown and lower well efficiency.

Poor well development upon completion of drilling is another factor causing well inefficiency. Without proper development, wells will not achieve maximum efficiency, and this results in increased pumping costs throughout the life of the system. It is important to first properly design and construct the well with quality materials, but it is equally as important to follow through with thorough development.

Short Well Life
Irrigation wells should be constructed for an expected useful life of at least 25 years. The cost of drilling two or three replacement wells that would be necessary to replace the service of one properly designed and constructed well is much greater than the one-time construction cost of a good well

capable of 25 years of service. In addition, maintenance costs are higher for poorly designed and poorly constructed wells.

Early irrigation well failure is often related to the improper selection of gravel pack and size of slot openings. Complete collapse of a well can occur due to excessive sand pumping. However, more well failures occur as a result of installing low-quality screening devices such as perforated or slotted pipe. These devices provide little open area and poor distribution of open area which causes water to enter the well at excessively high velocities. As the velocity of water moving into a well increases, the rate of corrosion or incrustation is accelerated at the screen and within the formation or gravel pack near the borehole. Incrustation can cause premature decrease of yield, while corrosion can cause early structural failure of the well.

Placement of the screen at elevations too near the static water level may result in premature well failure. If drawdown lowers the pumping level below the top of the perforated section, water entering the well above the pumping level will free-fall (cascade) to the pumping water surface. Cascading water reduces the life of an irrigation well because it accelerates corrosion and incrustation. In addition, cascading water causes air entrainment, which results in pumping of air, reduction in well yield, and erosion of pump components.

Chapter 9. Water Quality Parameters

by Brian Boman, Chris Wilson, and Esa Ontermaa

A basic knowledge of water quality is very useful to microirrigation system management and is an important consideration in the design and operation of the system. Therefore, the ability to read and understand a water quality analysis is important to the irrigation system manager. A careful analysis of the source water is prudent as a preliminary step to designing the microirrigation system. A microirrigation system requires good quality water free of all but the finest suspended solids, and relatively free of dissolved solids such as iron, which may precipitate out and cause emitter plugging problems. Neglecting to analyze the quality of source water and provide adequate treatment is one of the most common reasons for the failure of microirrigation systems to function properly.

Obtaining and Interpreting a Water Quality Analysis

It is important that a representative water sample be taken. If the source is a well, the sample should be collected after the pump has run for approximately a half hour. When collecting samples from a surface water source such as a ditch, river, or reservoir, the samples should be taken near the center and below the water surface. Where surface water sources are subject to seasonal variations in quality, these sources should be sampled and analyzed at various times throughout the irrigation season.

Glass or plastic containers are preferable for sample collection. For most analyses, the samples should be at least a pint. The containers should be thoroughly cleaned and rinsed before use to avoid contamination of the water sample. Sample bottles should be filled completely to the top (with all air removed), carefully labeled, and tightly sealed. Samples should be sent immediately to a water testing laboratory. For a microirrigation suitability analysis, the following tests should be requested from the laboratory: Electrical conductivity (EC) or total dissolved solids (TDS), pH, calcium, iron, alkalinity, and chloride, and if filtration information is needed, the quantity and size of suspended solids. Other parameters that are sometimes also needed to properly assess the suitability of the irrigation water are sodium, boron, potassium, manganese, nitrate, and sulfides (hydrogen sulfide must be measured at the wellhead).

Units Used in Reporting Water Analysis

One obstacle in interpreting a water analysis is that water testing laboratories report results in various units. For example, the concentration of chemical constituents may be reported as parts per million (ppm), milligrams per liter (mg/L), or milliequivalent per liter (meq/L).

A concentration reported in weight-per-weight is a dimensionless ratio and is independent of the system of weights and measures used in determining it. For many years, water analyses made in the U.S. were reported in parts per million (ppm). One ppm is equivalent to 1 mg of solute per kg of solution. Therefore, 1% is equal to 10,000 ppm. Parts per thousand (ppt, which is equivalent to 1 g solute per kg of solution) is frequently used to report salinity levels in sea water. In recent years, the development of more sensitive analytical equipment has led to terms such as parts-per-billion to report trace elements and pesticide concentrations.

Common laboratory analyses are made by initially measuring a volume of solution and then performing the appropriate procedure. The results are normally reported in units of weight per unit of volume (typically mg/L). To convert these values to ppm, it is generally assumed that 1 liter of water weighs exactly 1 kg. For most agricultural water, this assumption is close enough. Therefore, for agricultural water samples, ppm and mg/L can be considered the same. In some cases, water analyses

Table 9-1. Conversion factors for water quality data.

To convert-	To-	Multiply by-
grains/gallon	mg/L	17.12
mg/L	grains/gallon	0.0584
dS/m	uS/cm	1000
dS/m	ppm	700
uS/cm	ppm	0.7
ppm	uS/cm	1.429
ppm	dS/m	0.001429
Ca^{++}	$CaCO_3$	2.50
$CaCl_2$	$CaCO_3$	0.90
HCO_3^-	$CaCO_3$	0.82
HCO_3^-	CO_3^-	0.49
Mg^{++}	$CaCO_3$	4.12
Na_2CO_3	$CaCO_3$	0.94
NO_3^-	N	0.23
N	NO_3^-	4.43

(typically hardness) are presented in units of grains per gallon (1 grain/gallon equals 17.12 mg/L). Conversion factors for several water quality parameters are given in Table 9-1.

Sometimes analytical labs report results in terms of milliequivalents per liter (meq/L). This term recognizes that ions of different species have different weights and electrical charges. Conversions are made by dividing ppm (or mg/L) by the equivalent weight of the constituent ion (Table 9-1) as follows:

meq/L = (mg/L)/EW   **Eq. 9-1**

mg/L = meq/L x EW   **Eq. 9-2**

where,

> meq/L = milliequivalent per liter
> mg/L = milligram per liter
> EW = equivalent weight

Example:
How many meq/L is 120 ppm Ca?
From Table 9-2, EW for Ca = 20
meq/L = (mg/L)/EW = 120/20 = 6 meq/L

How many mg/L is 5 meq/L of CO_3
From Table 9-2, EW for CO_3 = 30
mg/L = 5 meq/L x 30 mg/L per meq/L
= 150 mg/L

If a water analysis reports 7 grains per gallon of Ca, how many mg/L of Ca are there?
From Table 9-1, the conversion factor is
17.12 mg/L per grain
7 grains x 17.12 mg/L per grain = 120 mg/L

Water Quality Parameters

Ions

Molecules may dissociate in solution into two or more ions which are theoretically free to move about independently. An ion differs from an atom or a molecule in that it carries an electrical charge. There are two kinds of ions: the cation, which is positively charged, and the anion, which is negatively charged. In water, the sum of the positive charges equals the sum of the negative charges, so that the solution remains electrically neutral.

Acid

An acid can be defined as a compound that releases hydrogen ions (H^+) in a solution. All acids contain hydrogen. In general, acids have more or less a sour taste, they change litmus paper red, and they react with bases to form salts and water. An example is acetic acid (vinegar), which is considered a weak acid because it releases only small amounts of free hydrogen ions into solution. Sulfuric acid (H_2SO_4) is considered a strong acid because it releases more hydrogen ions into solution.

Bases

Bases are substances which can release hydroxyl (OH^-) ions. Bases change litmus paper blue. As is the case with acids, bases demonstrate varying degrees of ionization. Those that ionize to a large extent are called strong, whereas those that ionize only slightly are known as weak bases.

Salts

A normal salt is a compound that is formed by the union of the cations of any base and anions of any acid. Other salts can be formed from acids or bases.

Table 9-2. Equivalent weight of common ions in irrigation water.

Cation (+ charge)	Equivalent weight	Anions (- charge)	Equivalent weight
Ammonium (NH_4^+)	18.0	Carbonate (CO_3^{--})	30.0
Calcium (Ca^{++})	20.0	Chloride (Cl^-)	35.5
Copper (Cu^{++})	31.8	Hydroxide (OH^-)	17.0
Magnesium (Mg^{++})	12.2	Nitrate (NO_3^-)	62.0
Manganese (Mn^{++})	27.5	Nitrate (NO_2^-)	46.0
Sodium (Na^+)	23.0	Sulfate (SO_4^-)	48.0
Iron (Fe^{++})	27.9	Bicarbonate (HCO_3^{--})	61.0
Iron (Fe^{+++})	18.6	Phosphate (PO_4^{---})	31.7
Potassium (K^+)	39.1	Phosphate (HPO_3^{--})	48.0
Sodium (Na^+)	23.0	Phosphate ($H_2PO_4^-$)	97.0
Zinc (Zn^{++})	32.7	Sulfide (S^{--})	16.0

Conductivity

Conductivity is a measure of the ability of water to pass an electrical current. Conductivity in water is affected by the presence of inorganic dissolved solids such as chloride, nitrate, sulfate, and phosphate anions (ions that carry a negative charge) or sodium, magnesium, calcium, iron, and aluminum cations (ions that carry a positive charge). Organic compounds like oil, phenol, alcohol, and sugar do not conduct electrical current very well and therefore have a low conductivity when in water. Conductivity is also affected by temperature: the warmer the water, the higher the conductivity. For this reason, conductivity is reported as conductivity at 25° C.

EC measurements are taken with platinum electrodes and presented in units of conductance. The drop in voltage caused by the resistance of the water is used to calculate the conductivity per centimeter. The SI unit of measurement is deci-Siemens per meter (dS/m), which is equal in magnitude to the commonly used conductance term of millimho/cm (mmho/cm). Both of these terms are generally in the range of 0.1 to 5.0 for waters used for irrigation. Conductivity is also reported in units 1000 times smaller: micromhos per centimeter (μmhos/cm) or microsiemens per centimeter (μS/cm). Most waters used for irrigation in Florida will be in the order of 100 to 5000 μS/cm. The conversion from electrical conductance to TDS depends on the particular salts present in the solution. The conversion factor of 700 x EC (in dS/m) is applicable for converting EC values to TDS for Florida soil extracts and irrigation waters (Table 9-1). In other areas, conversion values of 630-640 x EC (dS/m) are commonly used.

As with TDS, a relationship exists between electrical conductivity and chloride concentration. Chloride estimates, however, are less reliable. The Florida norm is 140 with 90% of all ratios measuring less than 250. The following equation is an approximate conversion from EC to chloride.

Cl (mg/L) = EC (dS/m) x 140   **Eq. 9-3**

Example:
A Floridian well was sampled and tested with a conductivity meter, with a reading reported as 2.55 dS/m. Estimate the TDS and Cl concentrations in ppm.

2.55 dS/m = 2550 μS/m = 2550 μmho/cm

2.55 dS/m x 700 ppm/dS/m = 1785 ppm TDS

Cl = dS/m x 140 = 2.55 x 140 = 357 ppm Cl

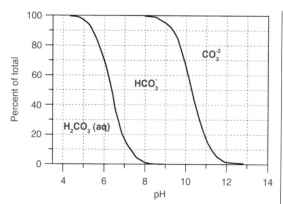

Figure 9-1. Percentages of total dissolved carbon species in solution for various pH ranges at 25° C and 1 atmosphere pressure.

Alkalinity

Alkalinity is primarily determined by the presence of bicarbonates ($-HCO_3^-$), carbonates ($-CO_3^-$), and hydroxides ($-OH^-$) in water. Alkalinity is a measure of the capacity of water to neutralize acids. Alkaline compounds in the water, such as bicarbonates (baking soda is one type), carbonates, and hydroxides, remove H^+ ions and lower the acidity of the water (which translates to increased pH). They usually do this by combining with the H^+ ions to make new compounds. Without this acid-neutralizing capacity, any acid added to a water source would cause an immediate change in the pH.

Total alkalinity is determined by measuring the amount of acid (e.g., muriatic acid) needed to bring the sample to a pH of 4.2. At this pH, all the alkaline compounds in the sample are "used up" (see Fig. 9-1). The result is reported as milligrams per liter of calcium carbonate (mg/L of $CaCO_3$). At the pH of most irrigation water, alkalinity is primarily a measure of bicarbonate in the water. Alkalinity expressed as mg/L of $CaCO_3$ can be converted to an equivalent concentration of HCO_3^- by dividing by 0.82 (Table 9-1).

Example:
Determine the equivalent expressed as concentration of HCO_3^- for an alkalinity of 80 mg/L $CaCO_3$.

80 mg/L x 0.82 = 65.6 mg/L as HCO_3^-

pH

The term pH is used to indicate the alkalinity or acidity of a substance as ranked on a scale from 0 to 14.0. Acidity increases as the pH decreases. The pH of water affects many chemical and biological processes in the water. For example, different organisms flourish within different ranges of pH (Fig. 9-2). The largest variety of aquatic animals prefer a range of 6.0-9.0. When pH is outside this range, diversity within the water body may decrease due to physiological stresses and reduced reproduction. Low pH can also allow toxic elements and compounds to become mobile and available for uptake by aquatic plants and animals. This can produce conditions that are toxic to aquatic life, particularly to sensitive species.

The pH scale measures the logarithmic concentration of hydrogen (H^+) and hydroxide (OH^-) ions, which make up water ($H^+ + OH^- = H_2O$). When both types of ions are in equal concentration, the pH is 7.0 or neutral. The pH value is the negative power to which 10 must be raised to equal the hydrogen ion concentration. Mathematically this is expressed as:

$$pH = - \log [H^+]$$

Below 7.0, the water is acidic (there are more hydrogen ions than hydroxide ions). When the pH is above 7.0, the water is alkaline, or basic (there are more hydroxide ions than hydrogen ions). Since the scale is logarithmic, a drop in the pH by 1.0 unit is equivalent to a 10-fold increase in acidity. So, a water sample with a pH of 5.0 is 10 times as acidic as one with a pH of 6.0, and pH 4.0 is 100 times as acidic as pH 6.0.

Generally, pH can be analyzed in the field or in the lab. If it is analyzed in the lab, you must measure the pH within two hours of the sample collection. The pH of a sample can change due to carbon dioxide from the air dissolving in the water.

A pH range of 6.0 to 9.0 appears to provide protection for the life of freshwater fish and bottom dwelling invertebrates in canals. Most resistant fish species can tolerate pH ranges from 4.0 to 10.0. At

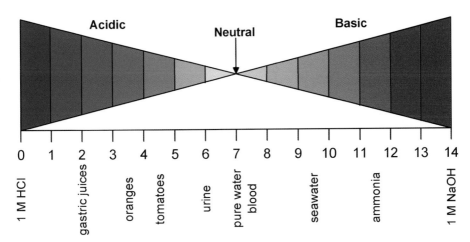

Figure 9-2. Range of pH for selected materials.

the extremes, fish eggs may hatch, but deformed young are often produced.

Hardness

The term "hardness" is one of the oldest terms used to describe characteristics of a water. In fact, Hippocrates (450-354 B.C.) used the "hard" and "soft" terms in a discourse on water quality. He states: "Consider the waters which the inhabitants use, whether they be marshy and soft, or hard and running from elevated and rocky situations.…" The hard term most likely refers to the condition of water that originated in the limestone formations of the upland regions. Over the years, hardness has come to be associated with the soap-consuming property of water or with the encrustations resulting from hard water when it is heated.

Hardness in water is caused primarily by calcium and magnesium, although iron and manganese also contribute to the actual hardness. Hardness may be divided into two types: carbonate and noncarbonate. Carbonate hardness is that portion of calcium and magnesium that can combine with bicarbonate to form calcium and magnesium carbonate. If the hardness exceeds the alkalinity (expressed as mg/L $CaCO_3$), the excess is termed noncarbonate hardness. The carbonate hardness is an indicator of the potential for calcium carbonate precipitation and scale formation. Total hardness (carbonate and noncarbonate) is customarily expressed as equiva-

lent of calcium carbonate ($CaCO_3$). Since the formula (equivalent) weight of $CaCO_3$ is near 100, hardness expressed in terms of mg/L of $CaCO_3$ can be converted to meq/L by dividing by 50. Hardness can be calculated from individual concentrations of Ca and Mg using Equation 9-4.

$$H = 2.5\ Ca + 4.1\ Mg \qquad \textbf{Eq. 9-4}$$

where,

H = total hardness (mg/L as $CaCO_3$)
Ca = calcium concentration (mg/L of Ca)
Mg = magnesium concentration (mg/L of Mg)

Example:
Determine the hardness of a water sample that contains 120 mg/L of Ca and 45 mg/L of Mg.

$H = 2.5 \times 120 + 4.1 \times 40 = 464$ mg/L of hardness
464 mg/L / 17.12 = 27.1 grains/gal hardness
464 mg/L / 50 = 9.3 meq/L

The terms hard and soft as applied to water are inexact, and several attempts over the years have been made to add more descriptive qualifiers. Table 9-3 lists typical ranges for hardness designations.

Total Solids

Total solids are dissolved solids plus suspended and settleable solids in water. In stream water, dissolved solids consist of calcium, chlorides, nitrate,

Table 9-3. Typical hardness ranges.

Hardness range		Descriptor
grains CaCO$_3$	mg/L of CaCO$_3$	
0-3.5	0-60	Soft
3.5-7.0	61-120	Moderately hard
7.1-10.5	121-180	Hard
>10.5	>180	Very hard

phosphorus, iron, sulfur, other ions, and particles that will pass through a filter with pores 0.45 microns (0.00045 mm) in size. Suspended solids include silt and clay particles, plankton, algae, fine organic debris, and other particulate matter. These are particles that will not pass through a 0.45-micron filter.

The concentration of total dissolved solids affects the water balance in the cells of aquatic organisms. An organism placed in water with a very low level of dissolved solids, such as distilled water, will swell up, possibly rupturing cells, because water will tend to move into its cells, which have a higher concentration of dissolved solids. An organism placed in water with a high concentration of dissolved solids will shrink somewhat because the water in its cells will tend to move out. This will in turn affect the organism's ability to maintain the proper cell volume.

Suspended solids can serve as carriers for organic compounds that readily cling to suspended particles. This is particularly of concern where the more water-insoluble pesticides are being used on irrigated crops. Where solids are high, pesticide concentrations associated with those solids may increase well beyond those of the original application as the irrigation water travels down irrigation ditches. Higher levels of solids can also clog irrigation devices and might become so high that irrigated plant roots will lose water rather than gain it.

Total solids also affect water clarity. Higher solids decrease the passage of light through water, thereby slowing photosynthesis by submersed aquatic plants. Water also heats up more rapidly

and holds more heat as total solids increase. This, in turn, might adversely affect aquatic life that has adapted to a lower temperature regime. Sources of total solids include fertilizers, runoff, and soil erosion. The measurement of total solids can be useful as an indicator of the effects of runoff from urban and agricultural areas. Concentrations often increase sharply during rainfall. Regular monitoring of total solids can help detect trends that might indicate increasing erosion in developing watersheds. Total solids are related closely to canal flow and velocity. Any change in total solids over time should be measured at the same site, and at the same flow rate.

Total solids are measured by weighing the amount of solids present in a known volume of sample. This is done by weighing a beaker, filling it with a known volume, evaporating the water in an oven and completely drying the residue, and then weighing the beaker with the residue. The total solids concentration is equal to the difference between the weight of the beaker with the residue and the weight of the beaker without it.

Primary Constituent Ions
Calcium
Calcium (Ca) is found to some extent in all natural waters and is very commonly a large component of the constituent ions in Florida water. Calcium combines with carbonate to form calcium carbonate scale, which can plug microirrigation system emitters. Calcium concentration is sometimes expressed as calcium hardness. To determine calcium concentration (mg/L), multiply calcium hardness by 0.4.

Magnesium
Magnesium (Mg) is also usually found in measurable amounts. Magnesium behaves much like calcium, but precipitates at higher pH levels and is not typically a problem in microirrigation systems. To determine magnesium concentration (mg/L), multiply Mg hardness by 0.24. Often laboratories will not separate calcium and magnesium but will report simply Ca + Mg in meq/L (as hardness).

Sodium

Sodium (Na) salts are all very soluble and as a result are found in most natural waters. High sodium in the irrigation water can impact both the soil and the plant. A soil with a large amount of sodium associated with a clay fraction has poor physical properties for plant growth and water infiltration. At high concentrations, Na can also be toxic to many plants. Sodium does not cause problems with the irrigation system. In order to prevent scale formation, domestic water supplies are often softened by replacing calcium with sodium.

Potassium

Potassium (K) is usually found in lesser amounts in natural waters. It behaves much like sodium in the water, but it is a major plant nutrient.

Iron

Iron is the fourth most abundant element, by weight, in the earth's crust. Natural waters contain variable amounts of iron despite its universal distribution and abundance. Iron in groundwater is normally present in the ferrous or bivalent form (Fe^{++}), which is a soluble state. It is easily oxidized to ferric iron (Fe^{+++}) or insoluble iron upon exposure to air.

Iron in water may be present in varying quantities depending upon the geological area and other chemical components of the water source. Ferrous (Fe^{++}) and ferric (Fe^{+++}) ions are the primary forms of concern in the aquatic environment. Other forms may be in either organic or inorganic wastewater streams. The ferrous form can persist in water void of dissolved oxygen and usually originates from groundwater wells. Black or brown swamp waters may contain iron concentrations of several mg/L in the presence or absence of dissolved oxygen, but this iron form has little effect on aquatic life. The current aquatic life standard is 1.0 mg/L based on toxic effects.

Iron in the soluble (ferrous) form may create emitter clogging problems at concentrations as low as 0.3 mg/L. Dissolved iron may precipitate out of the water due to changes in temperature or pressure, in response to a rise in pH, exposure to air, or through the action of bacteria. The presence of iron bacteria often results in the formation of an ochre sludge or slime mass capable of plugging the entire micro-irrigation system.

Manganese

Manganese (Mn) occurs in groundwater but is less common than iron, and it is generally in smaller amounts. Like iron, manganese in solution may precipitate out as a result of chemical or biological activity, forming a sediment that will clog emitters and other system components. The color of the deposits ranges from dark brown, if there is a mixture of iron, to black if the manganese oxide is pure.

Bicarbonate

Bicarbonate (HCO_3) is common in natural waters. Sodium and potassium bicarbonates can exist as solid salts; baking soda (sodium bicarbonate) is an example. Calcium and magnesium bicarbonates exist only in solution.

Carbonate

Carbonate (CO_3) is found in some waters at high pH (>8.0). At normal pH levels alkalinity is primarily in the form of bicarbonate.

Chloride

Chloride (Cl) is found in most natural waters. In high concentrations it is toxic to some plants. All common chlorides are soluble and contribute to the total salt content (salinity) of soils. The chloride content should be determined to properly evaluate irrigation waters if TDS is greater than 1000 mg/L.

Sulfate/Sulfide

Sulfate (SO_4) is abundant in nature. Sodium, magnesium, and potassium sulfates are readily soluble in water. Sulfate has no characteristic action on the soil except to contribute to the total salt content. The presence of soluble calcium will limit sulfate solubility.

Groundwater that contains dissolved hydrogen sulfide gas (H_2S) is easily recognized by its rotten-egg odor. As little as 0.5 ppm is noticeable. If the irrigation water contains more than 0.1 ppm of to-tal sulfides, sulfur bacteria may grow within the irrigation system, forming masses of slime that may clog filters and emitters.

Nitrogen

Nitrogen (N) is one of the most abundant elements. About 80 percent of the air we breathe is nitrogen. It is found in the cells of all living things and is a major component of proteins. Inorganic nitrogen may exist in the free state as a gas(N_2), or as nitrate (NO_3^-), nitrite (NO_2), or ammonia (NH_4^+). Organic nitrogen is found in proteins and is continually re-cycled by plants and animals.

Nitrogen-containing compounds act as nutrients in streams and rivers. Nitrate reactions (NO_3^-) in fresh water can cause oxygen depletion. Thus, aquatic organisms depending on the supply of oxygen in the stream may be at risk. Bacteria in water quickly convert nitrites ($NO_2^=$) to nitrates (NO_3^-). Together with phosphorus, nitrates in excess amounts can accelerate eutrophication, causing dramatic in-creases in aquatic plant growth and changes in the types of plants and animals that live in the stream. This, in turn, affects dissolved oxygen, tempera-ture, and other indicators. Excess nitrates can cause hypoxia (low levels of dissolved oxygen) and can become toxic to warm-blooded animals at higher concentrations (10 mg/L or higher) under certain conditions.

Nitrites can produce a serious condition in fish called "brown-blood disease." Nitrites also react directly with hemoglobin in the blood of humans and other warm-blooded animals to produce meth-emoglobin, which destroys the ability of red blood cells to transport oxygen. Nitrite-nitrogen levels below 90 mg/L and nitrate-nitrogen levels below 0.5 mg/L seem to have no effect on warm water fish.

Nitrates from land sources reach ditches and canals more quickly than other nutrients like phosphorus. This is because they dissolve in water more readily than phosphates, which have an attraction for soil particles. Water that is polluted with nitrogen-rich organic matter might show low nitrates. Decompo-sition of the organic matter lowers the dissolved oxygen level, which in turn slows the rate at which ammonia is oxidized to nitrite (NO_2) and then to nitrate (NO_3). Under such circumstances, it might be necessary to also monitor for nitrites or ammo-nia, which are considerably more toxic to aquatic life than nitrate.

Nitrate is not commonly found in large amounts in Florida waters. While beneficial as a plant nutrient, nitrate may indicate contamination from excessive use of fertilizers or from sewage. Nitrates have no effect on the physical properties of soil except to contribute slightly to its salinity, and nitrate is not harmful to the irrigation system.

Boron

Boron (B) occurs in natural waters. A small amount of boron is essential for plant growth, but a con-centration slightly above the optimum is toxic to plants. Some plants are more sensitive to a boron excess than others.

Phosphorus

Phosphorus (P) is one of the key elements neces-sary for growth of plants and animals. Phosphorus reactions in water are complicated. In nature, phos-phorus usually exists as part of a phosphate mol-ecule (PO_4). Phosphates exist in three forms: ortho-phosphate (inorganic phosphorus), metaphosphate (or polyphosphate) and organically-bound phos-phate. Each compound contains phosphorus in a different chemical formula. Phosphate that is not associated with organic material is inorganic. Inorganic phosphorus is the form required by plants. Animals can use either organic or inorganic phosphate.

Orthophosphates are produced by natural pro-cesses. Polyphosphates are used for treating boiler waters and in detergents. In water solutions, they change into the ortho form. Organic phosphates are important in nature. Organic phosphate consists of a phosphate molecule associated with a carbon-based molecule, as in plant or animal tissue. Their occurrence may result from the breakdown of or-ganic pesticides which contain phosphates. They may exist in solution, as particles, loose fragments,

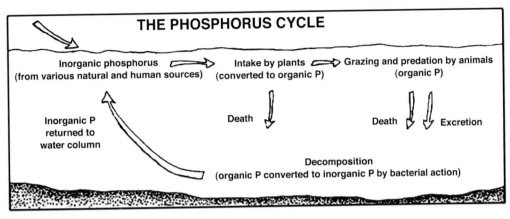

Figure 9-3. Phosphorus changes as it cycles through the aquatic environment.

or in the bodies of aquatic organisms. Both organic and inorganic phosphorus can either be dissolved in the water or suspended (attached to particles in the water column).

Rainfall can cause phosphates to wash from farm soils into ditches and canals. Since phosphorus is the nutrient in short supply in most fresh waters, even a modest increase in phosphorus can, under the right conditions, set off a whole chain of undesirable events in a stream including accelerated plant growth, algae blooms, low dissolved oxygen, and the death of certain fish, invertebrates, and other aquatic animals. The excessive growth and decay of algae and aquatic plants will consume large amounts of oxygen, and is indicative of the condition known as eutrophication or over-fertilization of receiving waters.

Phosphorus cycles through the environment, changing form as it does so (Fig. 9-3). Aquatic plants take in dissolved inorganic phosphorus and convert it to organic phosphorus as it becomes part of their tissues. Animals obtain the organic phosphorus they need by eating either aquatic plants, other animals, or decomposing plant and animal material.

As plants and animals excrete solid wastes or die, the organic phosphorus they contain sinks to the bottom, where bacterial decomposition converts it back to inorganic phosphorus, both dissolved and attached to particles. This inorganic phosphorus gets back into the water column when the bottom is stirred up by animals, human activity, chemical interactions, or water currents. It is then taken up by plants and the cycle begins again.

In a stream system, the phosphorus cycle tends to move phosphorus downstream as the current carries decomposing plant and animal tissue and dissolved phosphorus. It becomes stationary only when it is taken up by plants or is bound to particles that settle to the bottom of pools. In solution, the phosphate form depends on the pH (Fig. 9-4). The term "orthophosphate" is a chemistry-based term that refers to the phosphate molecule all by itself. "Reactive phosphorus" is a corresponding method-based term that describes what you are actually measuring when you perform the test for orthophosphate. Because the lab procedure isn't quite perfect, you get mostly orthophosphate but you also get a small fraction of some other forms. More complex inorganic phosphate compounds are referred to as condensed phosphates or polyphosphates. The method-based term for these forms is acid hydrolyzable.

Monitoring phosphorus is challenging because it involves measuring very low concentrations down to 0.01 mg/L or even lower. Even such very low concentrations of phosphorus can have a dramatic impact on streams. Less sensitive methods should be used only to identify serious problem areas. While there are many tests for phosphorus, only

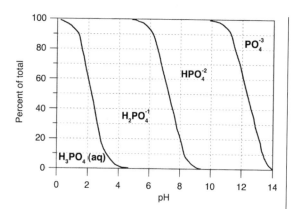

Figure 9-4. Percentages of total dissolved phosphate species in solution as a function of pH at 25° C and 1 atmosphere pressure.

the following are likely to be performed by growers in the field.

1. The total orthophosphate test is largely a measure of orthophosphate. Because the sample is not filtered, the procedure measures both dissolved and suspended orthophosphate. The EPA approved method for measuring total orthophosphate is known as the ascorbic acid method. Briefly, a reagent (either liquid or powder) containing ascorbic acid and ammonium molybdate reacts with orthophosphate in the sample to form a blue compound. The intensity of the blue color is directly proportional to the amount of ortho phosphate in the water.

2. The total phosphorus test measures all forms of phosphorus in the sample (orthophosphate, condensed phosphate, and organic phosphate). This is accomplished by first digesting (heating and acidifying) the sample to convert all other forms to orthophosphate. Orthophosphate is then measured by the ascorbic acid method. Because the sample is not filtered, the procedure measures both dissolved and suspended orthophosphate.

3. The dissolved phosphorus test measures that fraction of the total phosphorus that is in solution in the water (as opposed to being attached to suspended particles). It is determined by first

filtering the sample, then analyzing the filtered sample for total phosphorus.

4. Insoluble phosphorus is calculated by subtracting the dissolved phosphorus result from the total phosphorus result.

All these tests have one thing in common: they all depend on measuring orthophosphate. The total orthophosphate test measures the orthophosphate that is already present in the sample. The others measure that which is already present, and that which is formed when the other forms of phosphorus are converted to orthophosphate by digestion.

Sample containers made of either some form of plastic or Pyrex glass are acceptable to EPA. Because phosphorus molecules have a tendency to adsorb (attach) to the inside surface of sample containers, if containers are to be reused they must be acid-washed to remove adsorbed phosphorus. Therefore, the container must be able to withstand repeated contact with hydrochloric acid. Plastic containers, either high-density polyethylene or polypropylene, might be preferable to glass from a practical standpoint because they will better withstand breakage.

The only form of phosphorus recommended for field analysis is total orthophosphate, which uses the ascorbic acid method on an untreated sample. Analysis of any of the other forms involves adding potentially hazardous reagents and heating the sample to boiling, which requires too much time and equipment to be practical. In addition, analysis for other forms of phosphorus is prone to errors and inaccuracies in a field situation. Pretreatment and analysis for these other forms should be handled in a laboratory.

Dissolved Oxygen
Surface water systems both produce and consume oxygen. Oxygen is gained from the atmosphere and from plants as a result of photosynthesis. Moving water, because of its churning, dissolves more oxygen than still water. Respiration by aquatic animals, decomposition, and various chemical reactions consume oxygen.

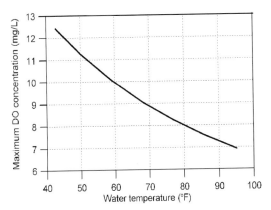

Figure 9-5. Relationship between water temperature and the maximum DO concentration possible in the solution.

Figure 9-6. Proper sampling for DO involves submerging the sample bottle, allowing to gradually fill, and capping while under water.

Organic materials in the water are decomposed by microorganisms, which use oxygen in the process. The amount of oxygen consumed by these organisms in breaking down the waste is known as the biological oxygen demand or BOD. Stormwater runoff can contribute large amounts of oxygen-consuming demand to surface water systems.

Oxygen is measured in its dissolved form as dissolved oxygen (DO). If more oxygen is consumed than is produced, dissolved oxygen levels decline and some sensitive animals may move away, weaken, or die. DO levels fluctuate seasonally, daily, and with water temperature (cold water holds more oxygen than warm water) (Fig. 9-5). Aquatic animals are most vulnerable to lowered DO levels in the early morning on hot summer days when stream flows are low, water temperatures are high, and aquatic plants have not been producing oxygen since sunset.

In contrast to lakes, where DO levels are more likely to vary vertically in the water column, the DO in canals changes more horizontally along the course of the waterway. This is especially true in smaller, shallower ditches. The DO levels below spillways are typically higher than those in pools and slower-moving stretches. A time profile of DO levels at a sampling site is a valuable set of data because it shows the change in DO levels from the low point just before sunrise to the high point

sometime at midday. Therefore it is important to note the time of DO sampling and to sample at similar times in order to make meaningful comparisons.

DO is measured either in mg/L or percent saturation. Milligrams per liter is the amount of oxygen in a liter of water. Percent saturation is the amount of oxygen in a liter of water relative to the total amount of oxygen that the water can hold at that temperature. DO samples are collected using a special BOD bottle: a glass bottle with a turtleneck and a ground glass stopper. You can fill the bottle directly in the stream, or you can use a sampler that is dropped from a bridge or structure deep enough to submerse the sampler. Samplers can be made or purchased. The sample bottle should be submerged and allowed to fill without allowing air to mix with the sample (Fig. 9-6). The bottle should be completely filled and held submerged until the cap is firmly in place.

Dissolved oxygen field kits using reagents and color wheels are relatively inexpensive. Field kits run between $35 and $200, and each kit comes with enough reagents to run 50 to 100 DO tests. Replacement reagents are inexpensive, and can be bought already measured out for each test in plastic pillows. If a high degree of accuracy and precision is needed for DO results, a digital titrator can be used. A kit that uses an eye-dropper-type or syringe-type titrator is suitable for most other purposes. The lower cost of this type of DO field kit might be attractive if you are relying on several teams of volunteers to sample multiple sites at the same time. Dissolved oxygen meters allow quick measurements, but they cost between $500 and $1,200, including a long cable to connect the probe to the meter. The advantage of a meter/probe is that you can measure DO and temperature quickly anywhere you can reach with the probe. You can also measure the DO levels at a certain point on a continuous basis. The results are read directly as milligrams per liter.

Total dissolved gas concentrations in water should not exceed 110 percent. Concentrations above this level can be harmful to aquatic life. Fish in waters containing excessive dissolved gases may suffer from gas bubble disease. However, this is a very rare occurrence. The bubbles (emboli) block the flow of blood through blood vessels, causing death. External bubbles (emphysema) can also occur and be seen on fins, on skin and on other tissue. Aquatic invertebrates are also affected by gas bubble disease but at levels higher than those lethal to fish.

Adequate dissolved oxygen is necessary for good water quality. Oxygen is a necessary element to all forms of life. Natural stream purification processes require adequate oxygen levels in order to provide for aerobic life forms. As dissolved oxygen levels in water drop below 5.0 mg/L, aquatic life is put under stress. The lower the concentration, the greater the stress. Oxygen levels that remain below 1-2 mg/L for several hours can result in large fish kills.

Biochemical/Oxygen Demand

Biochemical oxygen demand, or BOD, measures the amount of oxygen consumed by microorganisms in decomposing organic matter in stream water. BOD also measures the chemical oxidation of inorganic matter (i.e., the extraction of oxygen from water via chemical reaction). A test is used to measure the amount of oxygen consumed by these organisms during a specified period of time (usually five days at 20° C). The rate of oxygen consumption in a stream is affected by a number of

variables: temperature, pH, the presence of certain types of microorganisms, and the type of organic and inorganic material in the water.

BOD directly affects the amount of dissolved oxygen in rivers and streams. The greater the BOD, the more rapidly oxygen is depleted in the stream. This means less oxygen is available to higher forms of aquatic life. The consequences of high BOD are the same as those for low dissolved oxygen: aquatic organisms become stressed, suffocate, and die.

Sources of BOD include leaves and woody debris; dead plants and animals; animal manure; effluents from pulp and paper mills, wastewater treatment plants, feedlots, and food processing plants; failing septic systems; and urban stormwater runoff.

BOD is affected by the same factors that affect dissolved oxygen (see above). Aeration of stream water by rapids and waterfalls, for example, will accelerate the decomposition of organic and inorganic material. Therefore, BOD levels at a sampling site with slower, deeper waters might be higher for a given volume of organic and inorganic material than the levels for a similar site in highly aerated waters. Chlorine can also affect BOD measurement by inhibiting or killing the microorganisms that decompose the organic and inorganic matter in a sample.

BOD measurement requires taking two samples at each site. One is tested immediately for dissolved oxygen, and the second is incubated in the dark at 68° F (20° C) for 5 days and then tested for the amount of dissolved oxygen remaining. The difference in oxygen levels between the first test and the second test, in milligrams per liter (mg/L), is the amount of BOD. This represents the amount of oxygen required by microorganisms to break down the organic matter present in the sample bottle during the incubation period. Because of the 5-day incubation, the tests should be conducted in a laboratory.

Chapter 10. Water Quality Monitoring

by Brian Boman, Chris Wilson, and Esa Ontermaa

Monitoring Versus Sampling

Monitoring is the process of checking or testing in order to regulate or control something. This definition implies three elements: continuity, organization or systematic testing, and purpose. Water quality monitoring costs money, takes time, and results can be difficult to interpret. There are many reasons to monitor, some more desirable than others. In fact, it may be hard to find any other water-related issue that growers are more emotional about than monitoring, except perhaps regulation. Before starting a water quality monitoring program, it is important to determine how it can benefit you and how to design a proper monitoring program. The purpose of this chapter is not to attempt to list all possible uses of a monitoring program, but rather to describe basic principles on which one is able to build a valid program while minimizing investment in the field as well as the operational cost.

A single sample in time and space does not constitute monitoring. The difference between monitoring and a single sample (sporadic sampling) is that the former is able to describe the behavior of the water, while the latter only provides a snapshot of the same system at a specific time and place. Nevertheless, a single sample is useful and is widely used. However, it should be kept in mind that, at best, a single sample is indicative of the status of the water sampled rather than representative or absolute. Making a long-term commitment and/or decision based on a single sample is risky business and bound to lead to unexpected complications or waste of money.

A number of samples taken sporadically over time should not be considered monitoring, even if the samples are analyzed for the same set of parameters. In fact, even if the samples are analyzed for the same parameters but the sampling method is altered or the time interval is varied without a predetermined reason, the sampling cannot be called monitoring in the strictest sense of the term. Since random sampling lacks continuity or organization, the data provided may not hold up in the face of a challenge and therefore should not be considered as monitoring.

The purpose of a water monitoring program can vary widely, depending on each particular grove or operation. To assure that the design serves the intended purpose, the object of the program must be clearly stated and understood. The objectives based on the purpose are the foundation for the entire program. The objectives of a monitoring program determine the means of sampling, system design, parameter selection, analytical services selected, cost, and the applicability of the data. If the statement of purpose and objectives for a monitoring program are weak, the monitoring program results will also be weak. However, well-defined and thoroughly thought-out objectives, plus a program built around them, should result in a successful program and save money and effort.

Need for Monitoring

The need for monitoring can be divided roughly into three categories:

1. The monitoring one wants to do → Grower interests

2. The monitoring one should do → Market interests

3. The monitoring one must do → Governmental interests

It is obvious that a number of intermediate stages do exist, and it is feasible to expect one monitoring program to serve several interests.

Recent trends in agricultural technology have led the scientific community as well as growers to search for cause/effect relationships in crop management systems. Water constituents and characteristics are an integral part of that management scheme.

The governmental agencies concerned with human health and the environment require ever-increasing assurances of proper growing practices when related to food safety and are pushing towards no impact on the environment at or beyond the grove. In addition, those marketing fruit both within the US and abroad are under pressure to show that their merchandise has been produced using environmentally sound and safe practices. These requirements usually result in the transfer of the burden of proof to the growers. This translates to an increased need to monitor for the influence of the water on the crop, as well as for the environmental impact of the growing practices.

Voluntary Monitoring

In normal business environments, one wants to know what helps increase profits or minimize losses. This means obtaining information that must be economical and focused on:

• cost savings,

• cultural controls,

• crop improvement.

The cost savings can include such concerns as water quality impact on crop protection chemicals. For example, a number of materials in the present market are pH sensitive, and the efficacy of the spray or other application may depend on the proper control of the carrier water pH. A water pH monitoring program will allow the grower to avoid costly mistakes and minimize material waste.

Improving profitabilty of the cultural program may require information describing the impact of rain on the quality of surface water or shallow wells. Knowledge of the concentration variation of Calcium (Ca) and Magnesium (Mg) in the irrigation water can help in making proper changes to a liming program, or there may be a concern about

mineral components in the water in general, and a desire to assess their contribution to the overall growing practices.

Measurement of the irrigation water salinity may indicate a need to select different fertilizer materials or alter irrigation strategies in order to improve yield and crop quality. It can also warn the grower if a steady increase in the salinity reading is observed. Depending on the water source, the trend may warn of impeding saltwater intrusion or overirrigation, dissolution, and transport of the fertilizer into groundwater or ditches and canals.

In all of these cases, equipment and services needed are available at very economical costs. Good portable pH and salinity meters cost anywhere from $200 to $500 and many of the needed wells or sampling points can be hand installed using various sizes of PVC pipe. More detailed water analysis can be determined at the analytical laboratories commonly used for soil and leaf testing. The crux of a good monitoring program is not the equipment, but the continuous and systematic approach that must be established and adhered to in order to meet the intended goals of the program.

Monitoring That Should Be Done

While the food supply is safer than it has ever been, our ability to determine residual pesticides at the part per trillion (ppt) level allows extremely detailed analysis of citrus and citrus by-products. The dissemination of this information by the media and interest groups (oftentimes without adequate scientific scrutiny) has resulted in public concerns about the wholesomeness of their food supply. Recent interest in organically grown fruit and juice attest to the fact that citrus products perceived as environmentally and humanly safe can be sold at a premium in retail outlets. Globally the trend is directly correlated to technological advancements: the higher the level of industrialization, the higher the concern. Fortunately, these concerns are largely unfounded. However, once negative public concerns become highly publicized, growers must be able to refute them with hard scientific data.

The global market economy has set the stage where marketing organizations and retailers are transferring the burden of proof of the wholesomeness to the grower. Any negative news that associates with the wholesomeness of fruit initially results in scrutiny of the retail store, and then of the producer. Inability to show that the fruit was grown, harvested, packed, shipped, and handled properly will significantly increase financial losses. Unfortunately, negative news is often fueled by media and public hysteria, and scientific reason is forgotten. In these cases, both the retailer and the grower will suffer. To what degree depends on how well each of them are prepared to address the situation. Therefore, globalized concern has often necessitated that citrus growers monitor water quality parameters in order to assure that the crop is grown with sound practices in regards to:

- Environmental stewardship → Discharge water quality and its impact on the downstream system

- Inputs to the growing crop → Irrigation water quality

- Food safety → Bacterial and chemical quality of the water that comes in contact with the fruit

The environmental stewardship and inputs to the crop are some of the burdens buyers and marketing branches are transferring to the grower. Food safety is a primary concern of packing houses and processing facilities. An environmental stewardship monitoring program may include such elements as characterizing incoming and outgoing water quality and monitoring aquifers and the shallow water table under the grove. The primary concerns are N, NO_3, P, salinity, turbidity, certain pesticides, and some metals. With wise planning, the environmental stewardship monitoring program can aid in:

- characterizing crop inputs;

- developing a nutrient application program and aid in irrigation scheduling;

- assuring customers that the fruit is wholesome.

In essence, the monitoring can help by optimizing inputs. It also allows maximization of profits by eliminating unnecessary waste while tuning the application-irrigation relationships. A monitoring program can help meet criteria of environmentally sensitive buyers, and probably many regulatory requirements as well.

The Food Quality Protection Act (FQPA), which is enforced by the Environmental Protection Agency (EPA) in the US, will probably result in a reduction in the number of available crop protection compounds. To contradict these trends, growers need to show that their actions have minimal impact on the environment. Regionwide water quality monitoring programs can not only protect growers from environmental cleanup liability, but also ensure that the fruit they grow can be freely marketed.

This type of monitoring provides some distinct benefits. The cost of the sampling goes down as well as the cost of analysis. Maintenance of the equipment and data handling can be centralized, or each participant can supply his portion to the central organization or consultant. Therefore, participants get the best possible data at the lowest possible cost with little or no involvement on their part. This allows the grower to concentrate on growing while reaping the benefits of monitoring. It requires close contact between the grower and the monitoring entity so that the data provided is in such a form and is reported in such intervals, that it allows the grower to use the data for his/her growing purposes as well.

In its simplest form, a regionwide monitoring program may simply consist of a number of individuals that have chosen to allow a single group (typically associated with a cooperatived or marketing association) to administer the monitoring instead of each doing their own. This arrangement lends:

- reliability due to independence of the monitoring entity;

- uniformity due to single group administration;

• continuity since dropping of one grower from the group will not necessarily affect the other;

• validity since the work is done by professionals;

• confidentiality due to regionwide reporting without identifying any grower.

To design a regionwide monitoring system with net inputs and net outputs in mind is a complex task. In many cases, this may not even be possible unless the area is predominantly a monoculture. One must consider environmentally sensitive areas within the region and determine any potential connections to them. Other questions concern:

• What is the role of other land uses?

• What is the source water quality and where does it come from?

• What are the ecological interrelationships?

A number of professionals from several disciplines of science should be consulted to reach a satisfactory, defensible monitoring network that is economically feasible. The list of expertise should include chemistry, hydrogeology, ecology, biology, agriculture, and sociology.

Required Monitoring

Most water monitoring programs in place today fall under this category. These monitoring programs are initiated by governmental agencies (usually environmental) in order to cause a change in the downstream or surrounding environment. An example of this is the extensive monitoring and sampling program in the Everglades Agricultural Area, designed to change water quality going to the Everglades National Park. This program is administered by the South Florida Water Management District. Another program is related to the restoration efforts made by the St. John's River Water Management District in regards to St. John's River. The Florida nitrate bill also requires monitoring in sensitive areas. Today, monitoring is done by the state agencies. However, concerned citrus growers should consider water quality monitoring in order to establish long-term data that documents the effects of their practices on the environment.

Monitoring requirements vary from periodic grab samples to continuous discharge monitoring, where samples are proportional to flow or within certain time intervals.

Types of Monitoring

Monitoring can be either descriptive, warning, or controlling. A water quality monitoring program should provide information that can be used to:

• draw cause/effect relationships,

• warn of a change,

• automatically control the system when water characteristics change.

Descriptive Monitoring

Descriptive monitoring is most often based on physical and chemical characteristics of the water. The samples are taken systematically over time in a prescribed manner, allowing one to correlate system responses to other external events or vice versa and derive cause/effect relationships. This type of monitoring program seeks to determine what things are happening and why. Therefore, this type of monitoring program serves four important functions:

• It provides information about the system responses to various conditions.

• It establishes behavioral limits of the system.

• It shows the direction, speed, and magnitude of a potential change.

• It allows reliable assessment of impact on a secondary system such as soil and/or plants.

As a result, there is a two-directional process that considers the impact of inputs and environment on the character and behavior of the monitored water, and assesses the impact of the water on any system with which it comes in contact. The monitoring program types and their general function are displayed in Fig. 10-1. The dotted line from the "Water to environment" box to the "Change" box denotes a need for additional information besides that provided by the monitoring system alone.

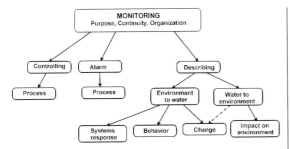

Figure 10-1. Types of monitoring and their general function.

Warning Monitoring

Monitoring where devices with alarms are used is a subcategory of the control monitoring. When a pre-determined value is reached, an appropriate action is required to take care of a problem. The alarm sensors are usually special systems measuring pH, salinity, temperature, turbidity, chlorine, nitrogen, and ammonia.

A potential use of these types of sensors can be where chlorination is used to maintain bacteria-free irrigation lines. When liquid fertilizer is injected, the chlorination should be turned off; a sensor that prevents fertilizer injection before chlorination is turned off, or sounds an alarm, is a good safety measure. If the chlorination is not turned off, there can be a number of problematic reactions at the point where the chlorine meets the fertilizer. Depending on fertilizer formulation, type of chlorine used, and concentrations of each compound, the reactions can range from overheating, precipitate formation, inactivation of chlorine, or an explosion.

These devices are commonly used in water, waste water, and chemical plants to help with process control. The technology is well established. However, adjustments and/or modifications to the technology may be needed in order to adapt them to agricultural operations.

Control Monitoring

Control monitoring is basically threshold monitoring with associated alarms or controls. Threshold monitors are devices commonly used in the chemical industry and water and wastewater treatment plants to monitor water pH, salinity, chlorine,

chemical concentrations, etc. These devices consist of a special sensor (characteristic to the parameter measured) that is installed into the water line or mixing container. The sensor passively measures the concentrations, color, turbidity, etc., while the main unit often has multiple options to select a value or a high-low window in order to automatically control chemical feed operations. Should preset values be exceeded, the units may be programmed to sound an alarm and adjust the chemical feed into the water stream, or to stop the process.

In agriculture, the use of such monitoring devices is generally limited to fertilizer or chemical feed systems in irrigation systems. However, chemical feed systems and controls are being used to alter the characteristics of the agricultural waters in reservoirs, canals, lakes, and ponds as well.

Costs of Monitoring

The minimum cost of any monitoring system is determined by the purpose of the program. This cost generally consists of the following:

- Equipment and maintenance →
  equipment, parts, man-hours
- Sample collection and shipment →
  man-hours, shipping cost
- Analysis →
  cost of analytical services
- Results →
  man-hours (data review, analysis, filing)

As a rule of thumb, once a program is in place, it is relatively inexpensive to expand the spectrum of analyses as compared to the field portion of the costs for the program. This is because usually the same effort that is used to collect samples for several parameters is needed to collect samples for one or two analyses. With this type of program expansion, care must be taken so that the initial purpose of the program is not compromised. The temptation to compromise or try to get by with the minimum is especially great when the pressure to monitor comes from outside. Inadequate planning often leads to vastly increased field time and cost because of continuous repairs, resampling, data retrofits, and eventual revamping of the entire sampling station design. If a company has several monitoring sites in various places, expenditures spent in this manner can be quite significant. To avoid this problem, one needs a clear focus and well-defined purpose combined with wisely spent money, experienced consultation, and proper equipment.

Some of the pitfalls that increase field cost are a lack of careful planning, no distinct objectives, inappropriate (cheap) equipment, erroneous programming or alteration of the programming, lack of systematic maintenance, change in focus that is not carried all the way to the instrumentation, and incompatible add-ons. All of the above also increase the data analysis time due to mismatched or missing data. While the system design directly influences the operational costs, the work required to analyze the data is often ignored in initial cost assessment. This may be due to the lack of a defined purpose for the program, or because the data collection program is meant to solely meet the needs of a government agency. In designing a monitoring program, this is not sufficient. The data resulting from the sampling should be the reason the program was initiated in the first place. In any monitoring program, the receipt and review of the data are the most important of all of the factors to be considered. Where the importance of the monitoring is acknowledged, it is not uncommon for a manager to spend a considerable amount of time reviewing and analyzing the data that has been collected.

Cost should not be the only consideration in a monitoring program. Careful systematic planning when it comes to water monitoring is essential. While the initial dollar figure is meaningful, one must remember that it will (in most cases) be the least of the total expenditures. If one decides that the equipment needed is too expensive, he probably should not begin a monitoring program with lesser equipment. Inappropriate, poorly constructed equipment may waste manpower and effort. To avoid some of the common pitfalls, the process of program design should consider the following:

1. Determine the general needs and have a written statement.

2. Determine whether monitoring can provide all or part of the necessary information. Describe the information needed.

3. Make a good faith estimate of the equipment and operational costs.

4. Determine the purpose of the monitoring program and prepare a written statement.

5. Determine the need for additional information to be combined with monitoring data such as purpose, collection method and intervals, etc.

6. Decide how the monitoring data are to be collected. Have a written statement that specifies the degree of automation, frequency, and sample sites.

7. Determine what data are needed and why they need to be collected. Record the specifics of analysis and characterization processes.

8. What are the requirements for data validity? Begin to establish a QA/QC (quality assurance/quality control) program, including sample preservation, storage time, and collection method.

9. Determine what equipment is needed, what it is supposed to do, and how the equipment functions. Determine type, cost, configuration, maintenance needs, and tolerance to the environment.

10. Speak to the analysis laboratory and other service providers to clarify their role, what they will provide, how they will perform the analysis, and how much it costs. Determine whether a contract is needed.

11. Develop a record system and a written description of the programs and their purposes.

12. Establish a maintenance schedule in writing.

13. Provide time for test runs and record problems, needed improvements, etc.

Assess Cost Benefit Ratio

There are many ways to assess a water quality monitoring program. Does the water respond to the environment as it used to? If not, what is changing and why? Is the water within acceptable or usual parameter limits? If not, where is the change coming from? Is there a value one can attach to the knowledge that the water one uses now is the same as it used to be five or ten years ago? Maybe something is being done right all along. Maybe material and effort are being saved because one is able to use the right adjuvant materials for the pest- and weed-control programs, and is not losing the primary product efficacy due to water quality.

Often the water quality impact on soil is measurable and should be factored into the formulation of cultural programs. Without a monitoring program, this cannot be done. Is there a way to assess the value of this type of information? The answer depends greatly on the purpose of the monitoring program and how good the past record keeping has been. For example, using monitoring to adjust fertilizer and irrigation programs in regards to N application may allow one to use less expensive mixes and gain higher application efficiency. Knowing the calcium and magnesium contribution of the irrigation water may allow skipping lime or dolomite application every few years.

Accuracy

No one wants to pay for something that is not useful. Since monitoring data will be relied upon and critical decisions will be based on it, the results must be valid. To ensure this, the monitoring program must have strict quality requirements. There must be sufficient information to ensure that the results are accurate.

The data quality and its accuracy are described by two variables: bias and precision. Bias is the measure of the systematic error in the process. Precision describes the repeatability of the process (Fig. 10-2).

Understanding these concepts while designing sampling regimes and intervals is essential. Faulty repair of a sampling device can skew results well

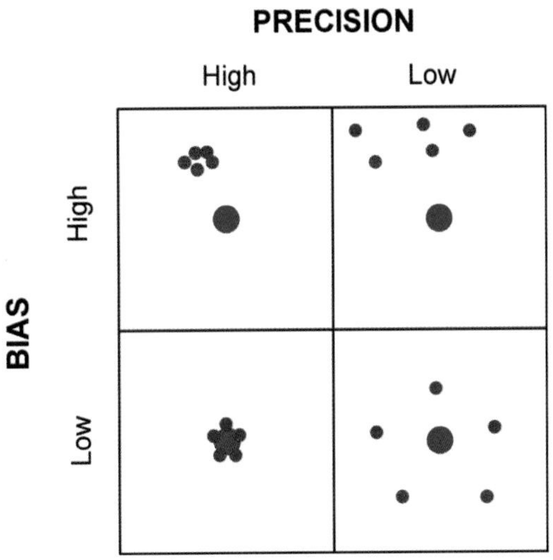

Figure 10-2. Illustration of how bias and precision affect accuracy.

off the baseline. For example, locating a sampling point close to the canal bank leads to consistently biased results due to sampling location within the stream. Use of soda cans, milk containers, and such introduce uncontrolled contamination to the samples and therefore decrease precision. Another factor to consider is the laboratory that is used for analysis. All laboratories have a bias and precision that is unique to their operation.

One other factor that should be considered is consistency. Consistent use of the same methods and laboratory permits reliable comparisons made with the same techniques and lab procedures. It makes comparative analysis of the observed trends and/or changes much easier since differences in procedures do not have to be factored in. This principle pertains also to the whole monitoring operation—consistent sampling methods, time intervals, protocols, analytical parameters, etc. Consistency in all of the above increases the reliability of the data.

What to Monitor

The decision as to what to monitor greatly depends on the initial question: What do we need to know? Possibilities are numerous and some broad direction can be pointed out, depending on the type of water one is dealing with. Table 10-1 identifies

Table 10-1. Parameters to include in a basic water monitoring program.

Surface Water	Well water	Soil water
Incoming water quality	Water quality at the well	Phosphorus
Outgoing water quality	pH	pH
pH	Electrical conductivity	Calcium
Electrical conductivity	Hardness (Ca, Mg)	Magnesium
N, P, K	Alkalinity (CO_3)	Copper
Turbidity (TSS)	Iron	N
Pesticides	Sulfur	

some of the parameters that may be included in typical monitoring programs.

The incoming or outgoing water quality may consist of a simple elemental set of N and P, and may be expanded to include metals such as Cu, Zn, Fe, Mn, or elements such as B and S. All of these can be obtained from one single acidified sample. Electrical conductivity and pH must be measured on site since they are subject to change in transport or are purposely changed for the preservation of the other elements in the solution. Turbidity, pH, and TDS requires monitoring equipment at the sampling site. Total suspended solids (TSS) may need to be measured at a lab. Another set of parameters consists of organic compounds: pesticides, herbicides, natural compounds, and their derivatives. Each compound may require specific sampling methods, and therefore an analytical laboratory should be consulted before attempting to determine if organic compounds exist in the samples.

Combining Data Sources

A monitoring program may focus on water alone, but to use the data for maximum benefit, a number of other parameters should be included and recorded along with the data from the monitoring program. Some of these are:

   Rainfall - duration, amount, distribution in large
      farms
   Fertilization - material, rate, application method
   Irrigation - time, duration, applied water
   Chemical applications - material, rate,
      application method
   Intake water analysis

   Soil analysis
   Plant analysis
   Crop survey - yields, quality, tree condition, etc.

Most of this additional data is routinely observed (if not recorded) by the grower. Systematic recording of the data generally does not necessarily increase the workload. The importance is to bring the different data sources together and sort them in a fashion that allows meaningful analysis to be performed. This method can be used to obtain useful information that otherwise might have been cost prohibitive to generate.

Constituents

Water samples can be analyzed for numerous chemical properties. The most important constituents concerning citrus growers generally include pH, electrical conductivity (EC), turbidity, hardness, chloride (Cl), iron (Fe), sulfate (SO_4), nitrogen (N), and phosphorus (P). Total-N and Total-P measurements account for all nitrogen and phosphorus, respectively, in organic and inorganic forms, measurable using standard chemical analysis procedures.

Nitrogen is an indispensable part of the life cycle. However, even though plants, animals, and most micro-organisms require some form of combined nitrogen for growth and reproduction, concentrations above certain levels can present problems. Total-N is the sum of Total Kjeldahl Nitrogen (TKN) and nitrate (NO_3^-). Total Dissolved Kjeldahl Nitrogen (TDKN) is the fraction of TKN that passes through a 0.45 micron filter. Ammonium

(NH_4^+) is another nitrogen parameter commonly reported. Inorganic nitrogen is normally found in soil water as ammonia, nitrate (NO_3^-), and nitrite (NO_2^-). Ammonia, a product of microbiological decay of plant and animal protein, is used in commercial fertilizers. Its excessive presence in raw surface waters usually indicates domestic or agricultural pollution. Above certain levels, it is toxic to fish. Excessive amounts of nitrate or nitrite in water can cause infant death and adult illness, and produce spontaneous abortion in cows. Fairly low levels of nitrite can be harmful to humans and aquatic life.

Phosphorus parameters commonly analyzed for are orthophosphate (ortho-P), soluble reactive phosphorus (SRP), total dissolved phosphorus (TDP), and total-P (TP). Particulate phosphorus is calculated as the difference between TP and TDP. Phosphorus occurs in natural waters as one of the forms of phosphate: ortho- or reactive phosphate, meta- or poly- (condensed) phosphate (requires hot acid digestion), and organic phosphate (requires severe digestion). Phosphates enter water supplies from soil runoff, cleaning operations, water treatment, and sewage. Although necessary for plant growth, too much phosphate causes excessive growth of aquatic plants and eutrophication of freshwater bodies. TP includes particulate and dissolved forms of phosphorus, the amount of which can be greatly influenced by sampling procedures. Therefore, it is advisable to also analyze samples for TDP.

Sampling Preservation and Analysis
Water sample collection devices vary widely in complexity and cost; samples can be manually dipped from a body of water, one at a time, using specially prepared containers. Sophisticated autosamplers are available for collecting a series of samples at specific time intervals, at preset times, or in amounts proportional to flow.

<u>Sample Collection and Storage</u>
Containers for collection of water samples for nitrogen and phosphorus analyses should be made out of nalgene, polyethylene, or some other inert material. Sample bottle sizes generally should

range from 250 to 500 mL to ensure that an adequate sample will be available for reanalysis when necessary. The cost of the special bottles will vary according to size and style. Prior to use, the sample bottles should be washed with a phosphate-free detergent, rinsed with distilled water, rinsed in a dilute hydrochloric acid (HCl) solution, rinsed with distilled water again, and dried. The bottles should be capped with foil, saran wrap, or bottle caps (washed and rinsed in the same manner as the bottles themselves) to protect against dust particles and other contaminants (i.e., free ammonia in the air, insects) entering the containers during storage. Once prepared for receiving water samples, nothing except the water to be sampled should come in contact with the inside of the bottles or caps.

The proper labeling of the sample bottles is critical to a successful water quality monitoring program. Bottles should be labeled in the field as the sample is being collected to ensure proper identification. Prelabeling bottles in the laboratory before taking them to the field can cause confusion for the person collecting samples as he or she searches for the appropriate bottle. It may be desirable to label a bottle for a site even if no water is available to sample at the site. This ensures that the receiving personnel in the laboratory know that the site was visited and no sample was available. The label should include the time, date, site, station, and name of the person collecting the sample. For best results, use waterproof, permanent marking pens and write directly on the bottle. The writing will be removed by the washing procedure. Alternatively, self-adhesive waterproof labels can be used. Removal of the labels, however, can be a time consuming, messy task.

<u>Grab Samples</u>
Grab samples are manually collected at the site. They may be taken from a water body randomly, systematically, or at regular intervals such as daily, weekly, or monthly. These samples are generally used for establishing nutrient concentrations at specific points in time. Grab samples are useful for establishing long-term trends or point-in-time concentrations. Grab samples can be dipped manually

or pumped from a body of water. If using a pump, a suction strainer assembly and a peristaltic pump are necessary. The suction strainer is simply a coarse filtering device with holes small enough to prevent the passage of material that is too big to pass through the suction hose. These strainers are commercially available from several different companies that manufacture water sampling instruments. Alternatively, suction strainers can be constructed out of noncorrosive, chemically inert materials. The strainer should be attached to a rigid shaft for ease in lowering the unit to the desired depth in the water body. A hose should be attached to the strainer and can be connected to the pump inlet hose. A battery-operated peristaltic pump is used to pump the sample into the bottle from the desired depth.

Where adequate water is available for grab samples (i.e., in canals or ditches full of water), the sample bottle should be rinsed with water from the source being sampled. To do so, simply collect a sample as would normally be done, and empty the bottle before collecting the sample for return to the laboratory. Do not just dip water from the surface for the rinse process since the surface of the water body may have a different chemical and biological makeup than the water at the desired depth.

Autosamplers

An autosampler is a device that automatically collects water at preset times, on preset intervals, and in preset volumes (Fig. 10-3). The instrument consists of a timer, controller, pump, sample distributor, and sample bottles. The timer unit can be programmed to initiate a sampling event and to continue taking samples at set time increments. Alternatively, electronic pulses from flow sensors can be used to trigger sample collection. Sample size typically can vary from 100 mL up to 1 liter. Samples are collected in custom-fitted polyethylene or glass bottles. A standard feature of most autosamplers is a bottles-per-sample or samples-per-bottle multiplexer, which allows several bottles to be filled at each sampling interval, or several samples to be placed in each bottle.

Figure 10-3. Typical programmable autosampler (right) and multiple-sample collection chamber holding 8 jars (left).

Autosamplers normally use a peristaltic pump system to transport the sample from the source to the sample bottle. The only materials in contact with the sample are the vinyl or teflon suction line, the inlet strainer, the silicone rubber pump tubing, and the polyethylene (or glass) sample bottles. Each sampling cycle includes an air prepurge and postpurge to clear the suction line both before and after sampling.

Samples can be based on either time or flow. The flow mode is controlled by external flowmeter pulses. This requires a separate flowmeter to be attached to the control box on the autosampler by a cable. The flowmeter is then programmed to trigger the collection of a sample after the passage of a preset flow volume (for example, a sample would be collected after every 1000 gallons passed the monitoring station), resulting in a composite sample that is composed of subsamples proportional to flow rates. Costs for a single unit with necessary accessories, but without shelter, start at approximately $2,500.

Soil Solution Sampling/Suction Lysimeters

The suction lysimeter is installed in the soil so that there is a tight fit between the porous cup and the soil. Without the tight fit, the unit will not extract water from the soil unless the cup is below the water table. The assembly should be left in the field for a suitable period of time prior to use for collecting samples for analysis. This allows the ceramic cup to chemically equilibrate with the soil water. It is also advisable to pump samples through the system to flush it prior to use.

Any section of the lysimeter tube that protrudes above the ground surface should be painted black. Sunlight will pass through the thin-walled white PVC pipe and allow the growth of algae within the tube. Painting the unit black will alleviate this problem. When sampling, the user must first apply a vacuum to the unit. The vacuum must be held in the lysimeter for at least two hours to ensure that an appropriate amount of water will enter the tube. If the cup is below the water table, then very little time will be needed. Samples may then be extracted from the lysimeter using the same type of peristaltic pump used for grab sampling.

The pump hose should be rinsed with distilled water before pumping a sample to avoid cross-contamination between lysimeters. It is important to make sure that all the rinse water is pumped out of the tubing prior to filling the sample bottle. Pumping some lysimeter water through the tube prior to filling the sample bottle will ensure that the sample constituent concentrations are not diluted with the rinse water. Alternatively, a vacuum flask and vacuum pump can be used to extract the sample without passing the sample through the pump tubing. This method is preferred if small sample volumes are available.

Suction lysimeters can be purchased already assembled or manufactured using commercially available parts. Each unit costs about $60. A service kit that includes a hand vacuum pump must also be purchased. This kit costs a little over $100.

Sample Preservation

Grab and soil solution samples are collected manually and, therefore, require little in the way of sample preservation in the field. Essentially, all that is required is that the samples be kept cool (40° F) during transportation from the field to the laboratory. This is easily accomplished by placing sam-ples in an ice chest or cooler filled with ice.

Autosamplers were originally designed to collect numerous samples automatically over a variable length of time. Each sample, therefore, has a different length of time that it remains in the sampler between collection and pickup times. Autosamplers generally collect up to 24 samples, enabling the sampler to operate over long time periods prior to pickup. Thus, autosamplers require more sophisticated means of maintaining the viability of collected samples. Preservation begins in the field with the appropriate shelter. The shelter should protect the autosampler from direct sunlight while ensuring adequate ventilation. Refrigerated shelters are available for instances where the time between sampling and pickup is considerable. In other cases, a ventilated shelter will suffice if ice is placed in the autosampler base where the sample bottles are stored. Some autosamplers have insulated bases and allow ice to be added to the collection chamber to keep sample temperature well below ambient for several hours. Samples that will be analyzed for ammonia should be acidified with sulfuric acid (H_2SO_4) to a pH less than 2.0 upon collection.

Once the samples are brought into the laboratory, preservation activities must continue. There are a number of procedures to be used, depending on the constituent parameters to be measured. The storage and preservation specifications for chemical parameters most often measured in agricultural water are listed in Table 10-2.

Laboratory Water Analysis Instruments and Procedures

Water samples can be filtered in the field, and should be if grab samples are being collected. Special equipment exists, enabling the filtration of the sample to occur as the sample is being pumped from the water body. Obviously, when numerous autosamplers are in use, field filtration techniques are not applicable nor practical.

When samples are brought in from the fields, they are filtered immediately. The filtration step allows for the determination of the concentrations of dissolved nutrient species (i.e., NO_3^-, NH_4^+, TDKN, ortho-P, and TDP) without interference from particulate matter. The filtration process involves

Table 10-2. Required water sample containers, preservation techniques, and maximum storage times suggested by EPA (1984).

Parameter	Container*	Preservation	Maximum storage time
Ammonia	P, G	Cool, 40° F H_2SO_4 to pH<2	28 days
Color	P, G	Cool, 40° F	48 hours
Hardness	P, G	HNO_3 to pH<2 H_2SO_4 to pH<2	6 months
pH	P, G	None	Analyze immediately
Kjeldahl and organic N	P, G	Cool 40° F H_2SO_4 to pH<2	28 days
Nitrate	P, G	Cool, 40° F	48 hours
Nitrate + Nitrate	P, G	Cool, 40° F	28 days
Nitrite	P, G	Cool, 40° F	48 hours
Phosphorus (Total)	P, G	Cool, 40° F H_2SO_4 to pH<2	28 days
Turbidity	P, G	Cool, 40° F	48 hours

*P = plastic, G = glass

passing the sample through a 0.45 micron chemically inert (polysulfone) filter membrane into a properly prepared vial. The filter membranes should be rinsed with deionized water prior to use. Once rinsed, to avoid contamination, they should not be touched directly by human hands. Each filter membrane should only be used once. A vacuum filtration apparatus can be purchased to accelerate the process, or a high-capacity vacuum filtration unit can be constructed out of readily available materials.

Sample Digestion

Water samples are digested prior to analyses for TDKN, TKN, TDP, and TP analyses. The Kjeldahl digestion process involves adding acid to the sample and heating it at 400° F for 1 hour. The heat is then turned up to 700° F for 1 1/2 hours to break down complex chemical compounds into ones suitable for colorimetric analyses. Primarily, the process converts complex compounds of nitrogen and phosphorus to NH_4^+ and PO_4^-, respectively.

pH

The pH of water is measured using a pH meter. The pH of a sample is defined as the logarithm of the reciprocal of the concentration of free hydrogen ions. Acid waters will have pH values below 7. Alkaline waters will have pH values ranging from above 7 to 14. A pH of 7.0 indicates that the water is neither acidic nor alkaline. Surface water samples will generally range in pH from 5 to 8. Due to the limestone composition of most aquifers in Florida, most well water is alkaline, with pH values ranging from 7 to 8.5.

Turbidity

The standard methods for the determination of turbidity are the nephelometric and visual methods. The turbidity of a water sample measured using a nephelometric method is reported in nephelometric turbidity units (NTUs). The visual Jackson Turbidity Method reports turbidity in Jackson turbidity units (JTU's). Meters for measuring turbidity range in cost from about $600 to $2,000.

Nitrogen and Phosphorus

Colorimetry is one method of analysis used to determine the concentrations of nitrogen and phosphorus in a water sample. The procedure involves the use of a colorimeter as a detection device. Essentially, upon chemical treatment in an explicit manner, a water sample will yield unique color traits, dependent on the concentrations of nutrients in the sample.

Colorimeters range from simple to extraordinarily sophisticated instruments. In situations that require analyses of a large number of samples in a limited amount of time, it is advantageous to use an auto-analyzer. An autoanalyzer is simply a colorimeter that has most of its functions automated, including the addition of the chemicals necessary to produce the measurable color. Autoanalyzers are capable of analyzing an extremely large number of samples while eliminating much of the potential for human error. The instruments are, however, subject to the usual maintenance problems associated with electronic and computerized equipment operating in wet chemistry environments.

Field Test Kits

Many field test kits (Fig. 10-4) are available for analyzing water samples for various constituent concentrations. Accuracy, range, and detection limits vary among kits. Field test kit data are useful as indicators of nutrient concentrations, with the general consensus being that the resulting concentrations will be ballpark figures. The user must also be extremely careful regarding what nutrient species are actually being measured. For example, a kit measuring total phosphorus, with no digestion procedure involved, is probably only measuring total soluble inorganic phosphorus. Acid reagents used in these kits will account for some, but not all, of the particulate organic and inorganic compounds. The test kits are reliable as long as the user is aware of their inherent limitations. Generally, field test kits should not be looked upon as inexpensive substitutes for laboratory analyses. However, several of the available test kits have been

Figure 10-4. Field test kit for acidity, alkalinity, CO_2, DO, hardness, and pH with color wheel and digital titrator.

certified by the EPA as accepted methods for determining and reporting some of the water quality parameters.

Quality Assurance/Quality Control (QA/QC)

Quality assurance is attained by employing adequately trained and experienced personnel, having good physical facilities and equipment, using certified reagents and standards, frequently servicing and calibrating instruments, and using replicate and known-addition sample analysis. It is desirable that QA/QC programs be acceptable by the Florida Department of Environmental Protection (FDEP) and/or the Florida Department of Health.

A good analytical quality control program consists of an organized plan for sampling procedures, sample custody, analytical procedures, calibration procedures and frequency, routine maintenance of equipment, quality control checks (matrix spikes, method blanks, standard calibration, check samples, laboratory duplicates, field quality controls, precision, accuracy), data reduction, data validation, and reporting. Each organization or laboratory involved with sample collection or analysis has the responsibility of implementing procedures that assure that the precision, accuracy, and comparability of the data submitted is of a known and documented level of quality.

Chapter 11. Physiological Response to Irrigation and Water Stress

by Ed Stover, Brian Boman, and Larry Parsons

Water stress develops because tree water loss exceeds the rate of water uptake for a sustained time period. Although citrus species are well adapted to conditions of moderate water stress, optimizing water management can provide significant benefit to the citrus grower. Water stress influences many components of citrus growth and development, with effects differing by stage of growth and severity of stress. Severe moisture deficits result in familiar symptoms of wilting, abscission of many leaves, and poor fruit quality. Although the effects are less dramatic, milder water stress in prolonged or repeated periods or at critical developmental stages may also significantly reduce productivity and profitability of citrus.

Root hydraulic conductivity, essentially the ability of roots to transfer water from soil into the plant, is quite low for citrus. This characteristic makes serious damage by drought stress less likely since soil water is conserved. However, it can also result in some water stress even when soil moisture is plentiful. Citrus canopies release so much water into the air that water transport within the root system and the rest of the tree cannot always keep pace. Stomates, the openings in leaves that allow carbon dioxide entry for photosynthetic production of sugars, then close to reduce loss of water, and tree photosynthesis and potential productivity are reduced. This property of citrus makes it impossible to prevent water stress completely. The challenge is to optimize efforts so that investments in water management provide good returns. Frequently, this means that we manage water so that stomatal closure is delayed until later in the day.

Diurnal and Progressive Changes in Water Relations

If there is adequate water in the plant, stomates remain open in the daylight. At night, stomates normally close and evapotranspiration (ET) decreases to a very low level, but water absorption continues throughout the night. By daybreak the next morning, water stress is usually at a minimum, and leaves have reached full turgor. This diurnal change in water loss and water uptake results in small changes in trunk, limb, and fruit diameters, even with well-watered plants. Sensitive instruments can be used to measure changes in fruit diameter and determine need for irrigation, but high labor requirements have limited this use. As the soil dries out further, less water becomes available for uptake by the roots. Midday wilt can occur, particularly on young, unhardened leaves. Initially, wilted leaves recover turgor through water absorption overnight. However, as soil drying continues, wilting of mature leaves starts earlier and lasts longer during the day. Eventually leaves do not recover from their wilt by the next morning. At this point, the tree is said to be in a permanent wilt and the soil is at the permanent wilting point.

Sensitivity of Physiological Processes to Water Stress

Physiological processes have different sensitivities to water stress as indicated in Fig. 11-1. Cell growth is most sensitive, and water stress of only a few bars can slow down or stop this process. Cell wall and protein synthesis are also quite sensitive to water stress. As stress continues to develop, synthesis of the hormone abscisic acid (ABA) increases, leading to earlier stomatal closure and a decrease in photosynthesis. With additional stress, respiration decreases and proline, an amino acid, increases. Eventually stomates close completely and photosynthesis stops. This stomatal closure reduces transpiration water loss greatly, but it also stops photosynthesis. As stress progresses even further, wilting and leaf curling occur, followed by

112

leaf abscission. ABA is implicated in many components of plant adjustment to water stress, inducing characteristics like earlier stomatal closure and culminating in the drastic step of dropping leaves to limit water loss. Prolonged severe water stress can cause limb dieback and eventual death of the tree.

Most commercial citrus operations irrigate, and severe water stress can usually be avoided. Moderate water stress in citrus groves reduces shoot growth and leaf expansion, slows canopy development and economic cropping in young trees, and reduces vegetative growth needed to support fruit production in mature trees. Stress during flowering and early fruit development can be especially detrimental since the level of cropping can be severely reduced. All of these effects will be discussed in more detail below. Twig and branch dieback may result from either moderately insufficient or excessive water stress over prolonged periods.

Root Systems

The root systems of citrus trees are relatively shallow compared to many deciduous fruit trees. Even so, studies on well-drained, sandy Florida Ridge soils have found citrus roots on trees with rough lemon rootstock at a depth of 11 feet, while roots on sour orange trees have been found at a depth of 9 feet. Water depletion studies confirmed that the roots were active at this depth, but most of the water extraction came from the top 2 feet. Citrus can also survive in very shallow soils, especially when the limiting factor is a high water table. Citrus trees grown in the coastal flatwoods areas of Florida typically have rooting depths of only 12 to 18 inches.

Citrus roots are sensitive to soil temperature and oxygen supply. Root hydraulic conductance is continuously reduced as soil temperature decreases from 95° F to 41° F. Drought stress further decreases the hydraulic conductance of roots. Even at low winter temperatures, citrus trees may suffer from drought stress when the soil is cold and moist but climatic conditions produce high evaporative demand due to high winds and low air humidity.

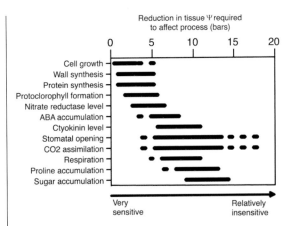

Figure 11-1. Generalized sensitivity to water stress for plant processes. (The length of horizontal lines represents the range of stress levels where a process is first affected. Dashed lines represent ranges where data is uncertain.)

As a result, severe leaf drop or scorch can occur. Similarly, low root oxygen during flooding can reduce hydraulic conductance, leading to wilting and other symptoms of drought stress.

Insufficient summer rainfall or irrigation reduces root development in citrus trees. This is especially true when the upper soil layers dry out between irrigations at long intervals. Long periods of drought or long intervals between rains or irrigation reduce growth and force the trees to increase water uptake from deeper soil layers.

Canopy Development

The vegetative development of young trees depends closely on the irrigation regime used. During dry periods, irrigation of young trees at long intervals can delay canopy development even though drought stress symptoms may not be evident. During the first several years after planting, there is generally a good relationship between canopy volume and yield. However, as trees reach full size, excessive growth induced by overirrigation and fertilization can decrease yields, mainly because of shading and the need for hedging. At this stage, reduced irrigation and fertilization practices can be used to limit canopy growth.

Measurement of growth of the trunk or main branches may be used to compare tree response to different irrigation regimes. However, various rootstocks develop different trunk sizes and there may be a poor correlation between trunk and canopy development across all rootstocks. For example, rough lemon develops a large canopy and small trunk as compared to Cleopatra mandarin, which has a large trunk but develops a small canopy during the first years of orchard development.

Fruit Set and Cropping

Citrus trees are very sensitive to water deficits during the flowering period. Stress during flowering can reduce fruit set, while continued water stress during the fruit set period is likely to cause a heavy physiological drop of small fruit. Water deficits during the early yield formation period (June) can result in further fruit shedding and reduce the rate of fruit growth. The increase in fruit size from June through December is highly related to water uptake. However, moderate water deficits in the late summer and fall can be desirable to maintain high juice quality. Severe water deficits in the summer, followed by rainfall or irrigation, often result in the development of a second fruit crop (which is typically of no value in Florida) and reduce the yields of the main crop in the following season. Fruit that matures while subject to water stress will usually have lower juice content and smaller size than fruit from trees receiving optimal water supply.

Effects of Irrigation on Fruit Quality

The amount of sugar in the juice is expressed as total soluble solids (TSS, usually measured with a refractometer), or as Brix (based on juice-specific gravity and usually measured with a hydrometer). Both of these measurements are essentially the same. Total soluble solids are an important parameter in determining the price of processed fruit. Acid concentration and the ratio between Brix and acid are important parameters in defining fruit quality and harvesting time. Grapefruit acidity usually responds more than orange acidity to irrigation, and decreases with reduced water stress. In general, water shortage causes increased concentration of Brix in the juice, while excessive rainfall or irrigation results in dilution of the sugars in the juice. Curtailing irrigation prior to harvest (Fig. 11-2) may increase Brix concentration in the juice, but may lower juice content.

Fruit size can be increased by irrigation. However, fruit enlargement is not always linear since better irrigation early in the season can increase the number of fruits per tree sufficiently to decrease fruit size due to higher crop load. Valencia oranges that develop under luxurious moisture conditions have a more tender peel compared with fruit from trees that are mildly drought stressed. Drought increases peel thickness and the peel-to-pulp ratio, and this decreases the juice percentage.

The effect of irrigation on color break is difficult to assess since irrigation can affect the nitrogen nutrition of the tree, and higher nitrogen will delay color break and enhance regreening. Usually color break is delayed when the irrigation is increased.

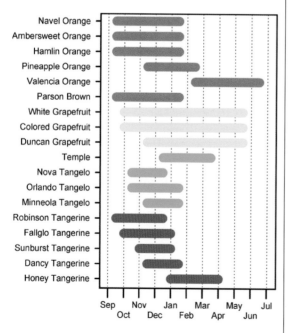

Figure 11-2. General maturity ranges for Florida citrus varieties.

Increased irrigation of 'Ruby Red' grapefruit has been shown to increase the chlorophyll and the red pigment (lycopene) content of the rind and cause more regreening, but it does not increase the yellow pigment (beta carotene). The effects of irrigation on citrus trees and fruit are shown in Table 11-1.

Diseases and Physiological Disorders

Conditions of high leaf and fruit turgor greatly increase susceptibility to oleocellosis, which is caused by the rupture of rind oil glands. Oleocellosis is avoided by delaying harvest when environmental conditions favor high fruit turgor, e.g., shortly after irrigation or rainfall, early in the morning, or under weather conditions that reduce water loss such as fog, dew, high humidity, or clouds.

Citrus foot rot, a fungal disease caused by *Phytophthora* spp., is closely related to conditions of high soil water content. The best preventive measure is good soil drainage. One should keep the soil around the trunk dry and avoid irrigation regimes that wet the trunk for long time periods.

Water (or flood) damage caused by asphyxiation of the roots from excessive water can cause severe damage or death to citrus trees planted in heavy, poorly drained soils. Some rootstocks, such as Carrizo citrange, may be very sensitive to this problem, while rough lemon and sweet lime have more tolerance. Citrus is more tolerant of flooding at the lower temperatures experienced in winter than in summer because ET demands are less in the winter. Flooding increases root resistance and reduces water absorption. Hence, high ET rates during warm summer days combined with reduced water uptake from flooding can promote damage sooner in warm weather.

Citrus blight is a wilt-like disease of unknown cause that is characterized by restricted water movement through the tree caused by vessel plugging. Studies of this disease fail to report a link

Table 11-1. Effects of irrigation on citrus.

Parameters		Increase	Decrease	No change
Tree size		X		
Fruit yield		X		
External quality	Fruit size	X		
	Fruit weight	X		
	Color		X	
	Green fruit	X		
	Peel thickness		X	
	Plugging		X	
Peel blemishes	Wind scar	X		
	Russeting			X
	Creasing			X
	Plugging		X	
	Scab	X		
Juice quality	Juice content	X		X
	Brix		X	
	Acid		X	X
	Brix/acid ratio		X	X
	Solids/box		X	
	Solids/acre	X		

between the disease and cultural practices, including irrigation. However, the disease causes drought stress symptoms. Poor uptake of water injected into the xylem is characteristic of blighted trees and is a simple way of diagnosing this problem.

Using Drought Stress to Induce Bloom

A period of dormancy (technically "quiescence" in citrus) markedly enhances flowering in citrus trees, and flowering generally increases with duration of dormancy. In subtropical citrus areas such as Florida, dormancy is primarily induced by low wintertime temperatures (less than 55° F). Water deficits can also be used to induce dormancy, and this is widely used in the tropics but can also be used to enhance flowering in the warmer citrus areas of Florida. An insufficient dormant period may result in a weak spring bloom, increasing the likelihood of significant off-bloom, and markedly reducing yield in many varieties.

Flower bud initiation occurs during this dormant period of minimal vegetative growth. Once the dormancy period has ended, an adequate water supply is necessary for the trees to bloom well. Severe water deficits following the break of dormancy will delay flowering, and when bloom does occur, flowering may be excessive, resulting in an overabundance of small fruit or even excess competition for resources leading to poor set.

Citrus trees growing in tropical climates can bloom continuously year round, and rainfall or irrigation after a period of drought can trigger a flush of flowering. In subtropical climates, some citrus species (notably lemon and lime) are capable of flowering year round. Irregular irrigation can also cause out-of-season flowering by inducing short periods of dormancy. The fruit that may set is usually undesirable, and heavy out-of-season flowering can reduce development of main season fruit.

The amount of drought stress necessary to induce flowering is important since excessive stress can be harmful. Studies have shown that a significant percentage of the flowers will abort after excessive drought stress. Excessive drought stress also harms the development of fruits already on the trees and will substantially reduce the yield of regular winter lemons. Lime trees seem to be less responsive than lemon to drought-induced flowering. Drought has been used in Mediterranean countries to stimulate flowering in lemons. By withholding water during certain periods, the timing of flowering can be altered so fruit can be harvested at a time when prices are higher.

In Florida, methods to increase citrus flower bud induction may be useful to enhance yield during winters with above normal temperatures. If little rain occurs, drought stress may be developed by discontinuing irrigation. Withholding irrigation should be continued until at least 30 to 40 days of stress have occurred.

Moderate drought stress may be developed and maintained on flatwood bedded groves by maintaining the water table just below the bottom of the water furrows. This should allow some daily water stress, but recovery will occur each night from water taken up by the deepest roots. If the water table is one or more feet below the water furrow, severe drought stress is likely if irrigation is stopped. Visual observation of leaf wilting is a good indicator of proper stress. Wilting by 10:00 or 11:00 AM and recovery overnight is ideal. Drought stress is more difficult to develop on deep sandy soils that have deeper root zones. Two or three weeks may be required to reach an adequate level of stress to begin flower bud induction. Drought stress should be used only on trees that have already been harvested. Trees bearing fruit are likely to experience excessive drop of the current crop if subjected to moderate water stress in the early spring.

If winter rain thwarts attempts to maintain water stress, the associated cool weather behind the front (a typical occurrence) often provides some cool temperature induction, but may not be sufficient to provide adequate bloom.

by Mark A. Ritenour, Wilfred F. Wardowski, and David P. Tucker

Irrigation and fertilizers are important components of commercial citrus production. Irrigation is necessary to adequately replenish soil water lost through evaporation and transpiration. Fertilizers replace nutrients removed during harvest and through leaching and maintain tree growth and vigor. Optimum management of both is critical for obtaining maximum yield. Irrigation and fertilization practices can also have significant impacts on fruit quality and quality retention during harvest, packinghouse operations, storage, and distribution. These include effects on fruit color, texture, disease susceptibility, juice composition, and the development of physiological disorders.

Although fruit quality usually improves as soil moisture and nutrient levels increase from deficient to optimum levels, those that produce maximum yield may not always correspond to levels that result in the highest fruit quality and maximum quality retention. Further, although the addition of nutrients above optimum levels may not reduce yields, they can have either negative or positive effects on aspects of quality that are not readily apparent. Other critical factors such as rootstock and scion selection, pest management, and environmental conditions will not be discussed in this section. Much of the information presented below was derived from research conducted more than fifteen years ago. Thus, new research is needed to understand irrigation and nutrient effects on fruit quality using the new rootstocks, varieties, cultural practices, etc., that have been adopted since then.

Water
Adequate moisture levels are critical for proper fruit set and to support optimum fruit growth and development through harvest. Moderate water stress in the winter may increase flower bud induction and be a useful management tool, especially during warmer winters with inadequate cold induction. Adequate soil moisture is important early in the season to ensure good fruit set. Water stress during spring fruit set can lead to the setting of a higher percentage of late-bloom fruit of inferior quality. For example, unirrigated trees were reported to produce up to 79% inferior, late-bloom fruit while irrigated trees had a maximum of only 9% of the late-bloom fruit.

Water stress anytime during the growth and development of citrus fruit can reduce yield and fruit quality compared to well-watered trees. Such losses cannot be completely recovered through proper irrigation during the rest of the season. Water stress may result in smaller, lighter fruit with thicker peel and reduced juice content. Excessive rainfall and/or irrigation immediately prior to harvest results in a dilution of soluble solids, whereas drought conditions concentrate soluble solids. Even though fruit from water-stressed trees may have higher total soluble solids and acids per fruit, solids per acre may be reduced because of lower total yields per acre.

Water stress also affects the fruit at harvest, with soft or dehydrated fruit experiencing more plugging. Conversely, irrigation or rainfall near harvest or harvesting with dew on the plant often results in fruit with a very turgid rind susceptible to oleocellosis (rupture of oil cells upon impact). Heavy rains, especially after a drought, can result in "zebra-skin" (areas of necrotic peel over the raised segments), which occurs mostly in early-season tangerines. Increased irrigation has also been reported to enhance the development of wind scar and scab.

Fruit harvested early in the morning, during rainy periods, or from trees with poor canopy ventilation have a higher risk of postharvest decay (i.e., green mold). Conversely, fruit grown in climates that are more arid tends to develop less rot during post-harvest handling, transportation, and marketing. If fruit must be harvested from trees during rainy periods, it may be best to avoid fruit from lower branches that may be exposed to more pathogens (i.e., brown rot). Increased irrigation has been reported to decrease postharvest incidence of stem-end rot but increase the incidence of green mold.

Nitrogen

Fruit quality responses to nitrogen, phosphorous, and potassium appear to be similar for both orange and grapefruit. Among other things, nitrogen (N) is an important constituent of proteins and plays a critical role in a cell's biochemical machinery. Adequate N is essential for optimal plant growth and development and is the mineral element used most by plants. Citrus fruit quality has traditionally been evaluated at N concentrations between 2% to 3% of leaf dry weight. In Florida, 2.5% to 2.7% leaf N is recommended for optimum citrus production. Of all the nutrient elements, nitrogen has the largest impact on fruit quality.

Probably the most pronounced effect of higher nitrogen levels on orange and grapefruit quality is on delayed time to color-break and reduced color development at harvest. However, levels of nitrogen that result in optimum color development may reduce yield. Thus, it is a balancing act to keep nitrogen levels high enough for maximum yield, but low enough to allow good color development. Using less nitrogen can also reduce re-greening of citrus. High nitrogen levels in young trees can result in coarser fruit with a thicker peel, while high nitrogen in mature trees can result in more and smaller fruit with a thinner peel. Although sometimes inconsistent, fruit grown under higher nitrogen levels tend to have a lower solids to acid ratio, increased acids, and lower ascorbic acid (vitamin C) content. In oranges, high N has been reported to lower juice content and increase rind staining. Orange fruit quality appears to be best if high N levels are avoided during the summer and fall.

On the other hand, higher levels of N can result in increased total soluble solids, soluble solids per box and per acre, juice color, grapefruit juice content, and observed reductions in the development of peel blemishes such as wind scar and russetting. Rind plugging during harvest also appears to be reduced with higher rates of N. Effects of N on creasing have been inconsistent.

Although high nitrogen has often been associated with increased postharvest decay in other commodities, very little is known about its effects on citrus quality retention during storage and marketing. The only reports thus far suggest that higher N may reduce the development of stem-end rot and green mold during storage.

Phosphorous

Phosphorous (P) is an important component of many plant compounds, including DNA, cell membranes, and energy-yielding intermediates of photosynthesis and respiration. Rates of P studied for fruit quality range between 0.10% and 0.21% per leaf dry weight, with the optimum for Florida citrus being between 0.12% and 0.16%. While P deficiencies are rare, when deficiencies do occur, they often result in fruit with hollow cores, high acids in the juice, and good external color but with thick peels.

High P levels have been reported to result in less color development and increased problems with re-greening. Such fruit often have lower acids and vitamin C content. Creasing can also become more of a problem as leaf levels rise between 0.10% and 0.14%. On the other hand, higher P can lead to fruit with thinner peels, better peel texture and a higher sugar to acid ratio. There have been inconsistent reports that increased P sometimes decreases fruit size and total soluble solids but increases juice content. There have been no reports of different rates of P affecting postharvest diseases (e.g., stem-end rot, green mold, or sour rot) and quality retention during citrus storage and marketing.

Potassium

Potassium (K) plays many important roles, including osmotic (water potential) regulation of cells and the activation of different enzymes in photosynthesis and respiration. In citrus, most research relating to K nutrition on fruit quality has been conducted within the range of 0.3% and 1.7% per leaf dry weight. Optimum K levels for Florida citrus are between 1.2% and 1.7%. For orange and grapefruit, one of the greatest negative effects of increasing K is a decrease in sugar to acid ratio. High levels of K can delay the time to legal maturity in grapefruit up to 83 days compared to fruit grown with low rates of K with associated delayed time to color-break and an increase in the number of green fruit at harvest. Re-greening of oranges may also be enhanced at higher K rates. Higher levels of K can reduce fruit total soluble solids, juice content, and juice color, increase acidity, and often, but not always, increase peel thickness and coarseness. Low levels of K have been associated with increased fruit splitting and fruit drop.

However, increasing K leads to increased fruit size, weight, and vitamin C content, and reduces the incidence of creasing. At harvest, fruit from trees with higher K tend to experience less plugging. Furthermore, foliar K applications have recently been reported to increase size by 0.1 to 0.2 inch without decreasing sugar to acid ratios, Brix, acid or juice contents, and with no increase in peel thickness. Other than one report suggesting that higher K may slightly decrease stem-end rot, there are no reports of it affecting other postharvest diseases (e.g., green mold, sour rot, etc.).

Calcium (Ca), Magnesium (Mg), and Micronutrients

Although Ca deficiency in other commodities commonly results in disorders (e.g., bitter pit of apple), Ca deficiency in citrus is very rare. Further, unlike many other commodities, there have been no reports of additional Ca enhancing fruit quality or quality retention during storage and marketing.

Magnesium usually has negative effects on fruit quality only if leaf content is deficient. In this case, increasing Mg can increase fruit size and weight, decrease peel thickness, and increase total soluble solids, sugar to acid ratio, solids per box, and solids per acre.

Micronutrient deficiencies usually do not impact fruit quality except for boron, which can result in brownish gum pockets in peel and sometimes pith areas.

Table 12-1. Effects of mineral nutrition on fruit quality (+ = increase, - = decrease, ? = unknown, and 0 = no change)[a].

	Parameters	N	P	K
External quality	Fruit size	-	0 or -	+
	Fruit weight	-	0	+
	Green fruit	+	+	+
	Peel thickness	-[b]	-	+
	Plugging	-	?	-
Juice quality	Juice content	+ or -	0 or +	-
	Soluble solids (SS)	+	0 or -	-
	Acid	+	-	+
	SS: acid ratio	-	+	-
	Color	+	0	-
	Solids/box	+	0	-
	Solids/acre	+	+	+

[a]Source: Adapted from Tucker and others, 1995.
[b]Fruit from young trees may have a thicker peel.

Table 12-2. Effects of mineral nutrition and irrigation on postharvest diseases (+ = increase, - = decrease, and 0 = no change)[a].

Parameter		Stem end rot	Green mold	Sour rot
Macro nutrients	N	-	-	0
	P	0	0	0
	K	-	0	0
Irrigation		-	+	0

[a] Source: Koo, 1988.

Chapter 13. Salinity Problems

by Brian Boman

When dealing with salinity problems, it is important to realize that all water is not equal. In fact, salinity management might be a major objective of irrigation management, even though the primary objective of irrigation is normally to maintain the soil matric potential in a range suitable for optimum crop growth. Irrigation with high salinity water requires irrigations to be more frequent and of greater amounts than when good quality water is used. During extended droughts, salinity levels will dictate irrigation scheduling.

All natural waters and soil solutions contain soluble salts. However, the amount and type of salts that are in water vary greatly. In some areas of the state, the groundwater can contain very high levels of salinity (Table 13-1). These salts are concentrated in the soil with the process of irrigation, evaporation, and transpiration. In addition, strong winds off the ocean can deposit salt spray many miles inland. Salt concentrations in rainfall can be as high as 40 ppm of total dissolved solids (TDS) along the coast.

In Florida, salinity problems are generally of concern only in flatwoods areas, as the irrigation water supply in Ridge areas is typically of excellent quality. Salinity problems have been documented on Indian River citrus since as early as 1900. More recently, problems with salinity have occurred in citrus groves in the Tampa Bay and Southwest Florida production areas.

In some coastal areas, high salinity levels in wells can be attributed to saltwater intrusion into the freshwater zone from the ocean. The effects of pumping rate in coastal areas on salt water intrusion are illustrated in Figure 13-1. Salt water has a density of about 1.027 as compared to 1.0 for fresh water. The Ghyben-Herzberg principle states that depth to the fresh: saltwater interface is 38 times

Table 13-1. Groundwater salinity by location (adapted from Wander and Reitz, 1951).

County	No. of samples	Average TDS (ppm)
Brevard	10	2,580
St. Lucie	38	1,100
Indian River	55	1,530
Manatee	26	1,045
Sarasota	14	1,315
Charlotte	11	2,485
Polk wells	2	195
Polk lakes	9	70

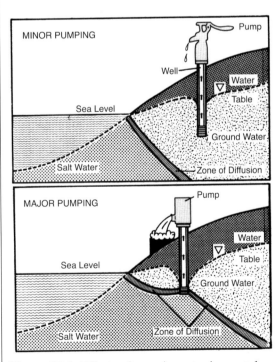

Figure 13-1. Effects of pumping rates in coastal areas on salt water intrusion into the freshwater shallow aquifer.

the distance between the static water table and mean sea level.

> **Example:**
> For a static water level of 15 ft msl, estimate the depth to the fresh: saltwater interface. If pumping resulted in a drawdown of 10 feet in the well (to 5 ft msl), determine the change in depth to the interface.
>
> Before pumping: Interface = 15 x 38 = 570 ft
>
> After pumping: Interface = 5 x 38 = 190 ft

In the Indian River area, irrigation wells normally are 600 to 1200 feet deep, and are in the upper Floridan Aquifer. Generally, deeper wells have higher salinity levels. The salts in these wells come from the highly mineralized limestone that is in the water-bearing strata. The salinity of these wells can vary from month to month and from year to year (Fig. 13-2). The quality in some wells deteriorates as the artesian pressure drops, while others remain relatively unaffected. (Note the change of over 1000 ppm in one well versus only 300 to 400 ppm in the other in Fig. 13-2.)

Surface water supplies in the Indian River area are also subject to periodic high salinity loadings (Fig. 13-3). As the dry season progresses, salinity levels of surface water in Indian River area ditches and canals increase as a combination of reuse of irrigation water and the augmentation of the surface water supply from the more saline Floridan Aquifer wells. Highest surface water salinity levels typically occur in April. When summer rains begin, canal salinity levels normally drop rapidly.

Groundwater can influence the salinity profile in the root zone if the net flow of water is upward for significant periods of time. High concentrations of salt may accumulate near the surface in the absence of sufficient irrigation or rainfall to maintain downward water flow. Typically, the salt concentration is usually higher in the soil than in the applied water. The increase in salinity results from plant transpiration and soil surface evaporation, which selectively removes relatively pure water and concentrates the salts. The average salt

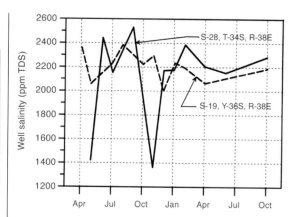

Figure 13-2. Salinity concentrations and changes by season for two Indian River area wells.

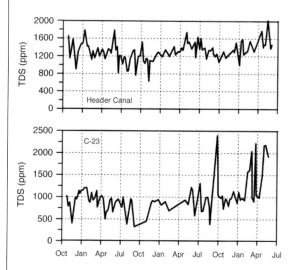

Figure 13-3. Salinity concentrations in Indian River area canals.

concentration of the soil solution in the root zone is often assumed to be about three times the salinity of the applied water.

Salt accumulations in the soil are generally only removed by leaching below the crop root zone. Therefore, the key to salinity control is to provide a net downward flow in the root zone. Even in well-managed groves, the soil water may be several times more saline than the irrigation water. With insufficient leaching and drying out of the soil, this ratio can easily increase tenfold or more, resulting in injury to the trees.

Accumulation of salts over the years is not a problem in most cases due to the abundant rainfall at sufficient rates to leach the salts from the root zones. Salts in typical sandy soils are generally leached out with the first inch of rainfall. However, in some poorly drained heavier soils, salt accumulation can be a problem. These soils require more careful monitoring of salinity-related problems.

Measurement

The quantity of salts in water is commonly reported in units of total dissolved solids (TDS) or electrical conductivity (EC). TDS are measured by evaporating a sample of water and weighing the residue. The results are reported in parts per million (ppm) or mg/L, depending on whether the calculation is on a weight or volume basis. For most practical purposes, ppm is equal to mg/L.

The EC of a solution is a measure of the ability of the solution to conduct electricity. When ions (salts) are present, the EC of the solution increases. If no salts are present, the EC is low, indicating that the solution does not conduct electricity well. The EC indicates the presence or absence of salts, but does not indicate which salts might be present. If the EC of a sample is relatively high, no indication from the EC test is available to determine if this condition was from irrigation with salty water or if the field had been recently fertilized and the elevated EC is from the soluble fertilizer salts. To determine the source of the salts in a sample, further chemical tests must be performed.

EC measurements are taken with platinum electrodes and presented in units of conductance. Handheld conductivity sensors (Fig. 13-4) are convenient for measuring conductivity in the field. They come in a variety of designs and can range in cost from $40 to several hundred dollars.

The SI (metric) unit of measurement is deci-Siemens per meter (dS/m), which is equal in magnitude to the commonly used conductance term of millimho/cm (mmho/cm). Both of these terms are generally in the range of 0-5. If the numbers reported are higher, in the range of 100 to 5000, the

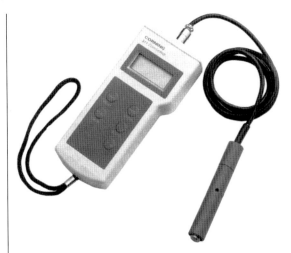

Figure 13-4. Typical handheld conductivity meter.

units are typically micro-Siemens per centimeter (μS/m), which is equivalent to micromho/cm (μmho/cm). The conversion from electrical conductance to TDS depends on the particular salts present in the solution. The conversion factor of 700 x EC (in dS/m) is applicable for converting EC values to TDS for Florida irrigation waters. Oftentimes commercially available meters will read directly in ppm. Care must be taken when using these meters so that results are reported consistently. Most of these type of meters will use conversion factors of 630 or 640 x EC to get ppm. These are common conversion factors that are used in many places throughout the world. However, some meters may use a factor as low as 500 or as high as 800 to convert from dS/m to ppm.

Conversions:
  1 mg/L = 1 ppm
  dS/m x 700 = ppm
  μS/cm x 0.7 = ppm
  μS/cm = μmho/cm

Example:
Determine the salinity in ppm for a water sample with EC of 2.3 dS/m.

2.3 dS/m x 700 = 1610 ppm

Example:

A meter that has a built-in conversion factor of 1 dS/m = 630 ppm has a reading of 2300 ppm. What would be the TDS if the factor of 700, suitable for most Florida water, was used instead?

Convert to dS/m using the meter factor of 630
2300 ppm / 630 ppm/dS/m = 3.65 dS/m

Then convert back to ppm using a factor of 700
3.65 dS/m x 700 ppm/dS/m = 2555 ppm

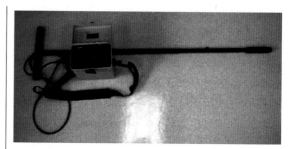

Figure 13-5. Soil salinity probe.

Soil Salinity Measurements in the Field

The EC of the soil has little direct detrimental effect on sandy mineral soils, but EC directly affects plants growing in the soil. As EC increases, more attention to water management is needed to prevent salinity from adversely affecting citrus. In Florida, the Extension Soil Testing Laboratory uses a 2:1 solution:soil ratio with which to determine EC. In most states and most literature, the saturated paste extract (EC_e) method is used. The saturated paste method is more time-consuming than the 2:1 extraction, and results in inadequate amounts of solution in Florida's sandy soils. The conversion from the 2:1 extraction result to the saturated paste is a factor of 8 ($EC_e = EC_{2:1}$ x 8). In general, when the soil $EC_{2:1}$ exceeds 0.25 dS/m (EC_e = 0.25 x 8 = 2.0 dS/m), citrus will begin to experience stress due to salts.

Example:

What is the equivalent EC_e for a soil with a measured $EC_{2:1}$ of 0.4 dS/m?

EC_e = 0.4 x 8 = 3.2 dS/m

Soil salinity probes are an effective means of tracking salinity levels in Florida's sandy soils (Fig. 13-5). Since most soil minerals are insulators, electrical conduction in saline soil is primarily through the pore water, which contains dissolved salts. The contribution of exchangeable cations to electrical conduction is relatively small in saline soils because these cations are less abundant and mobile than the soluble electrolytes. EC in soils is also affected by the number, size, and continuity of soil pores, as well as salt and water contents.

Figure 13-6. Salinity probe being used to measure apparent conductivity in citrus grove.

For given soil types where water content is at a standard level, apparent conductivity (EC_a) is related to soil salinity. The water content at field capacity is sufficiently reproducible to serve as the reference water content required to establish calibrations between EC_a and soil salinity (EC_e). In irrigated groves, the soil water in the wetted area comes to field capacity shortly after an irrigation.

Salinity probes can be used to measure soil electrical conductivity profiles within the soil. A succession of measurements (typically at 6-inch increments) is taken with the vertical sensor inserted into the soil via an access hole (Fig. 13-6). Average soil salinity (EC_e) may be determined from EC_a measurements once calibrations are established between EC_e and EC_a for a particular soil, provided the EC_a determinations are made at approximately the same water content as that for which the

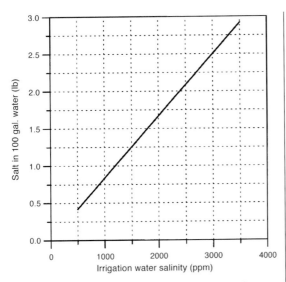

Figure 13-7. Pounds of salt in 100 gallons of water at various water salinity levels (TDS).

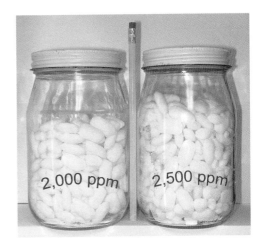

Figure 13-8. Salt load in 100 gallons of water at 2000 mg/L (1.7 lb) and 2500 mg/L (2.1 lb) concentration in irrigation water (1 qt jars).

calibrations were made. Separate calibrations for each soil are not usually necessary, since calibrations are similar enough for soils of similar water-holding capacities and textures.

Salt Load

Large amounts of salts can be deposited on the soil during continued irrigations with high salinity water (Fig. 13-7). For example, in water with 2000 ppm TDS, there is about 1.7 lb of salt in each 100 gallons applied (Fig. 13-8). The salts that are applied will remain in the soil unless they are leached out through excess irrigation or rainwater applied to the soil. Consider a block of citrus that receives the equivalent of 40 gal/tree/day of 2000 ppm TDS of irrigation water. In one week, each tree will have 4 3/4 lb of salt applied around it. As the drought continues, more and more salt will accumulate if adequate irrigation strategies are not employed. Without proper water and nutrient management, citrus irrigated with high salinity water can suffer the reduced growth, small fruit, and decreased yields that accompany salt stress.

Osmotic Stress

Salts in solution exert an osmotic effect that reduces the availability of free (unbound) water

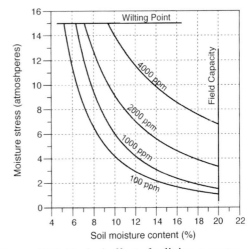

Figure 13-9. Typical effect of salinity on water stress levels within plants (adapted from Wadleigh and Ayers, 1945).

through both chemical and physical processes. Roots are therefore not able to extract as much water from a solution that is high in salts as from one low in salts. In effect, the trees have to work harder to move water into the roots. Figure 13-9 shows typical effects of salinity on the water stress within a plant. In Fig. 13-9, the stress level with 100 ppm water when half the water is depleted (going from 20% to 10% moisture) is about 4.0 atm. Remarkably, this stress level is similar to the 3.5 atm stress level that occurs at field capacity when

Figure 13-10. Tip burn caused by excess salinity.

Figure 13-11. Leaf-bronzing caused by excess salinity.

Figure 13-12. Canopy thinning resulting from excess salinity.

the soil solution has 2000 ppm TDS. (Field capacity is the moisture content that occurs in sandy soils a day or so after excess rains completely wet the soil.) In other words, even at field capacity the 2000 ppm salinity has significant water stress.

Therefore, for citrus irrigated with saline water it is essential that irrigations be frequent (daily) to minimize salinity stress.

There are distinct differences in the rate of chloride and sodium uptake among citrus rootstocks. The general decreasing order of salinity tolerance to chlorides for common rootstocks is (best to worst): Cleopatra mandarin → rough lemon → sour orange → Swingle citrumelo → Carrizo citrange. The range of some other rootstocks is given in Table 13-2. It is important to remember that growth and yield of trees on all rootstocks can be reduced by excessive salts.

Symptoms of Salt Injury
The critical salinity level varies with the buffering capacity of the soil (soil type, organic matter), climatic conditions, and the soil moisture status. Many salinity-induced symptoms such as reduced root growth, decreased flowering, smaller leaf size, and impaired shoot growth are often difficult to assess, but occur prior to ion toxicity symptoms in leaves. Chloride toxicity, consisting of burned necrotic or dry appearing edges on leaves (Fig. 13-10), is one of the most common visible salt injury symptoms. Na toxicity symptoms seldom distinctly appear, but rather an overall leaf "bronzing" appears along with reductions in growth (Fig. 13-11). As with Cl, high leaf Na can cause nutrient imbalances at much lower concentrations than those required for visible symptoms.

As salinity loads increase, trees will begin to shed leaves and a thinning of the canopy is evident (Fig. 13-12). The symptoms are usually most evident looking up into the top of the canopy. There will also be an abundance of leaves on the ground. Progressive salinity will lead to defoliated branches and twig dieback (Fig. 13-13).

In an Australian study, leaves were monitored on Washington navel trees irrigated with 300 and 1200 ppm TDS water (Fig. 13-14). Leaves on trees irrigated with the higher salinity level had significantly shorter lives. After 9 months, only about 15% of the spring flush leaves were still on the

Table 13-2. Citrus rootstocks ranked in order of decreasing ability to restrict chloride and sodium accumulation in scion (adapted from Mass, 1992).

Rank	Chloride	Sodium
1	Sunki mandarin	Sour orange
2	Grapefruit	Cleopatra mandarin
3	Cleopatra mandarin	Rusk citrange
4	Rangpur lime	Rough lemon
5	Sampson tangelo	Rangpur lime
6	Rough lemon	Sweet orange
7	Sour orange	Cuban shaddock
8	Ponkan mandarin	Savage citrange
9	Citrumelo 4475	Citrumelo 4475
10	Trifoliate orange	Troyer citrange
11	Cuban shaddock	Sunki mandarin
12	Sweet lemon	Grapefruit
13	Calamondin	Sampson tangelo Ponkan mandarin
14	Sweet orange	Calamondin
15	Sweet lime	
16	Savage citrange	
17	Rusk citrange	
18	Troyer citrange	
19	Carrizo citrange	

Figure 13-13. Defoliation and twig death resulting from excess salinity.

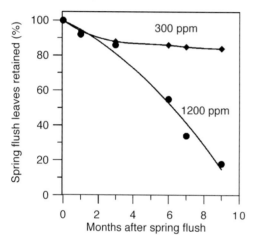

Figure 13-14. Retention of spring flush leaves for Washington navel trees irrigated with 300 and 1200 ppm water (adapted from Howie and Lloyd, 1989).

trees irrigated with the 1200 ppm water, as compared to nearly 85% of those with the 300 ppm water.

Wetting Foliage
Irrigation water that wets the foliage (partially or fully) can result in severe damage to the leaves in the skirt of the trees. There are reports where chloride and sodium concentrations of the lower leaves were about four times greater than those of the upper leaves (grapefruit, Valencia and Washington navel).

The lowest concentration of either sodium or chloride generally associated with leaf burn is about

0.25%. Controlled experiments showed that citrus leaves easily accumulate chloride and sodium from direct contact with water drops. The accumulation is greater from intermittent than continuous wetting, and from daytime than nighttime irrigation. Accumulation is a function of the rate of evaporation, which results in increased salt concentration of the water film on the leaves. The sensitivity of a citrus to injury through direct foliar contact bears no relationship to its general soil salinity tolerance. Unlike soil-applied salinity, trees on all rootstocks are about equally sensitive to injury through direct foliar contact. Young, tender shoots are especially vulnerable to salt burn. Young trees (1 to 2 years) on Swingle citrumelo rootstock seem to be more susceptible to spray on their trunks and often develop brown "blisters" of dead tissue on their trunks.

Fertilization
The frequency of injecting nutrients or of applying granular fertilizer has a direct effect on the concentration of TDS in the soil solution. A fertilization program that uses frequent applications with relatively low concentrations of salts will normally result in less salinity stress than programs using only two or three applications per year. Controlled-release fertilizers and frequent fertigations are ways to economically minimize salt stress when using high salinity irrigation water. Growers using surface water in high salinity areas generally see a marked improvement in water quality when the summer rains begin. Under these conditions, fertigations during the wet season should pose no problems.

Selecting nutrient sources that have a relatively small osmotic effect in the soil solution can help reduce salt stress. The osmotic effect that a material adds to a soil solution is defined as its salt index relative to sodium nitrate, which has a salt index of 100 (Appendix 5). Since sources of phosphorus (P) generally have a low salt index, they usually present little problems. However, the salt index per unit (lb) of N and potassium (K) should be considered.

The salt indexes of natural organic fertilizers and slow-release products are low compared to those of commonly used soluble fertilizers. High-analysis fertilizers may have a lower salt index per unit of plant nutrient than lower-analysis fertilizers since they may be made with a lower salt index material. Hence, at a given fertilization rate, the high-analysis formulation may have less of a tendency to produce salt injury.

> Example:
> Compare the salt index per unit plant nutrient of 100 lb of 8-0-8 solution made from ammonium nitrate (23.5 lb) and muriate of potash (13.3 lb) to a blend made with ammonium nitrate (16.9 lb) and potassium nitrate (17.4 lb).
>
> From Appendix 5, the salt indices for the materials are:
>
> Ammonium nitrate (34% N) = 105
> Muriate of potash (60% K$_2$O) = 116
> Potassium nitrate (13% N and 46% K$_2$O) = 74
>
> Ammonium nitrate + muriate of potash =
> 23.5 lb x 105 + 13.3 lb x 116 = 4010
> Ammonium nitrate + potassium nitrate =
> 16.9 x 105 + 17.4 x 74 = 3062
> 4010/3062 = 1.3

Although both solutions have the same analysis, the salt index of the ammonium nitrate + muriate of potash blend is 30% greater than that for the ammonium nitrate + potassium nitrate blend.

The Cl in KCl or Na in NaNO$_3$ materials add more toxic salts to the soil solution. High rates of salt application can alter soil pH, and thus cause soil nutrient imbalances. Some ions can also add to potential nutrient imbalances in trees. For example, Na displaces K, and to a lesser extent Ca, in soil solutions. This can lead to K deficiencies and, in some cases, even to Ca deficiencies. Such nutrient imbalances can compound the effects of salinity stress. Problems can be minimized if adequate nutritional levels are maintained, especially those of K and Ca.

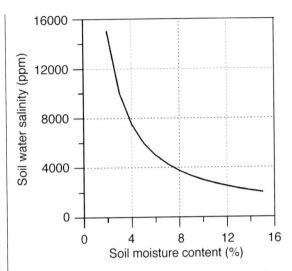

Figure 13-15. Increase in soil water salinity as the soil dries and the same amount of salts remains in solution.

Salt Buildup

As the dry season progresses, salts accumulate in the soil. Evaporation removes relatively pure water from the soil surface, leaving the salts behind. Trees also attempt to exclude salts from being taken up into the water stream. Evaporation from the soil surface and evapotranspiration (ET) results in a reduced volume of water in the soil. Since there is less water and only a slight drop in the amount of salts in the soil, the concentration (ppm) of salts in solution increases.

For example, consider flatwoods sandy soil that holds 15% moisture at field capacity (Fig. 13-15). If the soil solution salt concentration is 2000 ppm at field capacity, the roots will be seeing concentrations of 4000 ppm when half of the water is depleted. Again, the solution to manage the problem effectively is to keep the soil wet!

Once salts accumulate in the soil, the only way to remove them is to leach them below the root zone with excess irrigation or rainfall. This means that with each irrigation, enough water should be applied so that there is a net downward flow in the root zone. In areas with shallow water tables, salts

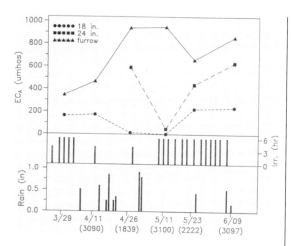

Figure 13-16. Salinity profiles in bedded citrus (values in parentheses (i.e., 3090) represent the irrigation water salinity in uS/cm).

irrigation (microsprinkler) necessary. Salinity levels at a depth of 18 inches dropped to near zero following rains beginning April 13. The rains on April 30 flushed out the salts from the 24-inch depth. Salts were flushed from the profile and were found to build in the water furrow. Irrigations every 2 to 3 days beginning on May 9 resulted in increases in soil salinity at both the 18- and 24-inch depths.

Young trees affected by salinity present a great challenge. In their first year, trees typically require less than 1gal/day per tree. However, frequent irrigations are even more critical on young trees when using high salinity water. With high salinity, young trees should be watered on a daily basis to minimize damage. During extended dry periods, salts will accumulate on the fringes of the wetted zone and move upwards as evaporation occurs. These salts will be put back in the soil solution with rainfall. If the rainfall amount is low, these salts will move back into the root zone and cause very high salinity levels. Therefore, it is a good practice to continue to irrigate until adequate rainfall is received to leach accumulated salts below the root zone (usually 1 inch of rain is sufficient on sandy soils).

that are flushed through the root zone can move back into the root zone if the surface and top of the root zone dry out.

Salt accumulations in most of Florida's sandy soils are flushed out fairly quickly following rainfall of 1+ inches. Figure 13-16 shows soil salinity during a 3-month period when drought conditions made

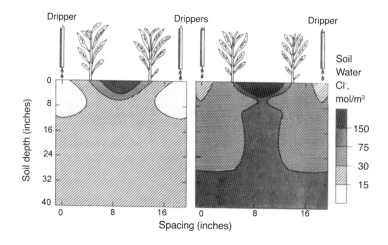

Figure 13-17. Chloride profiles for 17% (left) and 2% leaching (right) in a field plot study with drip irrigation (adapted from Hoffman et al., 1979).

Managing High Salinity

Managing irrigation and fertilization with high salinity irrigation waters requires routine evaluations of the water with an EC meter. Irrigate frequently to prevent concentration of salts. Excess irrigations to leach accumulated salt may become necessary, and should be made no less frequently than every other week during the peak irrigation season. Irrigation rates should be monitored to make sure that excess salts are leached below roots (Fig. 13-17). It may be necessary to initiate irrigations when small rainfall events occur. Rain will put salts that have accumulated near the surface back into solution. If there is insufficient rain (less than 1 inch), the salts may end up back in the root zone. Therefore, it is a good practice to continue irrigations until the salts are flushed from the root zone.

Keep poor quality water off of leaves, especially under conditions of high evaporative demand.

Irrigate at night whenever possible to minimize evaporative concentration of salts.

Choose fertilizer formulations that have the lowest salt index per unit of plant nutrients. Increase the frequency of fertilizations, thereby making it possible to reduce the salt content of each application and aid in preventing excess salt accumulation in the root zone. Maintain optimum, but not excessive, nutrient levels in soil and leaves with rates based on the long-term production from the grove. Fertilizer rates can usually be lower for trees with high salinity since production levels will probably be lower. Leaf tissue analysis should be used to detect excessive Na or Cl levels, or deficient levels of other elements caused by nutrient imbalances from the salt stress. Na levels greater than 0.2% and Cl levels over 0.5% indicate imminent problems.

Chapter 14. Soil And Water Relationships

by Thomas Obreza and Brian Boman

Floridian Plateau

Florida is probably the youngest part of the United States because it was the last to emerge from the ocean. The land area is the highest portion of the Floridian Plateau, which consists of a core of metamorphic rocks buried under more than 4000 feet of sedimentary rocks (mostly limestone) that represent the seaward extension of the coastal plain of Georgia and Alabama. Over the course of eons, it has been alternately dry land or covered by shallow seas. The Floridian Plateau has been one of the most stable parts of the earth's crust.

The plateau is broad and nearly level. The portion above sea level forms the state of Florida, while the submerged part (the continental shelf) slopes gently away from the land at less than 3 feet per mile. The slope is steeper in the Atlantic Ocean than in the Gulf of Mexico. Out beyond a depth of 300 feet, the level of the sea bottom deepens rapidly. The citrus industry exists on two of Florida's five natural topographic regions: the central highlands and the coastal lowlands.

Central Highlands

The Central Highlands (Ridge) follows the crest of the Floridian Plateau, but lies closer to the west coast than the east coast. The eastern and southern parts are known as the Lake Region, where the hills rise higher than anywhere else in peninsular Florida. The Lake Region is where Florida's citrus industry was traditionally located because the deep, sandy soils are naturally well drained. However, cold weather has caused a substantial portion of the industry to move south to the lowlands of Florida's east coast and southwest regions.

Coastal Lowlands

The Coastal Lowlands (Flatwoods) is the region most recently emerged from the sea, and is generally less than 100 feet above sea level. It consists of plains representing four marine terraces, which were sea bottoms during four previous high-water levels. They remain in their original, unbroken condition over large areas, but elsewhere have been modified by underground solution and surface erosion. A large part of these lowlands is covered by pine and palmetto forests, interspersed with cypress swamps. Many of these flatwoods areas were converted to citrus groves beginning in the mid 1980s. Although the soils are generally sandy, they require intensive artificial drainage due to flat topography and impermeable subsurface soil layers.

Geology of Southern Florida

Sedimentary rocks underlie all of South Florida. The surface strata include limestones, marls, and calcareous sandstones, which formed during the last ice age (when glaciers existed in the northern USA). Changes in the continental ice caps caused changes in sea level: high seas as ice melted, low seas as ice formed. In the melting stage, sea level rose and the waves cut bluffs. The waves, washing against the bluffs, formed terraces on the shorelines. When the ice formed again and the waters fell, the old shorelines were left far inland. Marine sands, marls, and shells form most of the mineral soils that overlie the harder rock strata. Organic soils also exist directly on top of hard limestone. As the sea levels changed, the land level changed very little. The marine sands left by these seas form the main sandy soils of Florida. The junction zones between the terraces are not often easily recognized escarpments. These zones are either hills and ridges of sand, or elongated depressions (sloughs, ponds, cypress swamps, or marshes).

Development of Soils

The deep sands of central Florida and the thinner sand, marl, and organic soils of southern Florida have had little time for development of characteristic profiles. Many of the subsoil layers, particularly the marls, are parts of the original geological formation. Most Florida citrus groves have been planted on sandy soils, but a few have been planted on marl. Almost none exist on organic soils.

Florida's sandy soils developed directly from marine sedimentation, while organic soils developed from decaying swamp and marsh plants. The flooded conditions prevented complete decomposition of these soils. As the seas receded, the layer of sand or marl became modified to form the present-day soils. Some soils developed on top of other sand deposits, some directly on top of rock. Dunes of the Central Florida sandhills region formed through wind action. Near Florida's southern tip, coverage of the limestone rock by marine sands was incomplete; thus, large areas of rock land were left without any soil cover after the last seas receded.

Physical Properties
Soil Texture
The texture of most Florida surface soils is dominated by sand. Texture is defined as the percentage by weight of sand (>0.02 mm diameter), silt (0.005 to 0.02 mm), and clay (<0.005 mm) in a soil. Textural classes describe a soil's particle size distribution. Broad textural groups include:

coarse-textured soils - sand and loamy sand;

moderately coarse-textured soils - sandy loam and fine sandy loam;

medium-textured soils - loam, silt loam, silt, and very fine sandy loam;

moderately fine-textured soils - clay loam, sandy clay loam, and silty clay loam;

fine-textured soils - clay, sandy clay, and silty clay.

With training and experience, texture can be determined in the field by feel. Texture is determined by rubbing the moist soil between the fingers and thumb and is commonly determined in the field. It is a subjective technique but can be mastered with some experience. Since it is a simple technique, it has an advantage over particle size determination,

which is a long, tedious process. Coarse sand particles can clearly be seen with the naked eye and have a gritty feel. Fine sand particles are less obvious, but also have a slight to moderately gritty feel. Silt particles cannot be individually detected with the naked eye, but can be seen with a 10X lens. They have a distinctive smooth silky feel, both wet and dry, but are only very slightly plastic. Silt-sized particles feel smooth like flour, are slightly plastic, and are barely visible. Clay-sized particles feel slick and sticky when wet. Individual clay particles cannot be seen with the naked eye. Clay soils are distinctly sticky and plastic when wet, but are usually compact and hard when dry.

Although different soils may have the same texture, they may not have the same particle size distribution. This is due mainly to variations in the amounts of organic matter, type of clay, shape of particles and degree of aggregation. It is usual for the upper horizons to contain varying amounts of organic matter. Large amounts of organic matter cause the soil to be smooth and to appear to have a higher silt content.

The shape of the particles can also be important. For example, when the sand grains are round, the grittiness will appear to be less than if they are angular. In some soils, the clay particles are cemented to form small aggregates that cause the soil to appear to have a higher silt content. Thus, a soil that has the particle size of a clay has the feel of a silty clay. The properties of the texture classes are as follows:

Sand:
extremely gritty, not smooth, not sticky or plastic, noncohesive balls that collapse easily

Loamy sand:
extremely gritty, not smooth, not sticky or plastic, slightly cohesive balls; does not form threads

Sandy loam:
very gritty, not sticky or plastic, slightly cohesive balls; does not form threads

Loam:
moderately gritty, slightly smooth, slightly sticky and plastic, moderately cohesive balls; forms threads with great difficulty

Sandy clay loam:
moderately gritty, not smooth, moderately sticky and plastic, moderately cohesive balls; forms long threads that will bend into rings with difficulty, moderate polish

Sandy clay:
moderately gritty, not smooth, very sticky and plastic, very cohesive balls; forms long threads that bend into rings with difficulty, high degree of polish

Clay loam:
slightly to moderately gritty, slightly smooth, moderately sticky and plastic, very cohesive balls; forms threads that will bend into rings, moderate polish

Silt loam:
not gritty to slightly gritty, very smooth and silky, slightly sticky and plastic, moderately cohesive balls; forms threads with great difficulty that have a broken appearance, no polish

Silt:
not gritty to slightly gritty, extremely smooth and silky, very slightly sticky and plastic, moderately cohesive balls; forms threads with difficulty that have a broken appearance, slight degree of polish

Silty clay loam:
not gritty to slightly gritty, moderately smooth and silky, moderately sticky and plastic, moderately cohesive balls; forms threads that will not bend into rings, moderate polish

Silty clay:
not gritty to slightly gritty, moderately smooth and silky, very sticky and plastic, very cohesive balls and long threads that bend into rings, high degree of polish

Clay:
not gritty to slightly gritty, not smooth, extremely sticky and plastic, extremely cohesive balls and

long threads that bend into rings easily, high degree of polish

Sandy soils generally have a low water and nutrient-holding capacity. Clays have a good nutrient- and water-holding capacity, but they may become waterlogged. The very silty soils tend to erode very easily. The most desirable soils for cultivation are the loams, particularly those containing about 5% to 10% organic matter, but these do not exist in Florida.

Soil Structure

When soil particles bind together into identifiable shapes, they form soil aggregates. The type of aggregate formed determines soil structure. Several different structures result from combinations of root penetration, wetting and drying, freezing and thawing, inorganic salts, and organic matter content. Commonly recognized structural types are spheroidal, platelike, blocklike, and prismlike (Fig. 14-1). Soils can also be structureless; particles can either exist in a single-grained state (as exemplified by sand where the single particles function separately) or in a massive state (large, irregular, and featureless). In a coarse-textured (sandy) soil, most of the particles may function separately, but the soil may also have some granular structure where two or more particles are bound into aggregates. Such a soil is loose and open with large individual

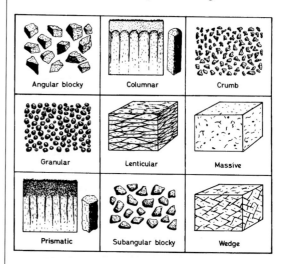

Figure 14-1. Examples of various types of soil structure.

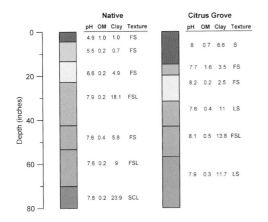

Figure 14-2. Example of differences in the profile of a Riviera fine sand series soil in an undisturbed location (Native) and that after citrus bed forming operations (Citrus Grove).

pore spaces that allow easy movement of air and water.

A granular structure can be developed in a properly managed, fine-textured soil. The individual granules (aggregates) are made up of many individual soil particles that are bound together. Organic matter plays an important role in adding stability to these aggregates. Such a soil may be loose, open, and friable, provided granules of the proper size and nature are developed and maintained. Improper management of such a soil often causes the granules to break down, resulting in a tight, less open condition. Favorable structure in a fine-textured soil is essential for satisfactory movement of water and air.

Medium-textured soils, such as loam, are likely to exhibit a combination of single-grained and granular structures. Such a soil may have some sand particles that act independently, but also have sand, silt, and clay particles bound into aggregates. This combination results in some large pores that allow easy movement of water and air and many smaller pores that retain water effectively.

Clay is the dominant binding agent in fine-textured soils. Organic matter modifies the effect of the clay as a binding agent because higher percentages of organic matter reduce the effect of the clay as a

binding agent. Since the pores in fine-textured soils are very small, they tend to hold water tightly, so these soils may have less water available for plants than medium-textured soils.

Some flatwoods soils may have significant change in profile properties when beds are formed. In sandy soils where the clay or spodic is below the depth of cut in the water furrow (i.e. Pineda and Immokalee series), changes in soil properties are related to the altered position of the topsoil and organic matter in surface horizon. In soils where substantial amounts of clay or spodic material is moved into the bed during bed formation, the changes in soil properties may be significant (Fig. 14-2). Subsoil moved to bed tops may increase soil water holding capacity and cation exchange capacity, but it also may result in lower drainage rates, increased soil pH, and reduced organic matter in surface layers.

Consistency

When soils are manipulated between the fingers and thumb, they exert varying degrees of resistance to disruption and deformation as determined by their mechanical composition, degree of aggregation, content of organic matter and moisture content. Generally, the pressure needed to disrupt a dry soil increases with the content of fine material. In general, sands are usually quite loose while clays form very hard aggregates. Moist sands have a small measure of coherence, whereas moist clays are plastic and become very sticky when wet. The presence of large amounts of organic matter in the soil is particularly important, for it increases the plasticity of sandy soils, but has the reverse effect on a clay by reducing the stickiness.

The consistence of soils of medium texture does not change very much with variations in moisture content. In either the dry or moist state, they are usually friable, i.e., firm with well-formed aggregates that crumble easily when pressure is exerted on them. When wet, these soils tend to be slightly sticky, but never to the same extent as clays. Some horizons that are massive and hard offer a considerable degree of resistance to disruption as a result of cementation by substances such as iron oxides and calcium carbonate. Resistance to disruption can result also from physical compaction. The consistence of most materials alters with a change in moisture status. For example, some clays may be hard when dry, plastic when moist, and sticky when wet.

Color

A very high proportion of the names of soils is based upon color, since this is the most conspicuous property and sometimes the only one that is easily remembered. Many inferences made about soils are based upon color. The color of the soil is usually determined by the nature of the fine material. Generally the color of a soil is determined by the amount and state of iron and/or organic matter. Grey, olive, and blue colors occur in wet soils and originate through the presence of iron in the reduced or ferrous state.

The color of the upper horizons usually changes from brown to dark brown to black as the organic matter content increases. Dark colors are produced also through the presence of manganese dioxide, or may be caused by elemental carbon following burning. Pale grey and white originate through the lack of alteration of light-colored parent materials, deposition of calcium carbonate, and efflorescence of salts. Pale colors also originate by the removal of coloring substances to form the distinctive leached horizons.

Some horizons have a color pattern that may be mottled, streaked, spotted, variegated, or tongued. Possibly the most common and important color pattern is yellow and brown mottles on a grey background, which is interpreted as resulting from seasonal wetting and drying of the horizon.

Classes of Soil Water

Water is held in a soil both as a film coating on the soil particles or aggregates, and in the pore space between individual particles or aggregates. As a soil is irrigated, the pore spaces fill with water. As the pores are filled, water moves through the soil in response to gravity and capillary forces. Gravity

moves water downward only, while capillary force (tension) can move water in any direction. The capillary force originates from the adhesion of water molecules to the pore walls and the attraction of water molecules to one another. Water move-

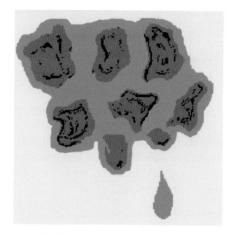

Figure 14-3. Gravitational water.

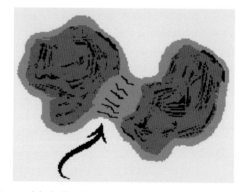

Figure 14-4. Capillary water.

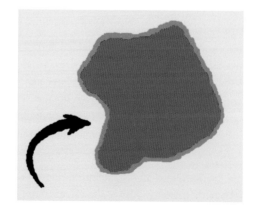

Figure 14-5. Hygroscopic water.

ment continues until there is a balance between the upward-acting capillary forces and the downward-acting gravity force. At this time, the entire wetted portion of the soil is at about the same moisture tension.

Soil water can be divided into gravitational (Fig. 14-3), capillary (Fig. 14-4), and hygroscopic (Fig. 14-5) classes. Gravitational water is the water subject to movement by gravity. This water drains downward; plants take up little of it because of the short period they have access to it. Capillary water is held by the soil in pore spaces and is the most important for crop production. The capillary and gravitational water that can be used by plants is termed "available" water. Hygroscopic water is held so tightly to individual soil particles that almost none is available for plant use.

In a medium-textured soil, the soil is usually at the upper limit of its available water range about 24 to 48 hours after irrigation. The water content of a well-drained soil at the upper limit of the available water range is called field capacity (Fig. 14-7). The lower limit of the available water range is called the wilting point. Water available for plant growth exists in the range between field capacity and wilting point in nonsaline soils.

Soil Water Storage

Water storage characteristics of a soil are very important in irrigation. The volume of water a soil can store for plant use (its available water capacity) is determined by the size and arrangement of the soil particles, organic matter, and the depth of soil considered (Table 14-1). The volumes of air and water within a soil can vary significantly, depending on soil moisture status (Fig. 14-7, Table 14-2). The volume of available water that a soil holds is important when irrigating because it determines the amount of water that should be applied at each irrigation, and it influences the time interval between irrigations.

Soil Water Status

The state of soil water is described in terms of its amount in the soil and the energy associated with

the forces that hold it there. The amount of water is defined by water content and the energy state of water is the water potential. Plant growth, soil temperature, chemical transport, and groundwater recharge all depend on the state of soil water. While

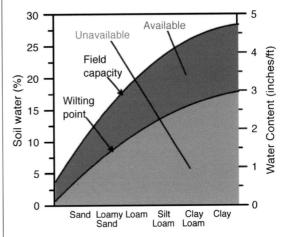

Figure 14-6. Water content ranges for soils of different textual class (Kramer and Boyer, 1995).

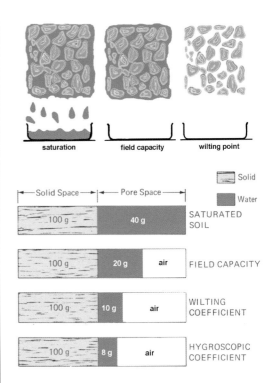

Figure 14-7. Volumes of solids, water, and air for a loam soil at various moisture levels (Brady, 1974).

Table 14-1. Typical amounts of available water by soil texture (in inches per inch of soil). The ranges of available water capacities are the adjusted water retention difference between 1/3-bar and 15-bar tension for the medium- and fine-textured soils, and between 1/10-bar and 15-bar tension for the moderately coarse and coarse textures).

Soil Texture	Surface (0-12 in.)	Subsoil (12-36 in.)	Lower horizons (36-60 in.)
	(in./in.)	(in./in.)	(in./in.)
Coarse sand and gravel	0.04-0.06	0.03-0.05	0.02-0.04
Sands	0.07-0.09	0.06-0.08	0.05-0.07
Fine sands	0.06-0.12	0.06-0.11	0.05-0.09
Loamy sands	0.10-0.12	0.09-0.11	0.08-0.10
Loamy fine sand	0.10-0.12	0.10-0.13	0.08-0.12
Sandy loams	0.13-0.15	0.12-0.14	0.11-0.13
Fine sandy looms	0.16-0.18	0.15-0.17	0.14-0.16
Loams/very fine sandy loams	0.20-0.22	0.17-0.19	0.17-0.19
Silt loams	0.22-0.24	0.20-0.22	0.20-0.22
Silty clay loams	0.18-0.23	0.16-0.20	0.16-0.20
Sandy clay loams	0.18-0.20	0.16-0.18	0.15-0.17
Clay loams	0.17-0.19	0.15-0.19	0.14-0.16
Silty clays	0.15-0.18	0.14-0.17	0.13-0.15
Clays	0.15-0.18	0.14-0.17	0.13-0.15

there is a unique relationship between water content and water potential for a particular soil, these physical properties describe the state of the water in soil in a distinctly different manner. It is important to understand the distinction when choosing a soil water measuring instrument.

Water content

Soil water content is expressed on a gravimetric or volumetric basis. Gravimetric water content (Θ_g) is the mass of water per mass of dry soil. It is measured by weighing a wet soil sample (m_{wet}), drying the sample to remove the water, then weighing the dried soil (m_{dry}).

$$\Theta_g = \frac{M_{water}}{M_{soil}} = \frac{m_{wet} - m_{dry}}{m_{dry}} \qquad \textbf{Eq. 14-1}$$

where,

Θ_g = gravimetric water content
m_{water} = mass of water in sample (g)
m_{soil} = mass of soil in sample (g)
m_{wet} = mass of wet soil (g)
m_{dry} = mass of dried soil (g)

Volumetric water content (Θ_v) is the volume of water per volume of soil. Volume is the ratio of mass to density (P), which gives:

$$\Theta_v = \frac{volume_{water}}{volume_{soil}} = \frac{\dfrac{M_{water}}{P_{water}}}{\dfrac{M_{soil}}{P_{soil}}} = \frac{\Theta_g * P_{soil}}{P_{water}} \qquad \textbf{Eq. 14-2}$$

Table 14-2. Typical soil moisture contents (percent moisture by volume) by texture and soil tension status.

Soil Texture	Field capacity (%)	Wilting point (%)	Available moisture (%)
Sand	9	2	7
Loamy Sand	14	4	10
Sandy Loam	23	9	14
Sandy Loam with Organic matter	29	10	19
Loam	34	12	22
Clay Loam	30	16	14
Clay	38	24	14

where,

Θ_v = Volumetric water content
volume$_{water}$ = volume of water in sample (cm^3)
volume$_{soil}$ = volume of soil in sample (cm^3)
P_{soil} = Soil bulk density (g/cm^3)
P_{water} = Density of water (normally 1.0 g/cm^3)

Soil bulk density (P_{bulk}) is used for P_{soil} and is the ratio of soil dry mass to sample volume. The density of water is close to 1 and is usually ignored. Another useful property, soil porosity (E), is related to soil bulk density as shown by the following expression.

$$\varepsilon = 1 - \frac{P_{bulk}}{P_{solid}}$$ **Eq. 14-3**

where,

E = soil porosity
P_{solid} = density of the soil solid fraction, which is approximately 2.6 g/cm^3.

Example:
A sample of known volume was weighed before and after oven drying at 105°C for 24 hours.

m_{wet} = 94 g
m_{dry} = 78 g
sample volume = 60 cm^3

Using Eq. 14-1,

$$\Theta_g = \frac{m_{water}}{m_{soild}} = \frac{94g - 78g}{78g} = 0.205 = 20.5\%$$

$$P_{bulk} = \frac{m_{dry}}{m_{soild}} = \frac{78g}{60\ cm^3} = \frac{1.3g}{cm^3}$$

$$\Theta_v = \frac{\Theta_g * P_{soil}}{P_{water}} = \frac{0.267 cm^3}{cm^3} = 26.7\%$$

$$\varepsilon = 1 - \frac{P_{bulk}}{P_{solid}} = 1 - \frac{\dfrac{1.3g}{cm^3}}{\dfrac{2.6g}{cm^3}} = 0.50 = 50\%$$

The porosity of 0.50 defines the maximum possible volumetric water content. The measured Θv value of 0.267 indicates the pore space is just over half-full of water. If the sample is from a 12-inch (30-cm) depth profile, there are 3.2 inches (8 cm) of water in the 12-inch profile. Water content indicates how much water is present in the soil. It can be used to estimate the amount of stored water in a profile or how much irrigation is required to reach a desired amount of water.

Water potential
Water flux (the movement of water) occurs within the soil profile, between the soil and plant roots, and between the soil and the atmosphere. As in all natural systems, movement of a substance depends on energy gradients. Soil water potential is an expression of the energy state of water in soil and must be known or estimated to describe water flux.

Water molecules in a soil matrix are subject to numerous forces. If no adhesive forces were present, the water molecules would move through the soil at the same velocity as in free air (minus delays from collisions with the solid matter) much like sand moves through a sieve. Soil water potential accounts for adhesive and cohesive forces and describes the energy status of soil water.

The fundamental forces acting on soil water are gravitational, matric, and osmotic. Water molecules have energy by virtue of position in the gravitational force field, just as all matter has potential energy. This energy component is described by the gravitational potential component of the total water potential. The influence of gravitational potential is easily seen when attractive forces between water and soil are less than the gravitational forces acting on the water molecule and water moves downward.

The matrix arrangement of soil particles results in capillary and electrostatic forces and determines the soil water matric potential. The magnitude of the forces depends on texture and the physical-chemical properties of the soil solids. Most methods for measuring soil water potential are sensitive only to the matric potential.

Soil water is a solution. The polar nature of the water molecule results in interaction with other electrostatic poles present in the solution as free ions. This component of the energy status is the osmotic potential. Methods for measuring soil water matric potential include tensiometry, thermocouple psychrometry, electrical conduction (generically Buoyucous blocks), and heat dissipation.

Soil Water Characteristics

Water movement in the soil is necessary for sustaining plant life and for removing surplus water that might be harmful. The infiltration rate (the rate at which a soil allows water to enter) is important because it influences the rate at which water should be applied. Water movement takes place through pore spaces and depends on the size, number, and continuity of pores. Texture, structure, organic matter, and chemical properties of the soil determine these characteristics. Soil drainage is important to remove excess water from the root zone so it does not become depleted of oxygen.

The force that holds water in the soil is the most important soil water characteristic for a growing plant. This force, known as suction force or moisture tension, is expressed as the energy required to remove the water from the soil. Moisture tension is usually expressed in terms of atmospheres of pressure or bars. An atmosphere is the average air pressure at sea level expressed in pounds per square inch (1 Atmosphere = 14.72 pounds per square inch = 1.014 bars). As soil moisture tension values increase, the energy with which water is held by the soil increases. Plants can extract water from a nonsaline soil at tensions from 0 to about 15 atmospheres. By definition, at this tension level the soil water has reached the permanent wilting point. The higher the soil tension, the harder it is for the plant to extract water from the soil.

Soil Water Characteristic Curve

There is a unique relationship between water content and water potential for each soil that is referred to as the soil water characteristic curve. These curves for four flatwoods soils and two Ridge soils are depicted in Figs. 14-8, 14-9, and 14-10. For a given water potential [presented in centibars (cb), which equal 1/100 of a bar], the finer the soil texture, the more water is held in the soil. Coarse-textured soils like sand consist mostly of large pores that empty when a relatively small force is applied. Fine-textured soils have a broader pore size distribution and larger particle surface area. Consequently, a larger change in water potential is required to remove the same amount of water. Greater soil particle surface area, which results from an increased proportion of fine particles in the soil, means that more water is adsorbed via electrostatic forces.

Capillary Rise from Water Table (Upward Flux)

Capillarity is caused by the attractive force of water to the walls of the channels through which it

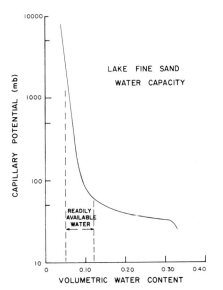

Figure 14-8. Soil water characteristic curve for Lake Fine Sand series soil (Ridge soil).

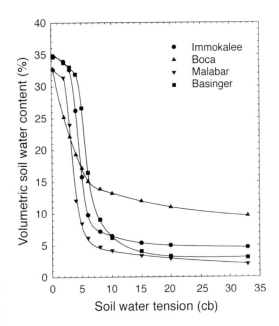

Figure 14-10. Soil water characteristic curve for four flatwoods soil series.

moves (adhesion) and the surface tension of water, which is largely due to the attraction of water molecules to each other (cohesion). Capillarity can be demonstrated by placing one end of a narrow tube in a glass of water. The water rises in the tube, and the smaller the tube bore, the higher the water rises.

Capillary forces are at work in all moist soils. However, the rate of movement and the rise in height are less than one would expect on the basis of soil pore size. One reason is that soil pores are not straight like glass tubes, but are crooked (tortuous). Also, some soil pores are filled with air, which may be entrapped, slowing down or preventing the movement of water by capillarity.

Upward flow of water by capillary forces from a water table can be an important water source to meet crop evapotranspiration (ET) requirements. This is in fact the principal basis of "seepage" or subirrigation. A water table near the root zone, which occurs in many Florida flatwoods locations, can supply a significant portion of the water to meet ET. If the water table, however, is too close to the root zone, a lack of aeration may develop that

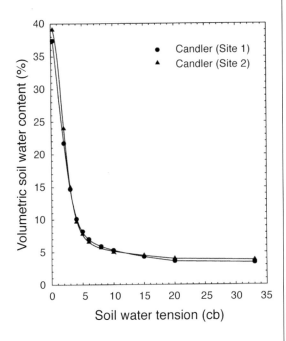

Figure 14-9. Soil water characteristic curve for Candler series soil (Ridge soil).

will limit production. The rate of water movement from a water table depends on soil characteristics and water potential gradients (differences). Irrigation management should take this contribution into account.

Nutrient Movement

Water is the key to the three mechanisms by which nutrients come in contact with plant roots: root interception, mass flow, and diffusion. The importance of each of these mechanisms differs widely depending on the nutrient. Likewise, the level of soil moisture affects each of these mechanisms differently.

As roots grow through soil pore spaces, they intercept nutrients in their path. The nutrients they contact may be present as soluble ions in the soil solution, or ions adsorbed to negatively-charged soil particles (clay or organic matter). Nitrate and sulfate ions are examples of the former, and exchangeable calcium, magnesium, potassium, and hydrogen are examples of the latter. The amount of interception depends on the volume of roots growing in the soil pore space.

Trees absorb large quantities of water and transpire most of it from their leaf surfaces. The water moves through the soil to the root surface where it is absorbed and moved up into the tree. This soil water contains nutrients such as nitrate, sulfate, calcium, magnesium, potassium, phosphorus, and small concentrations of micronutrients. In sandy Florida soils, many of these nutrients are transported to the root surface by mass flow of water. The significance of mass flow in transporting nutrients to roots depends on the relative mobility of a given nutrient in the soil, its concentration in soil water, the amount of water used by the plant, and the nutrient needs of the plant. Anything that changes these factors will change the importance of mass flow as a supplier of nutrients. Water is the carrier of the nutrients moving by this mechanism.

Plant roots may not receive enough of a nutrient by the combined processes of root interception and mass flow. When they don't, they absorb nutrients adjacent to their surface and thus reduce nutrient concentration between the soil at the root surface and the bulk soil. Nutrients then move slowly by diffusion along this concentration gradient toward the root, because nutrients move from areas of higher concentration to areas of lower concentration. All nutrients may move by any of the three mechanisms, if the appropriate conditions prevail.

Nutrients that are not held by clays or organic matter, like nitrate or sulfate, will reach roots by mass flow. If more nutrient is transported to the root than it can absorb, there is an accumulation near the root, or the nutrient may leach below the root zone. In a moist soil, nitrate moves to roots almost exclusively by mass flow; thus, in a dry soil it will be unavailable. Frequently the top inch or two of soil is dry. When this occurs, mass flow will not move nitrate or other soluble nutrients to roots. Obviously, water is very important for transporting plant nutrients to roots.

Figure 14-11. Example of shallow rooting of citrus planted on flatwood soils.

Mass flow can typically supply citrus with adequate nitrate, potassium, and sulfate, while calcium and magnesium will be supplied mainly by root interception. However, phosphorus is not supplied to roots by mass flow, and is only partially supplied by root interception. Diffusion in the soil solution is the principal transport means for phosphorus. Diffusion becomes important when root interception and mass flow do not adequately supply nutrients. The rate of diffusion is greater when more water is present, which makes phosphorus more available. Water is a key factor in all three mechanisms of nutrient uptake because roots intercept more nutrient ions when growing in a wetter soil than when growing in a drier one.

Rooting Depth
The root systems of citrus trees in deep, well-drained soils are 4 to 5 feet, with the main root system reaching only a depth of 2 to 3 feet. Studies have found roots on trees with rough lemon rootstock at a depth of 11 feet, while roots on sour orange trees have been found at a depth of 9 feet in well-drained sandy Florida Ridge soils. Water depletion studies confirmed that the roots were active at this depth but most of the water extraction came from the top 2 feet. Citrus can also survive in very shallow soils, especially when the limiting factor is a high water table. Citrus trees grown in the coastal flatwoods areas of Florida typically have rooting depths of only 12 to 18 inches as a result of shallow water table and hardpan layers (Figs. 14-11 and 14-13).

Soil Series
Example profiles of the most common citrus soils are presented in Figs. 14-14 through 14-26.

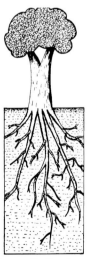  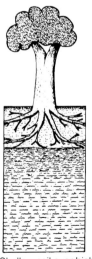 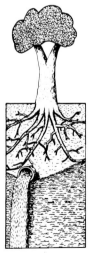

| Deep soil with adequate rooting depth and water holding capacity. | Shallow soil over rock. Limited rooting depth and low water holding capacity. | Shallow soil over high water table. Limited rooting depth and excessive water in root zone. | Shallow root zone increased by the use of tile drains. |

Figure 14-12. Effects of soil and water table on citrus rooting depth (Fitzpatrick, 1974).

Figure 14-13. Spodic horizon in Oldsmar Fine Sand series soil.

Figure 14-15. Bassinger.

Figure 14-14. Astatula.

Figure 14-16. Candler.

Figure 14-17. Felda.

Figure 14-19. Immokalee.

Figure 14-18. Holopaw.

Figure 14-20. Myakka.

Figure 14-21. Oldsmar.

Figure 14-23. Riviera.

Figure 14-22. Pineda.

Figure 14-24. Tavares.

Figure 14-25. Wabasso.

Figure 14-26. Winder.

Chapter 15. Soil Water Measuring Devices

by Brian Boman and Larry Parsons

Measurement of Soil Water

Good irrigation management requires that the status of soil water be accurately evaluated. There are direct and indirect methods to measure soil water content, and several alternative ways to express it quantitatively. There is no universally recognized standard method of measurement and no uniform way to compute and present the results.

Although there are many ways the status of soil water can be expressed, all can basically be classified into one of two categories: the amount of water (water content) held in a given amount of soil or the potential (tension) with which the water is held by the soil. These two properties describing the soil water status are related to each other throughout the entire soil water content range. This relationship is important because it describes the ability of a soil to hold water available for plant growth and the forces with which it is held by the soil as it is depleted. The amount of water that is held by a certain mass or volume of soil can be expressed as a percentage by weight or percentage by volume. Soil water content as a percentage by weight (gravimetric water content) is based on the dry weight of the sample and can be calculated as follows (see Eq.14-1):

$$\frac{\text{weight of water}}{\text{weight of oven-dry soil}} \times 100\%$$

Soil water content as a percentage by volume (volumetric water content) is based on the volume of the sample and can be calculated as follows (see Eq.14-2):

$$\frac{\text{volume of water}}{\text{bulk soil volume}} \times 100\%$$

The volume percentage is a more useful measure to the irrigator because it readily allows the amount of water available to a crop to be calculated. For example, a soil with a water content of 10% by volume contains 10% x 12 in./ft = 1.2 inches of water per foot of soil.

The water potential, or tension with which water is held by the soil, is a function of water content. Water molecules are attracted to soil particles (adhesive forces) and to other water molecules (cohesive forces) so that they are not readily removed from the soil by plants. Rather, the attractive forces between the soil and water must be overcome by the plant in order to obtain water for its use. As soil water is used by plants, that most readily available is removed first. Each succeeding increment becomes less readily available as water is first depleted from the large soil pores and is held more tightly by the smaller soil pores.

Water potential or tension is measured in a variety of units. Common units of measurement of water potential are the bar (1 bar = 14.5 psi = 0.986 atmospheres) or atmosphere (1 atmosphere = 14.72 psi = 1.014 bars). Subunits commonly used are centibars (cbar) and millibars (mb). One millibar is approximately equal to the pressure required to support a column of water at a height of one centimeter (0.39 inches). Other units that are commonly used are megapascals (MPa) or kilopascals (kPa). One bar = 0.1 MPa and 1 cbar = 1 kPa.

Water potentials are negative to reflect the fact that force must be exerted to extract water from soils. However, because it is implicit, the negative sign is often omitted. Water is available to plants at tensions from near zero to as much as 15 bars. However, water is readily available to plants (no growth reduction) over a much narrower range of tension. Commonly, irrigations are scheduled at tensions of less than one bar. Because of limited available water in sandy soils, they are usually irrigated when soil water tensions reach 10 to 30 cbar. Several methods can be used to provide an indication of the volume of water contained in a volume of soil. The most popular include gravimetric sampling, tensiometers, and neutron scattering.

Soil Water Measurement

Proper irrigation scheduling depends on many factors such as crop, climate, soil, crop age, cultural practices, water table depth and water quality. There are two basic approaches to irrigation scheduling: monitoring soil water status or maintaining a water budget. Often a combination of both approaches is used. The soil water content is the amount of water contained by soil on a weight or volume basis. Many soil water sensors have been developed for measuring soil water content and have advantages and disadvantages.

Several electromagnetic type devices are available. These devices typically measure an electrical property of the soil that is related to soil water content. In some cases, the device itself is affected by the surrounding soil water content. Electromagnetic sensors generally rely on an electronic circuit of known behavior that is influenced by the soil water content. Since the electrical properties of soils vary, calibration is required to relate the measured electrical response to the water content of the specific soil.

Feel and Appearance Method

The quickest and simplest way to estimate soil moisture in the field is the feel and appearance method. Although estimating moisture conditions by feel and appearance is not the most accurate method, with experience and judgement it is possible to estimate the moisture level to within 10% to 15%. To obtain a soil sample, dig or auger into the soil and examine samples in 6-inch increments of depth. Squeeze a handful of the soil sample very firmly and compare its appearance and feel with the description given in Figures 15-1 through 15-6,

or in Table 15-1, which describes moisture conditions as a percentage of total available water (TAW) for each soil texture. Samples should be taken from near the tree's dripline in an area watered by the irrigation system. One measurement should be made in the upper quarter of the root zone and additional measurements at lower levels.

Example:
Assume that soil samples were taken from the dripline of a citrus tree at depths of 6, 12, and 18 inches. The soil is a fine sand (Riviera Fine Sand Series). The sample from the 12- and 18-inch depths closely resembles Fig. 15-5, and the 6-inch depth lies somewhere between Figures 15-5 and 15-6. Estimate the soil depletion level in percent and inches for a root zone of 18 inches.

From Fig. 15-5, the TAW depletion would be somewhere between 25% and 50% at the 12- and 18-inch depths—assume it is about 35%. At the 6-inch depth, the depletion would be somewhere around 50% since the sample has characteristics of both figures.

Looking at the moisture depletion table (Table 15-1) for a fine sand with 25% to 50% depletion, we find that it takes 0.30 to 0.65 in./ft to bring the 12- and 18-inch depths back to field capacity. By experience, we think that it will take 0.40 in./ft to refill the 12- and 18-inch depths to field capacity. For the 6-inch depth, Table 15-1 indicates it will take about 0.65 in./ft to raise it to field capacity. However, experience tells us that for this soil it takes slightly less than the table value. We decide on a value of 0.60 in./ft. By multiplying each depth by the required depth to refill it, we can estimate the overall irrigation depth as:

Depth = 0.5 ft x 0.60 in./ft + 0.5 ft x 0.40 in./ft
+ 0.5 ft x 0.40 in./ft
= 0.80 inches

Table 15-1. Guide for judging how much moisture is available in soil by feel method. Ball is formed by squeezing a handful of soil very firmly.

Depletion of total available soil moisture (TAW) (%.)	Feel or appearance of soil			
	Medium sand	Fine sand	Loam	Silt loam and clay
75%-100%	Dry, loose, single grained, flows through fingers	Dry, loose, flows through fingers	Powdery dry, sometimes slightly crusted but easily broken down into powdery conditions	Hard, baked, cracked, sometimes has loose crumbs on surface
Depletion (in./ft)	0.70-0.90	1.0-1.3	1.5-2.0	1.6-2.2
50%-75%	Appears to be dry, will not form a ball with pressure	Appears to be dry, will not form a ball	Somewhat crumbly but holds together with pressure	Somewhat pliable, will ball under pressure
Depletion (in./ft)	0.45-0.7	0.65-1.0	1.0-1.5	1.1-1.6
25%-50%	Appears to be dry, will not form a ball with pressure	Tends to ball under pressure but seldom holds together	Forms a slightly plastic ball that will sometimes slick a little with pressure	Forms a ball, ribbons out between thumb and forefinger
Depletion (in./ft)	0.20-0.45	0.30-0.65	0.50-1.0	0.55-1.1
0% (field capacity)	Tends to stick together slightly, sometimes forms a very weak ball under pressure	Forms weak ball, breaks easily, will not slick	Forms a ball, is very pliable, slicks readily if relatively high in clay	Easily ribbons out between fingers, has slick feeling
Depletion (in./ft)	0.0-0.20	0.0-0.30	0.0-0.5	0.0-0.55
At field capacity	Upon squeezing, no free water appears on soil but wet outline of ball is left on hand	Upon squeezing, no free water appears on soil but wet outline of ball is left on hand	Upon squeezing, no free water appears on soil but wet outline of ball is left on hand	Upon squeezing, no free water appears on soil but wet outline of ball is left on hand

Fine sands

Characteristics:
Sample is moist to wet and it forms a flat mass with slightly defined finger marks when squeezed. A very light coating of loose and aggregated sand grains adheres to fingers. Sometimes moisture is visible and can be felt by touching the sample.

Figure 15-1. Characteristics of a fine sand with soil moisture content in the range of 0% to 25% depletion of TAW.

Characteristics:
Sample is moist and a weak ball is formed with defined finger marks when squeezed. A light coating of loose and aggregated sand grains adheres to fingers.

Figure 15-2. Characteristics of a fine sand with soil moisture content in the range of 25% to 50% depletion of TAW.

Characteristics:
Sample is slightly moist and a very weak ball is formed that will crumble easily. Some loose and aggregated sand grains will adhere to fingers. Sometimes moisture is visible and can be felt by touching the sample.

Figure 15-3. Characteristics of a fine sand with soil moisture content in the range of 50% to 75% depletion of TAW.

Medium sands

Characteristics:
Sample is moist to wet and forms a weak ball with defined finger marks when squeezed. A light coating of loose and aggregated sand grains will adhere to fingers. Sometimes moisture is visible and can be felt by touching the sample.

Figure 15-4. Characteristics of a medium sand with soil moisture content in the range of 0% to 25% depletion of TAW.

Characteristics:
Sample is moist and a weak ball is formed with defined finger marks when squeezed. A light coating of loose and aggregated sand grains will adhere to fingers.

Figure 15-5. Characteristics of a medium sand with soil moisture content in the range of 25% to 50% depletion of TAW.

Characteristics:
Sample is slightly moist. A very weak ball that will crumble easily is formed when squeezed. Some loose and aggregated sand grains will adhere to fingers.

Figure 15-6. Characteristics of a medium sand with soil moisture content in the range of 50% to 75% depletion of TAW.

Gravimetric

The gravimetric (oven drying) technique is probably the most widely used method for measuring soil water. The method is also the standard for the calibration of most other soil water determination techniques. The gravimetric technique requires removing a soil sample from the field and determining the mass of water contained in the soil in relation to the mass of dry soil. Although the use of this technique ensures accurate measurements, it also has a number of disadvantages. Laboratory equipment, sampling tools, and 24 hours of drying time are required. In addition, it is a destructive test (it requires sample removal). This makes it impossible to measure soil water at exactly the same point at a later date. Thus, measurements vary due to field variability from one site to another.

Samples are collected from the site, weighed, and oven dried at 105° C. The samples are periodically weighed until successive weights are the same (indicating all the water has been removed). The weight difference before and after drying is considered to be water removed. Water contents can be calculated on a weight basis (Eq. 15-1) or on a volume basis (Eq. 15-2) if the soil volume or bulk density is measured when the sample is taken. Sample volume or bulk density is difficult to measure in the field and subject to errors.

The gravimetric method itself is subject to errors because it depends on sampling, transporting, and repeated weighings. It is also laborious and time-consuming, since a period of 24 hours or more is usually required for complete drying. The standard method of oven drying is also arbitrary. Some clays may still contain appreciable amounts of adsorbed water even after drying. Conversely, some organic matter may oxidize and decompose at 105° C so that the weight loss may not be due entirely to the evaporation of water. The errors incurred with the gravimetric method can be reduced by increasing the size and number of samples. However, the sampling method is destructive and may disturb a location sufficiently to distort the results. Thus, it is primarily used to calibrate other soil moisture measurement methods.

Neutron Probe

Neutron scattering (Fig. 15-7) can be used to measure volumetric water content at several depths within the soil profile. The neutron probe operates on the principle of nuclear thermalization. Fast neutrons are emitted from a radiation source and these neutrons lose energy (and velocity) and become slow neutrons as they collide with hydrogen ions. The slow neutrons are counted by a detector on the instrument. The number of slow neutrons is an indirect measure of the quantity of water in the soil because of the hydrogen ions present in the water molecule (H^+, OH^-). Because only hydrogen ions (and not water molecules) are counted, the neutron probe should be calibrated for each specific soil type on which it will be used. This is especially true if the quantity of organic matter is variable between soil types, because readings will reflect the large number of hydrogen ions present in the organic matter.

Figure 15-7. Typical access tube location and zone for soil moisture measurements taken with a neutron probe.

A spherical volume of soil is sampled as neutrons are emitted in all directions and counted as they are reflected back to the detector. The radius of that volume varies from only a few inches in a wet soil to a foot or more in a very dry soil. Access tubes (typically aluminum or steel pipe with 1 1/2- to 2-inch diameter) are required for each soil moisture measurement site. Soil water readings are normally taken at several depths, with 0.5 to 2 minutes per reading required. Inaccurate soil water measurements may result if there is significant salinity in the soil.

This method offers the advantage of measuring a large soil volume, and also has the possibility of scanning several depths to obtain a profile of water distribution. However, it also has a number of disadvantages: the high cost of the instrument ($3,000 to $5,000), a radiation hazard and licensing requirement, insensitivity near the soil surface, insensitivity to small variations in water content at different points within a 12- to 15-inch radius, cost and complexities in calibration, and variation in readings due to soil density variations that may cause an error of up to 15%. Therefore, the neutron meter is used more often by researchers than by individuals in commercial agricultural enterprises.

Tensiometers

Tensiometers are devices that indicate the water status of a soil. They do not directly measure the amount of water in the soil, but measure the energy a plant must exert to extract water from the soil. The amount of energy a plant exerts to extract water is directly related to the soil water content. A tensiometer (Fig. 15-8) consists of a porous ceramic tip sealed to the base of a water-filled tube, usually plastic, which is sealed at the top with a removable airtight cap and a vacuum gauge connected to the upper part of the tube. In use, the porous tip and body are filled with deaerated water (water is deaerated by boiling it).

Soil moisture tension (SMT) values are periodically read from the vacuum gauge. Soil moisture tension is a direct measurement of the energy-holding water in the soil. When the relationships between SMT and available soil water are known, the tensiometer can be used to determine the amount of available water present in the soil at any time and hence can be used to schedule irrigations.

The vacuum gauges available on most commercial tensiometers are calibrated to read from 0 to 100 cbar. Some tensiometers may read in kilopascals (kPa) units, with the conversion of 1 cbar = 1 kPa. The useful range of the tensiometer is from 0 to 80 cbar. At SMT values greater than 80 cbar, air enters the porous tip and the tensiometer no longer functions properly. Most of the water available for plant growth can be detected between gauge readings of 0 and 80 cbar in sandy soils. Less can be measured with a tensiometer in medium-textured soils, while only a small amount can be detected with a tensiometer in fine-textured soils. Therefore, tensiometers are most applicable to coarse- and medium-textured soils.

When a tensiometer is filled with water and properly installed with the porous tip in the root zone of a growing crop, movement of water into or out of the porous tip occurs in order to maintain an

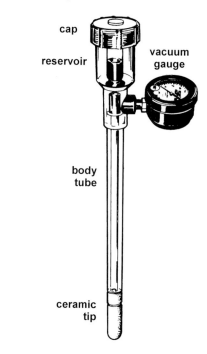

Figure 15-8. Components of typical tensiometer.

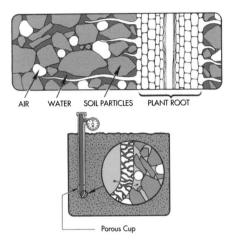

AIR   WATER   SOIL PARTICLES   PLANT ROOT

Porous Cup

Figure 15-9. Cross section of soil showing relationships among roots, soil particles, and soil water (top) compared to tensiometer pores in contact with soil (bottom).

equilibrium between water in the soil and water in the tensiometer (Fig. 15-9). As water moves out of the tip in response to soil moisture tension in the soil, it exerts a tension or vacuum on the vacuum gauge. As the soil dries, water moves from the porous tip into the soil and a higher SMT is registered on the gauge. On the other hand, when irrigation or rain water causes wetting of soil surrounding the tip, water moves from the soil into the tensiometer with a resultant decrease in SMT. Thus, the tensiometer indicates the soil water status for soils as they become drier in response to crop use and as they receive water from rainfall or irrigation.

Tensiometers must be prepared prior to their insertion in the soil. The cap should be removed and the tensiometer tip should be placed in a bucket of water and allowed to soak for at least 24 hours. Water will move through the porous tip and accumulate in the tube. A handheld suction pump (sold by the tensiometer supplier) should be used to remove entrapped air from the porous cup. Soak the tensiometer a few more hours and repeat the process of removing entrapped air. Water used to fill the tensiometer should contain an algicide (available from the manufacture or at waterbed suppliers). Fill the hollow tensiometer tube with water and seal the cap firmly. The tensiometer is now ready for insertion into the soil.

Installation of tensiometers in the soil is a simple process requiring only a hammer and a piece of steel rod having the same diameter as the body of the tensiometer. The rod, which should have a tapered point, is driven to the desired depth and extracted. In coarse-, moderately coarse-, and medium-textured soil, pounding the tapered rod into and extracting the rod from the soil is easily done. An alternative method of installation is to make the tensiometer hole using a soil sampling probe (coring tools are sold by tensiometer manufacturers) or a soil auger of the required diameter. After inserting the body of the tensiometer into the hole, the tip should be forced gently into the soil at the base of the hole.

In order to establish good contact between the porous tip and the soil (good contact must be established for the tensiometer to function properly), pour about one-fourth cup of water down the hole prior to inserting the tensiometer. The bottom of the gauge should be about one inch above the soil surface. Get a snug fit between the body of the tensiometer and the soil surface by compacting the soil around the tensiometer so that water will not run down along the body of the tensiometer. Minimal disturbance of the soil adjacent to the tensiometer is essential. Allow the tensiometer to remain overnight before taking the first reading.

Service each tensiometer frequently when air bubbles accumulate in the tensiometer tube. This involves recording SMT, removing the cap, adding deaerated water to refill the instrument, removing entrapped air bubbles with the handheld suction pump, and replacing the airtight cap. If the gauge indicates an unexpectedly low (or zero) reading following a previous high gauge reading, the tension is probably broken (a reading over about 80 cbar). This erratic behavior can be corrected by servicing the tensiometer after the soil is wetted by irrigation or rain. Servicing the tensiometer before irrigating the soil will be of no benefit because the SMT is so great that the tension will break the water column again.

Freezing temperatures can damage tensiometers, with the vacuum gauge being the part that is usually damaged. Unfortunately, it is also the most expensive part to replace. For this reason, remove tensiometers if freezing weather is imminent.

Tensiometers should be placed at representative sites or stations in a field. Since many groves have several different soil types, a specific recommendation on the number of tensiometers required per block is difficult to determine. More tensiometers per block means a better characterization of the grove soil water status, and a grower should invest in as many tensiometers as he can reasonably afford and maintain. Tensiometers should be placed in the root zone where water is taken up most rapidly. Since citrus roots are more abundant near the soil surface, the upper soil layers usually dry out sooner than the deeper soil layers.

There are various opinions on best depths for tensiometer placement. Tensiometer stations should consist of at least two tensiometers of different lengths, preferably 6 and 12 inches (Fig. 15-10). A third tensiometer in the station can be placed at 18 inches in the flatwoods or 18 to 24 inches in the Ridge to detect dry spots at deeper depths; but the deeper soil will not dry as rapidly as will soil near the surface.

Tensiometers must be placed at stations where they will not be damaged or broken during normal cultural operations. Typically, they should be installed in the tree row near the dripline of the tree.

Tensiometers need to be in an area wetted by the micro-irrigation system to provide meaningful information. For typical Ridge sands, irrigation in the spring should commence at 9 to 10 cbar (Table 15-2). In the fall and winter when water

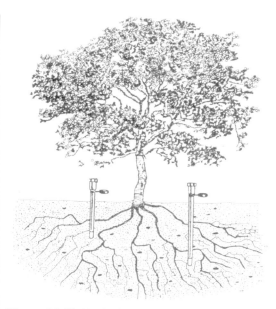

Figure 15-10. Typical location of tensiometers (in tree row) for mature citrus trees.

requirements are less critical, irrigation can start at 15 cbar. On better soils in the flatwoods, springtime irrigation can start at 15 cbar and fall and winter irrigation can start at 30 cbar.

Gamma Attenuation

The gamma ray attenuation method is a radioactive technique that can be used to determine soil water content. This method assumes that the scattering and absorption of gamma rays are related to the density of matter in their path. In addition, it is assumed that the specific gravity of the soil remains relatively constant as the water content increases or decreases. Changes in soil density are measured by the gamma transmission technique, and the water content is determined from this density change. The technique allows quick (<1 minute) measurements of soil volumetric water content.

Table 15-2. Tensiometer readings (cbar) for starting irrigation.

Period	Ridge	Flatwoods
February - June	10	15
July - January	15	20-30

The technique is restricted to soil layer thicknesses of 1 inch or less, but measurements are made with high resolution. Gamma attenuation is expensive and can be affected by changes in soil bulk density. As a result, there can be significant errors when used in highly stratified soils. The method can be used to determine the average water content within the profile. Due to the high cost and need for licensing, gamma attenuation in not generally used in commercial agriculture.

Resistive Sensors

Electromagnetic techniques include methods that depend upon the effect of water on the electrical properties of a soil. Soil resistivity depends on water content. Therefore, it can serve as the basis for a sensor. It is possible either to measure the resistivity between electrodes in a soil or to measure the resistivity of a material in equilibrium with the soil. The difficulty with resistive sensors is that the soil resistivity depends on ion concentration as well as on water concentration.

One of the most common methods of estimating matric potential is with gypsum or porous blocks (Fig. 15-11). The device consists of a porous block containing two electrodes connected to a wire lead. The porous block is typically made of gypsum or fiberglass (Fig. 15-12). When the device is buried in the soil, water will move in or out of the block until the matric potential of the block and the soil are the same. These blocks are placed in the soil at the location where the water content measurement is desired. The porous blocks absorb water from the surrounding soil, and electrical resistance in the block changes, corresponding to its water content. A lower resistance corresponds with a higher water content. The change in resistance can be measured with a simple ohm meter or a calibrated meter (Fig. 15-13).

Gypsum blocks are sensitive to soil water salinity, including fertilizer salts. They can be relatively insensitive to small water content changes. A calibration curve is made to relate electrical conductivity to the matric potential for any particular soil. Using a porous electrical resistance block system offers

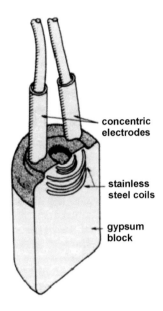

Figure 15-11. Gypsum block soil moisture sensor.

Figure 15-12. Watermark-resistance-type soil moisture sensor.

Figure 15-13. Resistance meter used in conjunction with watermark sensor.

the advantage of low cost and the possibility of measuring the same location in the field throughout the season. The blocks function over the entire range of soil water availability.

Gypsum blocks allow continuous soil water measurement, are economical, easily installed, and, once calibrated, require a minimum of calculations. Their main disadvantages are that they deteriorate with time and do not work well with the sandy soils that have low water-holding capacities on which most of Florida citrus is grown. They are also sensitive to temperature and salts in the soil. Another disadvantage of the porous block system is that each block has somewhat different characteristics and must be individually calibrated. In addition, the calibration changes gradually with time. Typically, resistance blocks require two to three hours to reach equilibrium with the surrounding soil.

Capacitive Sensors

Capacitive methods measure the dielectric constant of the soil using an electric circuit arrangement in which the soil is the dielectric media of one capacitor in a circuit. Capacitance-based sensors rely on the fact that water has a much higher dielectric constant than air or dry soil. Hence, changes in the water content of the soil are reflected in capacitance changes of the soil near the sensor.

Soil water content is determined by measuring the capacitance between two electrodes implanted in the soil. Where soil water is predominantly in the form of free water (such as occurs in sandy soils), the dielectric constant is directly proportional to the water content. The probe is normally excited at a high frequency to permit measurement of the dielectric constant.

Capacitive sensors provide a near instantaneous measure of volumetric soil water content. Periodic calibration may be required to maintain long-term accurate readings. These devices are not as sensitive as resistance sensors to salts. However, the relationship between the sensor response and the soil water content is not linear and is influenced by the type of soil. Thus, a more complex calibration method is required. Two types of devices are available: one is a portable sensor that is lowered into the soil through an access tube (Fig. 15-14), and the other is more permanently installed with numerous sensors connected to a data logger (Fig. 15-15).

Figure 15-14. Enviroscan portable handheld capacitance probe.

Theoretically, capacitive sensors can provide absolute soil water content. They also can provide water contents at any depth. Sensor configuration can vary in size so sphere of influence or measurement is adjustable. The sensors have a relatively high level of precision when ionic concentration of soil does not change. An advantage of capacitance sensors is that they are easily adapted to reading by remote methods. Other advantages of capacitive-type sensors are their high accuracy, the availability of excellent software for interpreting data, and the fact that there are no known hazards associated with their use. The primary disadvantage is their cost (in the $1,000 range for single portable units and more than $10,000 for permanent installations with several sensors).

Time-Domain Reflectometer (TDR)

Time-domain reflectometry (TDR) determinations involve measuring the movement (propagation) of electromagnetic (EM) waves or signals through soil. Propagation constants for EM waves in soil,

such as velocity and attenuation, depend on soil properties, especially water content and electrical conductivity. Generally with the TDR devices, a voltage pulse is propagated along a transmission line or rod that is driven into the soil (Fig.15-16). The lines act as wave guides along which the signal is propagated in the soil. The signal is reflected back at the end of the transmission line and is returned to the TDR receiver. The wave velocity and amplitude are measured and related to water content of the soil. The propagation of electrical signals in soil is influenced by soil water content and electrical conductivity. The dielectric constant, measured by TDR, provides a good measurement of this soil water content.

With TDR, the volumetric water content measurements are aided by propagation of electromagnetic wave measurements. Typical response time of TDR sensors is about 30 seconds. Although TDR systems are generally costly, they provide soil moisture measurements that are independent of soil texture, temperature, and salt content. It is possible to obtain long-term in situ measurements through automated TDR systems.

The advantages of the TDR method are that they can make relatively accurate measurements of water content, are not significantly affected by soil

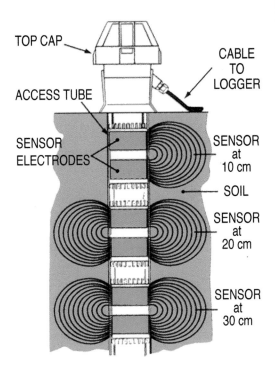

Figure 15-15. Schematic of permanently installed capacitance sensors at various depths in the root zone.

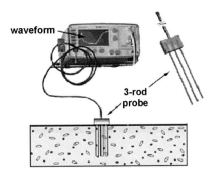

Figure 15-16. Typical TDR soil moisture measurement system with probe and receiver.

salts, can detect rather small changes in soil water content, and use relatively stable transmission lines or rods that do not degrade rapidly. Some of the disadvantages include the sensitivity and relative expense of the instrument that senses the changes in electrical characteristics, and soil profile changes that are not easily detected in a shallow root zone because the measurement is made over a distance rather than at a point.

A recent development in soil moisture sensing is the Aquaflex system, which consists of a 10-foot long sensor that is buried in the soil along with an electronic module to read the sensor (Fig. 15-17). The Aquaflex works using a standard transmission line process similar to TDR (time domain reflectrometry). An electrical pulse is sent down a transmission line and the evanescent field around this transmission line interacts with the surrounding soil. The speed and shape of this pulse is affected by the dielectric properties of the soil/water regime. Water has a much higher dielectric constant than the soil particles. Therefore, the presence of water has a profound impact on the electrical signal, and as a result, enables the measurement of moisture content.

Aquaflex differs from traditional TDR in several ways. The Aquaflex sensor consists of a transmission line, which extends to the end of the sensor ribbon cable and then returns back to the sensor electronics. In this way, the original signal pulse is detected by the electronics rather than a signal reflected off the end of the sensor as in traditional TDR. The pulse circulates many times through the sensor cable before the pulse delay is measured, resulting in a precise measurement of the time delay. In addition to water content, the Aquaflex sensor can sense soil conductivity. This is important to compensate the moisture measurement reading and maintain accuracy in soils that have higher salinity levels.

The Aquaflex Sensors are installed in the ground at an appropriate depth for the application and measurement required. The number and location of sensors are chosen to achieve a representative sample of soil moisture, taking local conditions and soil texture into consideration. The sensors are connected to the logger via a data cable providing continuous moisture and temperature readings at intervals adjustable from ten minutes to six hours. The logger stores the data until such time as it is transferred to a computer, either directly or via a handheld device.

Remote Sensing Techniques

Some types of remote sensing of soil water include satellite, radar (microwaves), and other noncontact techniques. The remote sensing of soil water depends on the measurement of electromagnetic energy that has been either reflected or emitted from the soil surface. The intensity of this radiation with soil water may vary depending on dielectric properties, soil temperature, or some combination of both. Infared thermometry (Fig. 15-18) can be used to detect stress in plants.

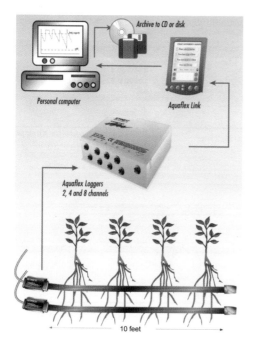

Figure 15-17. Components of Aquaflex soil moisture sensing system.

For active radar, the attenuation of microwave energy may be used to indicate the water content of porous media because of the effect of water content on the dielectric constant. Thermal infrared wavelengths are commonly used for this measurement. Remote sensing techniques measure soil surface water through the measurement of electromagnetic energy. The systems required to sense and evaluate the electromagnetic energy are often large and complex, and are generally used for measurements to be taken over large areas.

<u>Sap Flow</u>

Sap flow sensors (Fig. 15-19) measure the rate of sap movement in a tree, and thus reflect on the water consumption of the tree. These sensors measure the amount of heat carried by the sap, which is converted to a flow rate in units of weight per unit time. The trunk gauges are nonintrusive, and are not harmful to the trees. The trunk gauges typically have 4 to 8 pairs of differential temperature sensors spaced around the circumference of the tree to ensure that flow rates that vary around the tree are accurately monitored and averaged into a single value. Sensors can be used on both the main trunk and on branches of the tree to evaluate water movement within the tree.

Figure 15-19. Components of typical sap flow trunk gauge system.

Figure 15-18. Infared thermometer.

Figure 15-20. Thermal dissipation probe used to measure sap flow in tree.

Thermal dissipation probes (TDP) measure sap velocity, which is converted to volumetric flow rate (Fig. 15-20). The TDP has two thermocouple needles that are inserted into the sapwood of the tree. The upper needle contains an electric heater. The probe needles measure the temperature difference between the heated needle and the sapwood ambient temperature at the lower needle. The temperature difference variable at the measurement time and the maximum temperature difference at zero sap flow provide a direct and calibrated conversion to sap velocity. Because sap flow varies around the circumference of trees, multiple probes (typically 2 to 4, depending on diameter) need to be inserted into a single tree to make sap flow calculations accurate.

Further Reading

Cary, J.W. and H.D. Fisher. 1983. Irrigation decisions simplified with electronics and soil water sensors. Soil Sci. Soc. Am. J. 47:1219-1223.

Gardner, W.H. 1986. Water content, p. 493-544. In: A. Klute (ed.). Methods of soil analysis, Part 1. Physical and mineralogical methods. 2nd ed. Agronomy Series No. 9. Am. Soc. Agron.

Gutwein, B.J., E.J. Monke and D.B. Beasley. 1986. Remote sensing of soil water content. ASAE Paper No. 86-2004. St. Joseph, Mich.

Jackson, T.J. 1980. Profile soil moisture from surface measurements. J. Irrigation and Drainage. 106(IR2):81-92.

Kano, Y., W.F. McClure, and R.W. Skaggs. 1985. A near infrared reflectance soil moisture meter. Trans. ASAE. 28(6): 1852-1855.

Phene, C.J., G.J. Hoffman, and S.L. Rawlins. 1971. Measuring soil matric potential in situ by sensing heat dissipation within a porous body: I. Theory and sensor construction. Soil Sci. Soc. Am. Proc. 35: 27-33.

Stafford, J.V. 1988. Remote, non-contact and in situ measurement of soil moisture content: a review. J. Agr. Eng. Res. 41:151-172.

Chapter 16. Evapotranspiration

by Brian Boman and Larry Parsons

The primary reason for citrus irrigation is to provide water to the trees when the frequency and amount of rainfall is not sufficient to replenish water that has been used. The water requirements for citrus depend on growth and fruit development needs as well as environmental demands. The quantity of water stored in the tree for growth and fruit production is relatively small compared to environmental demands. Therefore, a knowledge of the parameters involved with the water use process will help the understanding of the driving forces associated with the tree water requirements.

An exchange process between incoming energy and outgoing water occurs at the crop canopy, which can be evaluated by an energy balance (Fig. 16-1). Water is transferred out of the plant system as a result of satisfying this balance. Water evaporates from soil and plant surfaces and transpires through the plant to satisfy the environmental demand. This transfer process is called evapotranspiration (ET). The process is a combination of two separate mechanisms whereby water is removed from the soil surface by evaporation (E) and from the trees by transpiration (T).

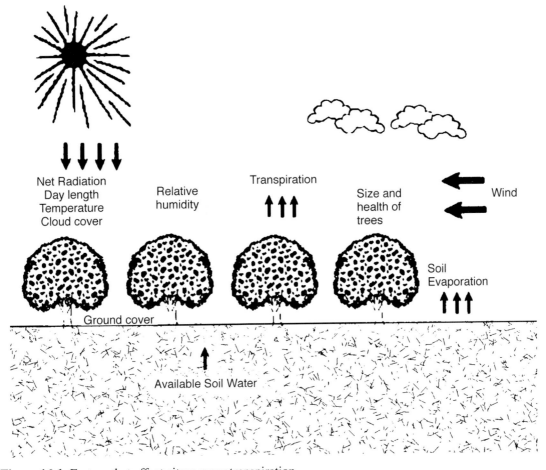

Figure 16-1. Factors that affect citrus evapotranspiration.

Evaporation

Evaporation is the process by which liquid water is converted to water vapor (vaporization) and removed from the evaporating surface. Water evaporates from a variety of surfaces, such as lakes, ditches, soils, and wet vegetation. Energy is required to change the state of water molecules from liquid to vapor. Direct solar radiation and, to a lesser extent, the ambient temperature of the air provide this energy. The driving force to remove water vapor from the evaporating surface is the difference between the water vapor pressure at the evaporating surface and that of the surrounding atmosphere. As evaporation proceeds, the surrounding air becomes gradually saturated, and the process will slow down and may even stop if the wet air is not transferred to the atmosphere. The rate of replacement of the saturated air with drier air depends on wind speed. The climatological parameters of solar radiation, air temperature, air humidity, and wind speed are the most important factors to consider when assessing the evaporation process.

Evaporation rate from the soil surface depends on the degree of shading from the crop canopy and the amount of water available at the soil surface. Frequent rains, irrigation, and water table upflux (water transported upwards from a shallow water table) can all wet the soil surface. When the soil is able to supply water fast enough to satisfy the evaporation demand, the evaporation from the soil is determined only by the meteorological conditions. However, with long intervals between rains or irrigation, the soil loses its ability to transport water to the soil surface. Therefore, the water content in the topsoil drops and the soil surface dries out. Under these circumstances, the limited availability of water exerts a controlling influence on soil evaporation. In the absence of any supply of water to the soil surface, evaporation decreases rapidly and may cease almost completely within a few days.

Transpiration

Transpiration consists of the vaporization of liquid water contained in plant tissues and the vapor removal to the atmosphere. Citrus trees lose water mainly through stomata. These are small openings on the plant leaf through which gases and water vapor pass (Figure 16-2). The water, together with some nutrients, is taken up by the roots and transported through the plant. The vaporization occurs within the leaf in the intercellular spaces, and the vapor exchange with the atmosphere is controlled by the stomatal aperture. Nearly all water taken up is lost by transpiration, and only a tiny fraction is used within the plant.

Transpiration, like direct evaporation, depends on the energy supply, vapor pressure gradient, and wind. The primary climatological factors affecting transpiration are radiation, air temperature, air humidity, and wind. Soil water content and the ability of a soil to conduct water to the roots also influence the transpiration rate. Other factors such as waterlogging and soil water salinity should be considered. The transpiration rate is also influenced by tree characteristics, environmental aspects, and cultivation practices, in addition to tree size and overall tree health.

Evapotranspiration

Evaporation and transpiration occur simultaneously, and there is no easy way of distinguishing between the two processes. Apart from the water

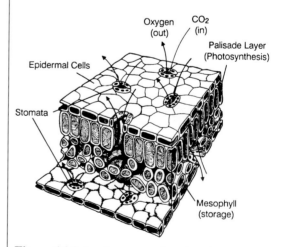

Figure 16-2. Leaf cross section showing stomata. In citrus, stomata are primarily on the lower leaf surfaces.

availability in the topsoil, the evaporation from a cropped soil is mainly determined by the fraction of the solar radiation reaching the soil surface. This fraction decreases as trees mature and the canopy shades more and more of the ground area. When the trees are small, water is predominately lost by soil evaporation, but once the trees mature and the canopy covers a larger portion of the soil, transpiration becomes the main process.

Units
The evapotranspiration rate is normally expressed in depth-per-unit time (in./day) or volume-per-unit time (gal/day). The rate expresses the amount of water lost from a cropped surface per unit of time. The time unit can be an hour, day, month, or even a year. Consider a one-acre mature citrus grove on a typical June day when the ET rate averages 0.18 in./day. Water lost with this ET rate is equal to 0.18 in. of water over the entire acre (0.18 acre-in.), which amounts to 0.18 ac-in. x 27,154 gal/ac-in. = 4888 gallons.

Sometimes it is necessary to express water use in terms of energy received per unit area (energy or heat required to vaporize free water). This energy, known as the latent heat of vaporization (λ), is dependent on the water temperature. For example, at 68° F, λ is about 580 cal/g. When ET is expressed as the latent heat flux (LE), units that can be used include langleys (cal/cm^2) per day and MJ/m^2 per day.

Crop Factors
The citrus variety and development stage should be considered when assessing the evapotranspiration from large, well-managed groves. Differences in resistance to transpiration, tree height, canopy roughness, reflection, ground cover, and rooting characteristics result in different ET levels under identical environmental conditions. Evapotranspiration under standard conditions refers to the evaporating demand from tree crops that are grown in large blocks under optimum soil water, excellent management, and environmental conditions, and achieve full production under the given climatic conditions.

Management and Environmental Conditions
Factors such as soil salinity, poor land fertility, poor nutrition, the presence of hardpan or impenetrable soil horizons, lack of disease and pest control, and poor soil management may limit the tree development and reduce the ET. Other factors to be considered when assessing ET are ground cover, tree planting density, and the soil water content. The effect of soil water content on ET is conditioned primarily by the magnitude of the water deficit and the type of soil. However, too much water will result in waterlogging, which might damage the root and limit root water uptake.

Additional consideration should be given to the range of management practices that act on the climatic and crop factors affecting the ET process. Cultivation practices and the type of irrigation method can alter the microclimate, affect the crop characteristics, or affect the wetting of the soil and crop surface. Outside rows reduce wind velocities and decrease the ET rate of the trees in the interior of a block. The effect can be significant under hot, windy, or dry conditions. Soil evaporation on a young block where trees are widely spaced can be reduced by using a drip system or top hats on microsprinklers that limit the amount of wetted soil. The use of mulches, especially when the trees are small, is another way of substantially reducing soil evaporation. Antitranspirants, such as stomata-closing, film-forming or reflecting material, reduce water losses from the trees and thus the transpiration rate.

Potential and Reference Crop Evapotranspiration
Energy Balance
Evaporation of water requires relatively large amounts of energy, either in the form of sensible heat or radiant energy. Therefore, the evapotranspiration process is governed by energy exchange at the vegetation surface, and is limited by the amount of energy available. Because of these limitations, it is possible to predict the evapotranspiration rate by applying the principle of energy

conservation. The energy arriving at the surface must equal the energy leaving the surface for the same time period.

When the water supply available to a crop system is limited, the ET component of the energy budget is reduced, and the heat flux terms are increased. However, for most crop systems, it is desirable to provide adequate water to the crop so that ET levels are at their maximum. This avoids water stress and helps achieve high levels of production. The term potential evapotranspiration (ETp) is used to describe the maximum or potential level of ET resulting from nonlimiting water conditions and satisfaction of the energy budget. ETp depends on the crop and other surface conditions. Therefore, the term reference crop evapotranspiration (ETo) is used to describe climate demand for a standard set of conditions. This permits climate conditions to be compared from site to site.

Evapotranspiration is a component of an energy budget of activities occurring at the crop surface. An energy budget is used to identify the individual components. These components are net radiation, sensible heat flux, soil heat flux, ET, and solar radiation stored as photochemical energy.

The primary energy input to the system is solar radiation, R_s (Fig. 16-3). However, some of R_s is reflected while the remainder is absorbed and converted to other energy forms (Fig. 16-4). The amount of R_s reflected depends on the albedo (α), which is the solar radiation reflectivity of the surface. Additional radiant energy transfers occur in the form of thermal radiation generated by the temperature of the surface (upward longwave radiation, R_u) and of the atmosphere (downward longwave radiation, R_d). All of these terms are combined to define the net radiation (R_n) received at the crop or soil surface.

Sensible heat flux (H) is the transfer of heat energy due to the temperature difference between two surfaces, such as the plant surface and the atmospheric environment. If the plant surface is warmer than the environment, heat energy flux will be from the plant surface to the environment, and vice versa.

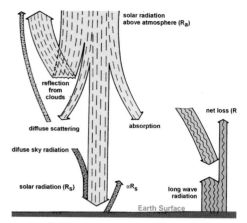

Figure 16-3. Components of radiation budget.

Similarly, soil heat flux (G) occurs due to a difference in temperature at the soil surface and layers below the surface.

Evapotranspiration is also known as latent heat flux (LE). The previous three components, R_n, H, and G, are used to calculate ET (Fig. 16-5). The last component, associated with photochemical energy storage, is very small when compared to the other components and can be neglected.

Reference crop evapotranspiration refers to ET from a uniform green crop surface, actively growing, of uniform height, completely shading the ground, and under well-watered conditions. A standard that has been accepted for use as a reference crop is grass maintained at a 3- to 6-inch height. Actual ET for a crop is calculated by multiplying ETo by a crop coefficient (Kc) that relates the water use properties of that crop to the reference level of ET.

$$ET = Kc \times ETo \qquad \text{Eq. 16-1}$$

where,

ET = Evapotranspiration of citrus trees (inch/day)
Kc = Crop coefficient
ETo = grass-based reference ET

The Penman equation for calculating ETo is probably the most widely accepted energy-based method. This approach combines two components to estimate ETo. These are a radiative component and an advective component.

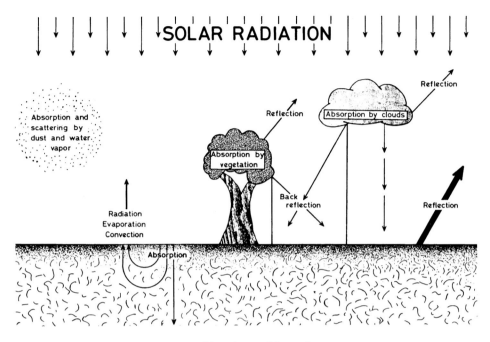

Figure 16-4. Utilization of solar energy reaching the earth's surface.

$$ETo = \frac{\Delta}{\Delta + \gamma} \times \frac{R_n}{L} + \frac{\gamma \times E_a}{\Delta + \gamma} \qquad \textbf{Eq. 16-2}$$

where,

ETo = Reference evapotranspiration

Δ = Slope of the saturated vapor pressure temperature curve of air

R_n = Net radiation at the crop surface

L = Latent heat of vaporization of water

γ = Psychrometric constant

E_a = A vapor pressure deficit and wind function term

= $0.263(e_a - e_d)(0.5 + 0.0062\ U_2)$

e_a = Saturation vapor pressure of air at the ambient temperature

e_d = Saturation vapor pressure of air at the dewpoint temperature

U_2 = Wind speed at a height of 2 meters (6.6 ft).

The driving forces of evapotranspiration (solar radiation and wind) are both influenced by the temperature and vapor pressure conditions of the air.

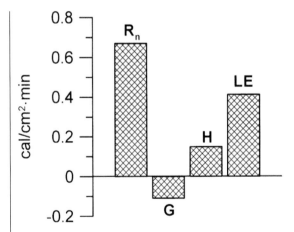

Figure 16-5. Energy balance at 2 pm on April 29 for a short grass pasture near Okeechobee, Florida (from Allen et al., 1978).

Net Radiation

Net radiation, R_n, is the net level of solar and thermal radiation energy at the crop surface. It can be broken down into the components of shortwave and longwave radiation. Net radiation is measurable and instruments are available for this purpose. However, direct measurements of net radiation are expensive and the reflected term depends on the albedo of the surface. Therefore, incoming solar radiation is normally measured as longwave radiation and is calculated from temperature measurements. Thus, R_n is normally estimated from the shortwave and longwave radiation components.

Radiation is measured in units of calories (cal) or joules (J). Instantaneous radiation levels on a surface are measured as an energy flux or rate per unit area such as calories per square centimeter per minute (1 cal/cm^2/min equals 1 langley/min) or watts per square meter (W/m^2) [an energy flux of 1 watt is equal to 1 joule per second]. Daily accumulated levels of radiation such as cal/cm^2/day or MJ/m^2/day are generally used in ET prediction equations.

Shortwave Radiation

Shortwave radiation is the radiation flux resulting directly from solar radiation. It is called shortwave radiation because it has wavelengths of only 0.3 to 4 micrometers. Shortwave radiation received by an object may be the result of direct, diffuse, or reflected solar radiation. Direct radiation comes directly from the sun and depends on the angle between the groves or crop surface and the sun. Radiation scattered and reflected by clouds has no specific direction and is called diffuse radiation. Additional reflection of shortwave radiation can occur from the ground surface or other objects. As with diffuse radiation, this terrestrial reflected radiation has no specific direction and is diffuse as well.

Rates of direct shortwave radiation range from 0 at night time to 150 W/m^2 (0.215 cal/cm^2/min) during very hazy or overcast daylight periods to 900 or 1000 W/m^2 (1.29 to 1.43 cal/cm^2/min) during bright sunny periods. Daily accumulated levels in Florida may range from 12 to 34 MJ/m^2 (300 to 800 cal/cm^2).

The reflectivity of a surface to shortwave solar radiation is known as the albedo (α) of the surface.

The higher the reflectivity of the surface is, the higher the albedo will be. Typical daily values of albedo are: open water, 0.05; dry soil (light color), 0.32; woodland, 0.16-0.18; and crop surfaces, 0.15-0.26 (0.23 is often used for complete canopy, green crops). The amount of reflected radiation is the albedo multiplied by the incoming shortwave radiation.

Longwave Radiation

Longwave (infrared) radiation is also called terrestrial radiation and has wavelengths in the range of 4 to 80 micrometers. Longwave radiation is emitted and absorbed by both the atmosphere and terrestrial objects. Atmospheric longwave radiation is emitted and absorbed by water vapor and carbon dioxide contained in the atmosphere. Similarly, terrestrial longwave radiation is emitted by objects on the earth's surface.

Longwave radiation depends on the temperature and emissivity of the emitting surface. As temperature increases, so does the emitted level of longwave radiation. Emissivity is a value that characterizes the efficiency of emittance of the surface. A black body has the highest efficiency of emittance and has an emissivity of 1.0. Clear-sky emissivity will depend on the levels of water vapor and carbon dioxide in the atmosphere, and ranges from 0.7 to 0.9 for Florida conditions. The higher levels occur at higher temperatures. Natural surfaces will generally have emissivities between 0.90 and 0.98.

Upward (R_u) and downward (R_d) rates will generally range from 300 to 500 W/m^2 (0.43 to 0.72 cal/cm^2/min). Of particular importance is the net longwave radiation. This value is the difference between R_u and R_d. Daily accumulated levels may range from 8 MJ/m^2 (200 cal/cm^2) net outgoing thermal radiation during clear cool weather to 2 MJ/m^2/day (50 cal/cm^2 /day) net incoming thermal radiation during warmer weather.

Latent Heat of Vaporization

The latent heat of vaporization (L) represents the amount of energy required for water to change from a liquid to a gas (water vapor). Water is in a liquid state in the plant and is changed to a vapor during ET. The value of L varies with temperature and is about 2425 joules per gram (580 cal/g) at 68° F.

Vapor Pressure

Air contains water vapor and can be either saturated or partially saturated. Saturation conditions exist when the air contains the maximum possible amount of water vapor for the existing temperature conditions. The air is partially saturated when the density of water vapor in the air is less than the maximum possible level. The amount of water vapor that can be held in the air is temperature dependent. As temperature increases, the amount of water vapor that it can hold also increases.

Vapor pressure (e) is the partial pressure exerted by the water vapor. Vapor pressure depends on vapor density and temperature (T) by the perfect gas law. It is measured in units of millibars (mb) or kilopascals (kPa), although inches of water or inches of mercury are sometimes also used. Saturation vapor pressure varies from 12 mb (1.2 kPa) at 50° F to 57 mb (5.7 kPa) at 95° F.

Vapor Pressure Deficit and Relative Humidity

The term $(e_a - e_d)$ represents the vapor pressure deficit (VPD) or vapor pressure gradient (VPG) of the air. VPD quantifies the amount of additional water vapor that the air can hold. The saturation vapor pressure at the ambient air temperature is designated by e_a, while the saturation vapor pressure at the dewpoint temperature is designated by e_d. These two values are related by the relative humidity. The relative humidity (RH) of the air is the ratio of e_a to e_d. A well-irrigated plant will generally have a moist surface in the leaf intercellular spaces, with a vapor pressure greater than the surrounding air. Water vapor will move from the plant surface to the surrounding atmosphere as long as these conditions exist.

Two additional terms related to vapor pressure and temperature are the slope of the saturation vapor pressure-temperature curve (Δ), and the psychrometric constant (γ). Δ is the slope of the curve defining the relationship between air temperature and saturation vapor pressure at the ambient air temperature. The psychrometric constant is the ratio of the heat capacity of the air to the latent heat of vaporization. Therefore, γ varies with temperature and pressure, but only a small amount so that it is considered to be a constant. The value of γ for Florida conditions is 0.37 millibars per degree F (mb/° F).

Temperature

Air temperatures are classified as dry bulb or ambient (T_a), wet bulb (T_{wb}), and dewpoint (T_{dew}). Ambient air temperature is the most common temperature reported, and it is measured using a standard thermometer shielded from direct solar radiation. Wet bulb and dewpoint temperatures characterize the moisture characteristics of the air.

As air moves over a wet surface such as a wet leaf, the temperature of the surface decreases until the air surrounding the surface is saturated, resulting in a temperature lower than ambient. This is called the wet bulb temperature. Measurement of wet bulb temperature is not difficult. A small sock or wick of material is placed over the end of a thermometer, the material is wetted with water, and air is blown over the wet material. The material cools to the wet bulb temperature and remains at that temperature until all of the water has evaporated.

The wet bulb temperature is the minimum temperature to which air can be cooled by evaporative cooling. An example would be a fan and wet pad systems used in greenhouses. Because Florida's humidity levels are relatively high, the air does not have much additional capacity for water, and thus wet bulb temperatures are only slightly lower than T_a.

Dew point temperature is the temperature to which air must be cooled for it to become saturated without adding water. For example, as ice is added to a glass of water, the glass cools. If the glass becomes cold enough, water condenses on the outside of the glass. This condensed water comes from the surrounding air. The dew point temperature is the temperature of the glass at which water first condenses on the outside.

Saturated air is at its maximum level of vapor density or pressure. At saturation, the relative humidity is 100%, and dry bulb, wet bulb, and dew point temperatures are all the same. This situation can be used to make ET calculations easier in humid climates where early, predawn atmospheric conditions are generally at or very near saturation, or 100% relative humidity. Predawn is normally the time of minimum daily temperature. As a result, measurements taken near dawn can usually be used for average daily dew point temperature and values e_d.

Conditions of partial saturation will have dry bulb, wet bulb, and dew point temperatures of different values. For example, air at a dry bulb temperature of 50° F and a RH of 50% will have a wet bulb temperature of 42° F and a dew point temperature of 32° F. As temperature and humidity increase, the differences between these different temperature parameters decrease. Air at 95° F and 70% RH has wet bulb and dew point temperatures of 86° and 83° F, respectively. The relationships among T_a, T_{wb}, and T_{dew} are given in Appendix 11.

Wind

Temperature, vapor pressure, and humidity dictate the movement of moisture from leaf surfaces into the atmosphere. The air surrounding leaves can approach saturation with water vapor from the tree. When air movement is zero, this saturated air mass moves very slowly away from the leaves, and the vapor pressure deficit is minimized. Therefore, air movement plays a major role in transporting water vapor transpired from a tree into the atmosphere. Wind can help to maintain a significant vapor pressure deficit around the leaf surface and increase the transpiration rate.

Wind speed varies with height above a crop surface. Wind will be at a minimum near the crop surface, and increases with height above the crop as surface friction tends to slow the wind (Fig. 16-6). The wind speed term in the Penman ETo equation uses wind speed (U_2) measured at a height of 2 meters (6.6 feet) and is generally measured in miles per day. In Florida's interior areas, average

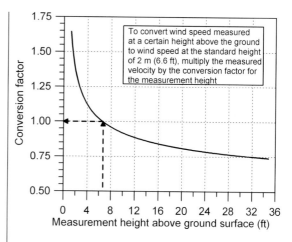

Figure 16-6. Factors for converting wind measurements to 2 m standard height.

wind run (windspeed) at a height of 2 meters ranges from 46 to 54 miles per day during the summer months and from 65 to 81 miles per day during the late winter and early spring months. The Penman equation will underpredict ET when using wind velocity measured at heights lower than 2 meters (6.6 feet) and will overpredict when wind is measured above 2 meters.

Instrumentation for Measuring Weather Parameters

Weather data collection can range from manually read instruments to sophisticated computer-activated systems. Most of the agricultural weather stations used to collect data for ET calculations utilize electronic data loggers to sense and record data (Fig. 16-7). These systems are typically battery-operated and are suitable for remote locations. They may be connected via phone modems or radio links to allow real-time data collection.

Radiation

Solar radiation is commonly measured with pyranometers. Pyranometers measure the shortwave incoming radiation in a solid angle in the shape of a hemisphere oriented upwards. In the most common glass-domed pyranometers, a thermopile is used within the instrument as the sensor, where thermal gradients are measured across hot and cold

Figure 16-7. Typical sensors used in an automated agricultural weather station.

areas (black and white). The radiation intensity is proportional to the temperature differences between the two sensing areas. Accuracy depends upon the sensitivity of the material used in the sensors, the response time, and the distortion characteristics of the material constituting the dome covering the sensors. A second type of pyranometer that is less expensive and that is gaining acceptance is the silicon diode instrument, where electric current is generated by a photo sensitive diode in proportion to solar intensity. Ordinarily, silicon diode pyranometers are not fully sensitive to the full spectrum of visible light, so that the calibration of the instrument is only valid for solar measurements when the instrument is pointed upwards.

When a pyranometer is oriented downwards, it measures the reflected shortwave radiation, and is thus called an albedometer. When two pyranometers are associated, one oriented upwards and the other downwards, the net shortwave radiation is measured. The instrument is then called a net pyranometer. A point of caution is that any instrument used as an albedometer or net pyranometer must have full sensitivity to all spectra of visible light. This is important since the composition of reflected light from vegetation is highly biased toward green. Therefore, most albedometers must be

of the glass-domed thermopile type and not the photo diode type.

Net radiation is measured by pyradiometers (or net radiometers), which sense both short- and long-wave radiation. They have two bodies, one oriented upwards and the other downwards, both covering a solid angle in the shape of a hemisphere. The sensors are made from several thermocouples, sensing heat generated by radiation from all wavelengths, and are protected by domes made in general of polyethylene treated in a specific manner. The black bodies can lose their sensing capabilities with time, so that these instruments require regular and frequent calibrations. Other net radiometers are comprised of ventilated differential thermopiles, but they are very seldom utilized. These radiometers transform the radiation energy into thermal energy, a portion of which is transformed into an electric voltage gradient that provides appropriate conditions for continuous recording dataloggers.

Sunshine duration is most commonly recorded with the Campbell-Stokes heliograph. A glass globe focuses the radiation beam to a special recording paper, and a trace is burned on the paper as the sun is moving. No records occur when no bright sunshine is sensed. Measurements are reliable when the recording paper is placed in the right position according to the relative position of the sun. Care is required to avoid accumulation of rainwater on the paper. The heliograph has to be oriented south in the northern hemisphere and north in the southern hemisphere.

Wind Velocity

Wind velocity is measured using anemometers normally located at a height of 6.6 feet (2 meters) or more above the ground. Most common are the three-cup anemometers. Also common are propeller anemometers. Measurements by both types are reliable, provided that maintenance ensures appropriate functioning of the mechanical parts. Older designs of anemometers utilize mechanical counters as the output device. Modern anemometers may be equipped with generators giving a voltage signal that is proportional to the wind speed. Other anemometers may be equipped with small, magnetic-reed switches or with opto-electronic couplers that generate electric impulses in proportion to the wind speed. The electronic devices are utilized in automatic weather stations. Accuracy of wind speed measurements depends on the upwind fetch as much as on instrumentation. A large upwind fetch that is free of buildings and trees is definitely required for representative measurements.

Temperature

The most commonly utilized sensor for measuring temperature is still the mercury thermometer. Maximum and minimum thermometers use mercury or alcohol. Bimetallic thermographs are the most common mechanical temperature recorders. They are easy to read and maintain. However, mechanical thermographs do require verification and adjustment of the position of the pen recorder. These instruments are installed in shelters that are naturally ventilated.

Modern temperature sensors, namely the thermistor and the thermocouple, provide very accurate analog measurements and are normally utilized in automatic weather stations. Thermistors provide independent measurements of air or soil temperature, whereas thermocouples require an additional base temperature reading, normally provided by a thermistor. To maintain the accuracy and representativeness of these instruments, they are installed in special radiation shields (shelters) having natural ventilation. Occasionally, the shields or shelters are artificially aspirated to reduce biases caused by heat loading from the sun.

Humidity

Dew point temperature is often measured with a mirrorlike metallic surface that is artificially cooled. When dew forms on the surface, its temperature is sensed as the dew point (T_{dew}). Other dew sensor systems use chemical or electric properties of certain materials that are altered when absorbing water vapor. Instruments for measuring dew point temperature require careful operation and maintenance and are seldom available in

weather stations. The accuracy of estimation of the actual vapor pressure from T_{dew} is generally very high.

Relative humidity is measured using hygrometers. Most frequently used in mechanically based field stations are the hair hygrometers that are normally operated as mechanical hygrographs. Measurements lose accuracy with dust and aging of the hairs. Modern hygrometers use a film from a dielectric polymer that changes its dielectric constant with changes in surface moisture, thus inducing a variation of the capacity of a condensator using that dielectric. These instruments are normally called dielectric polymer capacitive hygrometers. Accuracy can be higher than for hair hygrometers. These electronic devices are utilized in most modern automatic weather stations.

The dry and wet bulb temperatures are measured using psychrometers. Most common are those using two mercury thermometers, one of them having the bulb covered with a wick saturated with distilled water which measures a temperature lowered due to the evaporative cooling. When they are naturally ventilated inside a shelter, problems can arise if air flow is not sufficient to maintain an appropriate evaporation rate and associated cooling. The Assmann psychrometer has forced ventilation of the wet bulb and dry bulb thermometers. The dry and wet bulb temperature can be measured by thermocouples or by thermistors, the so-called thermocouple psychrometers and thermosound psychrometers. These psychrometers are used in automatic weather stations and, when properly maintained and operated, provide very accurate measurements.

Evaporation Pans

One of the simplest methods of measuring daily ET in the field is by measuring evaporation from a standardized free water surface. Studies have shown good correlation between ET and evaporation from free water surfaces. The standard water surface commonly used is a National Weather Service Class A evaporation pan (Fig. 16-8). The

Table 16-1. Typical ET calculated from long-term data and the Penman ETo equation (Jones et al., 1984) and citrus Kc values for citrus in Florida (Kc from Rogers et al., 1984).

Month	Kc	Tampa inches/day	West Palm Beach inches/day	Miami inches/day	Lakeland inches/day
Jan	0.90	0.09	0.10	0.10	0.07
Feb	0.90	0.12	0.13	0.13	0.11
Mar	0.90	0.14	0.16	0.16	0.13
Apr	0.90	0.19	0.19	0.19	0.18
May	0.95	0.20	0.19	0.19	0.20
Jun	1.0	0.20	0.18	0.18	0.21
Jul	1.0	0.18	0.18	0.18	0.20
Aug	1.0	0.17	0.18	0.17	0.19
Sep	1.0	0.16	0.16	0.15	0.16
Oct	1.0	0.14	0.14	0.14	0.12
Nov	1.0	0.11	0.12	0.11	0.09
Dec	1.0	0.08	0.10	0.10	0.06

Class A pan should be located in a well-watered and maintained grassed area with a grass height of 2 to 4 inches. Evaporation is calculated from the change in water level from one day to the next (taking into consideration any rainfall that has occurred). The ratio between potential ET and evaporation from a well-maintained Class A evaporation pan is usually about 0.7. Citrus ET can be estimated by multiplying pan evaporation by 0.7 and then multiplying by the crop coefficient (Kc).

ETo ≈ 0.7 x Pan evaporation   **Eq. 16-3**

Typical ET Rates
ET rates vary from day to day, depending on temperature, wind, humidity, and cloud cover. In general, winter ET rates are less than half of summer ET rates (Table 16-1). In many years, highest ET rates will occur in May when there are hot, dry, clear days.

Typically, summer rains will begin to increase humidity and increase cloud cover by mid-June. As a result, average daily ET rates will be somewhat moderated. Keep in mind, though, that days are longer in the summer, and when typical summer rains fail to materialize and hot, dry weather is prevalent, ET rates can be higher than long-term averages. Citrus ET can be calculated by using Eq. 16-1 and either pan evaporation or ETo calculated from environmental parameters such as with Eq. 16-2.

Figure 16-8. National Weather Service Class A evaporation pan with stilling well and hook gauge used to measure water level.

Example:
Calculate average May citrus ET using long-term Tampa Penman ETo values.

From Table 16-1, ETo for May = 0.20 in./day and the average Kc = 0.95.

Using Eq. 16-1,

ET ≈ Kc x ETo = 0.95 x 0.20 in./day = 0.19 in./day

Example:
Calculate average January citrus ET for a day with pan evaporation of 0.12 in.

Using Eq. 16-3,

ETo ≈ 0.7 x Pan evaporation = 0.7 x 0.12 in./day = 0.08 in./day

From Table 16-1, January Kc = 0.90

ET = Kc x ETo = 0.90 x 0.08 in./day = 0.07 in. day

Chapter 17. Citrus Water Use and Irrigation Scheduling

by Brian Boman, Larry Parsons, Thomas Obreza, and Ed Stover

Good irrigation must be practiced to obtain benefits from improved pest control, fertilization, and other cultural practices. When such improved cultural practices are combined with good irrigation, yield increases can be obtained. Merely irrigating in a routine fashion is not sufficient. Irrigation should be carefully supervised by knowledgeable, well-trained personnel employing modern ideas, techniques, and equipment. There are many factors (Fig. 17-1) to consider when determining irrigation schedules. The various factors are discussed in detail in other chapters and are summarized here to assist in understanding the two basic questions of irrigation scheduling:

• When should irrigations start?

• How much should be applied?

Soils

Soil consists of various-sized solid particles and the pores (spaces) between them. In most Florida citrus soils, pore volume is 40% to 55% of the total soil volume. These pores hold water and air that are needed by citrus roots. Water is held in the soil pores by attraction between the water molecules and the surfaces of the solid particles. At a low water content, water is spread as a thin film over the surfaces of all the soil particles. The thinness of the film is dependent upon the amount of water present and the total surface area of all the solid particles. Fine-textured (clay) soils have more particle surface area than coarse-textured (sandy) soils. Therefore, water molecules are held tightly to the particle surfaces. The amount of energy that must be expended to remove a unit of water from

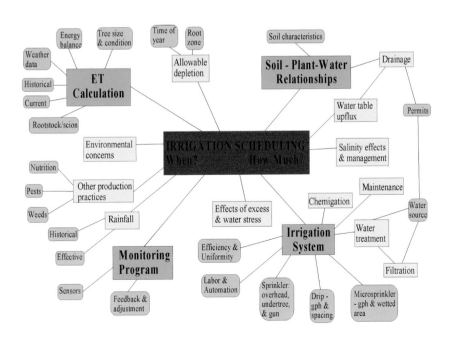

Figure 17-1. Relationships of factors involved in determining when and how much to irrigate Florida citrus.

the soil is measured in units of bars, kilopascals, or atmospheres. This force is called soil suction or soil moisture tension.

During irrigation or rainfall, water infiltrates the soil and is redistributed within the soil by gravity and soil capillary forces (attraction for water). When all of the pore space is filled with water, a soil is saturated. In this condition, gravitational forces dominate. Water drains downward through the soil, since the energy required to remove a unit of water from saturated soil is low. Drainage is initially rapid, but in most sandy soils it decreases to a very small rate after a day. For practical purposes in sandy soils, downward gravity drainage can be neglected two days after rainfall.

Drainage initially removes water from the largest pores, where it is held least securely. As the water drains, air space is created in these pores. The remaining water is somewhat closer to the surfaces of soil particles and is held firmly enough to prevent rapid drainage by gravity. This water condition, when measured in the field, is called field capacity (FC). Field capacity in Florida sandy soils usually occurs within a day after an intense rainfall because of the rapid movement of water in sandy soils. At this point, the soil water in the root zone will be removed primarily by plant transpiration or evaporation from the soil surface.

When water is used by plants, the medium-sized pores lose water first. As the soil dries, eventually the smaller soil pores lose water. In the water extraction process, water films on the soil particles gradually become thinner, and greater energy is required to remove each subsequent increment of water. At first, the change in energy requirement is slight and has little effect on the plant. Since citrus roots do not come in contact with every particle of soil in the root zone, water must move through the soil to reach root surfaces. The distance traveled may be a fraction of an inch or it may be several inches, depending on soil moisture status. Water moves more rapidly in thick films than in thin films. As water films become thinner, the rate of flow decreases and eventually becomes too slow to meet the needs of the tree.

Table 17-1. Typical available water (AW) for various soil types.

Type of Soil	Available Water (AW)	
	range (inches/ft)	average (inches/ft)
Sands and fine sands	0.40 to 1.00	0.75
Moderately coarse-textured: sandy loams and fine sandy loams	1.00 to 1.50	1.25
Medium-textured: very fine sandy loams to silty clay loam	1.25 to 1.75	1.50
Fine- and very fine-textured: silty clay to clay	1.50 to 2.50	2.00
Peats and mucks	1.75 to 2.75	2.25

The physical availability of soil water to a citrus tree is a continuously variable characteristic, largely depending on the water film thickness. At some point (normally at about 15 bars of soil tension), the soil water films become so thin that the tree wilts. When the tree remains wilted overnight, the soil water status has reached a level termed wilting point (WP). Citrus will generally suffer daytime wilting repeatedly before this condition is reached, and yield reductions typically occur long before WP is reached.

The difference between FC and WP is called the available water (AW). Table 17-1 lists typical values of AW for various soil types. Local soil surveys and irrigation guides available from the Natural Resources Conservation Service (NRCS) provide information on specific Florida soil types.

Once AW is known, the total water available (TAW) can be determined. TAW represents the maximum amount of water in the root zone that can be used by the tree. It is obtained by multiplying AW by the depth of the tree's root zone. For layered soils, TAW is calculated by adding the multiples of AW and depths of all soil layers contained in the root zone.

The effective root depths of Florida vary from 12 inches in some flatwoods areas to about 2 to 4 feet in many Ridge locations. The best way to determine effective root zone depths is by digging and observing where most of the roots are located. The effective root zone is where most of the roots actively involved in water uptake are located. In a humid area such as Florida, irrigations should be concentrated in this upper portion of the root zone where the great majority of the tree's roots are located.

Allowable Soil Water Depletion
The allowable soil water depletion is the fraction of the available soil water that will be used to meet evapotranspiration (ET) demands before irrigation commences. As ET occurs, the soil water begins to be depleted. As the soil dries, the remaining water is bound more tightly to the soil, making it more difficult for trees to extract it. For this reason, ET

will start to decrease long before the WP is reached. Lower ET generally results in smaller fruit, lower overall total soluble solids (TSS) production, and lower fruit yields. Therefore, irrigations should commence before the root zone water content reaches a level that restricts ET.

The critical soil water depletion level depends on several factors: crop factors (rooting density and tree age/size), soil factors (AW, effective root depth, and salinity concentration), and atmospheric factors (current ET rate). Therefore, no single level can be recommended for all situations. Allowable depletions up to 1/2 to 2/3 of the available soil water are commonly used in scheduling irrigations during the nonsensitive periods. Lower depletion levels should be used during critical stages such as bloom and fruit set. As a rule of thumb, soils should be allowed to deplete no more than 1/3 of AW from February through the "June drop," and 2/3 of AW during other times of the year. Lower depletion levels are required where high-salinity irrigation water is used (see chapter 13).

Roots
Soil water status is a major component of the root environment, affecting the growth and health of roots. Other components of the root environment affected by irrigation are soil aeration, temperature, and salinity. The ideal environment for citrus roots is a porous, medium-textured, well-drained soil, where water is easily available but not in excess. When such a soil is irrigated, water distributes itself throughout the soil profile in the root zone or drains away, leaving no excess. The growth and production of citrus may be impaired either by an excess of water in the root zone or by a lack of easily available water. A deficiency of water in the root zone produces detrimental effects on root growth. As the soil dries, root growth becomes slower and eventually ceases. However, if some of the roots are in soil containing easily available water, the roots in dry soil are not damaged. Roots do not grow through dry soil to reach moist soil, but remain healthy and ready to become active whenever the soil in which they exist receives

moisture. In groves that have chronic conditions of excess water, the most healthy roots are those existing in portions of soil that remain well drained during the rainy season.

Trees develop thicker feeder roots when repeatedly exposed to water stress. If soil conditions permit, trees subjected to regular drought stress may develop deeper root systems than those under trees maintained with more abundant soil moisture conditions. Excess water in the root zone exerts a much greater effect on the root system than a deficiency. The decline of citrus tree health resulting from excess water is nearly always due to decay of larger roots that support the feeder roots.

Citrus roots need good soil aeration if they are to remain healthy. It is therefore important to manage the drainage system to provide a well-aerated environment for tree roots. Low soil oxygen is most damaging to plants during hot weather because tree metabolism is high and increased temperature reduces the solubility of oxygen in water. Therefore, it is especially important to avoid waterlogging during hot weather.

The roots of citrus and the trees themselves respond to differences in soil temperature. One study showed that citrus roots grow best at soil temperatures of 79° F. As the temperature rose to 86° F, the growth rate dropped off 25%. At 93° F, the growth rate was reduced 90%. When the temperature dropped lower than 79° C, the growth rate again was reduced. Little root growth was found to occur below 68° F.

The water-supplying ability of citrus roots is also affected by the temperature of the root environment. In a study that measured the transpiration rate of rooted lemon cuttings grown in various root-temperature environments with similar aerial environment, the transpiration rate increased as root temperatures rose from 41° F to 77° F. The transpiration rate decreased at temperatures higher than 77° F. Low temperature can affect water absorption by roots as adversely as high temperature. When soil temperatures fall low enough during cold winters, the water absorption by the roots is impaired.

Stress
Water stress is the physiological condition to which a tree is subjected whenever the rate of water loss from the leaves by transpiration exceeds the rate of water absorbed by the root system. Stress can occur in citrus trees before signs of water shortage are clearly evident. Reduced growth and abnormal loss of leaves can result when citrus trees are allowed to go without irrigation until signs of water shortage are evident. Stress can be caused by the following conditions:

- On a hot, dry, or windy day, water may be removed from the leaves at so rapid a rate that roots are incapable of absorbing water fast enough to make up for the loss. This is especially found in flatwoods soils where the root zone is limited.

- A low soil moisture condition may reduce the availability of water, so that the roots cannot extract water effectively.

- As water is extracted from soil and it becomes drier, the hydraulic conductivity is generally significantly reduced. After the roots have absorbed the water with which they are directly in contact, the replacement of this water by movement from points some distance away (these distances need not be more than a few millimeters) is not rapid enough to meet the needs of the tree.

- When root zones are flooded, available oxygen may be depleted quickly. Since water uptake is reduced when oxygen levels are low, flooded trees may become stressed under high transpiration conditions.

- Unfavorable root temperature can result in insufficient water uptake on warm, windy winter days.

- Sudden root injury such as that caused by discing can reduce water absorption rates and create stress.

A citrus tree usually sets more fruit in the spring than it can reasonably bear to maturity. In late spring or early summer, abscission layers form at the base of the stem and some fruit is shed. To avoid excessive "June drop," the soil must be kept fairly moist during this period. Good irrigation is the most important operation in combating stress problems to which citrus trees are susceptible, such as fruit drop, leaf drop, and twig dieback. This is not merely a matter of timing a single irrigation just before the maximum danger period. Proper water management is necessary throughout the year to build up a dense, healthy root system and a vigorous tree.

Water Use

Water is used in a citrus grove in several ways. Water is assimilated into the plant and fruit. There is direct evaporation from the soil and plant surfaces. Transpiration moves water vapor from plant leaves into the atmosphere. There are also other uses of water, such as leaching of salts, use as a transport medium for chemigation, and use for freeze protection. Typically, less than 1% of the water used in crop production is assimilated into the plants.

Most of the water applied to meet the water requirements of a crop is used in evaporation and transpiration. As part of the photosynthesis process, CO_2 is taken in, and water vapor is lost through the open stomata. This water loss is known as transpiration, and usually cannot be reduced without also reducing photosynthesis. Transpiration can increase the rate of nutrient movement into and through the plant. Because it is an evaporative process, transpiration can cool the leaves. While this cooling can be beneficial in some situations, the cooling effect stops when the stomata close due to water stress. Therefore, transpiration is not a very reliable cooling system because it fails when water stress becomes too great.

Under similar climatic conditions, citrus trees are known to have lower transpiration rates than other crop plants. For example, the daily ET in the springtime may average about 0.25 inch/day for many field and vegetable crops in Florida, but only about 0.18 to 0.20 inch/day for citrus. The low citrus transpiration rate is due to low leaf and canopy conductance. Canopy conductance calculated from soil water depletion data and potential evapotranspiration is typically 0.09 to 0.11 inch/second compared with 0.8 to 1.3 inch/second for a number of field crops.

Citrus trees require water for transpiration throughout the year. Water requirements vary with climate, ground cover, cultivation practices between rows, weed control, tree size and age, scion, and rootstock. Water requirements of grapefruit are higher than orange or mandarin varieties. Water requirements for mature trees generally range from 35 to 50 inches per year, depending on tree size, fruit load, and tree condition.

The total seasonal amount of irrigation water needed by a fully grown citrus grove for optimum yield depends on the daily rate of evapotranspiration, the rainfall distribution, and the citrus variety grown. In Florida, there is usually a considerable amount of rainfall during the summer. However, oftentimes the timing of summer rains is insufficient to satisfy citrus water needs, and supplemental irrigation is beneficial. A study in Florida showed that irrigation was beneficial in eight out of nine years, and the average yield increased 22% when an annual application depth of 11 inches of water was added to the 43 inches of annual rainfall.

The amount of water needed by citrus is highly dependent on tree size and tree health. Typically, trees only average about one gal/day during the first two years. Large, vigorous, healthy trees require more water than young or nonproductive trees. Large trees at low planting densities (60 to 70 trees/acre) may use 20 to 30 gal/day during the winter months and 60 to 70 gal/day in July and August (Fig. 17-2). In a study conducted on a developing grove (with Bahia grass cover) over a 10-year period at Ft. Pierce, average annual ET was reported to be 48 inches. Daily water use by the trees in this study peaked at about 40 gal/day during the June-July period (middle curve,

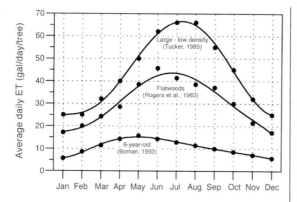

Figure 17-2. Water use for Florida citrus trees in gallons per tree per day for various ages and sizes of tree.

Fig. 17-2). In another study, water use from 6-year-old Valencia orange trees peaked at about 15 gal/day in the summer months (Fig. 17-2, lower curve).

As the soil dries during extended dry periods, it is important that irrigation water be supplied at the appropriate frequency and volume in order to minimize stress to the trees. With proper irrigation, most of the water used by the trees will come from the area wetted by the irrigation system. The volume of water that is available in the wetted area depends on the soil texture, organic content, and the emitter design. Typical microsprinklers wet an area of 12 to 16 feet in diameter, while drip systems frequently wet bands 2 1/2 to 3 feet wide. The water available within the area wetted by the microsprinkler can be calculated based on soil water-holding capacity, root zone depth, and the area that is managed as in the following example.

Example:
Calculate the amount of water for a microsprinkler with a 15-ft diameter wetting pattern in a Riviera fine sand that has 1.0 in./ft of available water. Assume a 1.5-ft root zone (refer to Fig. 17-3) and no rainfall.

TAW = 1.0 in./ft x 1.5 ft = 1.5 in.

= 1.5 in. /12 in./ft = 0.13 ft

Determine area wetted for a microsprinkler with a 15-foot pattern.

$$area = \frac{\pi d^2}{4} = \frac{3.14 \times 15^2}{4} = 177 \text{ ft}^2$$

Volume of water = Area x depth of water in root zone = 177 ft² x 0.13 ft = 23.0 ft³

Calculate the total gallons of water in the 15-ft diameter by 1.5 deep soil volume at field capacity (1.0 ft³ = 7.5 gallons)

23.0 ft³ x 7.5 gal/ft³ = 173 gal

• Calculate the amount of water available between FC and 1/3 depletion:

173 gal x 0.33 = 57 gal

Thus, after 57 gallons have been used by the tree, an irrigation should be made to replace it. This quantity of water will be used by typical mature citrus trees in 1 to 2 days during the normal spring irrigation season.

• Calculate the amount of water available between FC and 2/3 depletion

173 gal x 0.67 = 116 gal

During the noncritical times of the year (June through January), up to 116 gallons of water may be extracted by trees before irrigations will be needed.

When these calculations are repeated for other diameters using the same rooting depth and available water (Table 17-2), it is apparent that small wetted diameters require more frequent irrigations to adequately replace water that has been extracted by mature trees.

Table 17-3 provides an estimate of the volume of water stored within the area wetted by a 15-foot

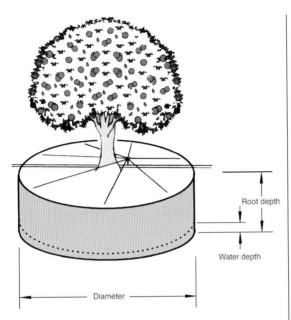

Figure 17-3. Schematic of the managed root zone beneath a citrus tree irrigated with a micro-sprinkler.

diameter microsprinkler based on soil texture and root zone depth. Available (useable) water is that portion of the total soil water that is held between soil tensions of 0.1 bar (field capacity) and 15 bars (wilting point).

Application Volumes

Water use rates and irrigation requirements are usually presented in terms of depth. Converting the depths to per-tree volumes is required to manage microirrigation systems that water only a portion of the ground surface. Detailed conversions from irrigation depth to irrigation volume require knowledge of tree planting density and tree size and health. As a rule of thumb, however, ET rates expressed as depths can be converted to volume by:

gal/tree/day = ET (in./day) **Eq. 17-1**
    x tree spacing (ft^2) x 0.622 gal/in.-ft^2

The conversion should only be used as a starting point with the actual water estimates based on the size and condition of trees in each block. The water use per tree for mature citrus trees in high density

plantings will be less than in low-density plantings due to tree size. Table 17-4 illustrates the effect of tree density and equal per-acre water use on individual tree water use.

The optimum interval between irrigation events depends on the design of the irrigation system, time of year, and soil characteristics as well as tree size and condition. Evapotranspiration rates of trees are highly dependant on climatic conditions. Higher evaporation and transpiration rates occur more often on clear, sunny days in May with low humidity and high winds than on cool, damp, overcast winter days. Therefore, irrigation frequency must be adjusted throughout the year. Table 17-5 provides estimates of the irrigation intervals required to replace water lost by evapotranspiration for flatwoods and Ridge soils (assuming no rainfall or upflux from the water table). Irrigation intervals would be greater for trees that use less water due to tree size, tree condition, dormancy, or insect and disease stress.

Example:
Estimate the irrigation interval in May for mature citrus trees planted at a 15 x 24-foot spacing in a Candler Series soil with a TAW of 0.75 in./ft and a managed root zone of 3 ft. Assume that the average daily evapotranspiration in May is 0.18 in./day.

Since the month is May, it is essential to minimize water stress to trees. Therefore, the allowable depletion of soil water will be set at 1/3. From Table 17-3, the available water for a sand with 0.75 in./ft-TAW and 3-ft root zone at 1/3 depletion is 84 gallons.

Using Equation 17-1:

ET = 0.18 in./day x 15 ft x 24 ft x 0.622 gal/in.-ft^2
    = 40 gal/tree/day

84 gallons/40 gal/tree/day = 2.1 days

If the root zone is only 2 ft, 1/3 depletion would be 56 gallons.

Table 17-2. Volume of water available in an 18-inch root zone for a soil with 1.0 in./ft of AW for various management diameters (1.0 ft^3 = 7.5 gal).

Wetted		Volume		Water Volume		
diameter (ft)	area (ft^2)	soil (ft^3)	water (ft^3)	field capacity (gal)	33% depletion (gal)	67% depletion (gal)
10	79	118	9.4	71	23	46
12	113	170	13.6	102	34	68
15	154	231	18.5	139	46	92
16	201	302	24.1	181	60	120

Table 17-3. Gallons of water available between FC and 1/3 or 2/3 depletion in the 180-ft^2 area under a microsprinkler with a 15-ft diameter wetting pattern based on soil type.

Soil Texture	Available water (in./ft)	Total available Root depth (ft)			FC to 1/3 depletion Root depth (ft)			FC to 2/3 depletion Root depth (ft)		
		1.5	2.0	3.0	1.5	2.0	3.0	1.5	2.0	3.0
Coarse sand	0.45	76	101	152	25	33	50	51	68	102
Sand	0.75	127	169	253	42	56	84	85	113	170
Fine sands	1.00	169	225	338	56	74	111	113	151	226
Medium-textured soils	1.40	236	315	472	78	104	156	158	211	317
Peat and muck	1.95	329	439	658	109	145	217	220	294	441

Coarse sand: Archibold, Satellite, St. Lucie
Sand: Astatula, Candler, Taveres
Fine Sand: EauGallie, Immokalee, Myakka, Oldsmar, Pineda, Riviera, Wabasso
Medium texture: Winder
Peat and muck: Chobee

Table 17-4. Water use in gal/tree for various planting densities assuming equivalent per-acre water use.

ET (in./day)	Tree spacing (ft x ft)	Tree area (ft^2)	Tree density (trees/acre)	ET (gal/tree/day)
0.10	8 x 22	176	248	11
0.10	10 x 24	240	182	15
0.10	15 x 25	375	116	23
0.15	8 x 22	176	248	16
0.15	10 x 24	240	182	22
0.15	15 x 25	375	116	35
0.20	8 x 22	176	248	22
0.20	10 x 24	240	182	30
0.20	15 x 25	375	116	47

The irrigation interval would be 56 gallons/ 40 gal/tree/day = 1.4 days (5 irrigations per week)

Assuming the same 0.18 inch/day ET rate, but occurring in August (when depletion can be set to 2/3), calculate the irrigation interval.

From Table 17-3, the available water for a sand with 0.75 in./ft-TAW and 3-ft root zone at 2/3 depletion is 170 gallons.

Using Equation 17-1:

ET = 0.18 in./day x 15 ft x 24 ft x
    0.622 gal/in.-ft^2
    = 40 gal/tree/day

170 gallons/40 gal/tree/day = 4.2 days

Soil moisture sensors such as tensiometers can be used to more effectively schedule irrigation since they measure water availability rather than rely on estimates. Good correlation has been found between tensiometer readings and leaf water potential. Tensiometers provide a valuable aid in irrigation monitoring, especially in systems that are irrigated frequently, as is common with drippers or microsprinklers. Electrotensiometers (with an adjustable switch) can be coupled to the irrigation system for automatic irrigation scheduling. Care should be exercised regarding the depth and placement of the tensiometers, and enough of them should be used to represent the whole orchard.

Young Tree Water Use

Newly planted trees require only a fraction of the water needed by mature trees. It takes a number of years (4 to 10, depending on planting density) to achieve maximum evapotranspiration for a citrus grove. The amount of water applied to a young grove should take into account the canopy size and root zone area. Average water use of about 1 gal/ day has been reported for newly planted Midsweet on Carrizo citrange rootstock trees over a 9-month period. However, irrigation systems that also water grassed areas in row middles and between trees can raise the water requirements considerably. One study found nearly double the water use by 2-year-old citrus trees with grass cover as compared to trees with bare soil.

In a Florida study, water use for 5- to 6-year-old Valencia orange trees on Swingle citrumelo rootstock peaked at about 15 gal/day in May. From November through January, ET rates averaged only about half the May rate. Water use was found to increase at a rate of about 20% per year for years 4 through 6. Typical spring applications for young trees are 2, 4, 7, 12, and 17 gallons per day from the first to the sixth year, respectively. Mature trees will generally require from 0.16 to 0.18 inch per day during the spring dry season.

Determining When to Irrigate

Oftentimes, the trees themselves are the best indicators of the need for irrigation. An indicator of plant water stress is the visual appearance of the plant. Unfortunately, however, growth processes may cease before visual wilting occurs, and yield and/or fruit quality reductions may have occurred for some time before wilting is seen.

Irrigation timing is often determined by calendar methods (for example, every 7 days in the winter and every 3 days in the spring) or based on long-term average irrigation requirements. However, these methods fail to consider the tremendous effect of climatic variability on tree water use. Therefore, the use of long-term average values may not be adequate during periods of hot, dry days. These values may result in overirrigation during periods of cool, overcast days, especially if rainfall is not considered. Day-to-day climatic conditions are highly variable during much of the year in Florida because of cloud cover and the random nature of rainfall occurrences.

Because of the these limitations, irrigations should be scheduled based on the soil water status. Scheduling techniques include using a water balance procedure based on the estimated crop water use rate and soil water storage, direct measurement procedures based on instrumentation to measure the soil

Table 17-5. Estimated irrigation interval (days) and net application volume (NAV) for healthy, mature trees planted at a density of 140 trees/acre demonstrating the effects of soil total available water (TAW), root zone depth, crop coefficient (Kc), and allowable depletion (AD). Intervals calculated based on estimated water that can be stored beneath the 180-ft^2 area wetted by a microsprinkler with a 15-ft diameter pattern (Table 17-3) and long-term Penman ETo for Lakeland (Table 16-1).

Month	Penman ETo (in./day)	Kc	Estimated citrus ET (in./day)	(gal/tree/day)	Allowable depletion (%)	Candler 0.75 in./ft 3.0 ft 253 gal NAV (gal)	Interval (days)	Riviera 1.0 in./ft 2.0 ft 225 gal NAV (gal)	Interval (days)	Winder 1.40 in./ft 1.5 236 gal NAV (gal)	Interval (days)
Jan	0.07	0.90	0.06	12	67	170	14	151	13	158	13
Feb	0.11	0.90	0.10	19	33	83	4	74	4	78	4
Mar	0.13	0.90	0.12	22	33	83	4	74	3	78	4
Apr	0.18	0.90	0.16	31	33	83	3	74	2	78	3
May	0.20	0.95	0.19	36	33	83	2	74	2	78	2
Jun	0.21	1.00	0.21	40	33	83	2	74	2	78	2
Jul	0.20	1.00	0.20	38	67	170	5	151	4	158	4
Aug	0.19	1.00	0.19	36	67	170	5	151	4	158	4
Sep	0.16	1.00	0.16	30	67	170	6	151	5	158	5
Oct	0.12	1.00	0.12	23	67	170	7	151	7	158	7
Nov	0.09	1.00	0.09	17	67	170	10	151	9	158	9
Dec	0.06	1.00	0.06	11	67	170	15	151	14	158	14

water status, and using soil water status instrumentation in combination with a water balance procedure. These procedures require a knowledge of the crop water requirements, effective root zone, soil water-holding capacity, and irrigation system capabilities in order to schedule irrigations effectively.

Water Budgeting for Irrigation Scheduling

Two questions must be answered in order to schedule irrigations: when to irrigate, and how much water to apply. A water budget (Figs. 17-4 and 17-5) accounts for irrigation and rainfall as well as water use by the trees. The water budget procedure can be used to answer both questions.

The crop root zone can be visualized as a reservoir where water is temporarily stored for use by the crop. Inputs to that reservoir occur from both rainfall and irrigation. If the capacity of the soil water reservoir (the volume of water stored in the crop root zone) and the daily rates of ET extraction from that reservoir are known, the date of the next irrigation and the amount of water to be applied can

be determined. Thus, ET and soil water storage in the plant root zone are the basic information needed to use the water budget method for irrigation scheduling.

The most significant crop factors that affect ET from a well-watered citrus grove are the size of tree, overall tree health, the time of year, and the leaf area with respect to the ground surface on which radiation is incident. Methods of expressing plant size and leaf area include the degree of ground cover or percentage of canopy coverage. ET rates are greatest when the entire soil surface is covered by the crop canopy. Exceptionally low relative humidity and high winds will increase ET rates above normal. Hot dry winds may raise the ET rates of isolated groves by 25% or more above normal, although such periods are usually brief.

Citrus trees do not shade the ground completely, especially during their early stages of growth, and evaporation from the dry soil surface between plants is low. When the tree canopy is not

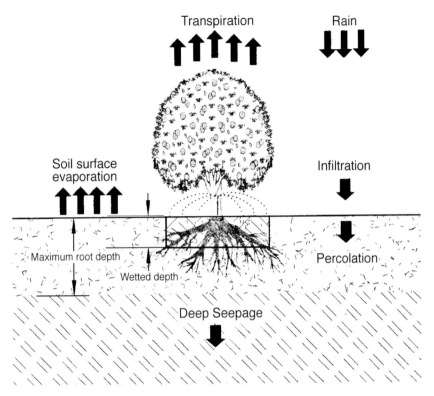

Figure 17-4. Components of citrus water budget for microirrigated Ridge citrus.

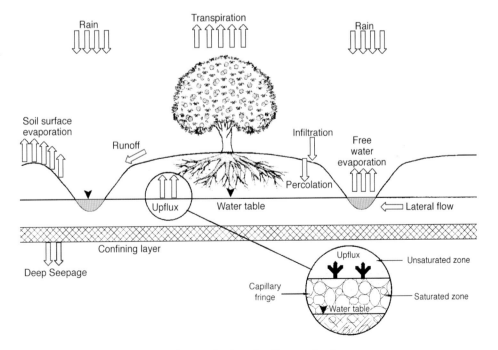

Figure 17-5. Components of water budget for flatwoods citrus.

complete, the ET rate is influenced strongly by the area of leaf surface that is intercepting sunlight (the percentage of soil surface shaded by the crop). For this reason, ET during early years is considerably less than the ET that would occur from a complete canopy. As growth increases, ET reaches its maximum at nearly complete ground cover. ET measurements indicate that when the percentage of ground covered by the canopy is above 60% to 70%, full ground cover and full ET rates can be assumed.

Immediately after an irrigation, evaporation from the wet soil occurs at approximately the same rate as full cover ET, but as the soil dries, rates of evaporation are quickly reduced. Thus, frequency of irrigation plays an important role in determining evaporation losses from the soil, especially when the entire soil surface is wetted.

<u>The Water Budget</u>

The water budget procedure is also called a water balance or bookkeeping procedure. It is similar to keeping a bank account balance. If the balance on a starting date and the dates and amounts of deposits and withdrawals are known, the balance can be calculated at any time. Most importantly, the time when all funds (water) would be withdrawn can be determined so that an overdraft is avoided (or an irrigation can be scheduled).

The water budget equation for irrigation scheduling on a daily basis can be written as follows:

$$S = R + I - ET - (D + RO) \qquad \textbf{Eq. 17-2}$$

where,

S = change in available soil water (inches)
R = rainfall measured at the field site (inches)
I = irrigation applied (inches)
ET = evapotranspiration estimated from pan
evaporation or other method (inches)
$D + RO$ = drainage (D) and runoff (RO): calculated as rainfall in excess of that which can be stored in the soil profile to field capacity (inches).

The soil water content on any day (i) can be calculated in terms of the water storage on the previous day (i-1), plus the rain and irrigation, and minus the ET, drainage, and runoff that occurred since the previous day as:

$$S_i = S_{i-1} + R + I - ET - (D + RO) \qquad \textbf{Eq. 17-3}$$

where,

S_i = soil water content on a particular day
S_{i-1} = soil water content on the previous day

The starting point for irrigation scheduling is often after a thorough wetting of the soil by irrigation or rainfall. This brings the soil reservoir to full capacity and S_i to TAW. If this does not occur, the initial available soil water must be determined by direct observation (measurement or estimation).

Daily measurements or estimates of ET are subtracted from the available soil water until the soil water has been reduced to the allowable depletion level. At that point, an irrigation should be applied with a net amount equivalent to the accumulated ET losses since the last irrigation. As a result, the soil reservoir is recharged to full capacity, and the depletion cycle begins again. Figure 17-6 illustrates the water budget calculations for a Florida sandy soil with a total available water of 1.5 inches in the tree's root zone. It was assumed that a management decision was made to irrigate when 1/3 of the available soil water (0.5 inch) was depleted. In this example, that level of depletion occurred after 3 days. At that time, an irrigation should be scheduled to replenish the 0.5 inch of soil water depleted.

The water budget procedure also accounts for rainfall. Rainfall is entered into the budget along with irrigation applications. If large rainfalls occur, only that portion required to restore the soil water content to field capacity will be effective. Greater amounts of rain will either run off of the soil surface or drain below the plant root zone.

The management decision concerning the level of allowable water depletion (AD) is one that will need to be made by each irrigation manager. It will vary depending upon soil, tree size, and climatic factors. Commonly it will vary during the growing season. For example, AD may be set at 2/3 from

July to January, but it should be decreased to 1/3 during critical growth stages during bloom through the June drop period. Decreasing AD increases the frequency of irrigation (but decreases the amount per irrigation) and provides a more favorable crop root environment to reduce water stress during critical growth stages. Decreasing AD will generally result in greater irrigation requirements because the soil will be maintained wetter and thus rainfall will be less effective. More frequent irrigations will also promote increased evaporation from the soil surface.

The soil depth to be managed for irrigation must be refined by field experience. In flatwoods areas, the managed root zone will typically be only 1 to 2 feet. Even on the Ridge, the managed citrus root zone should be much less than the 5- to 8-foot depths where roots exist. Rather, the irrigated zone should be the upper 2 to 3 feet of the root zone where the majority of the roots are located. This practice also has the advantage of allowing some soil capacity to store rainfall in the soil.

Daily ET values for specific water use periods should be estimated from pan evaporation or ET equations. If current daily ET estimates are not available, the use of soil moisture measurement instrumentation or the installation of evaporation pans should be considered. The use of long-term average ET data will result in scheduling errors because day-to-day ET rates are highly variable. Long-term average ET data can be used as a guide for daily ET estimates, but they will need to be modified for climatic variability. For instance, they will need to be increased during long-term hot, dry periods, and decreased during mild weather periods.

Soil Moisture Indicators for Irrigation Scheduling
Devices for monitoring soil moisture have been available for many years. Among them, tensiometers are the instruments most commonly used for scheduling irrigations. Gypsum blocks are also used on a limited basis. These devices register the water in the soil, in terms of soil water tension, at the depth at which the device is placed. They have

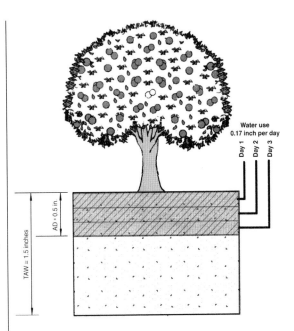

Figure 17-6. Illustration of water budget for citrus with 1.5-inch TAW and 1/3 allowable depletion.

the advantage of providing a measurement of soil water status rather than relying upon estimates of ET to calculate soil water content. When placed in the plant root zone, they indicate the soil moisture that the plants are experiencing. Disadvantages of soil moisture sensors include their cost, labor requirements for reading and servicing, and need for periodic calibration. They also measure soil moisture only in the area around the sensors. Thus, a number of instruments may need to be installed to accurately represent a given field.

No single soil water tension level can be recommended as indicating the need for irrigation when using tensiometers. For the same reasons that allowable soil water depletion is not constant for all crops and conditions, critical soil water tension also varies with soil and crop conditions and management objectives. The level also varies with depth of placement of the tensiometer. On Ridge sands, irrigation should be started at 8 to 12 cbar in the spring to avoid reduced fruit set. On flatwoods soils, irrigation should be started at 13 to 15 cbar. In the fall and winter when water stress is less critical, irrigation can be started at 15 cbar on

Ridge soils and 30 cbar on most flatwoods soils. Field experience is required to refine the interpretation of instrument readings for a given crop and management system. Tensiometers or any other soil moisture monitoring device are most effectively used in combination with ET data. The device is read to determine when to irrigate, and the ET data are used to calculate the volume of water lost since the last irrigation. From this, the volume to be replaced can be determined.

Durations of Water Applications

The optimum length of time to operate a microirrigation system at each irrigation depends upon the individual system owner's management procedures as well as the irrigation system design, the soil hydraulic properties, and the root zone. Therefore, the optimum time may be unique for each system manager because of his management constraints, even with identical soils and irrigation systems. There are, however, constraints on the maximum and minimum periods of time that the systems should operate.

<u>Minimum Time of Operation</u>

The minimum operation time for a microirrigation system should allow the water to move into that portion of the crop root zone where the majority of the roots actively involved in water uptake are located. Because the soil surface in the irrigated area will be wetted, nonproductive evaporation losses will occur. Those losses will be greatest immediately following each irrigation while the soil surface is wettest. Irrigation applications should be of sufficient duration to allow the applied water to penetrate into the soil to a depth where the bulk of the roots are located.

Since a microirrigation system is only capable of irrigating a small portion of the total tree root zone, it may be necessary to operate the irrigation system very frequently (often daily during extended droughts). The frequent water applications allow maintenance of nearly constant low soil water potentials (moist soil conditions) that minimize water stress in the trees.

<u>Maximum Time of Operation</u>

The maximum operating time for a microirrigation system depends on rooting depth, soil texture, and antecedent conditions. Generally, irrigations should be scheduled to thoroughly wet the soil to the depth of the effective (managed) root zone. Unless irrigating with high salinity water, care should be taken to not apply water volumes that will wet to depths below the root zone. Overirrigation increases the chances of nutrient leaching below the root zone.

<u>Wetting Depths</u>

The depth to which applied water will infiltrate depends on the soil texture, antecedent soil moisture conditions, rate of application, the presence of restrictive layers, channelizing actions of worms, ants, etc., plus factors related to the application method (i.e., rain, drippers, overhead sprinklers). Sandy soils typically have little lateral movement of applied water compared to soils with higher clay content (Fig. 17-7). In contrast, water applied to sandy soils will infiltrate to deeper depths more rapidly.

Water Table Considerations

Water table levels fluctuate widely in the coastal flatwoods areas of Florida during the rainy season due to the effects of variable soils and high-intensity rainstorms. Soil water drainage is especially important in these areas since root damage may occur during prolonged conditions of high

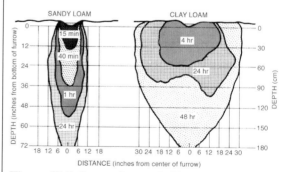

Figure 17-7. Rates of water infiltration into sandy loam (left) and clay loam (right) soils (redrawn from Cooney and Peterson, 1955).

water table. Effective water management, which includes both irrigation and drainage on these poorly drained soils, is essential for profitable citrus production.

Shallow water tables in bedded groves can significantly augment water available for root uptake. High water tables permit upflux of water from the water table into the root zone, and may decrease the need for supplemental irrigation. Capillary rise may be up to 12 inches above the free water surface in a loamy fine sand, but less than 6 inches in a fine sand. In sandy soils, when the water table depth is 20 inches below the mid root zone, there is generally insufficient upflux to meet the tree ET demand.

The rate of water table recession following heavy rainfall is dependent upon the antecedent conditions, soil series, bed height, drainage structures, and gradient of the water table. The time required for the water table to drop 6 inches following heavy rainstorms may be a few hours in well-drained soils to over 3 days in heavy depressional soils.

Therefore, whenever water table levels are elevated due to rainfall or irrigation, irrigations should be delayed until the water table subsides to where it cannot supply sufficient water to meet the tree needs. Observation wells with floating water table indicators can be used very effectively to track the water table position.

Field Water Balance
Even in well-designed systems, water supplied for irrigation must exceed actual crop needs. Water losses may occur because of inefficiencies in the conveyance system, evaporation and wind drift (if water is sprayed through the air), surface runoff, or percolation below the root zone. Though impossible to eliminate, these losses can be minimized through good management practices, and they must be considered when determining the total (or gross) irrigation water requirement.

The total irrigation water requirement is the total amount of irrigation water that is required for crop production (including ET) and other needs, such as freeze protection and leaching of salts, plus all losses incurred in delivering water to the crop. In humid areas such as Florida, a large part of the crop water requirement can be provided by rainfall. Effective rainfall is stored in the root zone and available for crop use. It proportionally reduces the amount of water that must be pumped for irrigation.

Runoff losses are normally minimal on sandy soils in Ridge groves. Application losses, including evaporation and wind drift, can occur during irrigation, especially from microsprinkler systems on hot, dry, windy days. These losses are, however, relatively small during periods of low radiation, low wind velocities, and high humidity. Also, water that evaporates during application, or that is intercepted and later evaporates from soil, plant, or other surfaces is not entirely lost. Rather, some evaporation during application promotes cooling. This reduces ET that would have occurred if the intercepted water had not been evaporated.

Evaporation and wind drift losses can be minimized by irrigating at night, early mornings, and late afternoons when wind and evaporation rate are generally lowest. However, the effects of prolonged wetting on disease development should be considered.

Deep percolation losses from well-designed irrigation systems can be minimized by good irrigation management. If water is applied uniformly and the water-holding capacity of a soil is not exceeded, water losses to deep percolation will be minimized. If saline water is used for irrigation, it may be necessary to leach excess salts from the crop root zone by adding water in excess of the soil water-holding capacity. However, excess irrigation for leaching should be required only during extended dry periods in Florida because rainfall normally leaches salts.

If the losses are kept to a minimum, most of the irrigation water applied will evaporate or transpire in proportion to the climatic demand. Irrigation can be reduced by anticipating rainfall and providing soil storage capacity (that is, irrigating to less than field capacity to leave room for rainfall storage) to increase rainfall effectiveness. However, in the spring, trees should not be water stressed because stress then can reduce fruit set.

Interactions with Other Horticultural Practices
Fertilization
Proper water management is essential to achieve high uptake and utilization of applied fertilizers. Research results have shown that if water stress is a limiting factor, economic returns from fertilizer applications will be reduced. Excess irrigation may result in leaching of nutrients, causing them to be unavailable to trees, and may move nitrates into the groundwater. Insufficient irrigation may result in low fertilizer uptake and lack of growth and yield resulting from water stress. Fertigation (injecting water soluble fertilizers with the irrigation water) can be an economical and effective method of fertilizer application. With fertigation, small fertilizer amounts can be applied at frequent intervals, thus making nutrients available to trees throughout the year.

Tree Density
Groves that have been planted at high tree densities have a higher per-acre demand for soil moisture during the first several years. As the trees mature and hedge rows are formed, per-acre water requirements for higher-density plantings are similar to lower-density plantings that have become hedge rows. Irrigation practices may need to be modified for initial years in high-density plantings to compensate for greater root densities compared to conventional tree spacings.

Rootstock/Scion
Large trees require more water than smaller trees. Due to their larger size, grapefruit trees typically require more water than sweet orange varieties. During the first several years, trees budded on fast growing scions, such as rough lemon, will require more water per tree than trees on slower-growing rootstocks. Therefore, irrigation rate and timing should be adjusted based on rootstock, scion, and tree planting density.

Some rootstocks (such as rough lemon) have an efficient and abundant fibrous root system that allows them to extract water more readily from the soil than less efficient root systems such as those on sweet orange. In addition, the lemon rootstocks (Milam, Volkamer, rough) and sour orange are most tolerant to flooding while Cleopatra mandarin, Carrizo citrange, and Citrus macrophylla are least tolerant. Consideration should be given to the adequacy of drainage before selecting the rootstock on which the trees will be planted.

Weed Control
An effective weed control program is essential for groves watered with microirrigation systems. Weeds and high grasses disrupt the wetting pattern of microsprinklers and lower the use efficiency of applied water. Lack of weed and vegetation control will increase irrigation system maintenance costs and reduce the effectiveness of fertigation and chemigation applications. Excessive vegetation control in bedded groves (wide herbicide bands) will result in erosion of soil from bed tops and reduce the effectiveness of water removal through water furrows and swales.

Water is important in the activation of the soil residual compounds in herbicides. Irrigation may be needed to move the applied materials into the zone of germinating weed seeds, or into the root zone of actively growing weeds. Too much irrigation, however, may leach the herbicide beyond the weed's root zone, resulting in ineffective weed control.

Disease/Insect Control
Without adequate disease and insect control, even proper and efficient irrigation may not improve profitability. Some nematicides and insecticides can be applied efficiently through the irrigation system (chemigation), providing a low-cost application method. In flatwoods groves, adequate drainage is essential to ensure that summer sprays for mites, insects, and fungal diseases can be applied at appropriate times. Without efficient drainage, application of insecticides and fungicides can be difficult or impossible on water furrow sides of trees during rainy periods. Many times,

spray applications made when water furrows are even marginally wet result in equipment damage and disruption of the flow patterns in water furrows. As a result, extra costs are incurred because of the need to reshape water furrows to make them drain properly.

Freeze Protection

Irrigation can be used effectively for citrus frost and freeze protection if it is properly managed.

Microsprinkler irrigation has been shown to be effective in providing good protection for young trees and partial protection of mature trees. In general, the higher the volume of water provided by microsprinklers, the more effective the cold protection will be. Approximately 2000 gal/acre/hr is the minimum required to provide reasonable cold protection.

by Larry Parsons and Howard Beck

Weather is one of the most important factors that affects citrus growth and production. Citrus can be grown in a variety of arid and humid climates, and can withstand temperatures ranging from 28° to 105° F. However, citrus performs best in the range between 60° to 86° F.

In addition to certain temperature requirements, citrus also requires a certain amount of water for optimum growth. In Florida, it generally takes 48 to 50 inches of water per year to grow a crop. The majority of this water comes from rainfall. Irrigation is a supplement, but if rainfall is deficient in the spring, it has a major impact on yield. In arid regions, most of the water needs are met by irrigation, with rainfall being a secondary provider of water.

When scheduling irrigation, it is important to have good rainfall and temperature data. In Florida, one of the best sources of up-to-date weather information is the FAWN weather network on the Internet. FAWN stands for Florida Automated Weather Network, and is available at http://fawn.ifas.ufl.edu. It provides weather information from a number of locations throughout the state at 15-minute intervals. Most of the weather collection points are at the various research and education centers (RECs) around the state. The sites are listed in Table 18-1. FAWN provides data on air and soil temperatures, rainfall, dew point, relative humidity, windspeed, wind direction, and radiation. Air temperatures are measured at 2-, 6-, and 30-foot heights, and soil temperature is measured at a 4-inch depth (Fig. 18-1). It also indicates minimum and maximum values of several parameters. The mission of FAWN is to provide accurate, real-time agricultural-based weather data to a variety of users.

FAWN will be enhanced with additional weather stations and more Web-related services. The next phase of expansion will involve placing more stations in the Panhandle. Decision modules, such as disease prediction, crop phenology, and impacts on crops based on climate forecasts, are also being added to the Web site. There are several ways to search the weather database from its present home page. The available weather data parameters are listed in Table 18-2. When searching the data, a grower can find daily, weekly, or monthly summaries. Of particular use is the daily summary that

Figure 18-1. Configuration of a typical FAWN weather station.

Table 18-1. FAWN Weather collection sites.

Alachua	Dover	Ocklawaha
Apopka	Ft. Lauderdale	Okahumpka
Avalon	Ft. Pierce	Ona
Belle Glade	Hastings	Peirson
Bradenton	Homestead	Putnam Hall
Brooksville	Immokalee	Tavares
Citra	Lake Alfred	Umatilla

gives the daily minimum and maximum temperature, rainfall, total radiation, calculated evapotranspiration (ET), and hours below certain temperatures. For scheduling irrigation, this daily summary gives much of the information needed to make good scheduling decisions.

Evapotranspiration (ET) is the combination of two processes: evaporation and transpiration. Evaporation is the vaporization of water from a free water surface, such as a lake or any moist surface such as soil. Transpiration is the movement of water vapor from the plant to the atmosphere. Various factors influence ET, including radiation, temperature, vapor pressure deficit (the difference between saturation vapor pressure and actual vapor pressure, or the dryness of the air), and windspeed. Most of these factors are included in the FAWN data set. FAWN also lists a calculated ET that can be used in irrigation management. Daily average ET varies from 0.07 to 0.20 inch/day from winter to summer. On certain clear, high-radiation, hot days, ET can be as high as 0.25 inch/day. During periods of prolonged high ET, irrigation schedules can be adjusted to replace the extra water lost by the plants.

Besides giving rainfall, temperature, and ET information that can help with irrigation management, FAWN weather data can be particularly valuable on freeze nights. Growers can follow changes in temperature throughout the state during the night. If cold air is blowing in from the north, one can watch the progression of the freezing weather as it moves down the state.

FAWN is not a weather forecasting system, but it provides timely information that growers can use to help make their own forecast. On predicted frost nights, the information from FAWN can be particularly useful, since humidity plays a major role in how cold it will get. FAWN provides both temperature and humidity (dew point and wet bulb) data. For definitions of terms used in frost protection, see Chapter 38: Microsprinkler Irrigation for Cold Protection.

If a grower knows the dew point and air temperature at sunset on a cold night, he can estimate the minimum temperature at his location. This method uses what is called the Brunt equation. If the sunset air temperature is 40° F and the dew point is 20° F, by using the chart shown in Fig. 18-2, a line can be drawn from the 20° F dew point to the diagonal sunset temperature at 40° F. A horizontal line is then extended from the point on the 40° F diagonal line to the left margin of the chart. The point where the horizontal line meets the left margin is the estimated minimum temperature for the next morning. In this case, the minimum temperature would be 22° F.

The method is simple and requires only two temperature measurements at sunset. A grower should measure the sunset temperature directly in his grove, because air temperatures can vary widely depending on elevation, topography, and height above the ground. He can determine the dew point by using a psychrometer. Since dew point temperatures do not vary greatly over a distance of a few miles, the grower can also use the calculated sunset dew point from the nearest FAWN site. While this FAWN dew point will not be the exact dew point temperature of the grove, it can be useful as an approximation.

Growers can either use the chart in Fig. 18-2, or they can use the FAWN site to estimate the

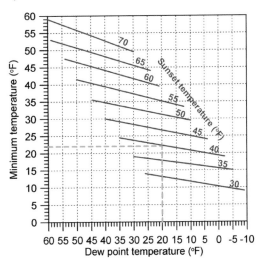

Figure 18-2. Minimum temperature prediction chart using air temperature and dew point temperatures observed at sunset.

Table 18-2. Weather data collected at FAWN sites.

Location	e.g. Lake Alfred
Date/Time	
Station	
Air Temperature	° F, 2, 6, and 30 feet
Soil Temperature	° F, 4 inches
Relative Humidity	%
Dewpoint	° F
Rainfall	inches
Wind Direction	°
Wind Speed	miles/hour (max. and min.)
Total Radiation	watts/m²

minimum temperature. To estimate this temperature, do the following:

1. Log onto the FAWN Web site – http://fawn.ifas.ufl.edu/

2. Obtain the dew point temperature at the site closest to your grove at the time nearest sunset.

3. Return to FAWN's main page and go to "Management Tools." Find the "Brunt Minimum Temperature Calculator."

4. Enter the dry bulb from your grove and dew point temperature obtained above. Click on "Calculate."

5. The estimated minimum temperature over sand and muck soils will appear on the calculator.

There can be a difference of a few degrees between the temperature calculated from the chart in Fig. 18-2 and the FAWN computer calculation, but either method will give an approximation of the minimum temperature.

This Brunt method for estimating the minimum temperature was designed to be used for a stable air mass of uniform moisture. If dry air moves in or winds increase noticeably, significant errors could be introduced.

Another advantage of FAWN is that it plots a calculated wet bulb temperature. The wet bulb temperature can be used as an indicator of when to turn the water off. There is always a risk when using water or microsprinkler systems for frost protection. Low humidities or wind can increase evaporative cooling, which can chill wetted trees below the air temperature. The wet bulb temperature is the lowest temperature to which air can be cooled by the addition of water. When the wet bulb temperature is 32° F or higher, the irrigation system can be stopped without danger of damage to the citrus trees. Considering the distance from the FAWN site and the estimation of the temperature, growers should wait until the wet bulb temperature reaches 34° F to be safe. In any case, it is not necessary to keep irrigating until all the ice has melted from the tree. When the wet bulb goes above freezing, growers can start to shut the irrigation systems down. By using the wet bulb temperature as a guide, growers can save substantial amounts of water and irrigation costs by stopping the irrigation when the wet bulb temperature goes above freezing.

In addition to the FAWN site, there are a number of other sites on the Internet that provide weather information. Several of these weather information sites come from television stations, industries, or government agencies. A listing of some of these sites is given in Table 18-3. Internet site addresses change from time to time, so a grower will need to update these site addresses. These Internet sites provide a wide range of information including weather forecasts, animated radar, and local conditions. Using this information, growers can better predict how to manage irrigation in their groves.

Table 18-3. Selected Internet sites for weather information.

Channel 2 NBC2 Online - Fort Myers Weather: http://www.nbc-2.com/weather/

Channel 9 WFTV - Orlando Weather: http://www.icflorida.com/partners/wftv/weather/index.html

Channel 10 NEWS WTSP - Tampa Weather: http://www.wtsp.com/weather/index.htm

Channel 13 FOX 13 WTVT - Tampa Weather: http://www.wtvt.com

The Ultimate Citrus Page: http://ultimatecitrus.com/index.html

Arapaho Citrus Management, Inc.: http://www.arapahocitrus.com

NBC Intellicast: http://www.intellicast.com

Accuweather: http://www.accuweather.com/adcbin/index

Florida Weather Center: http://www.weathercenter.com

NOAA Weather Site: http://www.goes.noaa.gov/

The Weather Channel: http://www.weather.com

National Climatic Data Center: http://lwf.ncdc.noaa.gov/oa/ncdc.html

Dept. of Atmospheric Sciences, University of Illinois at Urbana-Champaign: http://www.atmos.uiuc.edu

NOAA National Weather Service Homepage: http://www.nws.noaa.gov

Internet Weather Resources: http://tdc-www.harvard.edu/weather.html

Agriculture Weather: http://www.agriculturalweather.com

National Weather Service, Tampa Bay, FL: http://www.srh.noaa.gov/tbw

Chapter 19. Drainage

by Brian Boman and Dave Tucker

Water table levels fluctuate widely in the coastal flatwoods areas of Florida during the rainy season due to the effects of variable soils, nonuniform rainfall, and high intensity rainstorms (Fig. 19-1). Drainage of the soil water is especially important in the wet season because citrus root damage may occur under prolonged conditions of high water table. Effective water management, which includes both irrigation and drainage on these poorly drained soils, is essential for profitable citrus production.

Both surface and subsurface drainage are generally required for citrus grown in flatwoods areas.

Drainage systems in flatwoods groves consist of systems of canals, retention/detention areas, open ditches, subsurface drains, beds, water furrows, swales, and the pumps required to move the drainage water. These systems require continued good maintenance in order to minimize the chances of root damage from prolonged exposure to waterlogged soils following high intensity rains. Drainage systems should generally be designed to allow water table drawdown of 4 to 6 inches per day, which should be adequate to prevent root damage.

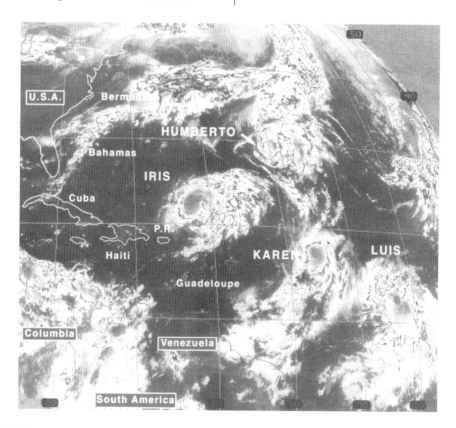

Figure 19-1. Satellite view of the "Parade of Storms" taken August 30, 1995 by the GOES-8 satellite in geosynchronous orbit 22,300 miles above the earth.

Flooding Damage

Research has shown that there is potential for water damage to citrus trees if roots are submerged in water for four days or more during frequent extended summer rains. During the cooler months of December through February, citrus trees can tolerate flooded conditions for longer periods than during the hot summer months.

In order to understand how citrus is damaged by flooding, one needs to understand the soil-water dynamics in a grove. In most flatwoods soils, a clay or organic layer within 20 to 48 inches of the surface acts as a barrier to downward movement of excess water. As a result, water must move laterally to be drained from saturated soils. The rate at which water moves through soil in units of distance/time (in./hr or ft/day) is called hydraulic conductivity. Sands typically have saturated hydraulic conductivity of 20 inches per hour or more, while the saturated hydraulic conductivity of many flatwoods soils with significant clay content is in the range of 0.1 to 0.2 inch per hour. Hydraulic conductivity varies tremendously with soil moisture content. For example, when saturated, the hydraulic conductivity of an Oldsmar fine sand soil is about 20 inches per hour. The rate drops to less than 0.05 inch per hour when soil tension reaches about 85 cbar. Porosity is the volume of pores (air space) divided by the total volume of soil. The higher the porosity, the more water that will drain out of the soil when the water table is lowered.

Gravity is the force that moves water in saturated soils. Therefore, water in saturated soils moves from a higher level to a lower level. The difference in elevation between two free water surfaces is called the hydraulic gradient. The steeper the gradient, the faster water will drain from the soil. Excessive rainfall will cause a perched water table to develop above the hardpan in flatwoods soils. Rainfall that infiltrates moves downward to the free water surface, and then must move laterally towards the water furrow for drainage to occur. As a result, a mound occurs in the water table between the water furrows.

The time required for this mound of water to recede back to level conditions varies greatly with soil texture, and the condition and quality of the drainage system. The time may vary from two days in coarse-textured soils such as Pineda, Myakka, and Immokalee series soils to over a week in heavier-textured soils such as Winder. Under normal conditions where drainage systems are adequately designed and maintained, water usually recedes at rates adequate to prevent root damage. Problems occur when heavy rainfall results in elevated water tables for several weeks. Once the soil is near saturation, it takes only a little rainfall to fill the available pore spaces in the soil, and the root zone becomes saturated. When the air is excluded, anaerobic conditions which promote root decay can develop.

If the surface runoff is removed quickly following rainfall events and the water furrows are free of water, the initial drawdown of the perched water table in the beds may be relatively fast. Since the difference in elevation between the free water surfaces in the bed middles and the water furrows is relatively large, maximum drainage rate will result. As the water drains from the beds, the height of the free water surface in the beds decreases, and the mounded water table becomes less pronounced. The rate of drainage from the beds will slow due to a decreased hydraulic gradient. A similar situation occurs when the water level in the furrows is relatively deep. The decreased gradient prevents water from draining from the beds at a high rate.

Observation wells are good tools for observing soil-water dynamics. They are the only reliable method for evaluating water-saturated zones in sites subject to chronic flooding injury. These wells can also be used to measure the rate of water table drawdown, which is the real key to how long roots can tolerate flooding. Observation wells constructed with float indicators allow water tables to be visually observed while driving by the well site. Local offices of the Natural Resources Conservation Service (NRCS) can assist with water table observation, well construction, and monitoring.

Short-term estimates during flooding stress can be obtained by digging into the soil and smelling soil and root samples. Sour odors indicate an oxygen deficient environment. The presence of hydrogen sulfide (a rotten egg odor) is an indication that feeder roots are dying.

Anaerobic bacteria (which grow only in the absence of oxygen) will develop rapidly in flooded soils and contribute to the destruction of citrus roots. In a field survey of poorly drained groves, toxic sulfides were formed by anaerobic sulfate-reducing bacteria at more than half the locations. Nitrites, formed by nitrate-reducing bacteria, and other organic acids that are toxic to roots were also found in these flooded soils.

Improper bed construction has been linked to areas with chronic root damage in several groves. Severe sulfide problems have often been found in grove areas that were developed over old swamps that were filled in before planting. Palmetto, cabbage palms, and other decomposable organic debris were frequently buried in these areas where land was leveled during preparation. It can take many years for Palmetto roots and stems to decompose in this environment. Certain organic acids in Palmetto, grass, and citrus roots provide a good source of energy for bacteria that require both energy and sulfates in order to reduce sulfates to sulfides. Thus, it is possible for citrus roots to contribute to their own destruction by acting as an energy source for these bacteria. Only small amounts of sulfur (3 ppm) are required for the bacteria to function at peak capacity. The forms of sulfur used by the bacteria can be elemental sulfur, thiosulfate, sulfites, or sulfates, which are usually present in all Florida soils.

Using topography alone as a diagnostic factor to assess potential for flood damage may be misleading. Flooding injury can occur in obvious spots such as poorly drained depressions, but it may also be present where least expected. It has been observed on hillsides, on relatively high ground, on isolated areas of flat land, and even on raised beds.

Hillsides may have pockets of clay. In flat areas, the problem may be impervious clay, marl, or organic-layered pockets that hold the water and prevent movement. Even beds in apparently uniform sandy areas can have buried palmetto roots and organic materials. These areas are subject to root damage, since the soils are able to support bacteria that can quickly generate toxic hydrogen sulfide if flooded. Old pond sites are prone to severe flooding injury. Trees on the periphery of old pond sites are often damaged as much as those in the middle.

Good drainage allows air to move into the soil and prevents oxygen-deprived conditions. Flooding stress is usually less when water is moving than when water is stagnant, for anaerobic bacteria cannot multiply if oxygen is present. Also, a higher subsoil pH may help to delay, for a few days at least, the death of citrus roots under flooded conditions.

With experience, flooding injury can be diagnosed during periods when groundwater levels are high. Even before there are visible tree symptoms, auguring and digging in the root zone may give an estimate of future tree condition. Indications of problems include high water tables with saturated soils in the root zones, sloughing roots, and sour odors in the soil. When the water table recedes, visible damage to the trees may become more obvious. New feeder roots appear and grow rapidly on trees that have survived and received adequate irrigation.

Symptom expression of damage may occur over a period of time depending on the severity of root damage. Symptoms usually start to show up after the water table drops and the soil dries out. Root damage symptoms include leaf yellowing, chlorosis, wilting, fruit drop, leaf drop, and dieback. Often root damage is so severe that trees may go into a wilt even though water furrows are still wet. Because the root system was pruned by the flooding, the full extent of damage may not be known for several months or until drought conditions occur.

Young trees are often more sensitive to flooding and may develop symptoms resembling winter chlorosis. More subtle symptoms include reduced growth and thinner foliage. This can occur at locations only a few inches lower in elevation than the surrounding area. Harvesting operations in a grove even after recent flooding may also further damage surface roots that have been injured by the flooding.

Hot, dry conditions following flooding will hasten the onset of stress and symptom expression. The reduced root system resulting from summer flooding is incapable of supporting the existing tree canopy. When this occurs, irrigation management becomes critical. Irrigation must provide moisture to a depleted (shallow) root system. Excessive water could compound existing problems. If root system damage is extensive and tree canopy condition continues to deteriorate with permanent wilt and foliage dieback, some degree of canopy pruning may be necessary to reestablish a satisfactory shoot/root balance.

Light, frequent irrigations will be required until the root zone has been reestablished. Subsurface moisture should be maintained to promote root growth into the lower root zone. If root damage is severe, frequent irrigation may even be required throughout the winter months, especially if dry winds persist. If irrigation water is high in salts, frequent irrigations are essential to prevent salt buildup, which will compound the flooding problem.

When trying to assess flood damage, *Phytophthora* foot rot problems may also need to be considered. However, if *Phytophthora* was not a problem before the flooding, excess flooding will not necessarily create one. Therefore, do not make costly soil or foliar fungicidal applications for the control of foot rot and feeder root rot unless soil propagule counts reveal such treatments are warranted. Soil and root conditions should be evaluated after the flooding has subsided and the potential for fungal invasion has been determined. If there are high propagule counts, *Phytophthora* root rot can accentuate the consequences of flooding injury. While

certain fungicides can help protect new roots during development if there is a *Phytophthora* problem, they will not bring dead roots back to life.

If flooding occurs, tree management must be intensified to minimize the effects of stress on water-damaged trees. Flooding will not always damage tree root systems, but trees should be closely monitored for symptoms. Duration of flooding conditions, rate of water table drawdown, presence of sulfur or organic matter in the soil, nature of the soil, tree age, rootstock, and root condition are all factors to be considered when trying to evaluate flooding injury and manage tree recovery. Other cultural practices should be adjusted to minimize stress on water-damaged trees. Fertilization rates and schedules may need to be adjusted for flood-damaged trees. Light ground applied or foliar fertilizer applications on a more frequent schedule are preferred until the root system becomes reestablished. Once the immediate drainage problem has been alleviated, the appropriate course of action is to wait, observe, and let tree response guide your actions.

Considerations for Plastic Drain Tiles
The initial consideration of suitability of a drainage system on a particular area requires a topographic survey. Since most of the flatwoods citrus areas are either nearly level or of basin-type topography, sufficient slope for gravity outfall probably won't exist. Therefore, a drainage pump for the outlet will most likely be needed. Aerial photographs, along with the survey, should facilitate selecting a suitable site for the drainage sump in a low area of the grove. Before constructing the drainage system, check with the appropriate water management district to ensure that the system will not impact wetlands.

A survey will also help determine if leveling of the land is required. Open ditches and water furrows may be required to remove the bulk of surface water. If shallow beds are to be constructed, they should be designed to minimize the quantity of earth moved in the leveling process. Drains should be installed in the direction of greatest slope.

There are several important soil factors that should be obtained from a soil survey of the area prior to the drain design. These factors include:

1. Soil types
2. Thickness of various strata
3. Continuity of strata
4. Position of strata with respect to ground surface
5. Hydraulic conductivity and porosity of various strata
   a. The hydraulic conductivity is the rate at which water moves through soil in units of distance/time (in./hr or ft/day). It varies tremendously with soil moisture content and is usually specified under saturated conditions.
   b. Porosity is the volume of pores (air space) divided by the total volume of soil. It determines how much water can drain out of the soil as the water table is lowered.

Several water factors need to be considered to ensure a proper design of the drainage system. Historical records can be examined to determine the relationship between rainfall and water table fluctuations. The source of all water coming into an area must be determined. If the water table is built up by irrigation, it may indicate that there is an improper system design, and that better management or a change in design may alleviate some of the water table problems. Seepage from reservoirs and ditches is often a source of high water tables in nearby areas of the grove. Money spent on perimeter ditches or throw-out pumps may decrease the drainage requirement. If there are free-flowing artesian wells, they should be capped or plugged.

Corrugated plastic drain tubing (Fig. 19-2) should meet the requirements of ASTM Standards F405 and F667, which, among other aspects, define quality classes. Standard tubing is satisfactory for most agricultural applications. Heavy duty tubing is recommended where wide trenches are required, where side support is poor, in narrow trenches

(where width is less than 3 pipe diameters wide), and where rocky soil conditions are expected.

Tubing normally used in agricultural operations is 4 inches or greater in diameter. The tubing is available in coils of various lengths, depending upon the diameter of the tubing. Water enters the plastic tubing through small openings located in the valleys between corrugations. The flexible drain tubing gains most of its load-bearing capacity by support from soil at the sides of the tubing. A load on top of the tubing causes the sidewalls to bulge outward against the soil. The soil resists the bulging, and the effect is to give the tubing a greater load-bearing capacity. The tubing must have sufficient strength to withstand the soil load without excessive deflection, collapse on top, or failure of the sidewalls.

Corrugated plastic drain tubing has several desirable characteristics. The tubing is light and flexible, weighing about 85 lb per 250 ft of 4-inch tubing. The tubing allows good alignment in unstable soils. The flexibility of the tubing allows long lengths on each roll with few joints required in the field, making good installations relatively easy Fig. 19-3.

Soils

Soils that are satisfactory for citrus in Florida but need profile subsurface drainage are listed in Table 19-1. The actual soil characteristics of soil profiles

Figure 19-2. Four-inch corrugated drain tubing with polyester sock.

Table 19-1. Depth and spacing for 4-inch plastic drains with synthetic envelope (adapted from Ford et al., 1985).

Series	Order	Minimum Depth (inches)	Minimum Spacing (ft)	Maximum Depth (inches)	Maximum Spacing (ft)	Spacing increase for each extra inch of depth (ft)	Ochre potential
Adamsville	Entisol	46	71	58	149	6	slight
Blichton	Ultisol	40	53	52	99	4	severe
Bradenton	Alfisol	40	40	52	79	3	slight
Broward	Entisol	28	57	34	90	5	slight
Charlotte	Spodosol	46	71	58	149	6	moderate
Delks	Spodosol	40	56	52	102	4	moderate
Delray	Mollisol	46	71	58	149	6	slight
Eau Gallie	Spodosol	46	60	58	130	6	severe
Elred	Spodsol	46	54	58	112	5	moderate
Farmton	Spodsol	46	63	58	115	4	severe
Felda	Alfisol	46	61	58	133	6	slight
Floridana	Mollisol	46	60	58	108	4	severe
Ft. Drum	Inceptisol	46	60	58	125	5	slight
Hilolo	Alfisol	46	32	58	64	3	severe
Holopaw	Alfisol	46	71	58	146	6	moderate
Immokalee	Spodosol	46	67	58	116	4	severe
Jumper	Ultisol	46	55	58	101	4	slight
Kanapaha	Ultisol	46	71	58	112	3	moderate
Lawnwood	Spodosol	46	43	58	78	3	severe
Lochloosa	Ultisol	46	56	58	104	4	slight
Malabar	Alfisol	46	101	58	162	5	severe
Micanopy	Alfisol	40	45	52	80	3	severe
Myaka	Spodosol	46	56	58	120	5	severe
Narcoosee	Spodosol	46	69	58	134	5	severe
Nettles	spodosol	46	66	58	125	5	severe
Nobleton	Ultisol	40	61	52	124	5	slight
Oldsmar	Spodosol	46	85	58	142	5	severe
Ona	Spodosol	46	85	58	142	5	severe
Parkwood	Alfisol	46	98	58	169	6	slight
Pendarvis	Spodosol	46	67	58	112	4	severe
Pepper	Spodosol	46	67	58	112	4	severe
Pineda	Alfisol	46	89	58	117	2	moderate
Pinellas	Alfisol	46	108	58	178	6	slight
Placid	Inceptisol	46	108	58	178	6	moderate
Pompano	Entisol	46	71	58	147	6	slight
Riviera	Alfisol	46	86	58	129	4	slight
Seffner	Entisol	46	72	58	126	4	moderate
Sparr	Ultisol	46	71	58	146	6	slight
St. Johns	Spodosol	46	74	58	125	4	severe
Susanna	Spodosol	46	73	58	110	3	severe
Tantile	Spodosool	46	86	58	131	4	severe
Tuscawilla	Alfisol	46	56	58	112	5	severe
Valkaria	Entisol	46	71	58	145	6	moderate
Vero	Spodosol	46	68	58	102	3	severe
Wabasso	Spodosol	46	70	58	115	4	severe
Wacahoota	Ultisol	46	91	58	121	2	severe
Wachula	Spodosol	46	85	58	128	4	severe
Waveland	Spodosol	46	56	58	86	2	severe
Winder	Alfisol	46	54	58	99	4	slight

Figure 19-3. Rolls of 4-inch corrugated drainage tubing.

have pronounced effects on depth and spacing of drains. It is common to find several different soil types in the same field so that adjustments in drain spacing may be necessary. The spacings at the minimum drain depth should result in a good rooting volume for the trees located midway between drain lines. Additional depth or closer spacing should result in additional rooting volume.

Drain Depth

The drain depth is often controlled by the depth of the outlet. Drains should be placed in the most permeable layer possible. However, the drain tubing should be installed at a constant slope, even if the tubing is located in a less permeable horizon for a portion of its length. The drains should have a minimum of 24 inches of cover to prevent collapse from traffic and heavy equipment.

Alignment

Changes in the horizontal direction should be minimized and made in such a way that the specified grade is maintained. Any change of alignment should be made with a gradual curve, the use of manufactured bends or fittings, or the use of junction boxes or manholes.

Drain Capacity

The drainage coefficient should be at least 0.5 to 0.75 inch per day for most groves Table 19-1 is

based on a drainage removal coefficient of 0.75 inch per day). This will allow water table drawdown of 4 to 6 inches per day, which should be adequate to prevent root damage. If surface water must also be removed, the drainage coefficient should be doubled to accommodate the extra water that needs to pass through the drains. Approximately 1 inch of rainfall entering the soil can raise the water table as much as one foot.

Spacing

The spacing between drains depends on the installed depth, hydraulic conductivity, and the amount of water to be drained. When two (or more) drain lines are installed, each one exerts an influence on the water table and the drawdown curves intersect at the midpoint between the drains. As the drains are moved closer together, the curves intersect at a lower level, at the midpoint between drains.

The grade (slope) at which the drain is installed should be based on site conditions, size of drain, and quality of installation. Minimum grades are:

  4 inch = 0.10% (1.2 inches per 100 ft)

  5 inch = 0.07% (0.8 inch per 100 ft)

  6 inch = 0.05% (0.6 inch per 100 ft)

The maximum grade should result in a flow velocity not exceeding 3.5 feet per second (fps) for sand or sandy loam soils and 7.0 fps for clay soils. Velocities exceeding these rates require the installation of protective measures such as air vents or relief wells on the drain tile. A gradual variation of up to 0.1 foot from the specified grade is allowable in most cases.

Connections

Manufactured couplers or fittings should be used at all joints. All connections must be compatible with the pipe. All fittings should be securely joined so they cannot come apart in the installation process, and the envelope material or "sock" should be taped or secured to provide integrity to the entire drain after installation.

Outlet

Outfall from the grove site is a first priority. Sufficient engineering surveys must be conducted to determine the existence of a natural water outlet from the grove site before considering a drainage system. Permits might be required by Water Management Districts or other agencies before large quantities of water can be removed rapidly from poorly drained wetlands. Drainage outlets that discharge into state waters are considered point source discharges by local and state pollution control authorities, and approval to discharge into such waters should be obtained during the planning stage of a drainage or water management system.

A sump and lift-pump-type outlet may be necessary for subsurface drainage, which could materially increase drainage costs. Sumps should be located at the low end of the collector ditch. A float-controlled pump resulting in automatic operation is preferred. The pump should be sized to remove the design capacity for the drained area. Drain outlets should be 6 inches above the normal water level in the ditch. In addition to sunlight weakening the plastic drainage tubing, it can be destroyed by fire or damaged by rodents or ditch maintenance procedures. Therefore, the discharge end should be rigid PVC (Fig. 19-4). At least 2/3 of the PVC pipe should be embedded in the ditch

Figure 19-4. Installing 4-inch corrugated drain tubing with PVC pipe discharge.

bank. The rigid outlet pipe may need an animal guard to keep rodents from entering and plugging the tubing.

Example:
Determine drain spacing for an Adamsville fine sand series soil with trees planted at a 25-ft across-row tree spacing on 50-ft wide beds. The tree spacing must be selected so that the drain trench will fall between rows or in the middles of double beds rather than in water furrows. From Table 19-1, the minimum spacing for Adamsville is 71 feet at a depth of 46 inches. The drain spacing can increase 6 feet for each inch the drain is installed deeper than 46 inches.

Since beds are 50-ft wide, a spacing of 100 feet results when drains are installed in every other bed.

Determine the required depth when drains are installed at 100-ft spacing.

Extra width greater than minimum
= 100 ft - 71 ft = 29 ft

$$\text{Extra Depth} = \frac{29 \text{ ft}}{6 \text{ ft / inch}}$$

= 4.8 inches > round to 5 inches

The trench would have to be lowered 5 inches. The trench depth in Table 19-1 should be applied closer to the upper end rather than near the drain outlet. It is better to be too deep than too shallow. Clay subhorizons could limit effective depth and drains placed directly in spodic horizons increase the risk of increased hydraulic entry resistance and biological clogging potential.

Installation Considerations

Most corrugated drainage tubing in Florida citrus groves is installed with plough-type machines (Fig. 19-5), which are generally capable of installing the tubing more economically than the slower chain-type machines. Tubing should be installed so that it does not deflect more than 20% of its nominal diameter. The plastic tubing has reduced strength at high temperatures. During installation on hot sunny days, the temperature of the tubing can reach over 120° F. The strength of 4-inch tubing is reduced 50% when temperature is raised from 70° F to 120° F. Therefore, precautions must be taken to prevent sharp impacts, heavy objects, or excessive pull on tubing during installation.

The strength of the tubing can also be reduced by stretching that may occur during installation. The amount of stretch during installation is influenced by the temperature, the amount and duration of drag encountered when the tubing is fed through the installation machine, and the stretch resistance

Figure 19-5. Laying corrugated drainage in citrus bed with plough-type drain installation machine.

of the tubing. Stretch should not exceed 5%, which reduces strength by about 11%.

Plastic tubing will float in water. During installation, it is essential that backfill material be placed around the tubing immediately and correctly if water is present so that the tubing does not get misaligned.

Care should be taken in all soils to prevent surface soil that contains organic material from being placed around the drain tubing. Likewise, the tubing should not be bedded in the organic layer of spodic soils. The organic material tends to trap iron and provide an energy source for the iron-reducing bacteria that cause ochre buildup.

Upon completion of installation, a map or aerial photograph showing drain locations should be prepared immediately to identify the precise location of the drain lines. Drain lines can be located using a small rod (3/8 inch) with a sharpened tip to probe for the tubing. The probe may penetrate the tubing when it is found, but it will not seriously damage it as the hole will tend to close when the probe is withdrawn.

Drain Clogging

The main types of deposits associated with bacterial activity in subsurface drains in Florida are ochre and sulfur slimes. Iron deposits (collectively called ochre or iron ochre) are the most serious and widespread. Ochre deposits and associated slimes are usually red, yellow, or tan in color. Ochre is filamentous (from bacterial filaments), amorphous (more than 90% water), and has a high iron content (2% to 65% dry weight). It is a sticky mass combined with an organic matrix (2% to 50% dry weight) that can clog drain entry slots, drain envelopes, and the valleys of the corrugations between envelope and inlet slots. Ochre often contains appreciable amounts of other minerals such as magnesium, sulfur, and silicon. Ochre can usually be detected at drain outlets or in manholes as a voluminous and gelatinous mass. Unfortunately, ochre may also be present in drain sublaterals, and not necessarily at the outlets. Under those conditions, it can usually be detected by excavation of poorly drained spots in a field. There are still disagreements concerning the physical, chemical, and biological factors contributing to ochre formation. For example, the gelatinous mass can trap fine soil particles so that ochre may contain more than 30% sand. In addition, old ochre can become crystallized and hard.

Sulfur slime is a yellow to white stringy deposit formed by the oxidation of the hydrogen sulfide that may be present in groundwater. Soluble sulfides are oxidized to elemental sulfur, predominately by the bacteria *Thiothrix nivea* and *Beggiatoa* sp., so that globules of elemental sulfur are deposited within the filaments of the bacteria. The fluffy masses of slime are held together by intertwining of the long filaments of the bacteria. Sulfur slime usually is not a serious problem in most agricultural drains. It is found most frequently in muck soils or in areas where the water used for irrigation contains hydrogen sulfide.

Ferrous Iron in Groundwater

Ferrous iron is a primary raw material for ochre formation, and it must be in solution in the groundwater rather than just located on soil particles. Ferrous iron will be present in groundwater of flooded soils only after the soil oxygen has been depleted. When that happens, certain iron-reducing bacteria attack and reduce insoluble ferric iron associated with mineral and organic soil particles. The biological action of the bacteria is energy intensive, so energy sources that can be utilized by bacteria must be present. This process cannot take place in a flooded soil without the action of specific bacteria. The bacterial bodies must be present and in direct contact with iron attached to soil particles.

There is often more ferrous iron in the groundwater of sandy soils and organic muck soils than in loamy and clay soils. Sandy soils usually have the most ochre problems. Flooding sandy soils excludes air rapidly and less energy is required for bacteria to reduce iron to the soluble ferrous form. Sandy soils may receive sufficient organic carbon from plant roots or organic residues. Iron is often

available from within sandy clay pockets and organic pans (spodic horizons). Organic and muck soils usually have sufficient iron and readily available organic carbon. Consequently, muck soils often have severe problems from ochre clogging. Clay soils, unless mixed with organic matter, have little if any ferrous iron in the groundwater, even when flooded for extended periods, since the suitable organic carbon level is often insufficient for strong iron reduction, and there are strong electrochemical attractions between the ferrous iron ion and the clay particle.

Soil pH is also a factor in potential ochre formation because the amount of ferrous iron is usually higher in the groundwater at pH values below 7.0. Soluble ferrous iron flowing in groundwater enters a different environment as it approaches the drain and passes through the drain envelope. If a low level of oxygen is present, certain filamentous and rod-shaped bacteria can precipitate some of the ferrous iron, forming insoluble ferric iron and incorporating it into ochre complex.

Ochre as a clogging factor may diminish or disappear over a period of 3 to 8 years if drains are maintained in a free-flowing condition. Many times ochre occurs rapidly and often can be detected at drain outlets within the first few months after drain installation. If drains can be maintained in working order, ferrous iron reaching them may diminish over a period of time.

Processes of Ochre Deposition

The minimum ferrous iron concentrations that can stimulate ochre formation are between 0.15 and 0.22 ppm. Iron-precipitating bacteria must be present for extensive clogging to occur, even when other conditions are just right for chemical precipitation of the iron. Iron alone does not have serious sticking properties. The reaction in drain tubes is a combination of bacterial precipitation and the incorporation of chemically precipitated iron into the sticky slimes of the bacterial masses involved in the ochre matrix.

There are several kinds of processes involved in ochre deposition. All of them do not occur under the same conditions. All of the reactions do require some oxygen to be present in the drain line.

- Oxidation of the iron by certain bacteria predominantly on the outside of the organisms, as shown by electron micrograph pictures

- Auto-oxidation (chemical change) and precipitation, with subsequent accumulation of the colloidal iron on the sticky surfaces of bacterial slimes

- Bacterially precipitated iron from complexed soluble organic-iron compounds

The soluble complex before precipitation may be either ferrous or ferric iron. The most effective iron-precipitating bacteria in drain pipes have been groups consisting of long filaments, such as *Gallionella*, *Leptothris*, and *Sphaerotilus*. They can grow quite rapidly, and the intertwined masses are capable of bridging small openings. There are certain rod bacteria, such as *Pseudomonas* and *Enterobacter*, that can precipitate iron, but the volumes of ochre produced are not as large as with the filamentous types.

Ochre can be found in drain filter envelopes, the zone abutting the envelope, the openings (slots or holes) in the drains, and within the drain tube itself. Most clogging in 4-inch-diameter corrugated polyethylene tubing can be traced to sealing of the inlet openings and accumulations within the corrugation valleys, particularly when synthetic drain envelopes are used. Within the tubing itself, the heaviest accumulation of ochre appears to be in the lower third of the drain length, although the lower third is usually not the region of maximum ochre formation.

Soil Conditions that Contribute to
Ochre Formation
Soils with the most potential for ochre formation are fine sands and silty sands, organic soils and soils with organic pans (spodic horizons), and mineral soil profiles with mixed organic matter. The least likely candidates for ochre hazard are silty clays and clay loams. When flooded, they are usu-

ally deficient in ferrous iron in the soil solution. It is possible to survey individual soil types for ochre potential. It is possible to estimate the maximum potential for ochre before installing drains, as well as to estimate whether specific soil types or profiles can be considered susceptible. The ferrous iron content of the groundwater flowing into a drain has been found to be a reliable indicator of the potential for ochre clogging. Analyzing the soils for total iron is of no value because the values do not indicate easily reducible soluble ferrous iron, nor the complex interactions between soil pH and soil type.

There are certain on-site observations that may give clues to potential ochre formation in advance of drainage. Surface water in canals may contain an oil-like film that is usually iron and may contain tothrix bacterial filaments. Gelatinous ochre may form on the ditch banks or bottoms of canals. Ochre may also form layers in the soil profile. In some locations, there may be iron concretions or so-called iron rocks. The presence of spodic horizons (organic layers) suggests ochre potential, and most organic soils, such as mucks, have some potential for ochre problems.

Measures to Minimize Drain Clogging
There is no known economical, long-term method for effectively controlling ochre clogging in drains. Although options are limited, the emphasis must be on living with the problem. It is necessary to follow certain practices to minimize the potential. All measures that minimize the development of anaerobic flooded conditions are acceptable. Closer spacings and shallower depths of drains may, for certain sites, be beneficial.

A drain envelope or filter is necessary for sandy soils. A graded gravel envelope is best, although it can become clogged under conditions of severe ochre potential. Thin synthetic fabrics are now used extensively in Florida. The principal materials installed are spun-bonded nylon, spun-bonded polypropylene, and a knitted sock. Surveys of selected drainage sites show that ochre clogging with the synthetic materials seems to occur first in the slots and valleys (the space between the envelope and slots) and can be present in amounts sufficient to cause drain failure. The spun-bonded fabrics also clog from ochre deposits in which the iron-precipitating bacteria grow across the voids in the fabrics.

Larger openings in the drains increase the period before drain outflow may be severely restricted. The ochre may adhere to the frayed plastic edges abutting the water inlet slots. Cleanly cut inlet slots are essential. Small slots also limit the effectiveness of jet rinsing as a method for cleaning drains installed with synthetic envelopes. Care must be taken to ensure that the size of the opening or slot is compatible with graded gravel envelopes or base soil.

The use of high- and low-pressure water jetting has been successful in cleaning many drains clogged with ochre. Most of the commercial cleaning has been on drains installed in gravel envelopes. Pressures as high as 1300 psi at the pump have been used, although pressures exceeding 400 psi in sandy soils may destabilize the sand around the drains and cause it to flow into the drain. Jetting nozzles should be designed for agricultural drains rather than municipal sewer lines. Jet cleaning has been unsatisfactory if delayed until the ochre has aged and become crystalline. Pressure requirements will exceed the 400 psi at the nozzle, which is suggested as the upper limit for sandy soils and synthetic envelopes.

Chapter 20. Water Table Measurement and Monitoring

by Brian Boman and Thomas Obreza

Effective water management under shallow water table conditions requires monitoring the water table position with enough precision to minimize pumping for irrigation and drainage. There is a wide range in the cost of the many methods that have been used to monitor water levels. The cost of monitoring devices is directly related to the frequency that measurements need to be taken and to the accuracy required. Methods range from a ruler used to measure water depth in an observation well to electronic sensors attached to data loggers. Even though growers often would like detailed information on water table position, the costs associated with installation and maintenance of recording devices, as well as collection and interpretation of data, are often too great to make such devices practical.

Upflux

The movement of water upward within the soil profile from the water table is called "upflux" (Fig. 20-1). As water is removed from the soil by the tree roots and by evaporation at the ground surface, water content of the soil decreases. By capillary action, water moves from the water table into the drier soil above. Water tends to adhere to soil particles due to surface tension between adjacent particles. Smaller soil particles have smaller inter-particle voids. The smaller particles provide greater surface areas upon which water can adhere. In addition, the smaller voids allow water to be retained at higher surface tensions. As a consequence, soils with smaller particles have the ability to move water greater distances by capillary action than coarser soils (Figs. 20-2 and 20-3). As a result, the

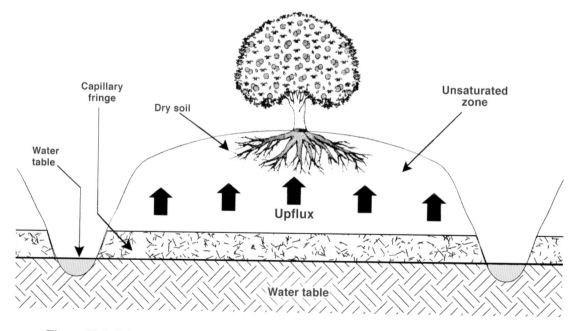

Figure 20-1. Schematic of upflux process in bedded citrus.

upflux process can move water into the root zone from a much deeper water table in clay soils than it can in sandy soils.

Excess water drains by gravity into the shallow water table after a saturating rain or irrigation cycle. The removal of soil water by evaporation and transpiration results in water movement upwards (upflux) by capillary action to replace some of the water in the root zone. The deeper the water table is, the farther the water has to travel upwards into the root zone. Therefore, the effectiveness of the water table for providing moisture to the roots decreases as the water table level drops. If the level is allowed to drop too low, capillarity is broken and upflux action ceases until another saturating rain or irrigation cycle refills the soil profile.

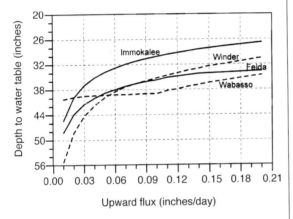

Figure 20-2. Typical upflux rates by water table depths for Immokalee, Winder, Felda, and Wabasso series soils (data from Obreza and Boman, 1992).

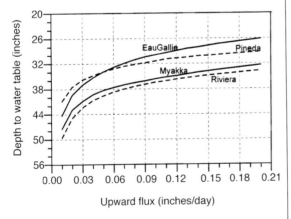

Figure 20-3. Typical upflux rates by water table depths for Eau Gallie, Myakka, Pineda, and Riviera series soils (data from Obreza and Boman, 1992).

Observation Wells

Often water table indicators that provide reasonably precise information quickly and easily are preferred over more detailed and labor-intensive systems. Knowledge of the water table level is essential to ensure that adequate drainage can be provided. The height of the water table is also important because a significant portion of the tree's water requirement can come from upward flux from the water table. Therefore, overdrainage should be minimized.

A groundwater level observation well is a porous casing buried upright in the ground. It permits the groundwater level to rise and fall inside it as the water level naturally changes. The changing water level moves an inserted float with an attached measuring rod. Therefore, the below-ground water level can be gauged aboveground. Water table observation wells installed in flatwoods soils usually penetrate only to the depth of the restrictive (argillic or spodic) layer. Typically, this layer is within 30 to 48 inches of the soil surface. Observation wells in these soils need water level indicators that respond to a minimal amount of recharge water so that even slight perching of the water table can be detected.

Well Construction

Observation wells with a simple float indicator can provide rapid evaluation of shallow water table depths. The float indicator components are readily available, durable, and economical. Installation is quick and calibration is quite simple. The float assembly provides a quick visual indication of water-table depth suitable for casual observation, or it can be used for detailed applications with more precise inspections. Use of the indicators to properly manage irrigation and drainage on shallow water-table soils can help conserve the quantity and quality of water resources.

The materials that are required to construct a water level indicator vary with the user's preferences. The basic components of the well itself include a short section of 3-inch perforated PVC pipe, a 3-inch PVC cap, screening material, a float, an indicator rod, and a small stopper (Fig. 20-4). The indicator rod can be a dowel, 1/2-inch PVC pipe (thin wall) or microsprinkler extension stake. Dowels are a poorer choice since they require painting and will rot out near the float within a few years. The float is typically a 2 1/2-inch fishing net float or a 500 ml (approximately 2 1/2 in. diameter x 6 in. high) polyethylene bottle with a 28-mm (1.1 in.) screw cap size. The float assembly can be constructed by inserting the microsprinkler extension stake into the fishing float or 1/2-inch pipe into the polyethylene bottle. The bottleneck provides a snug fit on the pipe and no sealant is required. The hole in the cap should be drilled slightly larger than the indicator pipe to serve as a guide for the float assembly. A rubber stopper (No. 2) can be placed in the upper end of the pipe to exclude rainfall. Fittings need not be glued so that the well can be inspected and components can be easily disassembled for cleaning or replacement.

Observation well casings are constructed from 3-in. diameter PVC pipe (PR-160). A circular saw or drill can be used to perforate the pipe prior to installation. Perforations should be staggered in rows around the pipe to allow flow into the well from the sides in addition to the bottom. Perforations totaling about 5% of the well's surface area have been adequate for the sandy soils encountered in the flatwoods. No perforations should be made within 12 inches of the surface in order to minimize the chances of ponded water from high intensity storms creating flow channels to perforations near the soil surface.

The pipe should be wrapped (sides and bottom) with a screening material to prevent soil particles

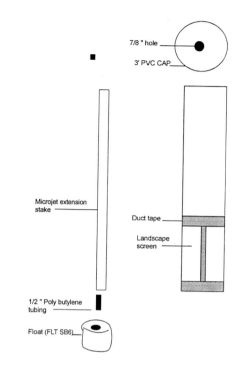

Figure 20-4. Typical components for shallow water table observation well.

from moving into the well. Materials such as cheesecloth, polyester drain fabric, and fiberglass screen have been used successfully as filters. The filter material should be taped in place with duct tape. A 3-inch soil auger can be used to bore holes for the wells. When possible, the observation wells should be installed when no water table is present in order to minimize chances of the well sides sloughing into the bore as it is dug.

When a water table is present, it is easiest to install the well by starting off with a larger diameter pipe. For a 3-inch observation well, a 4-inch installation pipe (Sch 40 preferred) will be needed. The installation pipe should be cut at least 6 inches longer than the intended depth of the well. Holes (1/2-inch diameter) should be drilled in the sides of the pipe opposite each other about 1 1/2 inches from the top of the pipe. These will be used to aid in removing the pipe from the soil after the observation well is installed. Auger a hole in the soil until it begins to slough in (when the water table is reached). The 4-inch pipe should then be forced into the hole. A 3-inch auger can then be used to remove soil from within the 4-inch casing. As soil is removed, the casing needs to be forced downward to keep the hole from sloughing. Continue to remove soil from inside the casing until the appropriate depth is achieved (typically when hardpan material begins to be excavated).

The well casing pipe should be cut to length and installed in the hole so that it extends 2 to 6 inches above the soil surface. Care should be taken to ensure that the casing is installed plumb to minimize binding of the float assembly. If a 4-inch installation pipe was used to excavate the hole, it needs to be removed. A 1/2-inch rod can be inserted through the holes that were drilled in the top of the 4-inch pipe. If the pipe cannot be removed easily by hand, a chain can be attached to the rod and a high-lift jack. Usually, after jacking the installation pipe up about a foot, the pipe can be easily removed by hand. The soil should be backfilled around the observation well casing and tamped to compact the soil and get a tight fit between the soil and the sides of the pipe.

A measurement should be taken of the distance from the bottom of the well to ground level. The float assembly can then be lowered into the well, making sure that the indicator rod and float do not bind against the sides of the observation well (Fig. 20-5). The well is now ready for calibration.

Calibration
Calibration is required to determine water table levels that are optimum for tree response, and to determine when to apply irrigation water to maintain water table levels. The measuring rod can be marked with divisions painted at appropriate intervals for the desired maximum and minimum water table depths.

The measuring rod in a water table observation has marks that indicate critical water table levels relative to ET rates. When the water table is within

these marks, the soil will provide sufficient moisture to meet tree demands. If a very slowly permeable spodic and/or argillic horizon is present, it may be necessary to use the depth to the restrictive layer as the lower depth limit. Observations over a period of time will help determine whether water perches on top of the restrictive layer. The upper depth is dependent on the depth of the citrus bed furrow and/or root depth. The upper depth should be a selected so that water does not pond in water furrows or result in root pruning. Controlling the groundwater table that is just below the root zone can be used as a tool to help fine-tune the amount of water used to produce a crop. This increases the efficiency and effectiveness of the water used.

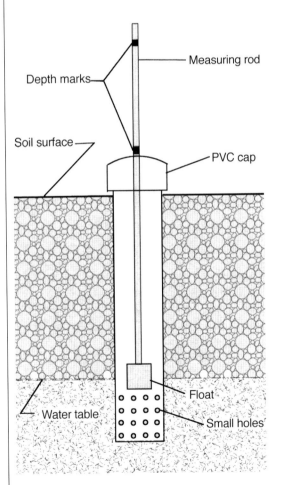

Figure 20-5. Schematic of completed observation well.

It is best if calibration is performed when the water table is above the bottom of the well. The weight of the float assembly and rod will affect how much of the float is submerged before it floats. A mark should be made on the measuring rod at the PVC cap on top of the well when the float is at the bottom of the well. A ring should be permanently painted on the indicator pipe at this level to serve as the reference mark for the well depth. The indicator pipe can then be marked with major divisions (feet) and minor divisions (inches) for easy reading of the water table depth. These rings can be painted at appropriate intervals using different colors for major and minor divisions. Marks painted at 2-inch increments provide enough accuracy for most users. Typically, the float assemblies can respond to water levels less than 1/2 inch above the bottom of the observation well.

The lower depth marks should be at points where the upflux of moisture is equal to the average daily evapotranspiration (ET). When the water table drops any lower, the tree will not receive the moisture it needs that day. If a spodic and/or argillic horizon is present that is very slowly permeable, it may be necessary to use the depth to the restrictive layer as the lower depth limit.

The upper depth is dependent on the depth of the citrus bed furrow and/or root depth. The upper depth should be a few inches below the furrow elevation to prevent furrow grass maintenance problems. The upper depth should be at least 6 inches below the bottom of the root zone to prevent root pruning. Choose the most limiting factor, furrow elevation or root depth, when selecting the maximum allowable upper water table depth.

Alternately, or in combination with the rings, zones can be marked on the float assembly. If it is desirable to maintain the water table within specified maximum and minimum depths, the corresponding zone can be painted on the float assembly. A quick glance at the observation well (Fig. 20-6) will indicate whether or not the water level is within desired limits.

Observations over a period of time will help to determine whether water perches on top of the restrictive layer. The key is not to hold the water table too high, which can cause root damage or prevent proper cover crop maintenance due to excessive wetness in the furrows. In contrast, if the water table is allowed to drop to the point where there is not enough water moving upward to supply the moisture needs of the tree, additional irrigations may be required.

Operation

The water table monitoring well is an excellent tool for determining when to irrigate or when the water table is too high. The goal of water table management is to maintain the water table at a level just below the root zone but not high enough to cause root damage. After a soaking rain or irrigation cycle, moisture adheres to the surface of soil particles until all available space is filled and the excess water settles to the water table. If the water table is close enough to the surface (as in the case of most flatwoods soils), this water may be available to the plant due to upflux, which generally lasts for several days as the water table gradually declines. If the water table is allowed to get too low, the soil dries out and the upflux continuity is broken. At this point, another rain or irrigation is needed.

Figure 20-6. Observation well installed in citrus grove.

Figure 20-7 shows the soil tensions at the 12-inch depth for a Wabasso series soil. During the afternoon hours, the soil water tension increases as water is taken up to meet the evapotranspiration requirements of the tree. In the evenings, the upflux from the shallow water table is equal to the ET rate, so the soil tensions remain fairly constant. In the early morning hours the trees are not transpiring; therefore, the soil tensions decrease at the 12-inch depth as upflux from the water table to the drier soil above. The 0.5 inch rainfall on Wednesday afternoon wets the soil thoroughly at the 12-inch tensiometer depth, causing the soil tensions to drop to near zero.

Observe and record the water table level over time and relate these observations to rainfall, soil moisture and plant response. Make observations of how rain or irrigation affects the water table level. This information can be used to develop an irrigation/drainage schedule that best fits your needs. The black mark on the indicator stick, when visible above the cap, represents the range of upflux values that will supply sufficient moisture to the root zone over the range of consumptive use requirements. This calibration is related to soil type and crop. This range can be modified based on the field observations.

Maintenance on the indicators is minimal. The painted rings may fade and require periodic repainting. The rubber stoppers can deteriorate with prolonged weathering and will probably need to be replaced after several years. If an indicator is suspected of erroneous readings, the assembly can be quickly removed to inspect the well. As routine maintenance, the observation wells should be closely inspected on an annual basis to ensure that they are functioning properly.

The purpose of the water table observation well is to assist the irrigation operator in determining the groundwater level in the field. Such information is helpful for setting a realistic irrigation schedule

Example:

Determine the water table depth required to supply typical citrus evapotranspiration requirements for a Winder series soil (using typical average values in Fig. 20-3).

Month	Typical ET (in./day)	Water table depth (in.)	Month	Typical ET (in./day)	Water table depth (in.)
Jan.	0.07	37	July	0.19	30
Feb.	0.09	36	Aug.	0.19	30
March	0.12	34	Sept.	0.17	31
April	0.15	32	Oct.	0.13	33
May	0.17	31	Nov.	0.09	36
June	0.18	30	Dec.	0.07	37

Marks should be made on the indicator to show depths of 30 inches (for summer) and 37 inches (for winter months). These will be used to signal the need to start supplemental irrigation throughout the year. Assuming a root zone depth of 18 inches, another mark should be made to indicate a depth of 24 inches. Whenever the water table is closer to the surface than the 24-inch depth, steps should be taken to drain the block and lower the water table.

that avoids harmful over- or underwatering. Using the wells is a learning experience. By faithfully monitoring the well, you can determine when to start and stop watering in order to maintain the desired water level for your soil type.

Water Table Behavior under Bedded Flatwoods Citrus

Most Florida flatwoods soils are poorly drained because of the low, flat landscape, and because a restrictive layer exists below the soil surface. Consequently, flatwoods citrus is planted on raised soil beds in order to create enough unsaturated soil volume for adequate root growth and development. As previously discussed, a shallow water table can exist close enough to the root zone to have a direct influence on the vigor and productivity of citrus trees even when planted on beds. Rainfall and subirrigation can quickly raise the water table, while topographical elevation, depth to the restrictive layer, and the ability of the artificial drainage system to remove water influence how quickly the water table declines.

Flatwoods citrus groves are normally planted with one to four rows per bed. Since the beds are crowned to aid surface water drainage to the water furrows, tree row configuration can influence water table behavior beneath individual rows. With single- and double-row beds, the tree rows are usually at the same elevation above the bottom of the water furrows, so the water table depth is about the same beneath each tree. However, the center row or rows on a 3- or 4-row bed are higher above the water furrow than the edge rows (Fig. 20-8), so there is potentially a greater vertical distance between the water table and the root zone under the middle rows. Even a difference of a few inches in water table depth can cause a large difference in root zone water regime, which can in turn cause long-term effects in tree growth and productivity. Water table depth differences between the middle and edge rows on 3- or 4-row beds typically differ by 4 to 6 inches throughout the year (Fig. 20-9). Upward flux in a sandy soil can vary widely with

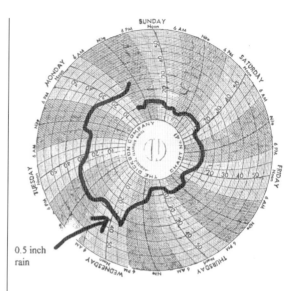

0.5 inch rain

Figure 20-7. Graph of soil tension from a recording tensiometer for a Wabasso series soil at 12-inch depth.

this magnitude of water table difference (Figs. 20-2 and 20-3), depending on water table position.

The water table under flatwoods citrus rises rapidly in response to either rainfall or subirrigation because sandy soils are highly conductive to water flow (Fig. 20-9). A general rule of thumb is that in sandy flatwoods bedded groves, 1 inch of rain will cause the water table to rise about 10 inches (Fig. 20-10). The rate of water table decline is also relatively rapid (although not as fast as water table rise) if the grove drainage system is working properly. It may take 4 to 6 days for the water table to return to its original position following rains of 1 inch or more (Fig. 20-11).

Because of the high porosity and conductivity of sandy flatwoods soils, it is possible for drip irrigation to raise the water table under a citrus bed if the irrigation system is run for a long enough time (Fig. 20-12). Lateral water movement outward from a drip emitter may be only 6 to 8 inches; thus, much of the applied water moves downward in a small cylindrical wetted pattern. It does not take long for the water to reach the free water surface below and cause the water table to rise. Even though the magnitude of rise is not as great as with flood (seepage) irrigation, it is measurable with a simple water table monitoring well.

If the intent of irrigation is to keep the irrigation water from moving below the root zone, then water table monitoring would be useful to determine if the irrigation system is being run for too long a time period. However, some irrigation managers use drip irrigation as a system for water table management. A water table monitoring device would also be useful in this situation, especially if it was used in conjunction with a float switch that turned the irrigation system on and off as the water table fluctuated between prescribed minimum and maximum levels.

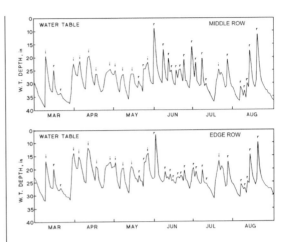

Figure 20-9. Water table fluctuations in response to seepage irrigation (i) and rainfall (r) under a 4-row citrus bed.

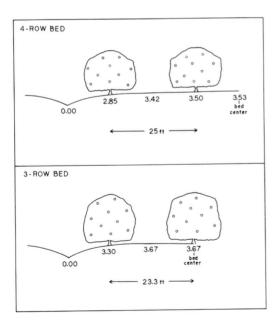

Figure 20-8. Examples of multirow bed configurations in flatwoods citrus. Numbers indicate the vertical distance in feet above the water furrow.

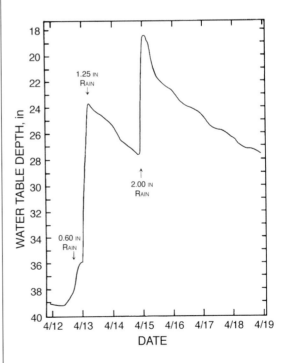

Figure 20-10. Example of water table depth response to rainfall in bedded flatwoods citrus.

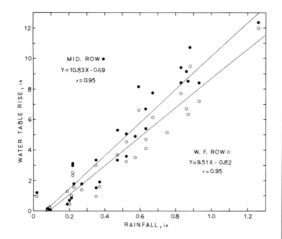

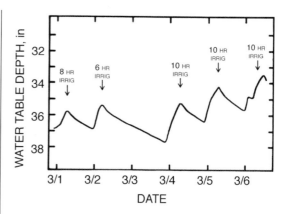

Figure 20-12. Example of water table depth response to extended periods of drip irrigation in bedded flatwoods citrus.

Figure 20-11. Magnitude of water table rise and fall with respect to rainfall volume under middle- and water-furrow-row positions in multirow bedded flatwoods citrus.

by Brian Boman

Hydraulic Principles

A couple of basic hydraulic concepts related to water movement through pipes are particularly important to the design and proper operation of microirrigation systems. Water weighs about 62.4 pounds per cubic foot (lb/ft³). This weight exerts a force on its surroundings, which is expressed as force per unit area of pressure (pounds per square inch or psi). The relationship between the height of a column of water and the resulting pressure it exerts is: 2.31 ft of water produces 1 psi.

$$Ht = P \times 2.31 \qquad \textbf{Eq. 21-1}$$

where,

Ht = elevation in ft
P = pressure in psi

The column of water does not have to be vertical. To calculate the static pressure between two points resulting from an elevation difference, only the vertical elevation distance between the two points needs to be known. However, other factors such as friction affect water pressure when water flows through a pipe.

> Example:
> Determine the height of water in a column that produces a gauge pressure of 10 psi.
>
> Ht = P x 2.31 = 10 (psi) x 2.31 (ft/psi)
> = 23.1 ft

Velocity

Velocity (v) is the average speed at which water moves through a pipe. Velocity is usually expressed in units of feet per second (ft/sec or fps). Water velocity in a pipe (Fig. 21-1) is greatest in the middle of the pipe (V_{max}) and smallest near the pipe walls. Normally, only the average velocity of water in the pipe is needed for calculations.

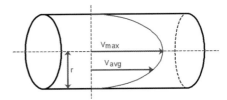

Figure 21-1. Typical velocity cross section profile for a full-flowing pipe.

To avoid excessive pressure losses due to friction and excessive potentially damaging surge pressures, most irrigation systems are designed to avoid velocities that exceed 5 ft/sec.

Flow

The relationship between flow rate and velocity is given by the equation of continuity, a fundamental physical law. The equation of continuity states that flow rate can be calculated from the multiple of the velocity times the cross-sectional area of flow.

$$Q = A \times V \qquad \textbf{Eq. 21-2}$$

or

$$V = Q/A \qquad \textbf{Eq. 21-3}$$

where,

Q = flow rate in ft³/sec
A = cross-sectional area of flow in ft²
$\quad$ (A = π x D²/4, π = 3.1416)
V = velocity in ft/sec

If pipe diameters change in adjoining pipe sections with no change in flow rate (Fig. 21-2), the relationship between flow and velocity can be calculated by:

$$A_1 \times V_1 = A_2 \times V_2 \qquad \textbf{Eq. 21-4}$$

where,

A₁ = cross-sectional area of flow for first section (ft²)

A_1 = cross-sectional area of flow for first section (ft²)

V_1 = velocity in first section (ft/sec)

A_2 = cross-sectional area of flow for second section (ft²)

V_2 = velocity in second section (ft/sec)

If the velocity is the same in both a 2-inch and a 4-inch diameter pipe, the flow rate with the 4-inch pipe would be four times as large as the flow rate from the 2-inch diameter pipe. Note that the cross-sectional area is proportional to the diameter

squared: $(2 \text{ in.})^2 = 4 \text{ in.}^2$, while $(4 \text{ in.})^2 = 16 \text{ in.}^2$. Therefore, doubling the pipe diameter increases the carrying capacity of a pipe by a factor of 4.

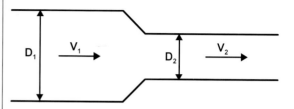

Figure 21-2. Diameter (D) and velocity (V) relationships for adjoining pipe sections with a constant flow rate.

Table 21-1. Maximum allowable flow rate and friction losses per 100 feet for PVC Class 160 IPS pipe. Flow rate is based on maximum velocity of 5 fps (friction loss calculated at maximum flow rate by Hazen-Williams equation).

Size	O.D. (in.)	I.D. (in.)	Wall thickness (in.)	Maximum flow (gpm)	Friction loss (psi/100 ft)
1.0	1.315	1.195	0.060	16	3.0
1.25	1.660	1.532	0.064	28	2.6
1.5	1.900	1.754	0.073	38	2.0
2.0	2.375	2.193	0.091	60	1.8
2.5	2.875	2.655	0.110	85	1.4
3.0	3.500	3.230	0.135	125	1.1
3.5	4.000	3.692	0.154	165	0.95
4.0	4.500	4.154	0.173	215	0.80
5.0	5.563	5.133	0.214	325	0.67
6.0	6.625	6.115	0.225	460	0.54
8.0	8.625	7.961	0.332	775	0.39
10.0	10.750	9.924	0.413	1200	0.30
12.0	12.750	11.770	0.490	1700	0.25

Table 21-2. Maximum allowable flow rate and friction losses per 100 feet for polyethylene (PE) SDR 15 (80 psi pressure-rated) tubing based on maximum velocity of 5 fps (friction loss calculated at maximum flow rate by Hazen-Williams equation, C = 140).

Size	O.D. (in.)	I.D. (in.)	Wall thickness (in.)	Maximum flow (gpm)	Friction loss (psi/100 ft)
0.50	0.682	0.622	0.060	4.5	8.0
0.75	0.884	0.824	0.060	8.5	6.6
1.00	1.119	1.049	0.070	13	4.5
1.25	1.472	1.380	0.092	23	3.4
1.50	1.717	1.610	0.107	32	2.9

Example:
Determine the flow rate (gpm) in a 4-inch Class 160 PVC pipe if the average velocity is 5 ft/s.

Solution:
From Table 21-1, the I.D. for 4-inch pipe is 4.154 inches

$D = 4.154$ in./12 = 0.346 ft
$A = \pi \times D^2/4 = (3.14 \times 0.346^2)/4 = 0.094$ ft^2
$Q = A \times V = 0.094$ ft$^2 \times 5$ ft = 0.47 ft^3/s (cfs)
$Q = 0.47$ cfs $\times$ 448 gal/cfs = 211 gpm

What would be the velocity if there was a transition to a 3-inch Class 160 PVC with the same flow rate?

From Table 21-1, $D_2 = 3.230$ inches
$D_2 = 3.230$ in./12 = 0.269 ft
$A_2 = \pi \times D^2/4 = (3.14 \times 0.269^2)/4 = 0.094$ ft^2
Rearranging Eq. 21-4 results in:

$$V_2 = \frac{A_1 \times V_1}{A_2}$$

$= (0.094$ ft$^2 \times 5$ ft/sec$) /0.057$ ft$^2 = 9.1$ ft/sec

Pressure Versus Flow

As water moves through any pipe, pressure is lost because of turbulence created by the moving water. The amount of pressure lost in a horizontal pipe is related to the velocity of the water, the inside diameter of the pipe, and the length of pipe through which the water flows. When velocity increases, the pressure loss increases. For example, in a 1-inch Sch 40 PVC pipe with an 8-gpm flow rate, the velocity will be 2.97 fps with a pressure loss of 1.59 psi per 100 ft. When the flow rate is increased to 18 gpm, the velocity will be 6.67 fps, and the pressure loss will increase to 7.12 psi per 100 ft of pipe.

Increasing the pressure in the system increases the flow rate. In Fig. 21-3, the flow rate in a 2-inch pipe increases by 100 gpm when the pressure is increased from 20 psi to 50 psi. Using a smaller pipe size does not increase the flow. Note that the smaller pipe sizes have considerably less flow at any given pressure. Since decreasing the pipe size does not increase the pressure at the source, the result of decreased size is reduced flow.

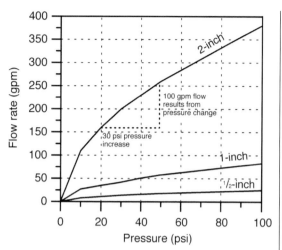

Figure 21-3. The relationships between pressure and flow through unrestricted 100-ft-long sections of pipe (with 4 couplings) for 1/2-,1-, and 2-inch Class 315 PVC pipe. Pressure losses include friction losses in pipe and couplings, as well as velocity head and entrance losses, but do not include exit losses.

Using a smaller pipe size does not increase pressure. In contrast, it will result in lower pressure because there will be greater pressure loss in the lines. In Fig. 21-3, a flow of 20 gpm would require about 9 psi pressure in a 1-inch pipe. In order to maintain a 20-gpm flow in a 1/2-inch pipe, over 50 psi would be required at the source. Smaller pipes result in greater pressure loss, not higher pressure.

Bernoulli's Theorem

At any point within a piping system, water has energy associated with it. The energy can be in various forms, including pressure, elevation, velocity, or friction (heat). The total energy of the fluid at one point in the system must equal the total energy at any other point in the system, plus any energy that might be transferred into or out of the system. This principle is known as Bernoulli's Theorem and can be expressed as:

$$H_{e1} + H_{p1} + H_{v1} = H_{e2} + H_{p2} + H_{v2} + H_f \qquad \textbf{Eq. 21-5}$$

where,

H_e = Elevation in feet above some reference point (ft)

H_p = Pressure (psi) divided by the specific weight of water to convert to ft

H_v = Velocity head (ft)

$$H_v = V^2/2g \qquad \textbf{Eq. 21-6}$$

where,

V = velocity in ft/s
g is the gravitational constant 32.2 ft/s^2
H_f = friction head (ft) usually calculated by Hazen-Williams equation

In irrigation systems, the amount of energy associated with velocity is usually small compared with elevation and pressure energy; thus, it is often ignored. If there isn't a pump (which adds energy) in the piping network, total energy remains the same at all points in the system.

Example:

The pressure at the pump of an irrigation system is 30 psi. A microsprinkler is located at another location, which is 10 ft higher than the pump. What is the static water pressure at the microsprinkler (assume no flow and thus no friction loss)? For convenience, we can use the pump as the elevation datum.

The total energy at the pump is determined:

$$H = P \times c + E \qquad \textbf{Eq. 21-7}$$

where,

H = energy head (ft)
P = pressure (psi)
E = elevation (ft)
c = conversion constant (2.31 ft/psi).

In this example, since there is no water flowing, the energy at all points of the system is the same. Pressure at the microsprinkler is found by solving this equation for P:

Energy head at the pump (H) =
30 psi x 2.31 ft/psi = 69.3 ft

Reorganize Eq. 21-7 => P = (H - E)/c

P = (69.3 ft -10 ft)/2.31 ft/psi = 25.7 psi.

Hazen-Williams Equation

Water flowing in a pipe loses energy because of friction between the water and pipe walls and turbulence. In the above example, when the micro-sprinkler is operating, pressure will be less than the 25.7 psi due to friction loss in the pipe and the microtubing. It is important to determine the amount of energy lost in pipes in order to properly size them. The friction loss calculations in Tables 21-1 and 21-2 are based on the Hazen-Williams equation, and they provide usable results from most pipe sizes and water temperatures encountered under irrigation system conditions. A more accurate equation, Darcy-Weisbach, is sometimes used for smaller pipes or when heated water is being piped; however, the computations are more difficult. The Hazen-Williams equation, with C = 150 (for plastic pipes), is generally suitable for irrigation systems and can be expressed as:

$$H_f = \frac{0.000977 \times Q^{1.852}}{D^{4.871}} \times L \qquad \textbf{Eq. 21-8}$$

where,

H_f = friction loss (feet)
Q = flow rate (gpm)
D = inside pipe diameter (inches)
L = length of pipe (feet)

Friction loss in pipes depends on flow (Q), pipe diameter (D), and pipe smoothness (C). The smoother the pipe, the higher the C value. Increasing flow rate or choosing a rougher pipe will increase energy losses, resulting in decreased pressure downstream. In contrast, increasing inside diameter will decreases friction losses and provides greater downstream pressure.

> Example:
> Determine the pipe friction loss in 1000 ft of 8-inch Class 160 PVC pipe if the flow rate is 800 gpm.
>
> Solution:
> From Table 21-1, the I.D. of 8-inch pipe is 7.961 inches
>
> $H_f = 0.000977 \times (800)^{1.852}/7.961^{4.871} \times 1000$
> $H_f = 9.5$ ft

Because of friction, pressure is lost whenever water passes through fittings, such as tees, elbows, constrictions, or valves. The magnitude of the loss depends both on the type of fitting and on the water velocity (determined by the flow rate and fitting size). Pressure losses in major fittings such as large valves, filters, and flow meters, can be obtained from the manufacturers. To account for minor pressure losses in fittings such as tees and elbows, refer to Appendix 9. Minor losses are sometimes aggregated into a friction loss safety factor (10% is frequently used) over and above the friction losses in pipelines, filters, valves, and other elements.

Hydraulics of Multiple Outlet Pipelines

If the pipeline has multiple outlets at regular spacing along mains and submains, the flow rate downstream from each of the outlets will be effectively reduced. Since the flow rate affects the amount of pressure loss, the pressure loss in such a system would be only a fraction of the loss that would occur in a pipe without outlets. The Christiansen lateral line friction formula is a modified version of the Hazen-Williams equation. It was developed for lateral lines with sprinklers or emitters that are evenly spaced with assumed equal discharge and a single pipe diameter.

$$H_{f(L)} = F \times \frac{L}{100} \times \frac{k \times \left(\frac{Q}{C}\right)^{1.852}}{D^{4.871}} \qquad \textbf{Eq. 21-9}$$

where,

$H_{f(L)}$ = head loss due to friction in lateral with evenly spaced emitters (ft)
L = length of lateral (ft)
F = multiple outlet coefficient (see Table 21-3)
$\qquad [1/(m + 1)] + [1/2n] + [(m + 1)^{0.5}/(6n^2)]$
m = velocity exponent (assume 1.85)
n = number of outlets on lateral
Q = flow rate in gpm
D = inside pipe diameter in inches
k = a constant 1045 for Q in gpm and D in inches
C = friction coefficient: 150 for PVC or PE pipe (use 130 for pipes less than 2 inches)

Example:
Determine the friction loss in a 3/4-inch poly-thylene lateral that is 300 ft long with 25 evenly spaced emitters. Each emitter has a discharge rate of 15 gph.

Solution:
Flow rate into the lateral is:
25 emitters x 15 gph each = 375 gph
or 375 gph/60 min = 6.3 gpm

From Table 21-2, the I.D. of 3/4-inch poly tubing is 0.824 inch

From Table 21-3, F = 0.355

C = 130 for 3/4-inch poly tubing

$H_{r(L)} = 0.355 (300/100)(1045 \times (6.3/130)^{1.852})$
$/0.824^{4.87} = 10.5$ ft = 10.5 ft/2.31 psi/ft
= 4.6 psi

Hydraulic Characteristics of Lateral Lines
The goal of uniform irrigation is to ensure, as much as feasible, that each portion of the field receives the same amount of water, as well as nutrients and chemicals. As water flows through the lateral tubing, there is friction between the wall of the tubing, and the water particles. This results in a gradual but nonuniform reduction in the pressure within the lateral line. The magnitude of pressure loss in a lateral line depends on flow rate, pipe diameter, roughness coefficient, changes in elevation, and the lateral length.

When a lateral line is placed upslope, the emitter flow rate decreases most rapidly. This is due to the combined influence of elevation and friction loss. Where topography allows, running the lateral line downslope can produce the most uniform flow since friction loss and elevation factors cancel each other to some degree.

Friction loss is greatest at the beginning of the lateral. Approximately 50% of the pressure reduction occurs in the first 25% of the lateral's length. This occurs because as the flow rate decreases, friction losses decrease more rapidly. Lateral length may have a large impact on uniform application. Lateral lengths that are too long, given the pipe diameter and the emitter flow rate, are one of the most commonly observed sources of nonuniformity in microirrigation systems. Flow rates are less uniform, in general, with longer lateral length.

Table 21-4 gives the fractional values (called the "F" value in Eq 21.6) of a system with multiple outlet pipes. A pipe with ten regularly spaced outlets, for example, would have 0.39 times the pressure loss of an equivalent pipe with no outlets.

Water Hammer
Water hammer is a hydraulic phenomenon that is caused by a sudden change in the velocity of the water. This velocity change results in a large pressure fluctuation that is often accompanied by a loud and explosive noise. This release of energy is caused by a sudden change in momentum followed

Table 21-3. "F" values of reducing pressure loss in multiple outlet pipes.

Number of Outlets	"F"	Number of Outlets	"F"
1	1	12	0.38
2	0.63	15	0.37
4	0.47	20	0.36
6	0.42	30	0.35
8	0.41	50	0.34
10	0.39	100 or more	0.33

by an exchange between kinetic and pressure energy. The pressure change associated with water hammer occurs as a wave, which is very rapidly transmitted through the entire hydraulic system. Severe or repeated water hammer events can lead to pipe failure.

The sudden change in velocity caused by the rapid closing of a valve can produce very high pressures in the piping system. These pressures can be several times the normal operating pressure and result in burst pipes and severe damage to the irrigation system. The high pressures resulting from the water hammer cannot be effectively relieved by a pressure relief valve because of the high velocity of the pressure wave (pressure waves can travel at more than 1000 ft per second in PVC pipe). The best prevention of water hammer is the installation of valves that cannot be rapidly closed and the selection of air vents with the appropriate orifice that do not release air too rapidly. Pipelines are usually designed so that velocities remain below 5 fps in order to avoid high surge pressures from occurring. Surge pressures may be calculated by:

$$P = \frac{0.028 \, (Q \times L)}{D^2 \times T} \qquad \textbf{Eq. 21-10}$$

where,

Q = flow rate (gpm)
D = pipe I.D. (inches)
P = surge pressure (psi)
L = length of pipeline (feet)
T = time to close valve (seconds)

Example:
For an 8-inch Class 160 PVC pipeline that is 1500 feet long and has a flow rate of 750 gpm, compare the potential surge pressure caused when a butterfly valve is closed (in 10 seconds) to a gate valve that requires 30 seconds to close.

Solution:
From Table 21-2, the diameter of 8-inch pipe is 7.961 inches.

Butterfly valve (P_{bv})

$P_{bv} = 0.028 \times (750 \times 4500) / (7.961^2 \times 5)$

= 297 psi

Gate valve (P_{gv})

$P_{gv} = 0.028 \times (750 \times 4500)/(7.9612 \times 30)$

= 51 psi

Head Losses in Lateral Lines

Microsprinkler field installations typically have 10- to 20-gph emitters. Emitters are normally attached to stake assemblies that raise the emitter 10 to 11 inches above the ground and the stake assemblies usually have 2- to 3-ft lengths of 4-mm spaghetti tubing. The spaghetti tubing is connected to the polyethylene lateral tubing with a barbed or threaded connector. The amount of head loss in the barbed connector can be significant, depending on the flow rate of the emitter and the connector inside diameter. The pressure loss in a 0.175-inch barb x barb connector is shown in Fig. 21-4. At 15 gph, about 1 psi is lost in the barbed connection alone.

In addition to the lateral tubing connection, there will be pressure losses in the spaghetti tubing. Fig. 21-5 shows the pressure required in the lateral line

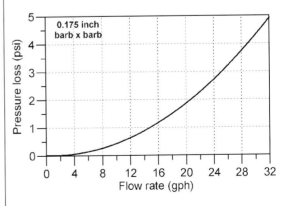

Figure 21-4. Pressure loss versus flow rate for 0.175-inch barb x barb connector.

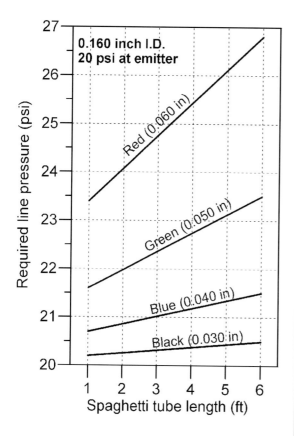

Figure 21-5. Lateral line pressure required to maintain 20 psi at emitter for various emitter orifice sizes and spaghetti tube lengths.

to maintain 20 psi at the emitter for various emitter orifices and spaghetti tubing lengths. Note that with the red base emitters (0.060-in. orifice), an additional 25% to 30% pressure is required in the lateral tubing to maintain 20 psi at the emitter.

It is very important to realize the hydraulic limits of irrigation lateral lines to efficiently deliver water. Oftentimes when resetting trees, two or more trees are planted for each tree taken out. If a microsprinkler is installed for each of the reset trees, the effects on the system uniformity and

system performance can be tremendous. Not only will friction losses increase and average emitter discharge decrease, but system uniformity and efficiency will decrease. Figure 21-6 shows the maximum length of lateral tubing that is possible while maintaining ± 5% flow variation on level ground with a 20 psi average pressure. The discharge gradient is calculated by dividing the emitter flow rate (gph) by the emitter spacing (ft).

The maximum number of microsprinkler emitters and maximum lateral lengths for 3/4- and 1-inch lateral tubing are given in Tables 21-4 and 21-5. Similar information for drippers with 1/2-, 3/4-, and 1-inch lateral tubing is given in Tables 21-6, 21-7, and 21-8. All calculations are based on ± 5% allowable flow variation on level ground. By knowing the emitter discharge rate, spacing, and tubing diameter, the maximum number of emitters and the maximum lateral length can be determined.

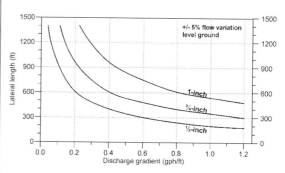

Figure 21-6. Lateral length allowable to achieve ± 5% flow variation for level ground with 22 psi inlet pressure (20 psi average pressure) for 1/2-, 3/4-, and 1-inch lateral tubing.

Table 21-4. Maximum number of microsprinkler emitters and maximum lateral lengths (ft) for 3/4-inch lateral tubing diameter (±5% allowable flow variation on level ground, based on Bowsmith 3/4 in.-50 tubing, I.D. = 0.818 in.).

Flow rate (gph)	Spacing on lateral (feet)									
	7.5		10		12.5		15		17.5	
	No.	Length	No.	Length	No.	Length	No.	Length	No.	Length
8	44	337	40	405	37	467	35	525	33	579
10	39	293	35	352	32	405	30	455	28	502
12	34	261	31	313	28	361	27	405	25	447
14	31	236	28	284	26	327	24	367	23	405
16	28	217	26	261	24	300	22	337	21	372
18	26	201	24	242	22	279	20	313	19	345
20	25	188	22	226	20	261	19	293	18	323

Example:
Using Fig. 21-6, determine the maximum allowable run length for 3/4-inch lateral tubing with 10-gph emitters spaced at 12 ft intervals.

Discharge gradient = 10 gph/12 ft = 0.83 gph/ft. From Fig. 21-4, the maximum run length would be about 380 ft (32 trees).

Example:
Using Tables 21-4 to 21-8, determine the maximum allowable run length for 3/4-inch lateral tubing with 10 gph emitters spaced at 12.5-ft intervals.

From Table 21-4, for 12-gph and 10-ft spacing,

Maximum number of emitters: 32

Maximum lateral length: 405 ft

Example:
Using Tables 21-4 to 21-8, determine the maximum allowable run length for 3/4-inch lateral tubing with 1.0 gph drip emitters spaced at 30-inch intervals.

From Table 21-7, for 1.0 gph at 30-inch spacing,

Maximum number of emitters: 227

Maximum lateral length: 568 ft

Table 21-5. Maximum number of microsprinkler emitters and maximum lateral lengths (ft) for 1-inch lateral tubing diameter (±5% allowable flow variation on level ground, based on Bowsmith 1 in.-45 tubing, I.D. = 1.057 in.).

Flow rate (gph)	Spacing on lateral (feet)									
	7.5		10		12.5		15		17.5	
	No.	Length	No.	Length	No.	Length	No.	Length	No.	Length
8	70	526	63	631	58	728	54	817	51	902
10	60	456	54	548	50	631	47	709	44	782
12	54	406	48	488	44	562	42	631	39	696
14	49	368	44	442	40	510	38	572	36	631
16	45	338	40	406	37	468	35	526	33	580
18	41	314	37	377	34	434	32	488	30	538
20	39	293	35	352	32	406	30	456	28	503

Table 21-6. Maximum number of drip emitters and maximum lateral lengths (ft) for 1/2-inch lateral tubing diameter (±5% allowable flow variation on level ground, based on Bowsmith P720P48 (1/2-inch) tubing, I.D. = 0.625 in.).

Flow rate (gph)	Spacing on lateral (inches)									
	18		24		30		36		42	
	No.	Length	No.	Length	No.	Length	No.	Length	No.	Length
0.5	266	400	240	481	221	554	207	623	196	687
1.0	171	257	154	309	142	356	133	400	126	442
1.5	132	199	119	239	110	275	103	309	97	341
2.0	110	165	99	199	91	229	85	257	81	284

Table 21-7. Maximum number of drip emitters and maximum lateral lengths (ft) for 3/4-inch lateral tubing diameter (±5% allowable flow variation on level ground, based on Bowsmith 3/4 in.-50 tubing, I.D. = 0.818 in.).

Flow rate (gph)	Spacing on lateral (inches)									
	18		24		30		36		42	
	No.	Length	No.	Length	No.	Length	No.	Length	No.	Length
0.5	424	637	383	766	353	883	330	991	312	1093
1.0	293	410	246	492	227	568	212	637	200	703
1.5	211	317	190	380	175	438	164	492	155	546
2.0	176	264	158	317	146	365	136	410	129	452

Table 21-8. Maximum number of drip emitters and maximum lateral lengths (ft) for 1-inch lateral tubing diameter (±5% allowable flow variation on level ground, based on Bowsmith 1 in.-45 tubing, I.D. = 1.057 in.).

Flow rate (gph)	Spacing on lateral (inches)									
	18		24		30		36		42	
	No.	Length	No.	Length	No.	Length	No.	Length	No.	Length
0.5	662	993	596	1192	549	1374	514	1544	486	1703
1.0	426	639	383	767	353	884	331	993	312	1095
1.5	328	493	296	592	273	683	255	767	241	846
2.0	274	411	246	493	227	569	213	639	201	704

Chapter 22. Water Measurement

by Brian Boman

Depth and Volume

Effective irrigation water management begins with accurate water measurement. Water measurement is required to determine both total volumes of water and flow rates pumped. Measurement of volumes will verify that the proper amount of water is applied at each irrigation and that amounts permitted by water management districts are not exceeded. Flow rate measurements help to ensure that the irrigation system is operating properly. For example, lower than normal flow rates may indicate the need for pump repair or adjustment, partially closed or obstructed valves or pipelines, or clogged drip emitters. Higher than normal flow rates may indicate broken pipelines, defective flush valves, too many zones operating simultaneously, or eroded sprinkler nozzles.

Pumping rate, application rate, or tree water use rate is measured in units of volume (or depth) per unit of time. Volume can be calculated from the depth applied if the area to which the water is applied is known. The volume can also be calculated by multiplying the flow rate (gpm) by the duration of flow (min). The depth of application is calculated by multiplying the application rate (in./hr) by the duration of application (hr). Velocity refers to the speed at which water flows (distance per unit time) as opposed to flow rate (volume per unit time). Thus, the terms flow rate and velocity cannot be used interchangeably. Both the velocity and cross-sectional area of flow must be known in order to calculate flow rates.

Volume (V) is commonly expressed as gallons, acre-inches (ac-in.), and acre-feet (ac-ft). An acre-inch is the volume of water that would be required to cover an area of 1 acre to a depth of 1 inch. The relationships among these units are:

1 acre-inch = 27,154 gallons

1 acre-foot = 12 acre-inches

1 acre-foot = 325,848 gallons

Depth (D) units are used to express soil water-holding capacity (inches of water per foot of soil depth) and irrigations are scheduled after a fraction of the soil water in the tree's root zone has been depleted.

> Example:
> Calculate the amount of water to be applied per irrigation for the following conditions: fine sand soil with 0.75 in./ft total available water (TAW, calculated as water available at field capacity - water available at wilting point)
>
> The root zone depth is 2.0 ft
>
> Irrigations scheduled at 50% depletion of TAW
>
> Total soil water storage is calculated as:
> D = 0.75 in./ft x 2-ft soil depth = 1.5 inch
>
> Amount to be applied per irrigation is:
> D = 50% x 1.5 inches = 0.75 inch

Note that the amount of water to be pumped must be greater than the 0.75 inch needed to be stored in the plant's root zone since some water will be lost during application. Application efficiencies are always less than 100% because of water losses due to such factors as evaporation, wind drift, and non-uniform water application.

Depth units are also convenient for comparison with rainfall depths. For example, a 1-inch rainfall would supply the same volume of water as a 1-inch irrigation. Evapotranspiration (ET) rates are also usually expressed in inches per day. For example, if the tree ET rate is 0.20 in./day, the root zone holds 1.2 inches, and allowable soil water depletion is 50%, an irrigation will be required after 3 days (1.2 inch x 50%/0.20 inch/day = 3 days). The irrigation would be required early on the fourth day to avoid depleting more than 50% of the available soil water in the root zone.

When irrigation amounts are expressed as depths, it means that the depth is to be applied over the entire area to be irrigated. For example, if the irrigated area is 1.0 acre and the depth applied is 1.0 inch, the volume to be applied is 1.0 ac-in. (V = 1.0 inch depth x 1.0 acre = 1.0 ac-in.) or 27,154 gallons. Likewise, if the irrigated area is 20 acres, the volume of water required is 20 ac-in. or 543,080 gal (V = 20 ac-in. x 27,154 gal/ac-in. = 543,080 gal). Thus, depth units are used interchangeably with volume units because it is convenient to use depth units when referring to soil water depletion, rainfall, and plant ET rates. However, when referring to amounts of water permitted or pumped, volume units are preferred.

Flow rate (Q) is defined as the volume of water per unit of time flowing past a point in the system. Flow rates commonly used are gallons per minute (gal/min or gpm) and acre-inches per hour (ac-in./hr). Water management district consumptive use permits often are expressed in units of million gallons per day (mgd) or cubic feet per second (cfs) when estimating storm water runoff. The relationships between these units are:

1 gal/min = 1 gpm = 0.00221 ac-in./hr

1 gpm = 0.00144 mgd

1 ac-in./hr = 453 gpm = 0.652 mgd

1 mgd = 694 gpm = 1.53 ac-in./hr

1 cfs = 448 gpm

Since flow rate is volume per unit time, the volumes of water applied during irrigation can be calculated if the flow rate and irrigation duration (time) are known.

Example:
Find the volume of water applied and depth of application for the following conditions:

Pump discharges of 800 gpm

Duration of irrigation is 4 hours

The volume of water applied is: V = 800 gpm x 60 min/hr x 4 hr = 192,000 gal or V = 192,000 gal / 27,154 gal/ac-in. = 7.1 ac-in.

The depth of water applied can be calculated by multiplying the application rate by the duration of irrigation. An application rate of 0.10 in./hr with a 6-hr irrigation duration results in a gross depth of water applied of 0.6 inch (D = 0.10 in./hr x 6 hr = 0.6 inch).

Note that the actual depth of water stored in the tree's root zone and available for plant use would be less than 0.6 inch because of water losses during application. A typical application efficiency for microsprinkler irrigation systems is 85% for Florida conditions. Thus, only 0.51 inch (0.6 inch x 85% = 0.51 inch) of the 0.6 inch pumped would be expected to be stored in the root zone and available for plant use.

Flowmeters

Flowmeters are one of the key components of a microirrigation system. Flowmeters are critical for managing irrigation efficiently and for monitoring system performance. Managing irrigation efficiently requires knowing how much water the crop has used since the last irrigation (irrigation scheduling) and operating the irrigation system to apply the required amount of water. Flowmeters provide the information necessary to apply the correct amount of water.

Monitoring the performance of an irrigation system makes it possible to identify changes in flowrate during the season (measured at the same pressure). This change may indicate problems such as clogging of emitters or filters, leaks in the system, or problems with the pump or well. The rate of water flow in an irrigation system depends on a variety of hydraulic factors (see Chapter 21), as well as pump and system constraints. Table 22-1 lists typical flow rate ranges for common pipe sizes.

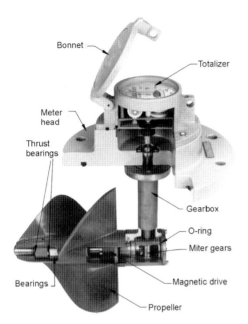

Figure 22-1. Components of a typical propeller meter.

Labels: Bonnet, Totalizer, Meter head, Thrust bearings, Gearbox, O-ring, Miter gears, Magnetic drive, Bearings, Propeller

Types of Flowmeters

Propeller meters are the most commonly used type of water measurement devices in irrigation systems (Fig. 22-1). Propeller meters need to be installed in straight sections of pipeline according to the manufacturer's recommendations. To minimize turbulence and the resulting inaccuracies caused by elbows, valves and other fittings, flowmeters should have at least a distance of ten times the pipe diameter upstream of the meter free of obstructions (straight pipe). In addition, there should also be five pipe diameters of straight pipe downstream of the meter (see Table 22-2). If the required upstream and downstream unobstructed distances cannot be

met, straightening vanes may be used to minimize the effects of turbulence caused by obstructions. For accurate measurement, the pipe must flow full. The readout of the flowmeter is through direct drive gears, cable connections, or magnetic drives. Models are available with both total and instantaneous flow rate measurements. Propeller meters are typically accurate to within 2% when new.

Example:

Determine the minimum distance downstream from an elbow required for installing a flowmeter in a 12-inch class 160 PVC pipe.

From Table 21-1, the diameter of 12-in. class 160 pipe = 11.770 in.

Upstream distance = 10 x 11.770 in. = 118 in.

Downstream distance = 5 x 11.770 in. = 59 in.

Note: Due to the high upstream distances required for large pipe diameters, most installations in citrus microirrigation systems utilize straightening vanes.

It is important to be aware that propeller meters have some potential problem areas. Debris in the water may entangle or damage the propeller and affect its operation. Many times the meter continues to register, but the rates registered are less than the true rate. If the debris is sufficient, the meter may fail to register completely. There is a small amount of pressure lost as water flows through the meter (typically 2 psi or less if the meter is the proper size). These drawbacks in propeller meters are minor. In fact, debris that might affect the propeller meter would also clog the

Table 22-1. Maximum and minimum flowrates (gpm) for various pipe diameters.

	Meter and nominal pipe size (inches)					
	4	6	8	10	12	14
Maximum flow rate	600	1200	1500	1800	2500	3000
Minimum flow rate	50	90	108	125	150	250

emitters and is therefore usually filtered out. In addition, pressure losses caused by the meter can be compensated for in the irrigation system design.

Magnetic flowmeters (Fig. 22-2) have the advantage of not causing obstructions in the pipe. This feature eliminates the problem of possible entanglement from debris in the water as well as any pressure loss across the device. Magnetic flowmeters measure the flow of conductive liquids flowing in filled pipes. The sensor creates a pulsating, alternating magnetic field on the inside of a pipe (Fig. 22-3). The liquid in the pipe will move through this magnetic field and generate a signal current proportional to its velocity. This information is collected by the electrodes and then processed by a microprocessor to provide the user with the desired flow information. Magnetic flowmeters also require less maintenance than propeller meters, have long-term accuracy, and can be

Figure 22-2. Magnetic flowmeter.

Figure 22-3. Schematic representation of magnetic flowmeter operation.

installed only 5 pipe diameters of straight pipe upstream from the meter. However, they have the disadvantages of a higher initial cost and the need for an external power supply.

Ultrasonic flowmeters measure flow velocity (and indirectly, flow rate) by directing ultrasonic pulses diagonally across the pipe both upstream and downstream. The difference in time required for the signal to travel through the moving water is measured and converted to flowvelocity. The operation of an ultrasonic flowmeter is based on the frequency shift (Doppler Effect) of an ultrasonic signal reflected by suspended particles or gas bubbles in the water stream (Fig. 22-4). This metering technique utilizes the physical phenomenon of a sound wave changing frequency when it is reflected by moving bubbles in the flowing water. In operation, ultrasonic sound is transmitted into a pipe with flowing liquids. The bubbles in the water reflect the ultrasonic wave with a slightly different frequency, the frequency difference being directly proportional to the flow of the liquid.

The transducer is mounted on the exterior of the pipe (Fig. 22-5). It receives electronic impulses from the transmitter through an interconnecting cable. The transducer generates an ultrasonic signal, which it transmits through the pipe into the flowing liquids. It simultaneously receives an ultrasonic signal reflected from the particles in this flowing liquid. The transmitter measures the difference between its output and input frequency and converts this different frequency into electronic pulses that are processed for an analog indication, and the pulses are scaled and totalized for flow quantity in gallons (Fig. 22-6).

Ultrasonic flowmeters are as accurate as propeller meters and, since they have no moving parts, require little maintenance. Because all attachments are external, these meters can be easily be moved to different locations. Ultrasonic flowmeters generally cost more than other types of meters.

Paddlewheel flowmeters (Fig. 22-7) can provide economical flow measurement for pipe diameters ranging from 1 inch to very large pipe with

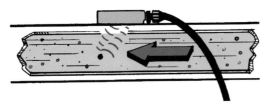

Figure 22-4. Schematic representation of ultrasonic flowmeter operation.

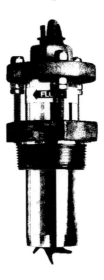

Figure 22-7. Example paddle wheel meter for insertion into 2-inch FPT fitting.

Figure 22-5. Transducers for ultrasonic flowmeter attached to 10-inch irrigation pipe.

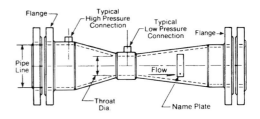

Figure 22-6. Transmitter and totalizing unit for ultrasonic flow.

Figure 22-8. Venturi flowmeter (bottom) with cross section view of features (top).

diameters of 36 to 48 inches or more. As water moves through the pipe, the paddle wheel rotates. The rotation velocity is proportional to the flow rate in the pipe. As the paddle wheel rotates past the base of the pickup coil, a voltage pulse is induced. An electronic unit converts pulses to units of flow based on the diameter and flow characteristics of the pipe.

Because the paddles contact only a limited cross section of the flow, the insertion depth of the rotor and the proper flow profile are critical to accurate readings. These meters are inserted to a specific depth into the pipe in which flow is measured. With proper calibration, the pulse rate represents an accurate measurement of velocity and, in turn, the flow rate. Paddle wheel flowmeters have an accuracy comparable to that of propeller flowmeters (within 2%).

Paddle wheel flowmeters normally need to be installed with at least 10 pipe diameters of straight pipeline upstream and 5 pipe diameters downstream from the sensor to ensure correct operation. However, if there are obstructions, longer upstream straight runs are recommended (Table 22-2). Some paddle wheel flowmeters can be installed in several sizes of pipe, which provides flexibility in system design. The applicability and cost are comparable to that of propeller meters.

Venturi flowmeters (Fig. 22-8) consist of a section of pipe with a restriction of a specific shape, across which pressure change is measured. The magnitude of this pressure change depends on the flow rate through the device. Venturi flowmeters have the advantages of unobstructed water flow, no moving parts in the main pipeline, little pressure loss across the meter, and good accuracy. Venturi flowmeters typically have low maintenance requirements; however, they cost slightly more than propeller meters.

Propeller Meters

Propeller meters are the meter of choice for the irrigation market for several reasons. They are intended to measure water flow rates mechanically, and typically indicate instantaneous flow rate and total volume. Because the reading is purely mechanical, it requires no power. The meters can be used in remote areas where power is nonexistent, or in temporary locations where power is inconvenient. Propeller meters can also be fitted with any kind of electronic readouts and outputs.

Propeller meters are relatively low cost and can be used economically even if the service need is short. With a typical accuracy of ± 2.0% of rate, the propeller meter gives irrigation users excellent control and accountability of their water. If needed, the meters can be made of high-grade stainless steels for harsh environments.

The propeller should be lightweight so it can respond to fast changes in the flow rates without inertial problems, and stiff enough to resist bending in fast-changing conditions. Friction in the drive system of the meter should be minimal; this will allow accurate measurement at low velocity ranges of the meter. The prop must also measure most of the pipe area (full bore measurement) to ensure accurate readings.

Propeller flowmeters normally have a propeller linked with a cable or shafts and gears to a flow indicator. These flowmeters can be installed in several different ways. They can be inserted into a short section of pipe, which is then either welded or bolted into the pipeline. Designs are available

Table 22-2. Recommended upstream distance of straight pipe to minimize effects of obstructions on flowmeter accuracy.

Obstruction	Pipe Distance
Concentric adapter	15 pipe diameters
1 Ell or Tee	20 pipe diameters
2 Ells	25 pipe diameters
2 Ells at 90°	50 pipe diameters
Ball, gate, or butterfly valve (wide open)	14 pipe diameters
Partially open valve	50 pipe diameters

that allow the meter to be clamped (Fig. 22-9), strapped, or welded onto the pipeline as a saddle-type meter or to be inserted into the pipe discharge. Propeller meters can also be installed as in-line meters in a short section of portable pipe. Couplings are welded onto each end of the flowmeter pipe section to connect sections of the portable pipe.

Propeller meters must be matched to the correct pipe size because the gear mechanism connecting the propeller to the indicator is based on the pipe diameter and the anticipated flow rate. Table 22-1 lists minimum and maximum flow rates for various pipe diameters. Most flow rate indicators report in gallons per minute or in cubic feet per second, while total flow indicators (totalizers) report in gallons, acre-feet, or cubic feet. Some indicators report in metric units. The flowmeter should be installed at a location of minimal water turbulence, since too much turbulence will cause the flow rate indicator to oscillate wildly, preventing reliable measurement.

Manufacturers often recommend that a section of straight pipe 10 pipe diameters long be placed immediately upstream from the flowmeter and a minimum of 5 pipe diameters downstream. However, longer pipe lengths may be required in locations where turbulence occurs due to obstructions in the pipeline (Table 22-2). A centrifugal sand separator or a series of elbows in the pipeline may cause a swirl or rotation to develop in the flowing water. The swirls may still be present even 100 pipe diameters downstream. The remedy is to place straightening vanes in the pipeline just in front of the flowmeter (Fig. 22-10). Reliable measurements can normally be made with even relatively short sections of pipe if straightening vanes are used.

A relatively stable rate indicator reading means that turbulence is minimal. Wide variations in indicator readings are a symptom of turbulence in the pipeline. If the readings are extremely erratic, air or gas may be present in the water.

Propeller flowmeters operate properly only if the pipe is flowing full. If the pipeline is partially full,

the flow rate measurement will not be accurate. In pressurized irrigation systems, flow will usually be full at the pump discharge, but in pumps with an open discharge, as into an irrigation ditch, pipe flow may not be full. This problem can be remedied by creating a slight rise in the discharge pipe, installing a gooseneck at the pipe discharge, or by installing an elbow (discharge end pointing upward at the pipe discharge).

Accuracy

Measurement accuracy is the difference between the true flow through a pipe and the flow measured with a meter. The measured flow should be as close as possible to the actual amount of water flowing in the pipe. Most irrigation meters should

Figure 22-9. Clamp-on type saddle propeller meter.

Figure 22-10. Propeller meter with straightening vanes.

have an accuracy of ± 2.0% of the true amount. The accuracy for a meter may be specified as a percentage of rate or as a percentage of full scale. In most instances, flowmeters with rate accuracy should be selected. If a flowmeter is operated below its recommended range, the accuracy may be reduced.

Inserting a propeller into the water flow will cause friction, resulting in pressure or head losses in the pipeline. The amount of pressure lost depends on the velocity and pipe diameter (flow rate). The higher the flow rate, the more pressure will be lost due to the flowmeter. Pressure losses from propeller meters are generally small. For example, with a 10-inch flowmeter, the pressure loss is typically less than 1 psi when flow rates are less than 2000 gpm.

Since propeller meters are mechanical devices and have moving parts, a scheduled maintenance program is recommended. Since most installations have no power or communication ability, a user must visually inspect his meters on site. If a propeller meter is inaccurate, it is often a mechanical problem that is easily discovered. Oftentimes, inaccurate flow measurements in propeller meters result from debris lodged inside the meter. If electronic registers are used with the meter, check the signal and all connections carefully. Small problems in electrical properties can make the measurement system inaccurate.

When checking an unfamiliar meter, it should be inspected to ensure that the meter is installed according to the manufacturer's recommendations. The instruction, operation, and maintenance manual should be reviewed for details regarding the requirement for straight pipe lengths before and after the meter. Recommended maintenance should be done according to the manual's suggestions. For propeller meters, the pipeline must remain full of water. If the pipe is sloped, errors will be introduced due to the lack of a completely full pipe under certain flow conditions.

If field checks cannot identify the cause of inaccurate measurements, a full calibration of the meter is required. This normally necessitates removing the meter from the line and sending it to a calibration laboratory. The propeller meters should be calibrated on a test rig that has been calibrated at the National Institute of Standards and Technology (NIST). The laboratory accuracy should be four times as accurate as the meter tested. For a meter with an accuracy of ± 2%, the laboratory should have an accuracy rate of at least ± 0.5%.

Shunt Flowmeters

Shunt flowmeters (Fig. 22-11) work on the principle that a fraction of the water in the irrigation main pipeline is shunted through a small diameter pipeline parallel to the mainline for a short distance where it is measured before it is returned to the mainline. The small diameter shunt pipeline is equipped with a totalizing flowmeter that measures only the shunted flow. Flow in the main pipeline is proportional to the amount of flow measured in the shunt pipeline.

A head loss is required in the main pipeline to create the pressure difference that causes water to flow through the shunt pipeline. Because the shunt flow rate depends on the head loss in the main pipeline, the relationship between the head loss and shunted flow must be determined by calibration. One commercially available shunt flowmeter operates by using a venturi constriction to create the head loss in the main pipeline. The constriction increases velocity and decreases pressure in the throat of the venturi. The venturi is constructed in tube diameters to be compatible with mainline pipe diameters from 6 to 10 3/4 inches. The venturi design has the advantage that much of the pressure loss in the throat of the venturi is recovered downstream of the constriction. The venturi is also a simple device whose hydraulic properties are well understood. Very importantly, there are no moving parts to wear out.

Flow is registered by a small, low-cost totalizing water meter on the shunt pipeline. The meter gears and register used are sized to indicate the total flow in the main plus shunt pipelines. This design has the advantage that all of the register components

Figure 22-11. Design of typical shunt flowmeter.

are contained on the shunt line where they are readily accessible for calibration, repair, or replacement. Calibration can be performed without removal of the meter. The meter register is a totalizer that does not directly measure flow rate. Rather, total volume of flow is measured in gallons, and this volume must be divided by the time of operation to determine the flow rate for any period of operation. The register components cost less than $100 if the entire register needs to be replaced.

Measurement accuracy of shunt meters can be as good as ± 2%. To improve accuracy, straightening vanes can be inserted in the mainline meter tube. However, as with all in-line flowmeters, certain minimal lengths of straight pipeline are required upstream of the meter. Because the straightening vanes are used, the minimum upstream distance of straight pipeline recommended by the manufacturer is only 2 pipe diameters as compared to the 6 to 10 diameters required for conventional impeller meters without straightening vanes.

Estimating Flow Rate with Trajectory Method
With a free-flowing horizontal pipe, the flow can be estimated by measuring the trajectory of the discharge (Fig. 22-12). The distance water flows horizontally past the end of the pipe is dependent on the velocity of the flow. The calculation requires measurements of the vertical drop in the trajectory compared to the horizontal distance the stream travels past the end of the pipe. The following steps can be used to estimate discharge from free-flowing pipes. (Adapted from: Pair, et al., 1983).

Step 1. Measure the inside diameter of the pipe, the freeboard, and the wall thickness of the pipe plus any fittings at the outlet. For pipes larger than 12 inches, measure the outside diameter instead.

Step 2. Choose one of the "H" values given in Table 22-4 (setting H equal to 13 inches makes the calculation simpler).

Step 3. Set the square so the inside vertical scale reads the "H" value plus the freeboard and the thickness of the pipe and fittings. Read the "X" distance on the lower horizontal scale at the end of the pipe. The long side of the square must be parallel with the pipe.

Step 4. From Table 22-3, read the "A" value.

Step 5. From Table 22-4, obtain the "K" value. Notice that an "H" value of 13 inches yields a "K" of 1.00.

Step 6. Multiply "A" times "X" times "K" to obtain the flow rate in gallons per minute. For full pipes, add ten percent to obtain a more accurate estimate.

Example:
Pipe inside diameter: 10 inches
Freeboard: 4 inches
Pipe Wall Thickness: 3/8 inches

Choose an "H" value of 9 inches.
"X" distance is measured as 16 inches with inside vertical scale reading 13 3/8 inches at the water surface (9 plus 4 plus 3/8).

From Table 22-3, obtain the "A" value under a pipe diameter of 10 inches and a freeboard of 4 inches. "A" = 49.5 square inches.

From Table 22-4, under an "H" value of 9 inches, read "K" = 1.20. Multiply "A" times "X" times "K" 49.5 x 6 x 1.20 = 950 gpm.

Estimating Volume/Flow in Ditches or Canals
Sometimes it is necessary to estimate the volume or flow rate in an open canal or ditch. Accurate

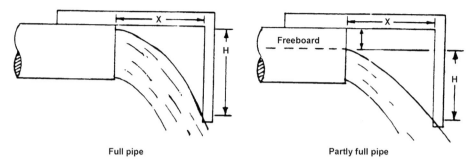

Figure 22-12. Estimating flow rates using the trajectory method.

Table 22-3. "A" values for determining flow rate (cross-sectional area of stream in square inches).

Freeboard (inches)	Pipe diameter (inches)					
	6 (I.D)	8 (I.D)	10 (I.D)	(12 I.D)	14 (OD)*	16 (OD)*
Full Pipe	28.3	50.3	78.5	113	138	183
0.5	27.5	49.0	77.0	112	137	181
1.0	25.0	47.0	74.5	109	134	178
1.5	23.0	44.0	71.5	105	130	174
2.0	20.0	40.5	67.5	101	126	169
2.5	17.0	37.0	63.5	96.0	121	163
3.0	14.0	33.0	59.0	91.5	116	158
3.5	11.0	29.0	54.0	86.0	110	152
4.0	8.5	25.0	49.5	80.5	104	145
4.5	5.5	21.5	44.5	74.5	98.0	138
5.0		17.5	39.5	68.5	92.0	131
5.5		13.5	34.5	62.5	86.0	124
6.0		10.0	29.5	56.5	79.0	116
6.5			24.5	50.5	72.5	109
7.0			20.0	44.5	66.0	101
7.5			15.5	39.0	60.0	93.0
8.0				33.0	53.0	86.0
8.5				27.5	47.0	78.5
9.0				22.0	40.5	70.5
9.5					34.5	63.0
10.0					28.5	56.6

*Based on a wall thickness of 0.375 inches

Table 22-4. "K" values for selected "H" values for trajectory estimates of flow rates.

H→	4 (in.)	5 (in.)	6 (in.)	7 (in.)	8 (in.)	9 (in.)	10 (in.)	11 (in.)	12 (in.)	13 (in.)	14 (in.)	15 (in.)
K→	1.80	1.61	1.47	1.37	1.28	1.20	1.14	1.09	1.04	1.00	0.97	0.93

measurement requires the installation of a weir or flume. However, estimates adequate for most purposes can be made from some simple measurements. Volume measurements require determining the cross-sectional area (Fig. 22-13). Typically, measurements should be taken at several points along the ditch to obtain the average depth profile. In addition, if the width is not uniform, several measurements should be taken and the average width determined. The average cross-sectional area is calculated by the average depth times the average width as:

$$A_{xc} = D_a \times W_a \qquad \textbf{Eq. 22-1}$$

where,

A_{xc} = Average cross-sectional area (ft^2)
D_a = Average depth (ft)
W_a = Average width (ft)

The volume of water in the ditch can be calculated by multiplying the cross-sectional area by the length of ditch (typically distance between structures). To convert to gallons, multiply ft^3 by 7.48, or to calculate ac-ft, multiply ft^3 by 0.00028.

$$Vol = A_{xc} \times L \qquad \textbf{Eq. 22-2}$$

where,

Vol = Volume of water (ft^3)
L = Length of ditch (ft)

Example:
For the following measurements related to locations in Figure 22-13, determine the cross-sectional area and volume for a ditch that is 1200 feet long. (Assume that the measurements represent the average for the entire length of ditch).

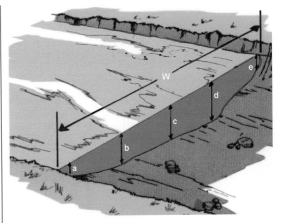

Figure 22-13. Typical measurement locations to determine cross-sectional area of ditch or canal.

Width: w = 10 ft
Depth measurements taken 2 ft apart (equally-spaced increments):

a = 2 ft b = 3 ft c = 4 ft d = 4.5 ft e = 2 ft

$$D_a = \frac{2 + 3 + 3.5 + 4 + 2.5}{5} = 3.0 \text{ ft}$$

A_{xc} = 10 ft x 3.0 ft = 30 ft^2

Vol = 30 ft^2 x 1200 ft = 36,000 ft^3

$$Vol = 36,000 \text{ ft}^3 \text{ x } \frac{7.48 \text{ gal}}{\text{ft}} \text{ x } \frac{1 \text{ ac-inch}}{27,158 \text{ gal}}$$

= 9.9 ac-inch

In order to calculate the flow rate in the ditch, the average velocity of water flow must be known. The velocity can be determined by measuring the time required for a floating object to travel a known distance. The velocity multiplied by the cross-sectional area will yield flow rate.

$$Q = A_{xc} \times V \qquad \qquad \text{Eq. 22-3}$$

where,

A_{xc} = average cross-sectional area (ft^3)
V = average velocity (ft/sec)
Q = flow rate (ft/sec)

Water along the banks of a ditch or canal flows slower than it does in the center. This is because the banks tend to resist the water flow. This resistance varies with shape and channel roughness. For most ditches and canals, a factor of 0.8 times the indicated velocity of flow can be used to estimate the average velocity.

> Example:
> Using the the cross-sectional area and length in the previous example, if the time required for a float to travel 100 ft in the center of the ditch is 70 seconds, estimate the flow rate of the ditch. Use a factor of 0.8 to convert measured velocity in the center of the ditch (V_c) to average velocity (V_{avg}).
>
> $V_c = 100$ ft / 70 sec = 1.43 ft/sec
>
> $V_{avg} = V_c \times 0.8 = 1.43$ ft/sec $\times$ 0.8 = 1.14 ft/sec
>
> $Q = A \times V = 30$ ft^2 $\times$ 1.14 ft/sec = 34.2 ft^3 /sec
>
> $Q = 34.2$ ft^3 /sec $\times$ 7.48 gal/ft^3 $\times$ 60 sec/min
>
> $\quad$ = 15,349 gal/min

Pressure Gauges

Measurement of pressure in key points of a network is of major importance for a system operator. Pressure gauges are effective and simple tools for controlling the performance of the system. A proper gauge and reasonably chosen measuring points help in discovering leakage and clogging, ascertaining an effective flow distribution, and verifying correct operation of different control devices in a system. Pressure measurements can also give a fair estimate of the flow rates in various zones of the irrigation system. Therefore, the reliability and accuracy of the pressure-measuring equipment is an important feature of the irrigation system.

Pressure gauges (Fig 22-14) should be installed in easily accessible places, so that they are convenient to read and maintain in proper working condition. They should be able to withstand large pressure fluctuations, overpressure, and vacuum. A pressure gauge must be selected according to the working pressure range and the local conditions of installation. A properly selected gauge has a scale that will measure up to at least twice the normal working pressure. This allows for pressure surges in the pipeline without damaging the gauge. It also keeps the pointer in the middle of the scale where the accuracy is best under normal operating conditions.

The main type of pressure gauge used in irrigation systems is the Bourdon gauge. This type of gauge uses a C-shaped tube made of copper or stainless alloy (Fig. 22-15). One end of the tube is connected to the inlet of the gauge and the other end is sealed. As the pressure increases, the coiled tube begins to uncoil slightly. This movement of the C-shaped tube is mechanically transferred to the pointer, which moves around a calibrated dial. The dial is typically calibrated in psi, with various ranges available.

The accuracy of the pressure gauge depends on the readability of the scale and the precision of the mechanism. Usually, gauge readings can only be made to 1/2 the span increment between scale measurements. For instance, readings for the gauge in Fig. 22-14 would be about 1 psi since the increments on the scale are at 2 psi increments. The accuracy of the gauge often changes along the range of the scale. Usually, the most accurate readings are in the range of 25% to 75% of the full scale reading. Accuracies of 1% to 2% in this region are normally adequate for irrigation systems.

The proper size of a gauge depends on working conditions. A larger dial permits finer gradation of

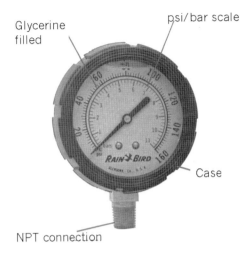

Figure 22-14. Liquid-filled 0 to160 psi.

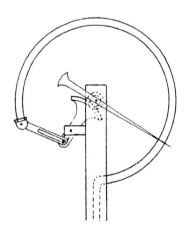

Figure 22-15. Bourdon tube and mechanical connections to pointer typical for irrigation system pressure gauges.

the scale and easier readout. Larger gauges (4- to 6-inch diameter) serve as master gauges for testing or as panel instruments in control rooms. Small dials should not be used for stationary control, as they cannot depict the system pressures with sufficient accuracy. Typically, 2 1/2- to 3-inch gauges are satisfactory for most irrigation system applications.

The materials that make up the gauge box, the transparent face pane, and the internal parts are of the utmost importance. For corrosive conditions,

the Bourdon tube and mechanism parts should be made of high-grade stainless steel or plastic. In working conditions where mechanical vibrations and pressure fluctuations occur frequently, the rack-and-pinion gear should be reinforced to resist rapid wear.

Special attention should be given to the transparent panes of gauges. They should be made of glass plate only. Gauges with plastic transparent panes tend to become cloudy quickly when used outdoors and exposed to the sun. Dials of pressure gauges for indicating water pressure are usually made of sheet steel. They should be painted white with black graduation marks and have large, easy-to-read digits. Pressure gauges with dials painted red, blue, or green are destined for fluids other than water (i.e., air, oxygen, coolants etc.) and are not usually suitable for irrigation system service. Pressure units used (typically psi) should be clearly marked on the dial, together with the name or trade mark of the manufacturer. The date of the last calibration and adjustment of the gauge should be recorded in ink on the dial.

Gauges for measuring water pressure usually have an externally threaded inlet at the bottom of the gauge, typically NPT in nominal sizes of 1/8 to 1 inch, depending on the size of the gauge. If necessary, seal the threads with Teflon tape, being careful not to cover the orifice. When installing a gauge, its base should be held firmly with a wrench at the square or hexagonal part of the nipple above the thread. Holding the gauge by the box and turning it when engaging the connecting threads is bad practice. It may impair the calibration and accuracy of the gauge, permanently damaging the gauge mechanism.

While handling and installing a pressure gauge, special care must be given to the small orifice at the end of the nipple. This part, which forms the inlet to the Bourdon tube, is extremely delicate, and careless handling may damage the orifice and plug the gauge inlet. In places where the gauge may be subjected to severe pressure fluctuations or water hammer, it should be protected with a valve that can be used to shut off the pressure. When installing the gauge directly in the outlet of a valve, the length of the threads must be considered. The gauge inlet orifice must never touch the rotating element of the ball valves. In systems where the gauge must constantly indicate the pressure and cannot be shutoff when not in use, vibration choking devices can minimize the effect of pressure fluctuations and water hammer. In these cases, the gauge may be filled with a high-viscosity liquid such as glycerin.

Another way of protecting a gauge from effects of surge is to connect it through a vertical loop of small diameter tubing. The upper part of the loop should contain air, which can compress and so compensate the pressure fluctuations. In such an arrangement, the presence of air in the loop should be checked periodically. If the air leaks and the loop fills with water, the device becomes ineffectual.

Where mechanical vibration in the piping may be a problem, such as near a pumping plant, a flexible connection of gauges, using small diameter, reinforced tubing, may prevent the transfer of vibration to the gauge mechanism, and so decrease wear of the parts of the mechanism. Such resilient connections may, in certain instances, even prevent the damaging effects of water hammer. The plastic piping must be appropriately chosen and adapted to the working pressure of the system, and to the possible pressure transients that may occur in the system.

Chapter 23. Pumps

by Brian Boman

There are two basic classes of irrigation pumps: positive displacement and kinetic, with each of these types having several subclassifications. The most commonly used pumps in irrigation systems are centrifugal pumps, a subclassification of kinetic pumps. Centrifugal pumps are generally either of the vertical and horizontal type, and each of these has many variations.

Centrifugal pumps convert mechanical energy into kinetic energy in the water by accelerating it to the outer rim of a rotating impeller (Fig. 23-1). In its simplest form, a centrifugal pump consists of an impeller, fixed on a shaft, that rotates in a volute-type casing. Water at the center of the impeller is picked up by the impeller vanes and accelerated to a high velocity by the impeller rotation. It is then discharged by centrifugal force into the casing where the high-velocity head is converted to a pressure head. The faster the impeller revolves, the higher the velocity of the water at the vane tip. Larger impeller diameter also results in higher water velocity at the vane tips. Higher velocities result in greater energy imparted to the water.

The kinetic energy transferred to water by an impeller is harnessed by creating a resistance to the flow. Resistance is created by the pump volute (casing), which catches the liquid and slows it down. When the liquid slows down in the pump casing, some of the kinetic energy is converted to pressure energy that can be read on a pressure gauge in the discharge line. The term "head" is used to quantify the kinetic energy that a pump transfers, and is related to the velocity that the water gains as it goes through the pump.

Suction or Lift
A centrifugal pump will operate either with the impeller submerged in water or with the impeller a limited distance above the surface. When located above the water surface (suction lift condition), the inlet side of the pump must be connected to the water source with an airtight pipe or hose. The pump does not lift the water. Rather, water must be pushed up to the pump by a higher pressure on the surface of the water source than exists at the center of the impeller.

All centrifugal pumps must be completely filled with water (primed) before they can operate. After the pump is primed, it is started and the rotation of the impeller throws the water that is contained in it toward the outside by centrifugal force. This creates a vacuum or an area of lower pressure at the center of the impeller. Since atmospheric pressure is pushing downward on the water surface, water is forced through the suction pipe to the lower

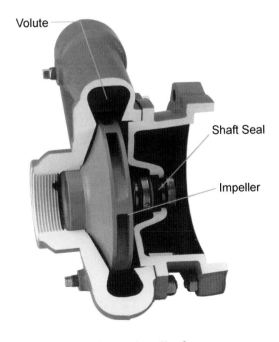

Figure 23-1. Semi-open impeller for centrifugal pump.

Table 23-1. Conversion from psi to equivalent head in feet of water.

psi	ft	psi	ft	psi	ft	psi	ft
1	12	30	69	55	127	80	185
10	23	35	81	60	139	85	196
15	35	40	92	65	150	90	208
20	46	45	104	70	162	95	219
25	58	50	116	75	173	100	231

pressure area at the center of the impeller to re-place the water being thrown outward. Thus, there is a continuous flow of water through the pump.

Head

A pump moves water from one point to another. The difference in elevation between the lower (pump) level and the higher (discharge) level is called the "static head" at which the pump oper-ates. It is usually expressed in terms of feet. How-ever, since 2.31 feet of elevation is equivalent to 1 pound per square inch (psi) of pressure (Table 23-1), the performance of a pump is often expressed in terms of discharge pressure in psi.

The total head that a centrifugal pump will develop depends on the impeller's diameter and speed. If a pump impeller rotates at its rated speed and a valve on the discharge side of the pump is closed, the pump will develop its maximum head that can be read on a pressure gauge. The gauge reading (tran-slated into feet) will register the height to which the pump is capable of elevating water. This head (pressure) is known as the "shutoff head." If the valve is slowly opened, the flow will increase and the pressure gauge reading will fall. This will con-tinue until the condition of maximum flow and minimum head is reached. If the total head devel-oped at any given rate of flow is plotted against the quantity of water delivered in terms of gallons per minute, the result will be a performance curve for this particular pump at this particular speed. If the power required to turn the pump is observed during this process, it will be noted that for a typical

centrifugal pump the power is at a minimum when there is no water being discharged from the pump, and it will gradually increase as flow rate increases and head decreases. The maximum efficiency will be somewhere about midway between zero and maximum flow.

Pump Curves

A typical centrifugal pump curve is given in Fig. 23-2. The pump curves can give a lot of informa-tion for a particular pump under various operating conditions. The size of the pump (2 1/2 x 3-7) is shown in the upper part of the pump curve illustra-tion. Note that the size number 2 1/2 x 3-7 indi-cates that the pump has a 2 1/2-inch discharge port, a 3-inch suction port, and a maximum nominal im-peller size of 7 inches. This type of nomenclature is common, with some companies putting the 3 in the first position instead of the 2 1/2. In either case, standard procedure is that the suction port is the larger of the first two numbers shown, and the larg-est of the three numbers is the nominal maximum impeller size.

Also, in the upper right hand corner, notice that the curve indicates performance at the speed of 3500 rpm (a common electric motor speed at 60 hz). All the information given in the curve is valid only for 3500 rpm. Generally speaking, curves that indicate rpm to be between 3400 and 3600 rpm are used for all two-pole (3600 rpm nominal speed) motors applications.

The pump's flow range is shown along the bottom of the performance curve. Note that the pump, when operating at one speed, 3500 rpm can provide various flows. The amount of flow varies with the amount of head generated. As a general rule with centrifugal pumps, an increase in flow causes a decrease in head.

The left side of the performance curve indicates the amount of head a pump is capable of generating. Notice that there are several curves that slope generally downward from left to right on the curve. These curves show the actual performance of the pump at various impeller diameters. For this pump, the maximum impeller diameter is shown as 6 3/4 inches and the minimum is 4 1/2 inches. Impellers are trimmed in a machine shop to match the impeller to the head and flow needed in the application.

The point on the curve where the flow and head match the application's requirement is known as the duty point. A centrifugal pump always operates at the point on its performance curve where its head matches the resistance in the pipeline. For example, if the pump shown above was fitted with a 6 3/4-inch impeller and encountered 160 feet of resistance in the pipeline, it would operate at a flow of approximately 280 gallons per minute and 160 feet of head. It is important to understand that a centrifugal pump is not limited to a single flow at a given speed. Its flow depends on the amount of resistance it encounters in the pipeline. For maximum efficiency, the pump discharge should closely match the irrigation system requirements. If a valve is used to control the flow of a centrifugal pump, the valve should be installed on the discharge pipe. Do not restrict a pump's flow by putting a valve on the suction line because it can damage the pump.

The brake horsepower lines (which slope upward from left to right) are also on this performance curve. These lines indicate the amount of

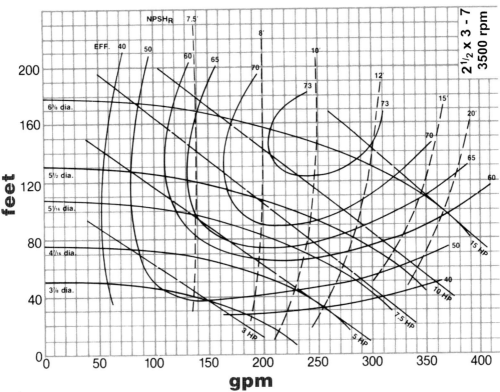

Figure 23-2. Example pump curves for 4 1/2- to 6 3/4-inch impeller diameters and 3- to 15-hp power requirement.

horsepower that is required to operate the pump at different points of the performance curve. The lines correspond to a BHP horsepower scale on the lower right hand corner of the page. For example, with an operating point of 280 gpm at 160 feet of head (for 6 3/4-inch impeller example), the corresponding BHP line equals about 15 horsepower.

When sizing a motor driver to fit a specific application, it is necessary to consider whether the pump will ever be required to operate at a flow higher than the duty point. The motor will need to be sized accordingly. If the pump may flow out to the end of the curve (if someone opens the restriction valve all the way, for example), it is important that the motor does not become overloaded as a result. Therefore, it is normal practice to size the motor not for the duty point, but for the end of curve (EOC) horsepower requirements. In the example, a 15 hp motor would adequately power the pump at a duty point of 280 gpm at 160 feet. However, consideration should be given to the need for additional power if more flow or pressure is needed under other operating conditions.

Performance Characteristics

The performance of a pump varies with the speed of rotation. The capacity or rate of flow varies directly in relation to the change in speed; that is, if the speed is increased by 10%, the capacity increases by 10%. The head that the pump will develop, however, varies as the square of the ratio of the new speed to the old speed. Thus, an increase of 50% in the speed means that head developed will increase $(1.5/1)^2$ or 2.25 times. Further, the horsepower required increases by the cube of the speed of change. This means that by increasing the speed by 50%, the required horsepower will be $(1.5/1)^3$ or 3.38 times that required at the lower speed.

Since vertical turbine pumps are installed inside the well, there are definite capacity limitations for a given diameter of well casing. It is necessary to increase the pump speed to increase capacity for a given well size. The maximum permissible speed depends on a number of factors but for 4-, 6-, and

even 8-inch pumps, a speed of 3600 rpm is not uncommon. This speed is not advisable for larger sizes. Since nominal electric motor speeds used in irrigation pumping are either 1760 rpm or 3450 rpm, intermediate speeds may be obtained only with right angle gears of suitable ratio or with belt and pulley drives.

Theoretically, varying the pump speed will result in changes in capacity, head, and brake horsepower according to the following formulae (affinity laws) that apply to all types of centrifugal pumps.

For speed (rpm) changes when impeller diameter remains constant:

$$\frac{Q_1}{Q_2} = \frac{rpm_1}{rpm_2} \qquad \text{Eq. 23-1}$$

$$\frac{H_1}{H_2} = \left(\frac{rpm_1}{rpm_2}\right)^2 \qquad \text{Eq. 23-2}$$

$$\frac{BHP_1}{BHP_2} = \left(\frac{rpm_1}{rpm_2}\right)^3 \qquad \text{Eq. 23-3}$$

When impeller diameter (Dia) changes and speed remains constant:

$$\frac{Q_1}{Q_2} = \frac{Dia_1}{Dia_2} \qquad \text{Eq. 23-4}$$

$$\frac{H_1}{H_2} = \left(\frac{Dia_1}{Dia_2}\right)^2 \qquad \text{Eq. 23-5}$$

$$\frac{BHP_1}{BHP_2} = \left(\frac{Dia_1}{Dia_2}\right)^3 \qquad \text{Eq. 23-6}$$

where,

Q = pump flow rate (gpm)
H = total head (feet)
BHP = brake horsepower required
rpm = pump speed in rpm
Dia = impeller diameter (inches)

If the pump is driven by a constant-speed electric motor connected directly to its shaft, the pump turns at the same speed as the motor. The pump speed can be changed by connecting it to the motor through a belt drive with pulleys of different diameters, or through a speed-changing gear. This arrangement, however, provides only a different fixed speed and not a variable speed.

When a pump is driven by an internal combustion engine, it is possible to vary its speed of operation within the speed range of the engine. However, this will often change the efficiency. Also, the horsepower capacity of an engine varies when the rpm of the engine is changed.

Example:
The head-capacity curve for a pump with a 6-inch impeller is given in Fig. 23-3. What will happen when the impeller is trimmed to 5 inches? From the 6-inch diameter curve, the following information is known:

D_1 = 6 inches
Q_1 = 200 gpm
H_1 = 100 feet
BHP_1 = 3 hp

Rearranging Eqs. 23-4 to 23-6 results in the solutions:

The new calculated data (Q = 167 ft, H = 69 ft) can be plotted on the graph (point B). Continue to calculate the Q-H relationships for different flow and head points to develop a new pump curve for the 5-inch impeller (dashed line). The new power requirement can be calculated as follows:

$$Q_2 = \frac{D_2}{D_1} \times Q_1 = \frac{5 \text{ in.}}{6 \text{ in.}} \times 200 \text{ gpm} = 167 \text{ gpm}$$

$$H_2 = \left(\frac{D_2}{D_1}\right)^2 \times H_1 = \left(\frac{5 \text{ in.}}{6 \text{ in.}}\right)^2 \times 100 \text{ ft}$$

$$= 69 \text{ ft}$$

$$BHP_2 = \left(\frac{D_2}{D_1}\right)^3 \times BHP_1 = \left(\frac{5 \text{ in.}}{6 \text{ in.}}\right)^3 \times 3 \text{ hp}$$

$$= 1.7 \text{ hp}$$

Calculating Impeller Trim
A microirrigation system design calls for a pump discharge rate of 225 gpm at 160 feet of head. The pump available that has the closest characteristics has the curve given in Fig. 23-2. Note that the required conditions fall somewhere between the 6 3/4-inch and the 5 7/8-inch impeller diameter curves. To determine the trimmed impeller diameter required to meet the head and discharge, draw a line perpendicular from the 225 gph, 160 foot coordinate point to the curve for the 6 3/4-inch impeller. At this point, the conditions are:

D_1 = 6 3/4 inches
Q_1 = 230 gpm
H_1 = 172 feet TDH

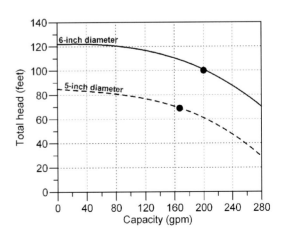

Figure 23-3. Pump head-capacity curve for 6-inch impeller and calculated curve for 5-inch impeller using affinity laws.

Rearrange Eq. 23-5 to solve for D2.

$$D_2 = D_1 \sqrt{\frac{H_2}{H_1}}$$

$$D_2 = 6.75 \sqrt{\frac{160}{172}} = 6.55 \approx 6\frac{5}{16} \text{ inches}$$

Determine whether the new impeller diameter will meet the required discharge rate by rearranging Eq. 23-4 and solving for Q2.

$$Q_2 = \frac{D_2}{D_1} \times Q_1$$

$$Q_2 = \frac{6.55}{6.75} \times 230 = 223 \text{ gpm}$$

Net Positive Suction Head (NPSH)

When a pump is operated with its impeller above the surface of the water supply, it is operating under suction lift. The permissible vertical distance between the center of the impeller and the surface of the supply is not equal for all centrifugal pumps. The allowable distance varies with pump design and operating conditions. If a pump successfully operates with a greater suction lift, it does not necessarily mean that it is a better pump. It merely means that the design characteristics are different. Usually, to design a pump that will operate under maximum lift means that some other desirable characteristics must be sacrificed.

Theoretically, if a pump could be designed to produce a perfect vacuum, and if it was operated at sea level, the acting atmospheric pressure of about 14.7 psi would allow water from a depth of 34 feet below the pump to be pumped. In practice, of course, this is impossible because a perfect vacuum cannot be created at the center of the impeller, and because there are losses from friction in the suction line and from turbulence at the entrance to the impeller. Usually a vertical suction lift of about 15 feet should be considered the maximum for reasonably efficient operation. A Net Positive Suction Head (NPSH) analysis on the suction side of a

pump is necessary to determine if the liquid will vaporize at the lowest pressure point in the pump.

The pressure that water exerts on its surroundings is dependent upon its temperature. This pressure (vapor pressure) is a unique characteristic of every fluid, and increases with increasing temperature. When the vapor pressure of water in a pump reaches the pressure of the surrounding medium, the water begins to vaporize or boil. The temperature at which this vaporization occurs will decrease as the pressure of the surrounding medium decreases. A liquid increases greatly in volume when it vaporizes. One cubic foot of water at room temperature becomes 1700 cu. ft. of vapor at the same temperature.

If a fluid is to be pumped effectively, it must be kept in liquid form. NPSH is simply a measure of the amount of head present to prevent this vaporization at the lowest pressure point in the pump. The NPSH required (NPSHr) depends on the pump design. As the liquid passes from the pump suction to the eye of the impeller, the velocity increases and pressure decreases. There are also pressure losses due to shock and turbulence as the liquid strikes the impeller. The centrifugal force of the impeller vanes further increases the velocity and decreases the pressure of the liquid. The NPSHr is the positive head (in feet absolute) required at the pump suction to overcome these pressure drops in the pump and maintain the liquid above its vapor pressure. The NPSHr varies with speed and capacity within any particular pump. Pump manufacturer's curves normally provide this information. Most manufacturers test their pumps to determine the suction lift under which they will operate, and plot this data on the pump performance curve as NPSH required by the pump. This information is available as a component of the pump characteristics curve.

Available NPSH (NPSHa) is a characteristic of the system in which the pump operates. It is the excess pressure of the liquid (expressed in feet of head) over its vapor pressure as it arrives at the pump suction. In an existing system, the NPSHa can be calculated with a vacuum gauge reading on the pump suction line by:

$$NPSHa = PB - VP \pm Hp + HV \qquad \text{Eq. 23-7}$$

where,

PB = Barometric pressure (feet)
VP = Vapor pressure of water at maximum pumping temperature (in feet)
Hp = Pressure head at the pump suction expressed in feet (negative value for vacuum and positive if flooded suction) corrected to the elevation of the pump centerline
Hv = Velocity head in the suction pipe (feet).
Hv = $V^2/2g$ where V = velocity of water in ft/sec and 2g = 2 times the acceleration of gravity (2 x 32.2 ft / sec^2)

Note: In using the formulas, it is important to correct for the specific gravity of the liquid and to convert all terms to units of feet of H_2O.

Example:
Calculate the NPSHa for a system pumping water with the following conditions:
Discharge rate of 1500 gpm
12-inch Schedule 40 PVC suction line (I.D. = 11.814 inches)
Barometric pressure is 29.8 in. Hg
Water temperature of 80° F

Table 23-2. Vapor pressure (VP) versus temperature for water.

Temperature ° F	Vapor Pressure	
	psi	feet H_2O
40	0.12	0.3
50	0.18	0.4
60	0.26	0.6
70	0.36	0.8
80	0.51	1.2
90	0.70	1.6
100	0.94	2.2
110	1.28	2.9
120	1.70	3.9

Vacuum reading of 12.5 ft H_2O on suction line.

From Appendix 2: inches Hg x 0.882 = ft H_2O

PB = 29.8 in. Hg x 0.882 = 26.3 ft H_2O

From Appendix 3: gpm/448.8 = ft^3/sec
Q = 1500 gal/min / 448.8 = 3.34 ft^3/sec
A = π x D^2/4
D = 11.814 in./ 12 in./ft = 0.98 ft
A = 3.14 x 0.98^2/4 = 0.81 ft^2
V = Q/A = 3.34 ft^3/sec / 0.81 ft^2 = 4.1 ft/sec
Hv = $V^2/2g$ = (4.1 ft/sec)^2 /(2 x 32.2 ft/sec^2) = 0.3 ft
From Table 23-2, VP = 1.2 ft H_2O
Hp = 12.5 ft H_2O

NPSHa = PB - VP $\pm$ Hp + Hv = 26.3 - 1.2 - 12.5 + 0.3 = 12.9 ft H_2O

Piping Size Considerations for Suction Pipe
There is always a compromise between reducing costs (smaller pipe size) and providing enough suction pressure, in terms of NPSHa, at the pump inlet (larger pipe size). The smaller the pipe size, the higher the friction, and consequently the greater the losses that take away from available NPSHa. Hydraulic losses are proportional to the velocity head ($V^2/2g$). The losses are inversely proportional to the pipe diameter to the 4th power for a given pipe flow.

$$H_{LC} = \left(\frac{D_1}{D_2}\right)^4 \qquad \text{Eq. 23-8}$$

where,

H_{LC} = change in friction loss
D_1 = diameter of first pipe (in.^{2})
D_2 = diameter of second pipe (in.^{2})

Example:
Given a cavitating/noisy pump that delivers 130 gpm with a 2-inch suction line, calculations indicate that the system provides the pump with approximately 20 feet of NPSHa. However, this is based only on the level from the water level in a ditch to the pump intake, but does not account for the hydraulic losses.

Although this sounds like an ample margin, it is not really what is available at the pump once the pressure drop in the suction line is considered. Calculations show that 14 feet of head are lost due to friction in the 2-inch suction line. This leaves only 6 feet of NPSHa for the pump to operate. It is evident from Fig. 23-4 that at 130 gpm, the required NPSHr of 9 feet exceeds the actually available NPSHa of 6 feet, so it is no surprise that the pump cavitates. If the 2-inch suction line is replaced with an 4-inch line, determine the change in friction loss in the line.

From Eq. 23-8,

$$H_{LC} = \left(\frac{2}{4}\right)^4 = \frac{1}{16}$$

Thus, there will be a 16-fold decrease in friction loss when the pipe size is doubled.

The friction loss with the 4-inch suction will be:

14 ft x 1/16 = 0.9 ft

NPSHa will be: 20 - 0.9 = 19.1 ft

There will be more than enough NPSHa with the 4-inch pipe. This could make the difference between a trouble-free installation and one plagued with problems. Historically, and as a rule of thumb, the suction pipe is selected for flow velocities of 5 ft/sec or less.

Cavitation

Cavitation is a term used to describe the phenomenon that occurs in a pump when there is insufficient NPSH available. The pressure of the liquid is reduced to a value equal to or below its vapor pressure and small vapor bubbles or pockets form. As these vapor bubbles move along the impeller vanes to a higher pressure area, they rapidly collapse. The collapse (implosion) is so rapid that it may be heard as a rumbling noise, as if you were pumping gravel. The forces during the collapse are generally high enough to cause small pockets of fatigue failure on the impeller vane surfaces. This action is

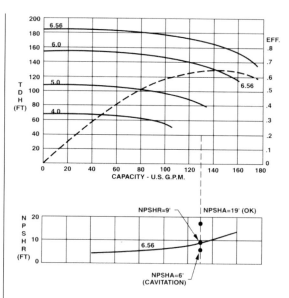

Figure 23-4. Pump curve with NPSHa requirements.

progressive, and under severe conditions can cause serious pitting damage to the impeller.

Suction-side cavitation (Fig. 23-5) is by far the most common form. Suction-side cavitation can be caused by restriction on the suction side of the pump system, which does not allow sufficient fluid to enter the pump to be discharged. Occasionally, discharge-side cavitation occurs. It is caused by a complete, or nearly complete restriction on the discharge-side of the pump. Since the water can't escape because of discharge side restrictions, it is recirculated in the pump casing. Oftentimes the water will boil and damage the impeller, guide vane, and casing.

Cavitation is relatively easy to recognize. In its mildest form, cavitation will be recognizable because of a sharp pinging noise that has often been described as pumping corn kernels or gravel through the pump. If you suspect cavitation in your pump system, but are not sure because you don't hear that noise, put the blade end of a screwdriver on the pump casing and the handle end up to your ear. This will enhance your ability to hear the noise within the pump casing.

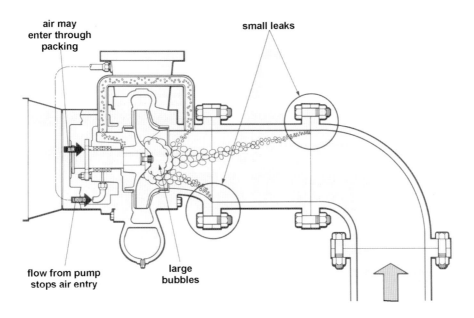

air may
enter through
packing

small leaks

AIR

LIQUID

flow from pump
stops air entry

large
bubbles

Figure 23-5. Small air leaks migrating to impeller eye, resulting in cavitation (courtesy of Cornell Pump Company).

Another sign of cavitation is that the discharge pressure gauge on the pump system will fluctuate wildly over a 5- to 10-psi range at a high rate of speed, indicating uneven discharge flow. One must be careful to put a new gauge on the system and check the gauge tap opening to ensure the gauge is operating correctly. A properly operating system will give a steady pressure gauge reading with little variation during pump operation.

Cavitation causes numerous undesirable side effects. Because the pump is not operating in its proper balance hydraulically, it is subject to internal stresses that can cause shaft deflection and premature bearing and seal wear. If bearings and seals constantly require replacement, severe misalignment or cavitation are probable causes. The most common reasons for cavitation are:

1. Suction-side supply blockage:
   Debris that blocks the pump suction intake will restrict the amount of water needed to operate at peak efficiency. Air leaks that develop in the suction line will also restrict flow of water into the pump.

2. Poor system design:
   Improper intake design or lack of attention to NPSH requirements may lead to pump cavitation.

3. Drop in water level:
   Water levels dropping below the design depth may result in insufficient NPSH available in the system.

4. Increased system demands:
   System flow rates can increase dramatically when resetting blocks with more trees (and emitters) than in the original system design. In addition, over time the orifices of micro-sprinklers tend to increase in diameter with wear. More water is allowed to flow through the emitters, which lowers the head pressure against the pump. The pump attempts to pump more and more liquid, but supply can't keep up with demand. As a result, the pump is no longer operating in its best efficiency range due to a change in the system performance requirements, and cavitation occurs.

5. High temperature combined with marginal suction supply:
   Partial blockage of intake screens and unusual swings in atmospheric conditions can result in cavitation occurring.

6. Reduced system demands:
   As discharge lines in the system corrode or plug, pump discharge output is restricted and discharge cavitation can occur. Check valves that may not operate properly on either the pump discharge or suction side as they also can cause cavitation.

7. The pump is oversized:
   Oversizing the pump may occur because of failure to conduct a detailed system analysis to determine the proper head pressure and flows required for the application or the tendency to "fudge" the numbers to be "safe."

An inspection of the impeller in a centrifugal pump will reveal the effects of cavitation. (Under proper operating circumstances, impellers do not wear out.) If centrifugal pump impellers have holes in the center of the impeller, there is suction-side cavitation. If there is damage around the outer diameter of the pump impeller and in the casing, this is probably discharge cavitation.

To determine if cavitation is occurring, install a combination gauge (reads in vacuum and psi) on the suction side of the pump and a discharge pressure gauge on the discharge side of the pump and take the readings. The discharge pressure, plus suction pressure or vacuum, will be the operating head at which the pump is performing.

Example:
A centrifugal pump (suction lift) has a vacuum reading of 5 inches of mercury on the suction side of the pump, and the discharge pressure gauge reads 50 psi.

Refer to the rotating shaft speed of the pump to find the pump's operating performance curve, then determine from these readings where the pump is operating on its

performance curve. Make sure that the pump performance curve is the same speed as the motor. Motors can be switched from one rpm to another.

To further refine these measurements, take amp readings on the motor inlet leads and convert them to brake horsepower. This will enable you to pinpoint the horsepower performance on the pump curve, which will be a double-check to the pressure readings taken earlier.

Correcting cavitation is fairly easy for suction-side cavitation. Close the valve on the pump's discharge side slowly until the cavitation noise that you heard disappears. If opening the valve to full open makes the noise disappear, this probably is discharge cavitation. However, other restrictions downstream may cause the problem to continue even with the valve open. Water in the system contains gas bubbles that may restrict the full flow of liquid. By returning the pump to its correct operating condition, you produce a steady stream of gas-free fluid that will render the most efficient flow of water from the pump.

In the case of discharge cavitation, it may be necessary to recirculate some of the water from the discharge side of the pump back to the supply of liquid. Do not recirculate water back directly to the pump's suction side, as this will not alleviate the problem. With discharge cavitation, it is necessary to bypass some of the water out of the discharge line so that the pump operates as if it is producing more flow than it really is.

While the above temporary fixes will work for centrifugal pumps, they will not work as well for positive displacement pumps. Any action to temporarily correct positive displacement pump cavitation should only be done by an experienced pump technician or field engineer.

The only way that cavitation can be corrected properly is to analyze your system precisely and determine the system head pressure and flow requirements. This evaluation produces a system head curve that can be used to determine the correct size and type of pump to do the job. Merely trimming the impeller or changing the speed of the pump corrects the problem in many cases. Sometimes it may be necessary to replace the pump with another pump properly suited for the existing system. Other times cavitation problems can be corrected by altering piping, reducing suction lift, and regulating the temperature. System changes can be made, including cleaning out the pipes, removing obstructions, or replacing worn components, which will solve the cavitation problem with little or no expense.

Energy and Efficiency

The efficiency of a pump is a measure of the degree of its hydraulic and mechanical perfection. It is defined as the ratio of the energy output to the energy input, and is expressed in percent. Horsepower is defined as the power required to raise the weight of 33,000 lb a vertical distance of 1 foot in 1 minute. The work performed by a pump (energy output in terms of horsepower) is determined by the weight of the water it delivers per minute multiplied by the total equivalent vertical distance in feet through which it is moved. For example, a pump delivering approximately 396 gallons per minute (3300 lb H_2O per minute) at a total head of 10 feet is performing work at the rate of 1 horsepower. If it were possible to achieve 100% efficiency, it would only be necessary in this instance to apply 1 horsepower to the pump shaft. Actually, of course, the energy input must be greater than 1 horsepower.

Some of the energy losses that result in a lower efficiency are the friction in the bearings that support the pump shaft, friction between the shaft and the packing in the stuffing box, unavoidable leakage between areas of high pressure and adjacent areas of low pressure inside the pump case, and the friction caused by the water moving across the metallic surfaces in the pump. There are also other losses of a more complex nature.

The efficiency of a pump is determined by actual test. Referring to the example above, if it is found that 1.25 horsepower must be applied to the input shaft when the pump is doing work equivalent to 1 horsepower, the pump efficiency would be 80% (1 divided by 1.25). It is often necessary for a pump designer to sacrifice some degree of efficiency in order to achieve some other characteristic that is desirable in providing a unit with maximum usefulness. The efficiency range to be expected varies with the pump size, type, etc., but usually it should be between 65% and 80%. Whenever feasible, a pump should be selected to operate near the point on its curve where maximum efficiency is obtained. This is referred to as the "design point."

Selection of Type Pump

Before a pump can be selected to perform a given job, three things must be known:

• the total dynamic head (TDH) against which it must operate

• the desired flow in gallons per minute

• the suction lift

Once this information has been established, selection of a proper pump can begin. The selection of an irrigation water pump is based almost entirely on the relationship between pump efficiency and the TDH the pump will provide at a specific flow rate. These parameters are the basis of the pump characteristic curve. Table 23-3 lists appropriate pump types for a broad range of flow rates and total dynamic heads. However, if the pump needs to lift water (suction lift), a centrifugal pump will have to be used.

Centrifugal Pumps (For surface supply and shallow wells):

If the source of water is a surface supply, such as a lake, canal, pond, or other body of surface water, the pump most commonly used is the horizontal centrifugal type (commonly referred to simply as a centrifugal pump). The horizontal pump has the pump shaft in the horizontal position (Fig. 23-6). This type is usually subdivided into two groups: single suction (end suction) and double suction (often called split case). Either of these may be single

Table 23-3. Characteristics of pumps for a range of flow rates and total dynamic heads (TDH).

Gallons per minute	TDH (feet)		
	50 or less	50 to 500	500 or more
0 to 300	Axial Flow Centrifugal	Centrifugal Vertical Turbine Submersible	Centrifugal Vertical Turbine Submersible
300 to 5000	Axial Flow	Centrifugal Vertical Turbine Submersible	Centrifugal Vertical Turbine Submersible
5000 or more	Axial Flow	Centrifugal Vertical Turbine Axial Flow Submersible	Centrifugal Vertical Turbine

Figure 23-6. Typical horizontal centrifugal pump used in citrus irrigation systems.

stage (only one impeller) or multistage, where several impellers are positioned so that the water travels from the discharge of one impeller to the suction of the second (Fig. 23-7). Thus, the total head is the head developed by a single impeller multiplied by the number of impellers in the pump. The most common pump and the lowest in cost is the end suction, single stage. Available sizes vary considerably with different manufacturers. In general, if the desired performance exceeds about 1000 gallons per minute of capacity and 150 feet of head, consideration should be given to the split-case type that is more rugged in construction and is capable of a much greater range of performance.

Vertical types of pumps (deep-well turbine) may also be used with a surface supply, and in many applications they offer advantages because they are inherently more flexible in performance and construction. If the source of water is a well and its pumping level is greater than the suction lift allowable for a centrifugal or shallow well jet (15 to 22 ft), a vertical turbine can be installed in the well.

Jet Pumps

For very low capacity requirements (5 to 20 gpm), one of the most common pump types used is the jet pump. This type consists of a small centrifugal pump located at ground level connected to a jet installed below the water level in the well. By circulation of part of the water from the pump back through the jet, water is forced up to the impeller in the pump and a continuous flow at reasonable pressure is provided.

Shallow well jet pumps operate on the same recirculation principle, but the jet is installed above ground and the allowable lift is limited to about 22 feet; in contrast, deep-well jet pumps have a maximum lift of about 85 feet. A jet pump requires almost twice the horsepower that a submersible pump would require to deliver the same amount of water from the same depth. Therefore, it is not

Figure 23-7. Five-stage horizontal centrifugal pump.

normally used in irrigation systems due to its low efficiency.

Deep-Well Turbine Pumps

The most widely used irrigation pumps for deep wells are vertical-type centrifugal pumps commonly referred to as deep-well turbines (Figs. 23-8, 23-9, and 23-10). This is a centrifugal type of pump designed for installation in a well. Deep-well turbine pumps are adapted for use in cased wells or where the water surface is below the practical limits of a centrifugal pump. Turbine pumps are also used with surface water systems. Since the intake for the turbine pump is continuously under water, priming is not a concern. Turbine pump efficiencies are comparable to, or greater than, most centrifugal pumps.

Turbine pumps are usually more expensive than centrifugal pumps and more difficult to inspect and repair. Because of their inherent flexibility, vertical turbine pumps are sometimes used in sumps, lakes, or other surface sources with a short column between the bowls and the discharge assembly.

The turbine pump has four main parts: the head assembly, the shaft and column assembly, the pump bowl assembly, and the drive unit. The head is normally cast iron and designed to be installed on a foundation. It supports the column, shaft, and bowl assemblies and provides a discharge point for the water. Also, it will support either an electric motor, a right-angle gear drive, or a belt drive.

The column and shaft assembly contains a pipe to suspend the bowl assembly in the well and convey the water to the surface. Inside this pipe (or column) is the shaft that connects the impeller to the driver (located at ground level). The shaft may be either water lubricated or oil lubricated. The water-lubricated pump has an open shaft. The bearings are lubricated by the pumped water. If there is a possibility of fine sand being pumped, select the oil-lubricated pump because it will keep the sand out of the bearings. The oil-lubricated pump has an enclosed shaft into which oil drips, lubricating the bearings.

The discharge assembly is often called the "head" or "base." It is normally cast iron and designed for installation on a foundation. It supports the column and shaft assembly, and the bowl assembly in the well, and provides a discharge for the water being delivered. It also accommodates the driver for the pump. The drive unit may be either an electric motor or a right-angle gear for connection to a power unit. When an electric motor is used, the usual type is a vertical hollow-shaft design that permits the pump shaft to come up through its center for securing at the top. The right-angle gear is also normally of the hollow shaft type for the same reason. It also has a horizontal shaft for connection to the engine drive or power takeoff. The internal gears are available in various ratios to accommodate an engine with an operating speed that is different from the desired pump operating speed.

Deep-well turbine pumps must have correct alignment between the pump and the power unit. Correct alignment is made easy by using a head assembly that matches the motor and column/pump assembly. It is very important that the well is straight and plumb. The pump column assembly must be vertically aligned so that no part touches the well casing. Spacers are usually attached to the pump column to prevent the pump assembly from touching the well casing. If the pump column does touch the well casing, vibration will wear holes in the casing. A pump column that is out of vertical alignment may also cause excessive bearing wear.

Figure 23-8. Typical vertical centrifugal pump used for citrus irrigation systems.

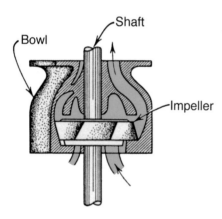

Figure 23-9. Detail of impeller and bowl of deep-well turbine pump.

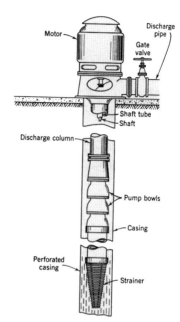

Figure 23-10. Schematic of vertical hollow-shaft electric motor connected to a deep well turbine pump.

The head assembly must be mounted on a good foundation at least 12 inches above the ground surface. A foundation of concrete (Fig. 23-11) provides a permanent and trouble-free installation. The foundation must be large enough to allow the head assembly to be securely fastened. The foundation should have at least 12 inches of bearing surface on all sides of the well. In the case of a gravel-packed well, the 12-inch clearance is measured from the outside edge of the gravel packing.

A well access pipe at least 1.5 inches in diameter must extend through the foundation into the well casing. The access pipe serves two purposes. The first is to measure both static and pumping water levels in the well, and the second is to allow chlorination of the well. A 3/4-inch diameter polyethylene tubing with the bottom end closed, inserted into the access pipe and extending to the pump level, will make measuring water levels easier. Small holes must be drilled into the tubing to allow water to move in and out of the tubing easily.

The operating characteristics of deep-well turbine pumps are determined by test and depend largely on the bowl design, impeller type, and the speed. Flow rate, TDH, BHP, efficiency, and rpm are similar to those given for centrifugal pumps. Vertical turbine pumps are generally designed for a specific rpm setting. The pump curve for a vertical

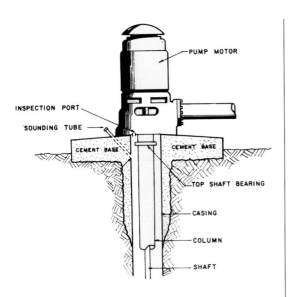

Figure 23 -11. Typical base and discharge assembly for turbine pump.

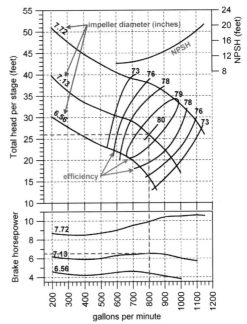

Figure 23-12. Typical deep-well turbine pump curve. Note that the brake horsepower and total head are for one stage. Total head and hp requirements are calculated by multiplying by the number of stages (gpm will stay the same no matter how many stages are added).

turbine (Fig. 23-12) is similar to the centrifugal pump curve, except instead of curves for various rpms, the curves are for different diameter impellers.

Because of the limited diameter of the impellers, each one develops a relatively low head. Thus, it is necessary for the average application to stack several impellers in series (one above the other) with each in its own bowl or diffuser housing (called staging). For instance, the three-stage bowl assembly in Fig. 23-10 contains three impellers all attached to a common shaft through the separate housing or bowls. The bowls are attached to the impeller shaft through the center of the pump column pipe. The shaft and column pipe must be long enough to position the bowl assembly below the water level in the well during pumping.

Pump curves for turbine pumps are normally shown for a single stage so the TDH obtained will be determined by multiplying the indicated head on the pump curve by the number of stages. The brake horsepower requirements must also be multiplied by the number of stages. Note that the flow rate will not change no matter how many stages are added.

Example:
Assume a well with 10-inch casing and a static water level 100 feet below the surface, and that it is desired to select a pump to deliver 500 gpm at a pressure of 40 psi measured at the point of discharge from the pump.

The static water level is 100 feet below the surface. However, when the well is pumped the level will fall (drawdown). The amount of drawdown varies with localities and the formation into which the well is located. Assuming the drawdown at a pumping rate of 500 gpm is 10 feet, then the total distance to the surface becomes 110 feet.

The 40 psi pressure need at the surface is converted to feet of head by multiplying by 2.31.

Pressure head = 40 psi x 2.31 psi/ft = 92 ft

Total head = lift + pressure head

Total head = 110 ft + 92 ft = 202 ft

To this must be added an allowance for friction loss within the pump column and the discharge assembly, which would normally be about 5 feet, assuming proper column size. Thus, the total dynamic head, or the total operating head, will be 207 feet.

The ideal pump would deliver 500 gpm at a head of 207 feet. If none are available, a pump that delivers a higher head at 500 gpm could be selected and modified. For instance, a 10-stage pump could be selected, with the diameter of the impellers trimmed to the point where the head developed by each would be 20.7 feet. The result would be 10 x 20.7 ft = 207 ft of total head.

Example:

The deep-well turbine pump curve in Fig. 23-12 is for a 5-stage pump with a 7.13-inch impeller supplying 800 gpm. What would be the TDH and BHP requirement?

Follow the dashed vertical line from 800 gpm up to where it meets the 7.13-inch impeller curve on the upper portion of the chart. Follow the dashed horizontal line left to where it shows 26 feet of TDH. Multiplying 26 by 5 gives 130 feet of TDH. Next, follow the dashed vertical line from 800 gpm up to the 7.13-inch impeller BHP curve on the lower portion of the chart and then follow the horizontal dashed line left to where it shows 6.5 BHP. Multiplying 6.5 BHP by 5 stages produces a 32.5 BHP requirement for this pump. Also note that the pump is operating at its peak efficiency of 80%.

Submersible Pumps

Submersible pumps are multistage vertical pumps connected directly to an electric motor that is designed to operate under water (Fig. 23-13). Both the pump and motor are suspended in the well below the water level by means of the discharge pipe that conducts the water to the surface. This type of

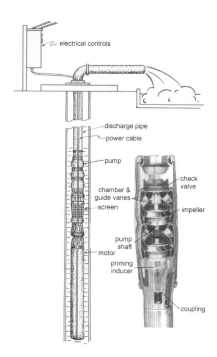

Figure 23-13. Components of typical submersible pump system.

pump is available in a wide range of capacities for 4-inch wells and larger. Submersible pumps have a relatively higher cost than centrifugal pumps, since making the motor waterproof is quite expensive.

Some advantages of a submersible pump are maximum efficiency of performance; it can be used in wells of any depth; no priming of the pump is necessary; well can be located any distance from point of water use; no possibility of air leaks; positive air charging system for pressure storage tank; and no waterlogging.

A submersible pump is a turbine pump close-coupled to a submersible electric motor. Both pump and motor are suspended in the water, thereby eliminating the long drive shaft and bearing retainers required for a deep-well turbine pump. Because the pump is located above the motor, water enters the pump through a screen located between the pump and motor. The submersible pump uses enclosed impellers because the shaft from the electric motor expands when it becomes hot and pushes up on the impellers. If semi-open impellers were used, the pump would lose efficiency. The pump curve for a submersible pump is very similar to a deep-well turbine pump.

Submersible motors are smaller in diameter and much longer than ordinary motors. Due to their smaller diameter, they are lower efficiency motors than those used for centrifugal or deep-well turbine pumps. Submersible motors are generally referred to as dry or wet motors. Dry motors are hermetically sealed with a high dielectric oil to exclude water from the motor. Wet motors are open to the well water, with the rotor and bearings actually operating in the water.

If there is restricted or inadequate circulation of water past the motor, it may overheat and burn out. Therefore, the length of riser pipe must be sufficient to keep the bowl assembly and motor completely submerged at all times. In addition, the well casing must be large enough to allow water to easily flow past the motor. Small submersible pumps (under 5 horsepower) use single-phase power. However, most submersible pumps used for irrigation need three-phase electrical power.

Electrical wiring from the pump to the surface must be watertight and all connections sealed. The electrical line should be attached to the column pipe every 20 feet to prevent it from wrapping around the column pipe. Voltage at the motor leads must be within plus or minus 10% of the motor nameplate voltage. If there is a 5% voltage drop in the submersible pump cable, voltage at the surface must not be less than 95% of rated voltage. Because the pump is located in the well, lightning protection should be wired into the control box. Lightning hits on wells with submersible pumps are a leading cause of pump failures.

Submersible pumps can be selected to provide a wide range of flow rate and TDH combinations. Submersible pumps more than 10 inches in diameter generally cost more than comparably sized deep-well turbines because the motors are more expensive. Many manufacturers make submersible booster pumps. These pumps are usually mounted horizontally in a pipeline. An advantage to using a submersible as a booster pump instead of a

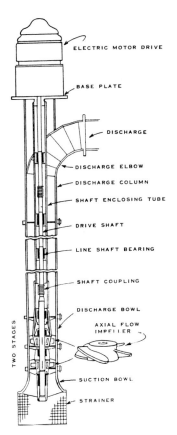

Figure 23-14. Cross section of axial flow propeller pump.

centrifugal is noise reduction. Submersibles have also been used as booster pumps in the suction lines of centrifugal pumps. This application is used in situations where the water level will fluctuate a considerable amount over the season. Having a submersible pump in the suction line will change the head at the inlet of the centrifugal pump from a suction to a positive head.

Propeller Pumps

Propeller pumps are used for low-lift, high-flow rate conditions. They come in two types: axial flow and mixed flow. The difference between the two is the type of impeller. The axial flow pump uses an impeller that looks like a common boat motor screw, and is essentially a very low head pump. A single-stage propeller pump typically will lift water no more than 20 feet. By adding another stage, heads from 30 to 40 feet are obtainable. The mixed-flow pump uses either semiopen or closed impellers similar to turbine pumps. In permanent installations, propeller pumps are mounted vertically (Fig. 23-14). For portable pumping platforms, they are mounted on trailers and commonly powered by the power-take-off (PTO) on tractors.

Axial-flow propeller pumps are an efficient means devised for either irrigation or drainage pumping at low heads of up to 50 feet and above 500 gallons per minute. Their efficiency is high, especially when the total head is in the range of 8 to 20 feet. Typical citrus drainage pumps (24 to 36 inch diameter) generally pump 20,000 to 30,000 gpm. The pumping element of an axial flow propeller pump consists primarily of a revolving propeller in a stationary bowl, which contains vanes above and below the propeller. Water enters the pump through the intake bell. It is discharged into the distributor section, then out the discharge elbow. Flow is essentially a straight line along the pump axis, keeping friction and turbulence to a minimum. Capacities of axial flow pumps vary from 500 gpm with a 6-inch diameter propeller up to 100,000 gpm using a 60-inch propeller.

Power requirements of the axial flow pump increase directly with the TDH, so adequate power must be provided to drive the pump at maximum lift. Axial flow pumps are not suitable where it is necessary to throttle the discharge to reduce the flow rate. It is important to accurately determine the maximum TDH against which this type of pump will operate.

Axial flow pumps are not suitable for suction lift. The impeller must be submerged and the pump operated at the proper submergence depth. The depth of submergence will vary according to various manufacturers' recommendations, but generally, the greater the diameter of pump, the deeper the submergence. Following recommended submergence depths will ensure that the flow rate is not reduced due to vortices. Also, failure to observe required submergence depth may cause severe mechanical vibrations and rapid deterioration of the blades.

Figure 23-15. Hand pump used to draw water through suction line into pump to prime it.

Optional Pump Features

Most pumps can be obtained with various features of construction to suit individual preferences and the various types of applications.

Priming

With the exception of some small self-priming models, centrifugal pumps must be primed prior to operation. With submersible and deep-well turbines, the pump bowls are located under the static water level so they remain primed. If horizontal centrifugal pumps are installed with flooded suctions, they will also remain primed.

There are several methods of priming, and various devices are available. Selection of the mechanism most suited for a particular application will usually add greatly to the convenience and dependability of the pumping unit. Almost all primers are quite simple in their design and construction, and their success or failure to provide reliable and efficient priming will depend primarily on their proper installation and operation.

Centrifugal pumps need to be primed by removing sufficient air from inside the pump and suction line

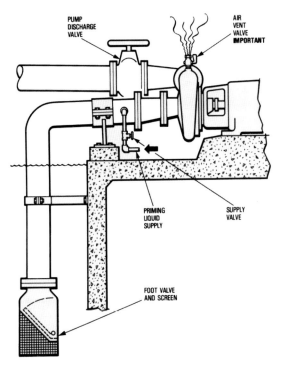

Figure 23-16. Supply tank is used to fill suction line and pump with water. Discharge valve must be closed and air vented off during line filling.

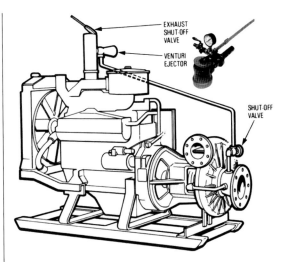

Figure 23-17. Exhaust system primer on diesel-powered pump.

to permit the atmospheric pressure on the outside of the system to cause water to flow into it. Air leaks, however small, will complicate the priming operation. Therefore, it is important that all possible sources of air leaks be checked carefully.

Removal of the air from the pump and suction line can be accomplished by simply displacing it with water, i.e., filling the pump and suction completely. This method requires a foot-valve on the suction line. One of the most popular methods of priming is to install a manually operated primer (Fig. 23-15) to pump the air out of the system. Such installations are relatively inexpensive and usually quite dependable.

Oftentimes, a water supply tank located at an elevation above the pump is used (Fig. 23-16). With this configuration, a valve on the pump discharge is required to prevent air from being drawn into the pump while priming. The tank can be filled from the pump discharge before each shutdown and later drained back into the pump to reprime.

Mechanically operated primers are used sometimes, and are available for electric motor or belt drive. Such installations are usually more expensive to install and maintain, but are advantageous where long suction lines of large diameter are used.

Some methods are applicable only to engine-driven pumps, such as exhaust primers and intake manifold primers. The exhaust primer consists of an air ejector with appropriate fittings for attachment to the engine exhaust and connection to the pump (Fig. 23-17). The function of the ejector is simply to draw air from the pump through the ejector suction when the exhaust gases are directed through the ejector nozzle by closing the exhaust valve. Exhaust primers should be selected according to the size of the engines on which they are to operate. They are available in standard sizes for engines of 4 to 200 hp and can provide suction lifts above 20 feet.

Intake manifold primers depend entirely on the vacuum developed in the engine intake manifold. While this varies in different engines, it is usually sufficient to provide suction lifts to at least 20 feet. The manifold vacuum will always be greatest when the engine is idling with no load. The primer is basically a sensitive float valve that, being mounted above the level of the pump, closes promptly when water is drawn into it by the engine air intake to which it is also connected. This type primer is usually adaptable to a wide variety of standard engines.

Shaft Lubrication

A water-lubricated pump column assembly is provided with fluted rubber bearings to permit lubrication of the shaft by the water being pumped. If the pump is to be operated at less than about 2200 rpm, these bearings, which are fixed in the column pipe couplings, are usually placed at 10-foot intervals. For higher speeds, the bearings are at 5-foot spacings.

An oil-lubricated pump column assembly includes a tube that encloses the shaft. Inside the tube are bronze shaft bearings that are threaded on the outside to serve as couplings for connecting the pieces of tube. The bearing spacing is usually on 5-foot centers. Lubricating oil is fed into the top of the tubing and passes by gravity over the surfaces of the bearings. At the bottom of the column is an opening that permits the oil to flow out. It then floats on top of the water in the well.

Water-lubricated turbine pumps are simpler, cheaper, and more commonly employed. If more than four or five of the rubber shaft bearings are above the water level and thus become dry when the pump is not operating, some means of prelubrication is required, such as a small prelube tank from which water can be spilled over the bearings prior to starting the pump. With smaller size pumps, a foot valve can be installed below the bowl assembly to keep the column pipe full of water.

When the water level is very deep, oil lubrication is normally used. While there is no definite point at which it becomes necessary, it is usually recommended for depths in excess of 150 to 200 feet.

Enclosed vs. Semi-open Impellers

Impellers used in turbine pumps may be either semiopen or enclosed (Fig. 23-18). Enclosed impellers have the vanes covered on both the top and bottom edges, while the semi-open impeller has

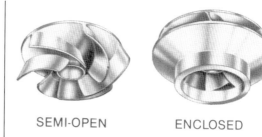

SEMI-OPEN　　　　　ENCLOSED

Figure 23-18. Enclosed and semi-open impellers.

only the top edge of the vanes covered. Both enclosed and semi-open impellers are used in vertical turbine and centrifugal pumps, but only enclosed impellers are used in submersible pumps. With enclosed impellers, performance and efficiency are not affected by small differences in their position, and constant performance can be maintained over a wider range of operating conditions. Adjustment is only needed to ensure that the bottom of the impeller does not touch the bottom surface of the bowl. Periodically, however, enclosed impellers need to be adjusted to compensate for wear between the throat of the impeller and the wear collar on the bowl.

The vanes on semi-open impellers are open on the bottom and they rotate with a close tolerance to the bottom of the pump bowl. This tolerance is critical and must be adjusted when the pump is new. During the initial break-in period, the line shaft couplings will tighten. Therefore, after about 100 hours of operation, impeller adjustments should be checked. After break-in, the tolerance must be checked and adjusted every three to five years or more often if pumping sand.

The bottom edge of the vanes runs at a close clearance with the pump bowl. By raising the setting of the semi-open impellers, the clearance between the vanes and the bowl seat is increased and water is allowed to circulate through this area. This makes a variation in the performance of the pump possible at any given speed. Thus, by adjustment of the impeller clearance, a specified performance can be maintained even though there is a change in the water level.

Impeller adjustments are made by tightening or loosening a nut on the top of the head assembly. Impeller adjustments are normally made by lowering the impellers to the bottom of the bowls and adjusting them upward. The amount of upward adjustment is determined by how much the line shaft will stretch during pumping. The adjustment must be made based on the lowest possible pumping level in the well. The proper adjustment procedure is often provided by the pump manufacturer. The distance between the bottom of the impeller vanes and the bowl should be as close as possible without touching when operating at the maximum head. The amount of adjustment must be carefully calculated.

Adjustments are made with the adjusting nut on the head shaft, and are based on line shaft stretch. The line shaft stretch is caused by the hydraulic thrust created by pumping. It is a product of the total pumping head, the impeller thrust constant (available from the pump curve or manufacturer), the line shaft diameter, and the number of bowls. Hydraulic thrust and line shaft stretch are calculated as follows:

$$S_{LS} = \frac{T_{HY}}{1000} \times \frac{L}{100} \times E_S + 0.004 \times N \quad \textbf{Eq. 23-9}$$

where,

S_{LS} = Line shaft stretch in inches
T_{HY} = Hydraulic thrust = total dynamic head (ft) x impeller constant (K)
L = Length of pipe column to bowls (ft)
E_S = Shaft elongation from Table 23-4 for proper shaft size (inches/100 ft per 1000 lb hydraulic thrust)
N = Number of bowls that make up the pump

Example:
Determine the line shaft stretch for a deep-well turbine pump with the following characteristics:

Length of pump column = 90 ft
Line shaft diameter = 1.5 inches
6 bowls on pump
Pumping TDH = 200 ft

Impeller constant (K) = 24

From Table 23-4 for 1.5-inch line shaft, the shaft elongation factor = 0.02341. The hydraulic thrust
= 200 ft x 24 = 4800

Using Eq. 23-9,

$$Stretch_{LS} = \frac{4800}{1000} \times \frac{90}{100}$$

x 0.02341 + 0.004 x 6 = 0.125 inch

The line shaft stretch will only be in a fraction of an inch. This is the distance the impeller must be lifted from the bottom of the bowl to keep it from rubbing when in operation. The adjustment is made by first lowering the impellers so they drag on the bottom of the bowl. Turn the adjusting nut counter-clockwise until the shaft cannot be turned by hand. Then lift the impellers the required distance (the amount of line shaft stretch) by turning the adjusting nut clockwise. The number of turns of the adjusting nut will depend upon the number of threads per inch on the head shaft. It is good practice to measure and count the threads per inch. The amount of adjustment obtained from the adjusting nut is calculated as follows:

$$Adjustment = \frac{1}{threads\ per\ inch} \quad \textbf{Eq. 23-10}$$

where,

adjustment is the vertical movement per 360° revolution of the nut

Example:
For a deep-well turbine pump with a 1.5-inch line shaft with 10 threads per inch, determine the adjustment required to raise the assembly 0.125 inch (hex nut).

Using Eq. 23-10, in./turn = 1/10 thread/in. = 0.10 in./turn

0.125 in./ 0.10 in./turn = 1.25 turns (clockwise)

When adjustment is completed, be sure to lock the adjusting nut into place with the locking device provided. Failing to do so can allow the pump to get out of adjustment with disastrous results. If you are not sure about making adjustments yourself, contact your pump supplier or someone more knowledgeable. Once the adjustments are made, obtain data again and calculate the efficiency for the pump. If the efficiency is not improved, check the pump further for the cause; pump repairs may be in order. An efficient pumping plant will help keep the pumping cost to a minimum.

Nonreverse Ratchet

At a small additional cost, either an electric motor or a right-angle gear drive can be provided with a ratchet to prevent pump rotation in the reverse direction. An electric motor will operate in either direction, and if it should accidentally be operated in the wrong direction by reason of phase reversal in the power supply, damage to the pump may result. While an engine cannot be operated in the wrong direction (except in case of backfire), it is preferable that the pump not be permitted to spin in the reverse direction, which will occur after stopping, because the water flowing back down the pump column through the impellers turns them like a water wheel. As the water in the column recedes, the shaft is turned in its bearings without lubrication, which causes undue wear.

Pump Selection Summary

It is important to select the proper pump for each application. The selection of an irrigation water pump is based almost entirely on the TDH the pump will provide at a specific flow rate. However, several other factors should be considered. Table 23-5 lists some of the advantages of each type of pump.

Table 23-4. Shaft elongation factors.

Shaft size (in.)	Stretch (in./100 ft per 1000 lb hydraulic thrust)	Shaft size (in.)	Stretch (in./100 ft per 1000 lb hydraulic thrust)
3/4	0.09371	1 15/16	0.01400
7/8	0.06865	2 3/16	0.01100
1	0.05626	2 7/16	0.0887
1 3/16	0.03736	2 11/16	0.00729
1 1/4	0.03376	2 15/16	0.00610
1 7/16	0.02505	3 3/16	0.00518
1 1/2	0.02341	3 11/16	0.00384
1 11/16	0.01850	3 15/15	0.00339

Table 23-5. Pump selection criteria.

Pump Type	Advantages	Disadvantages
Centrifugal	High efficiency over a range of operating conditions	Suction lift is limited; it needs to be within 20 vertical feet of the water surface
	Easy to install	Priming required
	Simple, economical, and adaptable to may situations	Loss of prime can damage pump
	Electric, internal combustion engines, or tractor power can be used	If the TDH is much lower than design value, the motor may overload.
	Does not overload with increased TDH	
	Vertical centrifugal may be submerged and not need priming	
Deep-Well Turbines	Adapted for use in wells	Difficult to install, inspect, and repair
	Provides high TDH and flow rates with high efficiency	Higher initial cost than centrifugal pump
	Can be used where water surface fluctuates	Repair and maintenance is more expensive than centrifugal pumps.
	Priming not needed	
Submersible	Can be used in deep wells	More expensive in larger sizes than deep-well vertical turbines
	Priming not needed	Only electric power can be used
	Can be used in crooked wells	More susceptible to lightning
	Smaller diameters are less expensive than comparable-sized vertical turbines	Water movement past motor is required
	Easy to install	
Axial Flow (propeller)	Simple construction	Not suitable for suction lift
	Can pump sand	Cannot be valved back to reduce flow rate
	Priming not needed	Intake submergence depth is very critical
	Efficient at pumping very large	Limited to low (less than 75 feet) TDH
	Electric, internal combustion engine, and tractor power can be used	
	Suitable for portable operation	

Chapter 24. Power Units

by Brian Boman

Irrigation power units commonly used include electric motors, internal combustion engines, and farm tractors. The power unit must match the irrigation pump requirements if an efficient operation is to be maintained. An engine or motor that is too small cannot provide the power necessary to deliver water at the rate and/or pressure required by the irrigation system.

Irrigation pumping generally is a constant load on the power unit. Therefore, the power unit must be suitable for continuous duty at the design load. Electric motors are designed and rated for constant loads. An internal combustion engine must either be rated for continuous duty at the design load, or derated from some other horsepower rating to a continuous horsepower rating. Either industrial internal combustion engines or electric motors are suitable for irrigation. Farm tractors should not be used for permanent installations because they are not designed for the 24-hour per day continuous loading required for irrigation. If a tractor is used on a temporary installation, it should be protected with safety controls to prevent engine damage and operated at less than maximum horsepower.

There are several factors that influence the selection of an appropriate power unit for a particular installation. When choosing a power unit, the following factors should be considered.

- The amount of brake horsepower required for pumping, including the amount of water to be pumped, elevation differences between the source and the delivery point, operating pressure, friction losses, power unit efficiency, and pump efficiency

- Hours of operation per season

- Availability and cost of energy or fuel (in case of electricity, availability of three-phase power may influence the selection)

- Availability of parts and service

- Depreciation

- Portability (if required)

- Labor requirements and convenience of operations

- Cost

It is very important to match the engine horsepower to the requirements of the pump. Previously used power units should be carefully checked and evaluated as to condition, available horsepower, and speed. Using an old power unit that may not fit the needs of the system often results in more costly operation than a more expensive new unit designed specifically for the system.

In order to determine the horsepower requirement for the power unit, the total dynamic head (TDH), pumping rate, efficiency of the pump, drive efficiency, and the type of power unit must be known. The components of TDH are illustrated in Fig. 24-1.

$$TDH = H_s + H_p + H_f + H_v \qquad \text{Eq. 24-1}$$

where,

TDH = total dynamic head (ft)
H_s = static head (ft)
H_f = friction head (ft)
H_v = velocity head (ft)
H_p = pressure head (ft of head = psi x 2.31)

Total Static Head is the total vertical distance the pump must lift the water. When pumping from a well, this would be the distance from the drawdown water level in the well to ground level plus the distance the water is lifted from ground level to the discharge point.

Pressure Head is the pressure required in the system to operate the emitters. The pressure (in

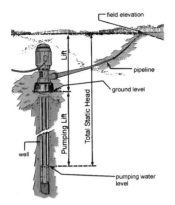 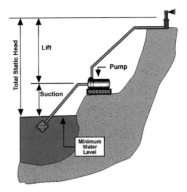

Figure 24-1. Total dynamic head (sum of suction head plus friction head plus pressure head) for deep well turbine (left) and centrifugal (right) pump systems.

psi) is converted to feet of head by: psi x 2.31 = H_p (in feet of H_2O).

<u>Velocity Head</u> is the energy imparted to the water to get it into motion. This is a very small amount of energy and is usually negligible when computing losses in an irrigation system.

<u>Friction Head</u> is the head loss due to friction when water flows through pipes. Friction losses occurring in fittings, valves, filters, or at changes in pipe size should also be considered. Values for these losses are obtained from friction loss tables (see Appendix 10).

The work required for a pumping plant to deliver water at the desired rate and pressure necessary for the system is called water horsepower (whp). Water horsepower is calculated from:

$$whp = \frac{Q \times TDH}{3960} \qquad \text{Eq 24-2}$$

where,

Q = flow rate (gpm)
TDH = total dynamic head (ft)

The water horsepower represents the power required to operate the pump if the pump and drive unit were 100% efficient. The brake horsepower (BHP) is the actual horsepower requirement at the drive unit connection and takes pump and drive efficiencies into consideration. The continuous horsepower rating of the power unit must equal this value and is calculated as:

$$BHP = \frac{whp}{Eff_p \times Eff_d} \qquad \text{Eq. 24-3}$$

where,

BHP = brake horsepower (continuous horse power rating of power unit)
whp = water horsepower
Eff_p = pump efficiency (%)
Eff_d = drive efficiency (%)

The pump efficiency must be obtained from the specifications for the pump. The drive efficiency can be estimated from Table 24-1. A well-designed irrigation pump should have an efficiency of 70%-80%. Typically, an estimate of 75% efficiency is adequate for preliminary planning.

Electric Motors
Electricity is often a very satisfactory power source. However, if freeze protection is a priority, alternate power sources should be considered.

Table 24-1. Typical efficiencies of various types of pump drives.

Type of Drive	Normal Efficiency
Direct	100%
Flat Belt (straight)	85%
Flat Belt (1/4 to 1/2 turn)	75%
V-Belt	95%
Right-Angle Gear Head	85%

Brownouts and rolling power outages make using electric-powered pumps for freeze protection of citrus risky. For routine irrigation, the dependability and long life of electric motors make them a desirable power source.

Single-phase motors are often used for loads up to and including 7.5 horsepower. However, three-phase motors are more efficient. Above 7.5 horsepower, single phase motors are not well adapted to irrigation pumps. Electric motors above 5 horsepower will generally have an efficiency of between 88% and 90%. Most squirrel cage induction motors are designed to operate satisfactorily under a continuous overload of 10% to 15%. However, it is not wise to plan on an overload.

If any adequate electric power supply is available, the electric motor may be the cheaper power source in many cases. Including the cost of the control, the initial investment will normally be less than that of an internal combustion engine drive. The cost of electric power for operation and standby charges may be higher than the fuel cost for an internal combustion engine, but the maintenance cost must also be considered. For an electric motor, maintenance is minimal, but it may be considerable for an internal combustion engine.

Properly installed and protected, electric motors will provide many years of service. The advantages of electric power include relatively long motor life, low maintenance costs, dependability, and ease of control and operation. An electric motor will deliver full power throughout its life and can be operated from no load to full load without damage. Disadvantages of electric motors include constant speed, an electric power supply required at each pumping spot, and normally a yearly minimum power cost.

Most large electric motors used for irrigation are squirrel-cage-induction-type, 60-cycle, 3-phase, 440-volt motors. Pumps may be connected to the motors by direct couplings, right-angle drives, or belts. The best design, if practical, is a direct coupling. Right-angle drives and belt drives are less than 100% efficient and so require more energy.

Most electric motors used on centrifugal pumps will have a horizontal shaft. On deep-well turbine pumps, either a vertical, hollow-shaft electric motor or a horizontal-shaft electric motor together with a hollow-shaft, right-angle drive must be used. The hollow-shaft unit is necessary so the pump impellers can be adjusted. The life of an electric motor is normally determined by the life of the motor winding insulation. Electric motor insulation is normally rated for a life of 20,000 hours. For example, Class A insulation has a 20,000-hour life at 105° C (220° F), Class B, a 20,000-hour life at 130° C (265° F), and Class F, a 20,000-hour life at 165° C (330° F). Electric motors for irrigation should be rated for continuous duty at 40° C (104° F) or higher maximum ambient temperature and have Class B or better insulation.

Overloading an electric motor causes internal temperatures to rise and drastically shortens expected motor life. Therefore, proper motor sizing is very important. If the calculated total pump load is close to the selected motor size, a service factor of 1.15 on the motor is desirable. A 1.15 service factor means the motor can safely carry a pump load of 115% of the rated electric motor horsepower. Unless specified otherwise, most electric motors come with a service factor of 100%.

Sometimes an irrigation pump must be served by a single-phase power supply. The alternatives are either a three-phase electric motor with a phase converter, or the use of a special single-phase motor. Phase converters convert single-phase electric power to three-phase power and may be either the static or rotary type. Auto-transformer-type static phase converters are appropriate where the load on the pump will remain constant.

Rotary phase converters are applicable on either single or multiple motor loads. They should be used where there may be a variable motor load (such as when the pump draws from a pond where the water depth varies). They also may be used where several motors will be run from one converter. The proper application of phase converters is critical and should only be done in consultation with the manufacturers, pump supplier, and electric power supplier. Always be certain all three leads to the three-phase motor are protected with the correctly sized overcurrent protection. This will protect the motor from any current imbalance that could result.

During start-up, electric motors draw more amperage than they do while running. Depending on location, this momentary high amperage draw may cause a nuisance to other persons receiving their electricity from the same electric distribution line. This high draw causes momentary low voltage that is normally seen in the dimming of lights. To minimize these starting problems, most electric power suppliers require that a reduced-voltage type of starting panel be used with large electric motors. These reduced-voltage starters limit the high amperage draw by the motors during start-up. Four types of reduced-voltage starters are available. The part-winding and wye-delta type require specially wound motors for their application. The primary-resistor type and the auto-transformer type may be used with any induction three-phase motor.

All electrical installations should be in compliance with the National Electric Code, along with state and local regulations. Motors, electrical enclosures, and other electrical equipment must be effectively grounded by a separate grounding conductor, and suitably connected to the power supply grounding system. Motors shall be supplied through a fused service disconnect.

A motor controller (magnetic or manual) and overcurrent protection should also be installed (Fig. 24-2). Overcurrent protection for single-phase motors should be of single-heater type. Three-phase motors require the three-heater type to provide overcurrent protection in each of the leads of the motor windings. Rating of the overcurrent protection devices should not exceed 115% of the motor name plate amperage. Ambient temperature-compensated overload protectors may be used to offset the effect of sunshine on control enclosures. Overload protection devices, consisting of temperature sensing elements buried in the motor

windings are highly recommended. However, they should not be used in lieu of overcurrent protection, unless the motor manufacturer fully guarantees the motor against locked-rotor and overload burnout with only the temperature-sensing element protection.

Motors should be drip-proof, and if operated with the shaft vertical, be designed to be drip-proof in this position unless protected from the weather by other suitable enclosures. Rodent screens should be installed either at the factory or in the field at the time the motor is installed. Secondary lightning arrestors should be installed and connected from ground to each ungrounded conductor in the supply. They should also be located on the secondary side of the transformer ahead of the service disconnect.

The operating costs of an electric-powered pump can be estimated from the following equation or from Table 24-2. Both fixed and operating costs should be considered before a final decision on the most appropriate unit is made.

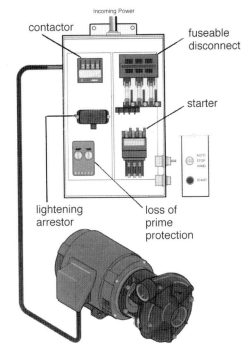

Figure 24-2. Minimum recommended components for electric pump system protection.

$$C = \frac{Q \times TDH \times c}{5310 \times Eff_p \times Eff_d}$$ **Eq. 24-4**

where,

C = Hourly pumping cost (in $)
Q = Discharge in gpm
TDH = Total dynamic head (pumping head, in ft)
c = Cost of electricity in $ per kwh
Eff_p = Pump efficiency
Eff_d = Efficiency of electric motor drive unit

Many electric motors can be safely overloaded to a small degree, depending on the service factor stamped on the motor. However, this is usually limited to about 10% of the horsepower rating stamped on the motor. If the permitted overload for an electric motor is exceeded, it will also operate inefficiently, wear rapidly, and fail prematurely. Underloading electric motors wastes energy. If a 100-hp electric motor were installed where a 50-hp motor would be adequate (percentage of continuous rated hp = 50%), about 18% of the electricity used would be wasted. If a 100-hp electric motor were installed where a 75-hp motor was needed (percentage of continuous rating = 75%), only about 5% of the electricity used would be wasted.

Internal Combustion Engines
Gasoline, diesel LP-gas, and electric power units are all used to drive irrigation pumps in Florida. Each has its advantages and disadvantages. Gasoline engines usually have lower initial cost. On the other hand, diesels are typically more durable and have a longer life. LP-gas engines require less maintenance than gasoline and the fuel may cost less. Power requirements for a particular installation can be estimated from Table 18-3.

Example:
Estimate the power requirement for a pump with 500 gpm at 140 ft of head.

From Table 24-3, the power unit would require 18 hp at 100% efficiency. If the pump is 75% efficient, 24 hp would be required. If the pump efficiency is 60%, the required power would be: 18 hp/60% efficiency = 30 hp.

The life expectancy of internal combustion engines is less than that of an electric motor. An electric motor provides only constant speed operation, but an internal combustion engine provides flexibility of pump performance through easy speed variation, which may be desirable. It is also necessary in selecting an engine power unit to make certain that adequate power is available at the desired speed at which the pump will be operated. A reserve horsepower allowance must also be made for wear, drive accessories such as electrical generators, and other factors that reduce the power an engine produces over time.

The most common fuels for internal combustion engines are gasoline, diesel, propane, and natural gas. Manufacturers have developed performance curves for each of their engines showing horsepower ratings at various speeds for use as a basis in engine selection. These curves are developed in a laboratory under conditions of 60° F, mean sea level elevation, and with a bare engine to produce the most horsepower per unit of engine weight. For field use, these curves must be corrected to reflect the power loss caused by the use of accessories, elevation differences, and air temperature.

Because of the characteristics of internal combustion engines, it is also necessary to further correct the horsepower curve to compensate for the continuous loading required in irrigation pumping. Corrections need to be made for the accessories, temperature, and for continuous operation for many hours at a time (Table 24-4). Selection of an engine for a power source for irrigation pumping should be based on the continuous service rating rather than the maximum BHP rating. For example, for the pump depicted in Fig. 24-3, the maximum permissible BHP load at 1600 rpm would be slightly less than 20 hp, as opposed to the 24.5 available at maximum performance.

Some manufacturers publish both the dynamometer curve and the continuous brake horsepower

Table 24-2. Theoretical per-hour pumping cost for single- (1) and three-phase (3) electrical service based on approximately 75% efficiency.

| Pump horsepower | Average kWh motor use | | Approximate cost of operation at kWh rate | | | | | | | | |
| | | | $ 0.02 | | $ 0.04 | | $ 0.06 | | $ 0.08 | | $ 0.10 | |
Phase →	1	3	1	3	1	3	1	3	1	3	1	3
1/2	0.50	0.52	0.01	0.01	0.02	0.02	0.03	0.03	0.04	0.04	0.05	0.05
1	0.76	0.77	0.02	0.02	0.03	0.03	0.05	0.05	0.06	0.06	0.08	0.08
1 1/2	1.00	0.96	0.02	0.02	0.04	0.04	0.06	0.06	0.08	0.08	0.10	0.10
2	1.50	1.42	0.03	0.03	0.06	0.06	0.09	0.09	0.12	0.11	0.15	0.14
2 1/2	1.98	1.83	0.04	0.04	0.08	0.07	0.12	0.11	0.16	0.15	0.20	0.18
3	2.95	2.70	0.06	0.05	0.12	0.11	0.18	0.16	0.24	0.22	0.30	0.27
5	4.65	4.50	0.09	0.09	0.19	0.18	0.28	0.27	0.37	0.36	0.47	0.45
8	6.90	6.75	0.14	0.14	0.28	0.27	0.41	0.41	0.55	0.54	0.69	0.68
10	9.30	9.00	0.19	0.18	0.37	0.36	0.56	0.54	0.74	0.72	0.93	0.90
15		12.8		0.26		0.51		0.77		1.02		1.28
20		16.9		0.34		0.68		1.01		1.35		1.69
25		20.8		0.42		0.83		1.25		1.66		2.08
30		25.0		0.50		1.00		1.50		2.00		2.50
40		33.2		0.66		1.33		1.99		2.66		3.32
50		41.3		0.83		1.65		2.48		3.30		4.13
60		49.5		0.99		1.98		2.97		3.96		4.95
75		61.5		1.23		2.46		3.69		4.92		6.15
100		81.8		1.64		3.27		4.91		6.54		8.18

kWh = Hours x 0.746 x motor hp/motor efficiency

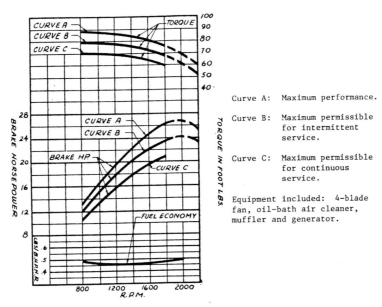

Curve A: Maximum performance.

Curve B: Maximum permissible for intermittent service.

Curve C: Maximum permissible for continuous service.

Equipment included: 4-blade fan, oil-bath air cleaner, muffler and generator.

Figure 24-3. Typical engine performance curve.

Table 24-3. Theoretical pump horsepower requirements for various head and flow rates (at 100% and 75% pump efficiency).

| Head (ft): → | 60 | | 80 | | 100 | | 120 | | 140 | | 160 | | 180 | |
| Press (psi):→ | 26 | | 35 | | 43 | | 52 | | 61 | | 69 | | 78 | |
gpm Eff. →	100	75	100	75	100	75	100	75	100	75	100	75	100	75
25	0.4	0.5	0.5	0.7	0.6	0.8	0.8	1.0	0.9	1.2	1.0	1.3	1.1	1.5
50	0.8	1.0	1.0	1.3	1.3	1.7	1.5	2.0	1.8	2.4	2.0	2.7	2.3	3.0
75	1.1	1.5	1.5	2.0	1.9	2.5	2.3	3.0	2.7	3.5	3.0	4.0	3.4	4.5
100	1.5	2.0	2.0	2.7	2.5	3.4	3.0	4.0	3.5	4.7	4.0	5.4	4.5	6.1
150	2.3	3.0	3.0	4.0	3.8	5.1	4.5	6.1	5.3	7.1	6.1	8.1	6.8	9.1
200	3.0	4.0	4.0	5.4	5.1	6.7	6.1	8.1	7.1	9.4	8.1	11	9.1	12
300	4.5	6.1	6.1	8.1	7.6	10	9.1	12	11	14	12	16	14	18
500	7.6	10	10	13	13	17	15	20	18	24	20	27	23	30
750	11	15	15	20	19	25	23	30	27	35	30	40	34	45
1000	15	20	20	27	25	34	30	40	35	47	40	54	45	61
1250	19	25	25	34	32	42	38	51	44	59	51	67	57	76
1500	23	30	30	40	38	51	45	61	53	71	61	81	68	91
1750	27	35	35	47	44	59	53	71	62	82	71	94	80	106
2000	30	40	40	54	51	67	61	81	71	94	81	108	91	121

Notes:

hp = gpm x head (ft)/3960

The hp figures at 100% efficiency must by divided by the pump efficiency to estimate power requirements.

Table 24-4. Correction factors to derate laboratory-measured engine horsepower.

For each 1000 feet above sea level	deduct 3%
For each 10° F above 60° F	deduct 1%
For accessories (generator, etc.)	deduct 5%
For radiator and fan	deduct 5%
For continuous operation	deduct 20%

curve in their literature. When only one curve is shown, that curve will generally be the horsepower determined by a dynamometer under laboratory conditions. The best operating load for an internal combustion engine is at or near the continuous brake horsepower curve. Running an engine under much lighter loads usually results in poorer fuel economy for the water delivered, since too much horsepower is used in overcoming engine friction and throttling losses. Running at maximum engine horsepower invites engine trouble as well as excessive fuel consumption.

Heat exchangers should meet the size requirements based on engine size and established by the manufacturer. In some instances, however, heat exchangers may be used on installations in sheltered areas where air movement around the unit is very poor and where the source of water is reasonably warm. This situation will require the use of a larger heat exchanger than is normally recommended. The addition of an auxiliary fan to move hot air away from the engine may increase fuel consumption slightly, but will eliminate safety switch shutdowns during extremely hot weather.

Operating costs pumps with for internal combustion power units can be estimated by using the following equation or by using Tables 24-5 and 24-6.

$$C = \frac{Q \times TDH \times Fc \times d}{3960 \times Eff_p} \qquad \textbf{Eq. 24-5}$$

where,

C = Hourly pumping cost (in $)
Q = Discharge in gpm
TDH = Total dynamic head (pumping head, in ft)
Fc = Fuel consumption in gallons per hp-hour
d = Cost of fuel in $ per gallon
Eff_p = Pump efficiency

Diesel power units operate at highest efficiency when fully loaded or loaded near their maximum continuous horsepower ratings. The efficiency of an irrigation power unit is reduced if the power unit is either too large or too small for the pump it is powering. When a power unit is overloaded, it will waste fuel, wear rapidly, and fail prematurely. Oversizing (underloading) power units by up to 20% will result in little loss in performance or fuel waste. When a gasoline engine is loaded at only 50%, it typically wastes about 27% of its fuel compared with a power unit of the proper size. Diesel engines loaded at only 20% of their continuous horsepower rating levels waste about 50% of the fuel compared to 60% for gasoline engines at 20% load.

The consequences of overloading a power unit are severe: both inefficient operation and power unit damage may be expected to occur. An internal combustion engine should never be operated for long periods of time under a load that exceeds its continuous horsepower rating for the speed at which it is being operated. When an internal combustion engine is overloaded, it will waste fuel, wear rapidly, and fail prematurely.

The general relationship between performance rating and percentage of continuous horsepower rating is not linear. Instead, performance ratings are high when power units are almost fully loaded; they decrease slowly for small underloads, then decrease more rapidly for severe underloads. All of the power units operate at high efficiency (above 95%) when loaded at more than 80% of their continuous horsepower rating. Diesel engines generally have performance ratings above 90% if they are loaded at more than 60%. The performance of gasoline engines drops most rapidly, yet is still above 90% when the engines are loaded at 70%. Below 60% to 70% of their continuous horsepower rating, performance ratings of all power units drop 10% to 12% for each 10% drop in loading.

Energy Costs
The fuel use performance rating (Table 24-7) is the ratio of the power unit's performance under a given partial load to its performance when fully loaded, expressed as a percentage. Thus, the performance rating is less than 100% for a partially loaded power unit and ranges up to 100% for a fully loaded one. The use of performance ratings rather than efficiency measurements allows the effects of loading to be compared. Performance standards are measured in units of horsepower-hours per gallon (hp-hr/gal) of fuel for internal combustion engines and horsepower-hours per kilowatt-hour (hp-hr/kWh) for electric motors. These standards rate the effectiveness of a typical power unit in converting fuel or electrical power to mechanical power. They are based on the assumption that the power unit is in good repair and fully loaded.

The fuel use rate (Table 24-8) is the inverse of the performance standard. To estimate the fuel use rate for a specific power unit, multiply the power unit size (continuous horsepower rating) times the fuel use rate from Table 24-8. The result is the fuel use

Table 24-5. Estimated fuel cost per hour by load and cost per gallon for diesel-powered pumps (based on 14.6 hp-hr/gal with 5% drive loss).

Pump load (hp)	Fuel cost per gallon ($)					
	0.80	0.90	1.00	1.20	1.40	1.60
10	0.55	0.62	0.69	0.82	0.96	1.10
20	1.10	1.23	1.37	1.65	1.92	2.19
30	1.65	1.85	2.06	2.47	2.88	3.29
40	2.19	2.47	2.74	3.29	3.84	4.39
50	2.74	3.09	3.43	4.12	4.80	5.49
75	4.12	4.63	5.14	6.17	7.20	8.23
100	5.49	6.17	6.86	8.23	9.60	10.97

Table 24-6. Estimated fuel cost per hour by load and cost per gallon for gasoline-powered pumps (based on 11.5 hp-hr/gal with 5% drive loss).

Pump load (hp)	Fuel cost per gallon ($)					
	0.80	0.90	1.00	1.20	1.40	1.60
10	0.69	0.78	0.87	1.04	1.40	1.39
20	1.39	1.56	1.37	2.08	2.43	2.77
30	2.08	2.34	2.60	3.12	3.64	4.16
40	2.77	3.12	3.47	4.16	4.85	5.55
50	3.47	3.90	4.33	5.20	6.07	6.93
75	5.20	5.84	6.50	7.80	9.10	10.40
100	6.93	7.80	8.67	10.40	12.13	13.86

Table 24-7. Typical performance standards for fully loaded irrigation power units.

Type of power	Units	whp	BHP required at 75% pump efficiency			
			Direct drive	V-belt	Right-angle gear or flat belt (straight)	Flat belt (1/4 or 1/2 turn)
		Drive efficiency →	100%	95%	85%	75%
Diesel	hp-hr/gal	11.0	14.7	13.9	12.5	11.0
Gasoline	hp-hr/gal	8.6	11.5	10.9	9.7	8.6
Propane	hp-hr/gal	7.0	9.3	8.9	7.9	7.0
Electric	hp-hr/kWh	0.89	1.19	1.13	1.01	0.89

Table 24-8. Typical fuel use rates for fully loaded irrigation power units.

Type of power	Units	whp	BHP required at 75% pump efficiency			
			Direct drive	V-belt	Right-angle gear or flat belt (straight)	Flat belt (1/4 or 1/2 turn)
		Drive efficiency →	100%	95%	85%	75%
Diesel	gal/hp-hr	0.0909	0.0682	0.0718	0.0802	0.0909
Gasoline	gal/hp-hr	0.1163	0.0872	0.0918	0.1026	0.1163
Propane	ga/hp-hr	0.1429	0.1071	0.1128	0.1261	0.1429
Electric	kWh/hp-hr	1.1236	0.8427	0.8870	0.9914	1.1236

rate in gallons (or kilowatts) per hour. Fuel or power use can also be estimated using Figure 24-4.

Example:
Estimate the fuel use rate for a 100-hp diesel power unit that is fully loaded that uses a right-angle drive (assume a 75% pump efficiency and 85% efficiency for right-angle drive).

From Table 24-8, the fuel use rate = 0.0802 gal/hp-hr for a diesel with a right-angle drive.

Fuel use = 100 hp x 0.0802 gal/hp-hr = 8.02 gal/hr.

Example:
Estimate the electric power use rate for a 50-hp electric motor that is fully loaded and uses a direct drive.

From Table 24-8, the fuel use rate = 0.08427 kWh/hp-hr for an electric motor with a direct drive.

Energy use rate = 50 hp x 0.8424 kWh/hp-hr = 42 kWh/hr.

Example:
Estimate the hourly fuel and power costs for a system requiring 1500 gpm at 125 ft of total dynamic head (TDH).

From Fig. 24-4, enter the vertical axis at 1500 gph and move right until the line intersects the 125 ft of TDH line. Read the diesel fuel requirements on the horizontal axis (directly below this intersection) at about 4.2 gph. Doing the same procedure results in about 5.3 gph for a gasoline engine and 52 kWh/hr for an electric motor.

The actual cost of operating an irrigation power plant depends on the fuel or electricity price and will vary by locale. Estimating the operating cost requires the water horsepower (whp) requirements and the total number of hours pumped during the year. The hours of operation depend on the pumping rate, acres irrigated, and the amount of water to be added per acre.

Example:
An 80-acre grove has an irrigation system that requires 800 gpm at a head of 120 feet. Average annual irrigation is 15 acre-inches per acre per year (240 hours of operation with 15 gph emitters). Calculate the total energy consumption for both diesel and electric power units. Both the motor and diesel engine will use right-angle gear drives with 75% pump efficiency.

From Eq. 24-2,

$$whp = \frac{800 \text{ gpm x } 120 \text{ ft}}{3960} = 24.2 \text{ whp}$$

From Table 24-7, a diesel power unit requires 12.5 hp-hr/gal and the electric power unit requires 1.01 hp-hr/kWh for a right-angle gear drive (85% efficiency) with 75% pump efficiency.

Diesel: $\dfrac{24.2 \text{ whp}}{12.5 \text{ hp-hr/gal}} = 1.9 \text{ gal/hr}$

Electric: $\dfrac{24.2 \text{ whp}}{1.01 \text{ hp-hr/kWh}} = 24 \text{ kWh/hr}$

Diesel: 240 hr x 1.9 gal/hr = 456 gal of fuel

Electric: 240 hr x 24 kWh/hr = 5760 kWh

After a pump and power unit have been installed and are running, an estimation of the actual power requirement can be made if the fuel consumption is known. To determine horsepower output from the power unit, the following formula is used:

$$BHP = FC \times Eff_{fuel} \qquad \textbf{Eq. 24-6}$$

where,

BHP = Brake horsepower
FC = Fuel consumption
Eff_{fuel} = Fuel efficiency (Table 24-8)

Example:

For a diesel power unit with a right-angle drive that uses 4.75 gallons per hour of fuel, what is the horsepower developed?

From 24-7, the performance standard for a diesel engine with right-angle drive is 12.5 hp-hr/gal

4.75 gal/hr x 12.5 hp-hr/gal = 59 BHP

Example:

What is the horsepower developed by an electric motor that is consuming 55 kWh per hour?

From 24-7, the performance standard for an electric motor with right-angle drive is 1.01 hp-hr/kWh

55 kWh/hr x 1.01 hp-hr/kWh = 56 BHP

The electric consumption may be determined by checking the electric meter. On meters that record electrical demand, the pointer will indicate consumption in kWh per hour. On electric meters without a demand pointer, the electrical consumption can be determined by counting revolutions of the disk and multiplying that by the Kh factor of the meter. The Kh factor indicates how many watts of electricity are consumed each time the disk revolves.

Safety Controls

Safety devices and procedures for protecting the pump and power plant equipment from damage by natural causes, sudden loss of water or power, and errors by operating personnel should be included with every irrigation system. Pumping plants should be equipped with safety devices that shut off the electric motor or engine when there is a break in the suction on centrifugal pumps, or a loss of pressure in the main pipeline. Pumps having water-lubricated bearings will be damaged by prolonged operation without water. Engines used to power sprinkler system pumps should be equipped with safety devices that stop the engine before damage occurs from overload, run away (if the pump becomes disconnected or loses its prime), loss of oil pressure, or overheating.

A ratchet coupling should be used between turbine pumps and electric motors or engines to prevent rotation of the power plants on high head sprinkler systems if there is a power failure while these systems are in operation. Automatic valves that permit pumps to be started or stopped without water damage to high head mainline pipes due to water hammer or surge should be installed as a safety feature.

Any type of pump is a valuable piece of precision equipment, and in order to operate satisfactorily over its normal life expectancy, it should be properly and carefully installed by competent personnel. Further, like any machine, it should be maintained and operated in accordance with the recommendations of its manufacturer.

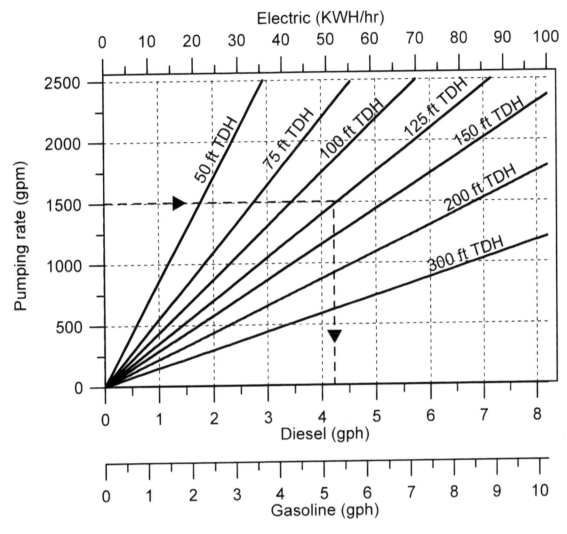

Figure 24-4. Estimated energy requirements for an efficient, well-maintained irrigation pumping unit.

Chapter 25. Pump and Power Unit Evaluation

by Brian Boman

Pump Station Evaluation

Inefficient irrigation pumping systems waste fuel and increase the cost per unit of water delivered. As power costs increase, the cost of operating an inefficient pump increases even more. Efficiency of a pumping system is defined as a ratio of the work being done by the system to the power or energy being supplied to it. If, for example, 50 hp is being put into a pump system and the actual work being done only requires 25 hp, then the pumping efficiency is 25 hp/50 hp = 0.5 or 50%.

Normally, power input is measured in terms of fuel (or electrical power) consumption. Work output is measured in terms of the flow rate and pressure produced by the system. Therefore, the efficiency computed is that of the entire pumping system. Overall efficiency includes the efficiency of fuel conversion in an internal combustion engine or electric motor, friction losses in belt or gear drives, the pump efficiency, and friction losses in the suction line, discharge column, and discharge head. The efficiency calculated including all of these components is often referred to as the pumping plant efficiency.

The pumping plant is designed to deliver water at the desired rate and pressure necessary for application. The work output (water horsepower) is calculated from Equation 25-1. Water horsepower depends on the rate that water is lifted and the height to which it is lifted. (For more details on pump performance factors, refer to Chapter 24).

$$\text{Water Horsepower} = \frac{Q \times TDH}{3960} \qquad \textbf{Eq. 25-1}$$

where,

Q = flow rate (gpm)
TDH = total pumping head produced (ft)
(1 psi = 2.31 ft of water)

The horsepower required to be delivered to the pump to produce the required water horsepower is called brake horsepower (BHP).

$$BHP = \frac{\text{Water Horsepower}}{\text{Pump Efficiency}} \qquad \textbf{Eq. 25-2}$$

Pumping Plant Tests

The performance of irrigation pumping systems can be checked in the field by measuring pumping plant performances. Standards have been developed based upon the water horsepower (whp) produced by the pump. Performance standards are shown for the different fuels most commonly used for irrigation systems in Table 25-1. These performance data reflect those that can reasonably be expected from a well-designed, well-maintained, pumping plant. Power unit performance standards are given in terms of power produced (in horsepower hours, hp x hr) per unit of fuel consumed (in gallons or kilowatt-hours). These figures represent the effectiveness of a typical power unit in converting fuel or electrical power to mechanical power.

Pumping system performance standards are given in units of water horsepower-hours (whp-hr) per gal or kWh. They include allowances for normal pump efficiencies, drive losses, and friction losses in the discharge column and discharge head. Pumping system performance standards are expressed in terms of units of fuel consumed because they can be easily measured. In contrast, mechanical power input to a pump can be measured only with specialized instrumentation.

A pumping-plant test requires that the physical properties that determine pumping-plant performance be measured. Pumping rate, lift, pressure at the discharge outlet, and the amount of fuel consumed over a period of time must be measured

Table 25-1. Performance standards for irrigation pumping plants.

Fuel	Power unit performance standards	Pumping plant* performance standards
Diesel	14.75 hp-hr/gal	11.06 whp-hr/gal
Gasoline	11.30 hp-hr/gal	8.48 whp-hr/gal
Propane	8.92 hp-hr/gal	6.69 whp-hr/gal
Electricity	1.18 hp-hr/kWh	0.885 whp-hr/kWh

*Based on 75% pump efficiency.

while the pump is operating at its normal load. The engine and pump speed should also be measured to ensure that the manufacturer's recommendations are being followed.

Water Horsepower Determination

To determine water horsepower, the flow rate (Q) must be accurately measured and converted to gpm. An impeller meter, orifice meter, or other measurement device may be used. Total pumping head (TDH) must also be accurately measured (Fig. 25-1). This includes a measure of the pressure produced at the pump discharge (pressure head) and the vertical distance the water is lifted from its source (aquifer, ditch, lake, etc.) to where the pressure is measured (pumping lift).

Pressure head must be measured with a reliable pressure gauge. A calibrated gauge should be installed for purposes of this measurement. A gauge that has been in the field for several years may not be reliable for a pump efficiency test. The gauge should allow pressures to be read to within ± 1 psi. Pressure head must be converted to feet of head by multiplying gauge pressure (psi) by 2.31 ft/psi.

Pumping lift must be measured as a component of the total pumping head. Pumping lift is measured as the vertical distance from the pumping water level at its source to the discharge point at which the pressure gauge is located. It is important that

this elevation be measured while the pump is operating because the water level in the aquifer may be considerably lower during pumping. Water elevations may be measured by any of several methods. One method consists of lowering a rusty steel measuring tape between the well column and casing. A watermark will remain on the rusty tape and allow it to be read immediately upon retrieval.

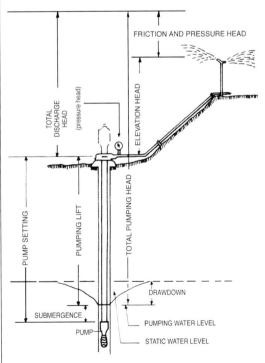

Figure 25-1. Components of total dynamic head.

Another commonly used technique involves locating the water elevation by allowing it to complete an electrical circuit (Fig. 25-2). A battery and ammeter are connected in series. One electrical lead is connected to the well casing; the second is lowered into the well. When the lead contacts the groundwater, the electrical circuit is completed and a current flow is indicated on the ammeter. Depth to water is obtained by measuring the electrical lead length. Similar results can be obtained by using a portable ohmmeter and measuring resistance changes when the circuit is completed.

Pumping lift should be measured to the nearest foot. Total pumping head is calculated as the sum of the pressure head and elevation head. Water horsepower is calculated by multiplying the measured flow rate (gpm) by the total head (ft) and dividing by 3960.

Fuel Consumption
To assess pumping plant performance, fuel use must be determined. For electric power units, this can be accomplished with any of several types of portable instrumentation to measure voltage, current flow, and power factor while the system is

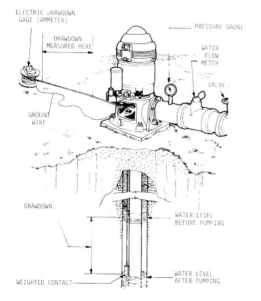

Figure 25-2. Measuring drawdown depth in a well with a vertical turbine pump.

operating. It can also be measured with acceptable accuracy using the electric company's kilowatt-hour (kWh) meter. The number of kilowatt hours consumed during an accurately measured period of time can be used to calculate the rate of power consumption in kWh/hr. To increase accuracy, a period of time of at least one hour should be used to evaluate electrical power consumption. The kWh meter should be read before and after the test period. The rate of power consumption is then calculated as the difference in kWh meter readings divided by the elapsed time.

For internal combustion engines, gallons of fuel consumed must be measured during an accurately monitored pumping period. For diesel and gasoline engines, fuel consumption can be monitored using a three-way valve connected to the engine's fuel line. The time for a given amount of fuel to be consumed can be accurately measured with a stopwatch. The amount of fuel used can be weighed from a smaller container (Fig. 25-3) or with a specially constructed graduated cylinder. It is important that the engine be fully loaded and at the correct operating temperature if this technique is to be used. During warm-up, fuel is consumed directly from the bulk fuel tank at the well site. To begin the test, the fuel source is switched from the main tank to the test supply. Fuel use rate is calculated from the known volume of fuel consumed divided by the elapsed time. If performed with care, this method is the most accurate method of determining the fuel use rate. Care must also be taken to avoid allowing dirt to enter the fuel line, as this can cause engine operational problems, especially with the injectors of diesel engines.

If it is not possible or desirable to inject directly into the fuel lines, fuel consumption rates can be measured by depletions from the bulk storage tank. If the fuel tank is symmetrically shaped, the consumption rate can be obtained by measuring fuel levels before and after an extended pumping period. Volume of fuel consumed will be obtained from the differences in depth before and after the pumping period, and from a knowledge of the

Figure 25-3. Weighing fuel used during pump test.

cross-sectional area of the tank. Such a tank can be calibrated by measuring fuel depth before and after an accurately measured volume is added to the tank.

Fuel tanks are often not regularly shaped (i.e., a cylindrical tank laying horizontally), and in such cases it is difficult to estimate volume of fuel consumed from measured reductions in liquid level in the tank (see Appendix 7). If a flowmeter is used to measure the amount of fuel pumped into the tank when it is refilled, that reading can be used to evaluate the fuel consumption rate, if pumping times are also accurately recorded.

Fuel consumption rates can also be calculated monthly or annually based upon fuel purchase records. Engine operating time must still be accurately recorded, preferably with an hour meter on the engine. Long-term record evaluations such as these are not as desirable as using previously described techniques, because long-term fuel storage efficiency becomes a factor to be considered. Vented tanks lose some fuel to vaporization, which results in an apparent decrease in the calculated pumping plant performance rating.

Fuel consumption rates for LP-gas (propane), which must be stored in tanks under pressure, can best be measured from the supplier's records. Pressure gauges indicating percentage of fuel remaining in the tank may not be sufficiently accurate to evaluate the pumping plant performance. Rather, the fuel supplier's record of gallons required to

refill the tank, and an accurate record of pumping time, must be used in the calculation of performance ratings.

Example Performance Calculation:
The following field illustrates the procedure for calculation of pumping plant performance:

Pump discharge rate, Q = 800 gpm
Pumping lift, Le = 40 ft
Discharge pressure, P = 50 psi
Pump speed = 1750 rpm
Fuel consumed (diesel) = 4.0 gal
Pump test duration = 1.0 hr

Measure pump speed (Fig. 25-4) using a portable tachometer to ensure that the pump is operating according to its specifications. The design pump operating speed should be stamped on a plate attached to the pump discharge head. If the speed is not close to the nameplate-required pump operating speed, adjustments should be made before continuing.

Total dynamic head (TDH):
TDH = pumping lift (ft) + discharge pressure (ft)
TDH = 40 ft + 50 psi x 2.31 ft/psi = 155 ft

Water horsepower, wph:

$$wph = \frac{Q \times TDH}{3960} = \frac{800 \text{ gpm} \times 155 \text{ ft}}{3960} = 31.3 \text{ hp}$$

Figure 25-4. Measuring engine rpm with digital tachometer.

Pumping Plant Performance:

$$(whp\text{-}hr/gal) = \frac{whp \times \text{Test Duration (hr)}}{\text{Fuel Consumed (gal)}}$$

$$\text{Performance} = \frac{31.3 \text{ hp} \times 1.0 \text{ hr}}{4.0 \text{ gal}}$$

$$= 7.83 \text{ whp-hr/gal}$$

Pumping Plant Performance Rating (PR):

$$PR = \frac{\text{Pumping Plant Performance} \times 100\%}{\text{Performance Standard}}$$

$$PR = \frac{7.83 \text{ whp-hr/gal} \times 100\%}{11.06 \text{ whp-hr/gal}} = 71\%$$

Fuel Wasted per Hour:
Fuel Wasted/Hour = Current Fuel Consumption Rate x (1 - PR) x 100%

Fuel Wasted/Hour = 4.0 gal/hr (100% - 71%) = 1.16 gal/hr

In this example, the actual pumping plant performance of 7.83 whp-hr/gal is only 71% of the performance standard for diesel-powered pumping plants. For the size of unit described, 1.16 gal/hr of diesel fuel is wasted because the pumping plant is not operating efficiently in its current condition. Whether this reduction in performance is significant enough to justify having the pumping unit repaired depends upon the expected repair cost and the number of hours of pump operation per year. In general, if the repair cost can be regained by savings in operating costs over a 2- to 3-year period of time, then it will be economically feasible to have the repairs made. The actual repayment time can only be calculated using a detailed economic analysis, including the expected performance increases, fuel cost, and the repair costs amortized over that period of time.

Causes for Substandard Performance
Substandard performance of a pump can be caused by several factors. One common cause is a mismatch of the pump with the present operating conditions. Originally, the pump may not have been

properly selected, or the operating conditions may have changed. A drop in the water table pumping level changes the pumping head requirement. Additional discharge required when replacing emitters with larger discharges, or resetting more trees than in the original system design, can lead to excess flow requirements. Sometimes, the engine or motor may not be operating at the specified speed (rpm) for maximum efficiency, resulting in poor performance.

If impellers are out of adjustment, a qualified repairman should be retained to adjust the impeller clearance with the bowl for the greatest efficiency. If the impeller is badly worn or corroded, adjustment will not help. Often, worn impellers are a result of cavitation, which occurs in pumps operating at flow rates greater than the well can supply. The cavitation pits the impellers and ruins them.

The engine or motor overloading may result in poor performance. An internal combustion engine operates most efficiently at 75% to 90% of its continuous horsepower rating at its design speed. Electric motors operate best at 100% to 110% of their nameplate rating. Overloading an internal combustion engine can seriously shorten engine life as well as increase fuel costs.

Oftentimes, an engine tune-up can improve performance. The ignition, timing, and carburetor should be adjusted on sparkignited engines. Diesel engines require fuel injection timing. Adjustments should be made by a qualified specialist to ensure maximum efficiency under the operating conditions. Electric motors generally do not need adjustment over their lifetime. Parts that are excessively worn should be replaced. Compression tests can be run to check for the need to overhaul an internal combustion engine.

Poorly designed pumping systems result in low performance ratings. These could be caused by such factors as an undersized suction pipe, restrictions in the intake strainer, or an improperly sized discharge column. Excessive wear, which is a sign of misalignment of the drive shaft, also decreases performance ratings.

Chapter 26. System, Pump, and Engine Maintenance

by Brian Boman

System Checks

Before the dry season begins, microirrigation systems need to be prepared for the anticipated use. Temporary repairs made during drought periods will probably need attention. Microirrigation systems are more than just a method to deliver water to the crop. They are becoming a management tool. A properly designed and maintained system allows the grower to supply precise amounts of water, nutrients, and other materials to the crop.

If the system is used to apply fertilizers or chemicals, varying pressure problems will cause uneven distribution of materials throughout the field. These problems can be due to clogged filters, regulators, or emission devices. Also, problems such as leaking barbs, tubing, emitters, and end plugs can allow water to collect at the base of the plant and invite disorders such as crown rot. Careful management and preseason maintenance can allow the grower to realize the full benefits of a microirrigation system.

Pumps

A pump that has been sitting idle for a few months needs to be checked for rodent activity and nests that could cause a short in the windings. A thorough cleaning is important, especially for pumps operating in dirty conditions. A pump dealer or manufacturer will be able to provide specific instructions for the care of the pump and motor.

The oil levels should be checked and filled at this time, and turbine pumps with automatic oilers should be checked to see that they are functioning properly. After a long layoff, it is also a good idea to start the oiler 24 hours before a deep-well pump is started. If you suspect that the efficiency of your pump has declined, a pump test is a quick and reliable way to assess its performance.

Filters

Several items need to be checked on both screen and media filters prior to start-up. On filters that flush automatically, the controller and valves should be checked for proper operation. If the controller is equipped with a pressure differential switch, the setting should be checked against the manufacturer's specifications. A differential can be created by removing one of the leads to simulate a high differential. If the differential switch is operating correctly, this will initiate a flush cycle.

Some media filter manufacturers and dealers recommend a minimum flush cycle at two- or three-hour intervals to prevent fine contaminants from becoming embedded in the media, even though the water may be relatively clean. The media filter should be opened to inspect the level of sand in each filter. Ideally, the level should be the same in each filter. Any difference could indicate trouble, such as a faulty valve or problem with the filter cake or the underdrain. Also note the condition of the media itself. If the media is channeled or caked, this could represent other problems, such as inadequate flush cycles during irrigation.

When setting the backflush time, it is important to allow for travel time (time required for the valve to move and fully seat). Most manufacturers recommend backflush times between 60 and 90 seconds. After the filter and controller have been checked and repaired, the backflush volume needs to be checked and adjusted. If the backflush volume is too high, the result is a loss of media. If it is too low, this can result in improper cleaning.

Screen filters need to be opened and inspected. The element, whether fabric, plastic, or steel, needs to be inspected for damage. The conditions of the seals and O-rings are important in isolating the incoming unfiltered water from the filtered water going to the system.

Laterals and Emitters

Once the pump and filters have been checked and repaired, lateral lines, emitters, and peripheral equipment need to be inspected. A thorough flushing of the system is the first priority, and this should be done in steps.

1. Flush the main line. Depending on the system and pump capabilities, a portion of the system may need to be shut off to increase the pressure and velocity.

2. Flush lateral tubing. Open the ends of lateral tubing and flush out accumulated debris. A portion of the laterals may need to remain closed to ensure good pressure and velocity for a thorough flushing of the tubing. After the entire system has been flushed, the system needs to be checked lateral by lateral.

3. Measure pressure and flow. If the system is equipped with water meters, the flow can be compared to previous evaluations. Reductions in flow can indicate problems such as obstructed lines, clogged emitters, or partially closed valves. If it is determined that the emitters have reduced flow rates, it could indicate a need for chemical treatment.

4. Walk the field. Check to see if there are emitters for reset trees. Are there areas in the field where the tree growth is not what it should be? Spot checks of pressure in weak or troubled areas may indicate changes that are needed.

Centrifugal Units

Engine: There should be a visual inspection every time the engine is started.

•   Radiator: Check fluid level, hose condition, corrosion and condition of mounts, and the condition of fan and belts (Figs. 26-1 and 26-2).

•   Engine: check oil level, look for fuel and oil leaks, check oil cooler and fittings, check for rodent damage and corrosion. Note vibration, oil pressure, and coolant temperatures.

•   Exhaust should be routed out of the pump house. The condition of the exhaust should be checked; there should be gray-black or white smoke.

PTO:

Apply maintenance grease to PTO shaft zerks as necessary. Check for alignment and adjustment of shaft and bearings; note vibration or wobble. Make sure all safety shields are in place (Fig. 26-3).

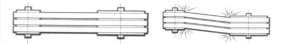

Figure 26-1. Align shafts so that they are parallel. Sheave grooves need to be directly in line to minimize wear and vibration (left).

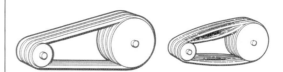

Figure 26-2. Proper (left) and improper (right) belt tension.

Figure 26-3. Properly shielded and placarded PTO for centrifugal pump station.

Control Gauges:

Confirm that all gauges are operational. Check to make sure that wiring and terminals are in good shape, that there is no damage from rodents or birds, and that there is no corrosion. Do not bypass shutdown controls, as they are very important from both safety and cost standpoints. Check to ensure all switch gauges are properly functioning. Standard shutdown switches should include low oil pressure and high coolant temperature. In addition, coolant level and pump pressure switches should be considered.

Coupling:

On close-coupled drives, the drive plate and rubber block should be checked. On units with shaft-to-shaft coupling, check alignment. If vibration occurs, check for cracks or damage in the rubber

element (Fig. 26-4). Where flexible couplings are used on engine or electric motor-driven pumps, alignment should be checked. Most flexible couplings will allow only about 0.020-inch misalignment. Alignment should be checked with feeler gauges and not judged by sight. Special consideration should be given to alignment on PTO-driven pumps. PTO shafts have a small amount of flexibility and should be connected in accordance with instructions supplied by the manufacturer.

Pump:

Perform a visual inspection to observe general performance and note any vibration (see Tables 26-1 and 26-2). Check packing and wear rings (Fig. 26-5). The running clearance for new pumps is about 0.010 inch on each side. If wear increases the clearance to over 0.030 inch on each side, the

Table 26-1. Troubleshooting positive displacement pumps.

No liquid delivered	Less than rated capacity pumped	Loss of prime while pump is operating	Pump is noisy	Rapid pump wear	Pump takes too much power
Pump not primed	Air leak in suction line or pump seal	Air leak in suction line	Cavitation	Pipe strain on pump casing; pump running dry	Speed too high
Insufficient NPSH	Insufficient NPSH	Liquid level falls below suction intake	Misalignment	Corrosion	Shaft packing too tight
Suction line strainer clogged	Suction strainer too small or clogged	Liquid vaporizes in suction lines	Foreign materials inside pump	Grit, sand, or abrasive material in liquid	Misalignment
End of suction line not in water	Increased clearances and slip caused by wear		Bent rotor shaft (rotary pumps)		Liquid more viscous than specified
Relief valve jammed open	Relief valve set wrong or stuck open		Relief valve chattering		High pressure from obstruction in discharge line
Pump rotates in wrong direction	Suction or discharge valves partially closed				Discharge line too small
Bypass valve open	Speed too low				Discharge valve partially closed
Suction or discharge valve open	Bypass valve partially open				

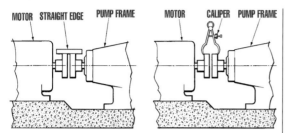

Figure 26-4. Correct alignment of pump and power unit (courtesy of Cornell Pump Company).

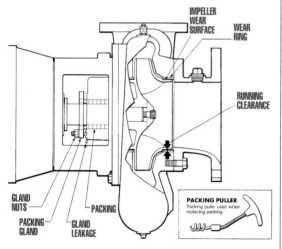

Figure 26-5. Wear rings should be replaced if clearance exceeds about 0.030 inch on each side (courtesy of Cornell Pump Company).

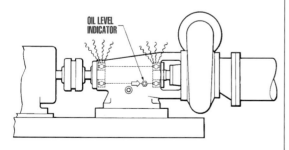

Figure 26-6. Normal bearing temperature should be about 160° F. Temperatures above 200° F may result in premature failure (courtesy of Cornell Pump Company).

wear ring should be repaired or replaced. Wear may be caused by abrasives in the water that is pumped (typically sand) or deflections in discharge piping. The packing gland should be checked to see if it is dripping, spraying, or running too dry. Tighten the gland nut 1/4 turn every 10 minutes until a leakage of 40 to 60 drops per minute is achieved. Do not overtighten. Add a ring when the gland bottoms out, and change packing if adjustment is ineffective. Use graphite grease or a grease cup to recoat dry packing. Check to ensure that bearings are not overheating, making noise, or vibrating (Fig. 26-6). Grease with standard grease, being careful not to overgrease. Brush, prime, and repaint rusted areas to control corrosion. Track and record the pump's performance, including rpm, pressure, and vacuum under normal operation.

Pump Frame:
Make sure it is structurally sound, since as corrosion attacks, it will cause vibration. Pumps powered with internal combustion engines should be installed in a level position due to cooling and lubrication requirements. Suction and discharge lines should be supported to eliminate strain on the pump and running gear. Pipe strain can damage the running gear, and cause cracked pump volutes, bearing failures, and misalignment.

Pump House:
If the pump is housed indoors or under a roof, provision should be made for proper cooling and circulation of air. A shelter with open sides is adequate except where extremely dusty conditions are encountered. Louvered sides can be used the entire height and length to provide proper circulation of the cooling air. Where it is absolutely necessary to enclose the engine and pump, a heat-exchanger cooling system should be used instead of the fan and radiator.

Electric Motors
Pay close attention to electric motors, as they pull dirt and debris into the motor itself. This material should be cleaned out before starting the motor any time the units remain idle for several months. The

Table 26-2. Troubleshooting for centrifugal pumps.

Problem	Probable Cause	Remedy
No water delivered	Pump not primed	Reprime pump. Check that pump and suction lines are full.
	Suction line clogged	Remove obstructions.
	Wrong direction of rotation	Change rotation to concur with direction indicated by arrow on bearing housing or pump casing.
	Foot valve not opening or suction pipe not submerged enough	Consult factory for proper depth. Use baffle to eliminate vortices.
	Suction lift too high	Shorten suction pipe.
Pump not producing rated flow	Air leak through gasket	Replace gasket.
	Air leak through stuffing	Replace or readjust packing or mechanical seal.
	Impeller partly clogged	Backflush pump to clean impeller.
	Worn suction side plate or wear rings	Replace part if defective.
	Insufficient suction head	Ensure that suction line shutoff valve is fully open and unobstructed.
	Worn or broken impeller	Inspect and replace as necessary.
Pump starts and then stops pumping	Improperly primed pump	Reprime pump.
	Air or vapor locks in suction line	Rearrange piping to eliminate air pockets.
	Air leak in suction line	Repair leak.
Bearings run hot	Improper alignment	Realign pump and driver.
	Improper lubrication	Check lubricant for suitablility.
Pump is noisy or vibrates	Improper pump/driver alignment	Align shafts.
	Partly clogged impeller causing imbalances	Backflush pump to clean impeller.
	Broken or bent impeller or shaft	Replace as required.
	Foundation not rigid	Tighten hold-down bolts of pump and motor.
	Worn bearings	Replace.
	Suction or discharge pipes not anchored or properly supported	Anchor properly.
	Pump cavitates	System problem.
Excessive leakage from stuffing box	Packing gland improperly adjusted	Tighten gland nuts.
	Stuffing box improperly packed	Check packing and repack box.
	Worn mechanical seals	Replace worn parts.
	Overheating mechanical seal	Check lubrication and cooling lines.
	Shaft seal scored	Remachine or replace as required.
Motor required excessive power	Head lower than rated - pumps too much water	Install throttle valve, trim impeller.
	Stuffing, packing too light	Readjust packing. Replace if worn.
	Rotating parts bind	Check internal wearing parts for proper clearances.

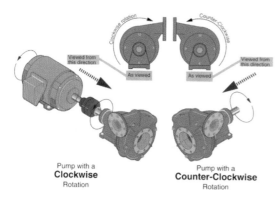

Pump with a
Clockwise
Rotation

Pump with a
Counter-Clockwise
Rotation

Figure 26-7. Rotation directions for centrifugal pumps.

electric motor housing makes an excellent home for rodents and insects. If ants, mice, wasps, etc., build a nest that works into the motor when it is started up, the result is a burned-out motor. Always disconnect power prior to inspecting or working on electric motors, and be sure to check for debris both above and below the electric motor's rotor.

When starting the motor after long periods of nonuse, lubricate all bearings. Enough lubricant should be used so it starts oozing out. This displaces the water which may have collected in the bearings. On larger irrigation units, fittings for applying new lubricants are provided, as well as a fitting for draining out the old lubricant. Some units have permanently sealed bearings, but most have the fittings. Pull the drain and the grease drain fitting, and push new grease through the bearing. Let it run long enough to expel any grease the motor doesn't want.

Blow compressed air into the control panel to clean out dirt and water that may have collected during the wet season. Check the contact points, and if they look corroded, redress them with a file. If the points are in bad shape, replace them. Be sure to check all electrical connections for possible discoloration. If the connection is dark or discolored from the normal color, take it apart and clean it. If there is oxidation on the connections, restrictions in current flow can cause heating; use emery cloth or commercial electrical cleaners.

Periodically check the motor and control panel to determine if there is proper insulation in the motor and any leaks to ground in the panel. Make sure the motor is rotating in the proper direction for the pump (Fig. 26-7). Start the motor and observe the direction of rotation. If the motor does not turn in the proper direction, refer to the owner's manual for proper electrical connections.

Turbine Pumps

With an oil-lubricated enclosed line shaft, fill the oil reservoir. If a manual dripper is used, make sure it is working properly. With a water-lubricated line shaft, make sure the prelube water tank is full and ready for startup. If a manual valve is used, open it before starting the pump. Remove the motor/gear drive bonnet and set the lateral adjustment according to the instructions from the pump installer. In general, this means:

1. Back off the top shaft nut until the impellers are on the bottom.

2. Tighten the top shaft nut until the shaft turns.

3. Accurately measure the stickup from the nut to the top of the top shaft.

4. Tighten the nut until the stickup has been increased by the amount specified by the installer. Caution: the bowl unit can be damaged if the impeller shaft is pulled up too high.

5. Replace the nut's locking device.

Double-check all valve positions, safety guards, switch positions, and all critical-run mode devices. After starting the pump, observe it carefully. If the inner column uses an electrically operated valve, be sure it is metering the oil satisfactorily. If a manual prelube is used, close the valve when the tank is full. Also, check the top shaft packing. It should be tight enough so that leakage is not excessive, but there must be a small amount of flow (a trickle). If necessary, repack the shaft after the pump is shut down.

Chapter 27. Filtration

by Brian Boman

Water used in microirrigation systems in Florida nearly always requires filtration, and frequently other forms of treatment are also needed. Solids in water can come from various sources, either naturally, accidentally, or by the action of man. Particulate matter may originate in wells in the form of silt or sand. In surface waters, solids can come from algae, aquatic weeds, silt, leaves, small animals, insects, and a wide variety of other debris. Solid matter can come from the system itself in the form of rust scales, PVC shavings, springs from check valves, and plastic parts from meters and valves. Particulate matter and organic growths that can plug microirrigation systems must be removed from the water supply using proper filtration. Often, chemical treatment is required to prevent organic growths and/or chemical precipitation in the irrigation system.

A variety of filters are available for the removal of physical clogging agents from irrigation water. The choice depends on the type and amount of contamination anticipated in the system, as well as the size of the irrigation system. Filters range from small, simple, single-unit filters to large batteries of filtering elements connected together by complex manifolds and activated by computerized solenoid systems.

Filtration units can be divided into groups according to their principle of operation. These groups include screen filtration, disc filters, media (sand) filters, and kinetic filters (sand separators). Sometimes, more than one method is combined to achieve higher filtration efficiency. Screen filters should be an acceptable primary choice when water is pumped from a well, and the only filtration requirements are to remove mineral particulate matter. They will normally be adequate when there is no organic contamination. Screen filters are typically less expensive and easier to maintain than media filters. Media filters and self-flushing screen filters have the ability to trap large quantities of contaminants. Media filters have historically been used when there are organic particles. However, recent advances in self-flushing screen filters make them a viable alternative to media filters in many cases. Kinetic filters are not used for organic loads, but are very effective for sand problems. Proper sizing and design of filters provide adequate filtration and good performance of the filter.

The nominal size of filters is designated by the size of their connections to adjoining piping. These connections are threaded in smaller sizes and flanged in sizes greater than 3 inches. Filters with cylindrical screen elements are generally available from manufacturers in sizes up to 12 inches (with 14 inches available from some manufacturers). Media filter tanks generally range from 18 to 48 inches in diameter.

For higher filtration rates, several filtration units may be constructed in parallel with a common inlet and outlet (Fig. 27-1). The filter elements in these multiple units can be independently cleaned and maintained. In large systems, they may be

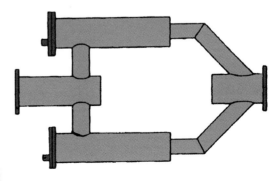

Figure 27-1. Typical layout for 2-barrel screen filter unit.

connected in batteries by manifold pipes on both the inlet and outlet sides. The design of these filter units must ensure equal distribution of flow among all of the elements. Unequal distribution of flow provides inefficient filtering and premature clogging in elements subjected to higher flow rates.

The optimum flow rate for any particular filter is usually specified by the manufacturer. Such specifications, however, should only serve as a starting point in determining needed capacity. The flow rate through any filter depends greatly on the quality of water flowing through it and on the frequency of maintenance and cleaning.

Suspended particles in irrigation water can be classified as being of either organic or inorganic in origin. Organic particles originate as algae, bacteria, plants, insects, fish, spiders, etc. Inorganic sources are generally sand, silt, and particles from within the irrigation system (such as PVC shavings, valve springs, and parts from plastic or metal components).

The permissible amount and size of particles in the irrigation system depends not only on the size of the emitter orifices, but also on the flow velocity in the mains, submains, and laterals. Water that moves slowly may deposit enough solids in the pipes to block them. On the other hand, abrasive solids carried by rapidly flowing water may result in premature failure of pipes and system components. In systems, there are usually portions of the distribution with flow velocities less than 1.5 fps, and other sections with velocities greater than 5 fps. Therefore, removal of even very fine particles is important.

The total amount of solids to be removed must also be considered when selecting a filtration method. Filtration problems tend to multiply when the amount of solids increases, and extra efforts must be employed to ensure filters do not clog. In severe cases, step filtration may be required. In these cases, water passes through a series of filters, each one removing finer particles until the desired rate and quality are achieved.

Filters will cause head losses in the irrigation system. Unlike valves or pipe, head loss in well-designed filters is generally independent of the flow rate. Head losses increase as the filter clogs. Manufacturers specify a head loss coefficient for filters in a totally clean state, which is seldom seen in practice. In a practical sense, this coefficient has little meaning since the head loss changes as the filter clogs and is cleaned.

Cleaning and Maintenance

Solids and sediments separated by the filter must be removed periodically to prevent clogging and to make room for sediments of subsequent filtration. This removal may be done infrequently, or may be required several times per day or even per hour in severe cases. The removal of the filtered particles may be manual or automatic, depending on the necessary frequency of cleaning, type of filter and required water quality.

In some systems, filters may need to be cleaned only at infrequent intervals (weeks to months). The filters used in these systems generally are screen filters of the simplest construction. In these cases, cleaning is best performed manually. The removal of debris from filters can normally be accomplished by opening drain valves and flushing out the accumulated debris. Periodically, the screen elements need to be removed and rinsed or brushed clean.

When the filters require frequent cleaning, automation can be used to achieve self-cleaning filters. Automatic cleaning can be accomplished by a number of methods. The most commonly used methods are: flushing debris using the filter's operating pressure, removal of debris by backflushing the filter element, or removal of accumulated debris by simultaneously flushing and cleaning the filtering element.

Working Pressure

When the working pressure method of filter cleaning is used, the control command opens a hydraulically operated drain valve. The drain valve allows water to flow across the filter at high velocity, flushing out any sediment that has accumulated at the bottom of the filter body. Oftentimes, a portion of the debris may adhere to the filter element, even after a thorough flushing. This debris needs to be removed by manually cleaning the filter.

A variation of this method of cleaning is the use of a constant small flow rate through the drain to remove debris from the filter. The flowing water constantly removes particles separated by the filter so they do not build up. If larger particles are in the water, they tend to reduce the flow of bleed water, and may limit the effectiveness of the self-cleaning action of the filter. With this type of filter, the cleaning is continuous, but periodic manual cleaning of the filter is necessary.

Backflushing

Backflushing is the most effective method of cleaning for volume filtration systems, since mechanical cleaning of the filter is complicated, and the filtering element consists of fine, granular materials. This method requires a valve system that directs the water flow from the downstream side into the filter, through the filtration element, and then drains it upstream. During backflushing, the regular flow of water through the filter must be stopped and the upstream piping shut off.

The best procedure for backflushing is to use clean water that has already passed through the filter. This is possible with a filtration system consisting of a battery of elements that can be actuated separately. The battery is then divided into two parts that alternate between filtering and backflushing. Such an arrangement prevents clogging of the filter from the downstream side by the debris that may flow in the flushing water.

Simultaneous Cleaning and Flushing

Flushing of the filter with a simultaneous cleaning of the screen may be done either mechanically or by a pressure differential. The mechanically cleaned filter features a brushing arrangement that rotates on the inside of the straining cylinder and is actuated either by an electric or a hydraulic motor. The drain valve opens with the beginning of the brushing, ejecting the debris removed from the screen by brushes.

In hydraulic self-cleaning filters, the cleaning device consists of a hollow tube equipped with suction arms that revolve inside the filtering element. A hydraulic motor operating on a jet principle incorporated in the lower part of the tube drives the suction arms. Opening the drain valve reduces the pressure in the suction tube, causing the debris accumulated on the filter screen to be removed by suction. During each of the cleaning procedures, the drain valve opens by a command from the control panel, allowing the removal of sediment from the filter body.

Automatic Control of Filters

Most automatic filters are activated by a pressure differential across the filtration element. The desired differential level can be preset according to local conditions, water quality, and other system requirements. When the filter clogs, the pressure differential rises, and if the preset differential level is reached, a pressure sensor activates the cleaning and flushing mechanisms.

In some cases, time-based backflushing is preferred, especially where water quality is constant throughout the season. In these cases, a mechanical or electric clock is installed in the activation panel to initiate the backflush operation at preset intervals. The actuating can be electrical, hydraulic, or a combination of both. The parameters that regulate automatic flushing of filters are:

- pressure differential across the filters,
- flushing period,
- interval between flushes.

Each is closely related to the other parameters and each influences the others. In many types of filters, flushing requires interrupting the flow to the irrigation system. Therefore, short flush cycles and long intervals between flushes are generally desirable. Unfortunately, these conditions are seldom possible in irrigation systems.

As solids accumulate in the filter, the pressure differential between the inlet and outlet rises, causing a loss of energy. In order to restore the filter to its clean, low, head-loss state, the solids must be removed by flushing from the filtering element and the sump. The flushing period depends on the amount of solids accumulated. A long interval between flushing causes accumulation of more sediments, prolonging the flushing period. If the flushing is not done thoroughly over a sufficient period of time, the loss of energy will not drop to its minimal level. Therefore, in time-controlled filters, both the flushing and interval periods must be adjusted in order to reduce head loss in the filter and minimize interruptions of water flow to the system. A timer (with double adjustment controls) initiates the flush start and the duration of flushing. The time period between the end of flushing and the next start is also programmed.

In pressure-differential-controlled systems, the triggering pressure must be adapted to the maximum permissible loss of energy in the filter. The preset high differential initiates the start of flushing, and a timer generally stops the flushing. In these systems, the time period between flushings is not fixed, but rather depends on the amount of the solids in the water and the system flow rate. Flushing intervals can range from a few minutes to several hours. The pressure differentials activating the flush cycle are generally in the range of 5 to 10 psi. The flush duration should be no longer than necessary in order to minimize interruptions to the system and prevent wasted water and flooding in the vicinity of the filter. Care must be given to ensure proper operation of valves controlling self-cleaning filters, since rapid operation (opening or closing) can result in surges that may damage the filter or other components of the system.

Screen Filters

Screen filters are recommended for the removal of fine sand or larger-sized inorganic debris. Unless the filters have automated backflush mechanisms, it is normally not effective to use screen filters for the removal of heavy loads of algae or other

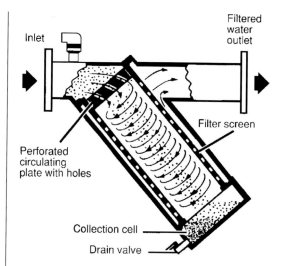

Figure 27-2. Cross section of a typical screen-type filter

organic matter. With heavy filtering loads, manual flush filters may clog rapidly, requiring too frequent cleaning to be practical.

Screen filter bodies can be made of metal or plastic, and the filter is based on the principle of a sieve. The filtration element consists of a perforated screen or mesh through which water flows (Fig. 27-2 and 27-3). The solid particles remain on the upstream side, either adhering to the screen or falling down into a sump. One of the more popular filters for irrigation systems with well supplies is the Thompson filter (Fig. 27-4). This type of filter has a simple design that results in a small pressure drop through the filter when clean (typically 1 psi or less). Water flows through the Thompson filter from bottom to top. As water flows up through the conical-shaped filter screen, any suspended sediment is trapped by the screen and forced downwards. Debris accumulated in the sediment reservoir at the bottom of the filter can be removed by opening the valve on the flush port.

The filtration level depends on the size and shape of openings, which may be round, rectangular, square, or of some other shape (Fig. 27-5). The screen is usually stainless steel or plastic. Sometimes fabric is used to cover perforated screens to allow finer filtration and easier cleaning.

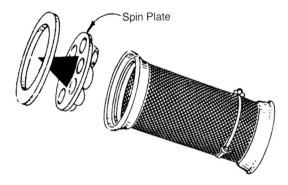

Figure 27-3. Components of a spin-clean screen-type filter.

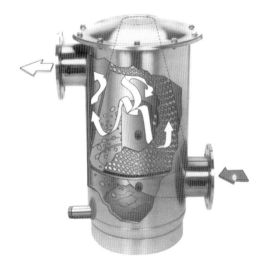

Figure 27-4. Water flow pattern in Thompson screen-type filter.

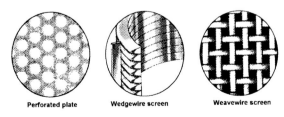

Figure 27-5. Typical types of screen filter elements.

When surface water is used for microirrigation, screens are often used as secondary filters after organic matter has been removed with media filters. In this capacity, they prevent washed-out media from entering the irrigation system. When well water is used for irrigation, a screen filter may be used as the primary filter, or it may be secondary to a vortex sand separator, depending on the mineral particle load in the water.

Filtering screens are classified according to the number of openings per inch (mesh), with a standard wire size given for each screen size (Table 27-1). The minimum size of particle retained by a screen filter with a certain mesh can be determined from Table 27-2. Recommended mesh sizes generally range from 50 to 60 mesh for large-orifice microsprinklers to up to 200 mesh for some drip irrigation systems. Water filtered with a 200-mesh screen will contain only very fine sand or smaller particles (Table 27-2).

A comparison of the relative openings for various stainless steel wire screen mesh sizes is given in Figs. 27-6 through 27-15. Various manufacturers may use different wire sizes to achieve the same mesh sizes. Thus, the percentage of open area may vary slightly from one manufacturer to another for the same mesh size.

Generally, it is recommended to remove particles down to a size that is 1/7th of the emitter's passageway so that grouping and bridging of particles will not cause clogging (Fig. 27-16). The maximum tolerable particle size for a given emitting device should be provided by the manufacturer. Organic particles with a density approaching the density of water tend to group and bridge more easily. Particles heavier than water—typically mineral particles—may settle and collect in the low flow zones of the irrigation system.

When estimating the appropriate size of filter for a specific application, one should consider the quality of water needed, volume of water required to be passed through the filter between consecutive cleanings, filtration area, and allowable pressure

Table 27-1. Representative screen mesh numbers and the corresponding standard opening size equivalents.

Mesh size	Opening size			Mesh size	Opening size		
	microns	inches	mm		microns	inches	mm
4	4760	0.1870	4.760	50	297	0.0117	0.297
6	3360	0.1320	3.360	60	250	0.0098	0.250
8	2380	0.0937	2.380	70	210	0.0083	0.210
10	1880	0.0787	1.880	80	177	0.0070	0.177
12	1410	0.0661	1.410	100	149	0.0059	0.149
14	1190	0.0555	1.190	120	125	0.0049	0.125
18	1000	0.0394	1.000	140	105	0.0041	0.104
20	841	0.0331	0.841	170	88	0.0035	0.088
24	813	0.0320	0.813	200	74	0.0029	0.074
25	707	0.0280	0.707	230	63	0.0024	0.063
30	595	0.0250	0.595	270	53	0.0021	0.053
35	500	0.0197	0.500	325	44	0.0017	0.044
40	420	0.0165	0.420	400	37	0.0015	0.037
45	350	0.0138	0.350	550	25	0.0009	0.025

The conversion values in Table 27-1 mesh sizes depend on the size of wire from which the screen is manufactured. The mesh designation is the number of wires in 1 lineal inch of screen. If different size wire is used, screens can have the same mesh designation but not have the same openings between wires. Therefore, the micron size (a measurement of opening size) is a more accurate indication of filtering capacity.

Table 27-2. Soil particle size classification and corresponding screen number.

Soil Classification	Particle Size (mm)	Screen size		
		microns	inches	mesh
Very coarse sand	1.00 - 2.00	1000 - 2000	0.0393 - 0.0786	18 - 10
Coarse sand	0.50 - 1.00	500 - 1000	0.0197 - 0.0393	35 - 18
Medium sand	0.25 - 0.50	250 - 500	0.0098 - 0.0197	60 - 35
Fine sand	0.10 - 0.25	100 - 250	0.0039 - 0.0197	160 - 60
Very fine sand	0.05 - 0.10	50 - 100	0.0020 - 0.0039	270 -160
Silt	0.002 - 0.05	2 - 50	0.00008 - 0.0020	— *
Clay	< 0.002	<2	< 0.00008	— *

*Screens are not normally used to remove particles of these sizes.

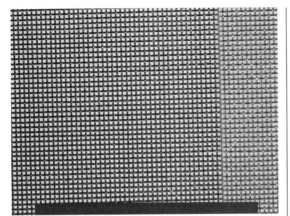

Figure 27-6. 200-mesh (74 micron) screen with 33.6% open area and 0.0021-inch wire diameter (bar length is equivalent to 0.25 inch).

Figure 27-9. 100-mesh (140 micron) screen with 30.3% open area and 0.0045-inch wire diameter (bar length is equivalent to 0.25 inch).

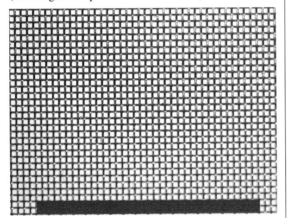

Figure 27-7. 150-mesh (104 micron) screen with 37.4% open area and 0.0026-inch wire diameter (bar length is equivalent to 0.25 inch).

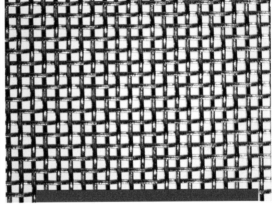

Figure 27-10. 80-mesh (178 micron) screen with 31.4% open area and 0.0055-inch wire diameter (bar length is equivalent to 0.25 inch).

Figure 27-8. 120-mesh (117 micron) screen with 30.7% open area and 0.0037-inch wire diameter (bar length is equivalent to 0.25 inch).

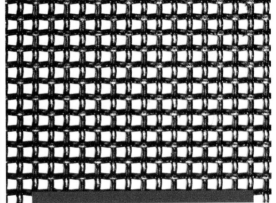

Figure 27-11. 60-mesh (234 micron) screen with 30.5% open area and 0.0075-inch wire diameter (bar length is equivalent to 0.25 inch).

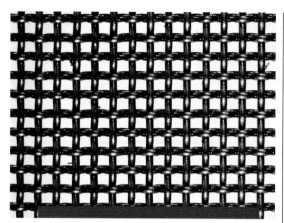

Figure 27-12. 50-mesh (279 micron) screen with 30.3% open area and 0.009-inch wire diameter (bar length is equivalent to 0.25 inch).

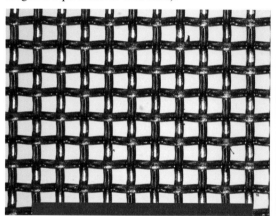

Figure 27-13. 40-mesh (381 micron) screen with 36.0% open area and 0.010-inch wire diameter (bar length is equivalent to 0.25 inch).

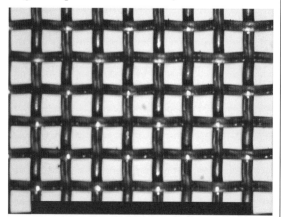

Figure 27-14. 30-mesh (516 micron) screen with 37.1% open area and 0.013-inch wire diameter (bar length is equivalent to 0.25 inch).

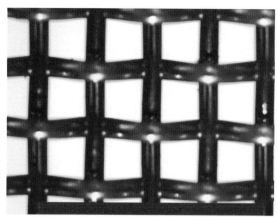

Figure 27-15. 16-mesh (1003 micron) screen with 39.9% open area and 0.023-inch wire diameter (bar length is equivalent to 0.25 inch).

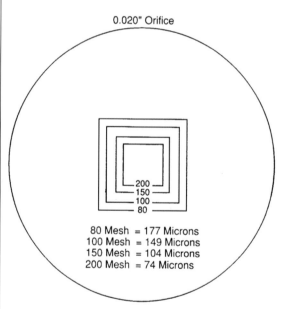

Figure 27-16. Relationship between 0.020-inch orifice diameter and various mesh sizes.

drop through the filter. A screen filter can handle a large range of discharges. However, discharges that are large in relation to the filtering surfaces will result in greater pressure losses, shorter life of the filter, and the requirement for frequent cleaning.

The size of the filter is specified by its effective area, which is the area of the openings in the screen. The size is specified in relation to the cross-sectional area of the main pipe. A desirable ratio is to have twice the area of screen openings as the cross-sectional area of the pipe. The mesh size of the filter (opening size) will depend on the smallest particle size to be removed from the irrigation water. If the water contains large amounts of sand or other inorganic contaminants, it is advisable to use vortex sand separators before the screen filters. Vortex separators remove heavier particles before they reach the fine mesh screen. The separate particles collect at the bottom of the filter where they can be periodically flushed away.

Disc Filters

The filtration element in disc filters consists of a stack of plastic or epoxy-coated metal discs placed on a telescopic shaft inside the housing (Fig. 27-17). When these rings are stacked tightly together, they form a cylindrical filtering body that has some resemblance to a deep tubular screen. Water flows through the discs from the outside inwards, along the radii of the discs. Particles suspended in the water are trapped in the grooves of the discs, and clean water is collected in the center of the discs.

Disc elements can be manually or automatically cleaned. During manual cleaning, the housing is removed, the telescopic shaft is expanded, and the compressed discs separate for easy cleaning. The filter is normally cleaned by rinsing with a water hose. The telescopic shaft prevents the individual discs from falling off the shaft during rinsing. On automatic flushing systems, the backflush is triggered by a preset pressure differential. This

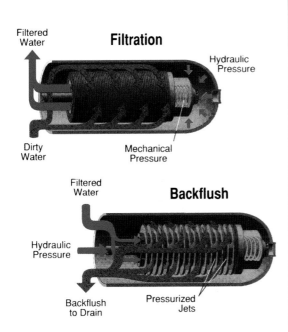

Figure 27-17. Cross-sectional view of disc filter during normal filtering and backflushing cycles.

Figure 27-18. Grooves in disc filters demonstrating the filtering process as water moves from outside disc to interior of filter.

pressure differential opens the exhaust valve and water flows backwards through the discs, removing trapped particles from the grooves.

The grade of filtration depends upon the number of grooves in the individual grooved rings. Typical filters are equivalent to 40-, 80-, 120-, 140-, and 200-mesh screens. Rings are commonly color coded for identification purposes. The grooves of each disc are randomly positioned relative to the adjacent discs, so a matrix of various screens is formed (Fig. 27-18). Water passing through the cylinder will encounter 12 to 32 groove intersections, depending upon the disc mesh size.

Disc filters are often a good filter choice for small flow rates (less than 25 gpm) because they have a larger filtration capacity than screens, and media tanks are more expensive for low flow rates. The small disc units are often composed of a single filter element that must be disassembled and cleaned manually.

Larger units are available in batteries of parallel cylinders (Fig. 27-19). Automatic backflushing may require a higher pressure (40 to 50 psi range) than is available from the irrigation pump, so a special booster pump to supply backflush water may be needed. Also, a small disc control filter is used for the control mechanism and must be manually cleaned. Typical differential pressures at the

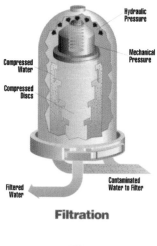

Filtration

Automatic Backflush

Figure 27-20. Water flow paths in a disc filter during normal operation and backflush cycles.

time of backflushing are 5 to 10 psi. When the disc filter is clean, the pressure differential should be less than 5 psi if the units are sized properly.

During backflushing, the discs are separated. In the case of automatic flushing (Fig. 27-20), multijet nozzles provide tangential spray on the loosened discs. As the discs spin from the spray, the retained debris is flushed outward. The volume of water used for backflushing is typically much less than that used for media tank filtration. Disc filters are generally not recommended for removal of large sand loads, as sand tends to become lodged between the discs during backflushing. Also, problems have been reported when they are used with stringy algae.

Figure 27-19. Battery of automatic backflush filters.

Figure 27-21. Electrically operated self-cleaning screen filter.

Figure 27-22. Components of self-cleaning screen filter.

The rules for pretreatment and adjustment of frequency, duration, flow rate, and dwell time are similar to those found for media tanks. Typical disc filters are made almost entirely of plastic, so corrosion is not a major concern. Periodic lubrication is required on some filter parts.

Self-Cleaning Screen Filters

Some tubular screen filters have a rotating mechanism inside them that can "scrub" the contaminants off the surface of the screen when it gets dirty (Figs. 27-21 and 27-22). There are several different designs for these types of filters. Self-cleaning screens have been available for many years, but have seen limited use for citrus irrigation. The new self-cleaning designs are generally superior to previous ones and can be used effectively with many surface water sources. Some designs use a rotating cleaning (flushing) mechanism, which is hydraulically driven, and other designs use an electric motor to rotate the cleaning mechanism. Advantages of the self-cleaning screen filters include a small "footprint" (i.e., they do not require a large concrete pad for installation) and a very efficient backflush (i.e., little water is needed for cleaning). They have automatic flushing that is triggered by a differential pressure switch. These filters have been used with surface water containing high organic loads in Florida.

In the hydraulically driven self-cleaning filter (Fig. 27-22), water enters via the inlet and is prefiltered through a coarse screen and then by fine screen. Water continues through the main body of the filter and exits through the outlet. As dirt accumulates on the inner face of the fine screen, the pressure differential between the inlet and outlet increases. At a preset pressure differential, the rinse controller opens a flushing valve and creates a strong backflush stream through the flushing port. The backflushing action creates a suction effect at the nozzles, which draws accumulated debris off the face of the fine screen directly in front of the nozzle openings. As the backflush water flows through the collector pipe via the hydraulic rotor to the flushing valve, the dirt collection assembly begins to spin. As the assembly spins, the collector assembly moves laterally. Thus, there is a combination spinning and lateral motion for the suction nozzles. This movement causes the nozzles to sweep over the entire fine screen area and remove accumulated debris from the screen. When the flushing collector assembly reaches the end of the screen, the cycle reverses, and the unit returns to its original position. Typically, the complete cycle takes only 10 to 15 seconds, and the backflush volume is 20 to 50 gallons.

Media Filters

Media filters are well suited for removal of either organic or inorganic particles. Due to their

Figure 27-23. Irrigation pump station with media filters.

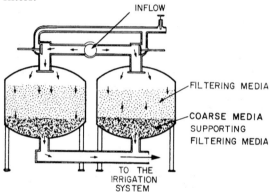

INFLOW

FILTERING MEDIA

COARSE MEDIA
SUPPORTING
FILTERING MEDIA

TO THE
IRRIGATION
SYSTEM

Figure 27-24. Typical pressure-type media filters (installed in parallel setup).

three-dimensional nature, media filters have the ability to entrap large amounts of contaminants. They do not seal off as easily, and therefore will not clog as often as screen filters. Media filters used in microirrigation systems are the pressure type (Fig. 27-23). They consist of fine gravel and sand of selected sizes placed in pressurized tanks (Fig. 27-24). The main body of the tank contains sand, which is the active filtering ingredient. The sand is placed on top of a thin layer of gravel that separates it from an outlet screen.

Sharp-edged sand or crushed rock is recommended for the filtering media because the edges catch soft organic tissue such as algae. Crushed granite or silica graded into specific sizes for a particular system are commonly used media. The size of particles is very important. Particles that are too coarse result in poor filtration, while particles that are too fine result in frequent backwashing of the filter.

Two factors describe the media used in the filter: the uniformity coefficient and the mean effective size. The uniformity coefficient is a measure of the range of sand sizes within a particular grade of sand. It is desirable to keep the sizes of sand particles as uniform as possible. The uniform size of filtering media assures better control of filtration, since only particles large enough to clog the emitters are retained by the filter. The uniformity coefficient is calculated as the ratio of the size of screen opening that will pass 60% of the filter sand to the size of the screen opening that will pass 10% of the same sand. For irrigation purposes a uniformity coefficient of 1.5 is adequate.

The mean effective sand size is the size of the screen opening that will pass 10% of the sand sample. It is a measure of the minimum sand size in the grade and therefore an indicator of the particle size that will be removed by the media. The quality of filtration increases with a smaller effective size of filtering media. Some examples of commercially available grades of sand media are shown in Table 27-3.

Table 27-3. Sand media grades and corresponding filtration quality.

Media number	Media material	Effective sand size		Filtration quality
		(mm)	(inches)	(mesh)
8	crushed granite	1.50	0.059	100-140
11	crushed granite	0.78	0.031	140-200
16	crushed silica	0.66	0.026	140-200
20	crushed silica	0.46	0.018	200-230
30	crushed silica	0.34	0.013	230-400

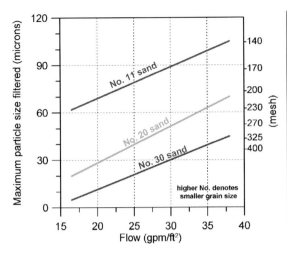

Figure 27-25. Effects of media size on filtering capacity.

Various kinds and sizes of sand media are available. For surface water that contains a high level of algae and organic materials, 70% to 75% of the sand particles should be 0.05 to 0.09 inch. The other sand should not be larger than 0.10 inch or smaller than 0.03 inch. For water with high levels of iron, the size of the sand media should be smaller, with 70% to 75% of the sand in the range of 0.03 to 0.07 inch, and the rest no smaller than 0.02 inch or larger than 0.08 inch.

Flow Through Media Filters
The design flow capacity of a media filter is expressed in gallons per minute of flow per square foot of surface area of the media. The filter should be sized so it can handle the poorest water quality at a given site, and still provide the required flow functioning for the irrigation system. The quality of irrigation water before filtration may vary with seasons and weather conditions. This is especially true for surface water, but can also apply to well water in some cases.

Media filters normally operate in the range of 15 to 30 gpm per ft² of filtration area. The standard 48-inch media tank provides 12.5 ft² of filtering area. Therefore, each 48-inch tank is able to accommodate 188 to 375 gpm of flow. A typical design flow rate for microirrigation systems is about 25 gpm/ft². Most systems are designed to

Table 27-4. Design flow rates through media filters in gallons per minute (gpm) per tank.

Flow capacity gpm/ft²	Tank diameter (inches)				
	18	24	30	36	48
15	27	47	74	106	189
20	35	63	98	141	251
25	44	79	123	177	314
30	53	94	147	212	377

allow additional filters to be added in parallel to increase the filtering capacity. Table 27-4 gives data on media filter flow rates for a range of design flow rates per square foot of sand surface.

Filtration Effectiveness
The effectiveness of a filter is a measure of its ability to remove particles of a certain size. As shown in Fig. 27-25, the effectiveness of filtration increases with a decreasing size of filtering media grain size (indicated by a larger designation number). However, smaller media size requires more frequent cleaning. Filtration effectiveness is also inversely proportional to the flow rate through the filter. The higher the flow rate, the lower the effectiveness of the media filter. For example, Fig. 27-25 shows that a filter with No. 11 media at a flow rate of 23 gpm/ft² will effectively remove particles of 75 microns (200 mesh). However, the increase in flow rate to 35 gpm/ft² will decrease the effectiveness of filtration to 100 micron-sized (140-mesh) particles.

Figure 27-26 shows the increase in pressure differential with time for a typical system. Notice that the pressure loss through a clean filter varies between 3 and 8 psi depending on the size of the media and flow rate used. The pressure differential increases with time as contaminants accumulate and partially plug the filter. Figure 27-26 was developed from field data at one location, hence for a particular quality of water, and should not be used as a guide for filter selection. The slope of all three lines presented on the graph will change, depending on the quality of water. Therefore, the filter manufacturer's specifications and water quality

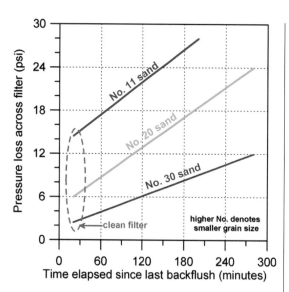

Figure 27-26. Typical effects of media size on pressure loss following backflushing.

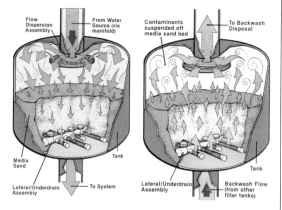

Figure 27-27. Water flow in media during normal filtration and during the backflush cycle.

samples should be used to select a filter for a specific application.

Cleaning of Media Filters

Media filters are cleaned by backwashing (Fig. 27-27). This operation consists of reversing the direction of water flow in the tank. Clean water is usually supplied from the second tank. The upward flow saturates the media and flushes collected contaminants. The backflush water is discharged and does not enter the irrigation system. The backwash flow must be carefully adjusted in order to provide sufficient cleaning without accidental removal of

Table 27-5. Recommended backwash flow rates for different designation numbers of filtering media in gallons per minute (gpm) per tank.

Media designation number	Tank diameter (inches)				
	18	24	30	36	48
8	51	91	141	201	360
11	38	68	106	150	270
16	32	57	89	126	225
20	26	48	74	105	188

the media. If flows larger than those recommended are used to backflush the media, sand can be flushed from the filter by the backflow water. Flow rates that are normally sufficient for backwashing are given in Table 27-5 based on media size and tank diameter.

Most media filters are backflushed at prescheduled time intervals or by using automatic devices based on the pressure loss across the filter. For systems with low-quality water requiring frequent backflushing, automatic cleaning is necessary to avoid problems. Since the filter is backflushed every time the pressure differential exceeds a predetermined value, large pressure drops in the irrigation system are avoided, maintaining the irrigation system's uniformity and efficiency. The pressure differential that triggers backflushing depends on the pressure required for proper functioning of the irrigation system (typically around 10 psi). Automatic backflush systems eliminate sudden changes in water quality, which can create problems if a filter is washed only at regular intervals. This is especially important in systems using surface water supplies with changing contamination levels.

The irrigation system pump should be able to supply enough pressure to compensate for the pressure drop through the filter just before the filter is washed. Pump capacity must also be adequate to supply enough pressure and flow rate to flush the filter. In large irrigation systems, several smaller tanks are typically used instead of a few large ones to minimize the effects of the high flows required to backwash large tanks.

Filtering Standards

The quality of water necessary for successful operation of an irrigation system depends on the type of emitter used. Emitter manufacturers' recommendations should be used to determine filtration requirements. Manufacturers' specifications may be given in terms of mesh equivalent or opening size with dimensions of inches, millimeters, or microns. In the absence of manufacturers' recommendations, a rule of thumb is to filter the water to a size that is 1/7th the diameter of the emitter orifice. The choice of media size will depend on the emitter passage size and flow rate, water quality and economic considerations.

Example:
Determine the filtration requirement for emitters with 0.040-orifice diameters.

Filtration diameter = 1/7 x orifice diameter
  = 1/7 x 0.040 = 0.0059 inch

From Table 27-1, 100 mesh equivalent filtration is required.

From Table 27-3, No. 8 crushed granite is recommended.

System Start-up Procedure

When starting up a media filter system after periods of nonuse, the following procedures should be used:

1. Check power to the controller. Be sure voltage is appropriate.

2. All grooved couplers should be fully tightened.

3. Check manway covers to be sure gaskets are in place and lids are properly tightened.

4. Power to the controller should be in the "OFF" position.

5. Open field valves to allow passage of water from filters to field.

6. Open the backwash manifold throttle valve to full.

7. Place one backwash valve in manual backwash mode by placing the manual solenoid value override screw in the override position.

Over pressurization can cause serious damage to any hydraulic system. The maximum pressure rating of the pump supplying the filtration unit should be checked. Several conditions during start-up and operation can stop water flow, producing a full head of static pressure within the filter unit. To protect against damage from overpressurization, a pressure relief valve with sufficient discharge capacity should be installed.

Before start-up, the start-up check list should be completely reviewed. Proceed only after pressure capacities of the pumping and filter systems are known. Initial cleaning can be expedited by closing the outlet manifold throttle valve. Begin by starting the pump (or opening the valve supplying pressurized water to filters) and set the first filter in the manual backwash mode. A backwash cycle should begin immediately. Wait approximately two minutes, then rotate to a second tank by closing the override screw on the first solenoid and opening the flush valve on the second filter. Repeat this process until all system tanks have been backwashed two or three times. Backwash only one tank at a time! Do not open the second solenoid until the first solenoid is closed. All tanks normally contain debris prior to initial backwash. Flow through the backwash manifold will be sluggish until some cleaning occurs. If after two or three backwash rotations the tanks are not clean, continue with additional cycles until cleaning occurs.

Open the regulating valve and allow clean water to enter the irrigation system. Once the system is fully pressurized, the pressure gauges will indicate a differential across the filter. Pressure gauge readings are not accurate until the system is fully operational. Readings are accurate only after the field regulators are set properly, the entire system is flushed, field leaks are repaired, and all emitters are operating properly. When the irrigation system is fully operational, final adjustments to the filter should be made.

Preset the controller for the recommended backwash and dwell times (typically 90 sec. backwash and 30 sec. dwell time). Turn on controller power and initiate a flush cycle. (See instructions accompanying the controller.) Adjust the backwash throttle valve as follows: during the flush cycle, check the backwash water through the view tube, or catch a backwash sample using a fine sieve. Adjust the backwash throttle valve downward until only a trace of filter media is being lost through the backwash outlet. A small loss of media is considered optimum. If no media is being lost and the backwash valve is fully open, use the outlet manifold throttle valve to direct additional flow away from the field for backwash. Adjust backwash flow until a small loss of media begins to occur. In systems where backwash flow is inadequate, the throttle valve must be adjusted in this manner each time a backwash cycle is initiated.

Experience is the best teacher for establishing a program of backwash cycles suitable for a particular filtration application. The source water quality most directly determines the frequency of backwash. A backwash cycle should be initiated when the pressure differential increases 4 to 6 psi above the optimum system differential (the differential reading when all tanks are clean). Automatic systems are normally shipped with the pressure differential switch preset at about 10 psi. If the desired differential range is not met by this setting, an adjustment can be made.

The pressure differential (PD) switch measures differential pressure between incoming (high) water pressure and outgoing (low) water pressure. When the difference between these pressures increases to a preset level, the switch signals the controller to begin a backwash cycle. To check any adjustment, manually initiate a backwash cycle. Observe the gauges until the switch automatically initiates a second backwash cycle. If backwash occurs at the desired PD as indicated by the gauges, the setting is correct. If not, readjust and repeat the test until the desired PD is achieved. Once the PD adjustment is made, the remaining adjustments are relatively simple.

Backwash flush time is adjusted by the filter controller. Start with a 90-second setting. Observe water quality at the beginning and end of a backwash cycle through the view tube or at the outlet of the backwash manifold. If the backwash water is still dirty near the end of the cycle, increase flush time. If the water appears clean several seconds before the end of the cycle, decrease flush time. In addition to the water cleanliness, check the pressure differential after backwash. If the PD has returned to the optimum level, the flush duration is adequate. If not, increase the backwash duration until optimum differential is achieved.

As the backwash valve closes, water from the backwash manifold is mixed with incoming water from the inlet manifold. Generally, during the last 20 to 30 seconds of the backwash cycle, the flow will be diluted. Therefore, it should not be considered representative of actual backwash water quality.

The dwell between tanks enables the irrigation system pressure to recover before the next tank flushes. This delay provides the system with a full head of pressure with which to initiate flushing of the next tank. To adjust the dwell time, observe the backwash cycle of the last tank in the battery. Calculate the length of time from the end of the cycle (when water stops flowing from the backwash manifold) until maximum system pressure is regained. This is the approximate time that should be set as dwell time.

The periodic start function is used when water quality does not require a backwash cycle for several hours. Set the periodic start to initiate a flush cycle at least every 6 hours. In situations where pressure differential operation is not desired, the PD switch can be disconnected. Backwash initiation then will depend entirely on the periodic start function, which can be set on the controller.

Maintenance
Like any mechanical device, media filters require periodic maintenance to ensure proper and efficient operation. Maintenance should be scheduled during periods when adequate rainfall allows the

system to be fully inspected and any problems repaired. Table 27-6 lists some of the causes and remedies for common problems. On a monthly basis, the following should be performed:

- Check all system fittings for leaks and make repairs as needed. Leaks in the 1/4-inch poly tubing or fittings can cause unneeded backwash cycles or prevent needed cycles.

- Remove, clean, and reinstall the screen in the end of the inlet manifold.

- Test backwash cycle and PD gauges. Initiate a backwash cycle by pushing the manual start button in the clock. Check to see that all tanks backwash and that no tanks continue flushing after the cycle is complete. Check PD reading after backwash, and make any required flush timing adjustments. Check pressure gauges during operation to see that the readings are reasonable and that gauges return to "0" when the system is off. Replace the pressure gauges if necessary.

- Check manual backwash throttle valve for proper setting. If it has been tampered with, readjust to the proper setting.

- Check sand level in tanks.

- To keep tanks looking good in corrosive atmospheres, wash with soap and water, dry, and coat with a light oil.

Shutdown Procedure

- Manually initiate a normal backwash cycle. Do not add cleaners or additives of any kind to the water. Shut down upon completion of cycle.

- Disconnect power to controller.

- Shut the outlet control valve and drain filter tanks.

- Open manways and check sand level. Add media, if required, and replace covers.

- Wash stainless tanks with soap and water. Rinse and allow to dry. Spray coating of a light oil on all stainless surfaces.

- Touch up with paint any carbon steel part showing rust (legs, controller box, etc.).

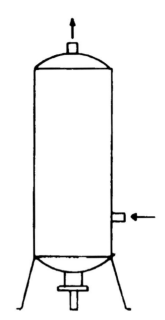

Figure 27-28. Sedimentation tank.

Kinetic Filtration (Sand Separators)

Kinetic filters are designed to separate out solid particles that have a specific weight greater than water. Therefore, they are used only for water that is carrying inorganic matter such as sand. Kinetic filters will not be effective in filtering out particles that can float in water, such as algae, leaves, or colloidal matter.

The simplest form of kinetic filter is the sedimentation tank (Fig. 27-28). Sedimentation tanks consist of a vertical cylinder with an inlet connection about a third of the way up and the outlet at the top. The diameter of the tank needs to be significantly larger than the adjoining pipe so that the velocity of water entering the tank decreases considerably. When the water enters the sedimentation tank, debris suspended in the turbulent, high-velocity flow of the pipeline settles out. The debris is removed from the sedimentation tank through a drain at the bottom. This simple method of filtration provides clean water, but flow rates are generally low. However, it can be an effective means of providing clean water for hydraulic-actuated control devices. In some models of sedimentation tanks, a screen is used to ensure better filtration

Table 27-6. Symptoms, causes, and solutions for media filter problems.

Symptoms	Cause	Solutions
Poor filtration	Filtering media too thin	Add more media to the recommended level
	Wrong media	Exchange for correct media
	High pressure differential forcing contaminants through the filter	More frequent backflushing
Poor filtration with cone forming in the middle of the sand layer	Too high flow rate	Add more tanks or reduce the flow rate per tank
	Air in tank causing bed disruption	Install automatic or manual air bleeds
Increased frequency of backwash	Change in quality of water source	More filters required
	Filtering media too thin	Add media to the recommended level
	Insufficient backwashing	Increase duration or flow of backwashing
Constant high pressure differential	Filters sealed - not enough water for backwashing tanks until clean	Remover the cover, scrape top layer, flush
	Filtering media too thin	Add media to the recommended level
Media downstream (in irrigation)	Too high flow	Reduce flow rate
	Wrong media	Exchange for correct media
	Air in the tanks, sand disrupted	Install air bleeds, manual or automatic

efficiency. Due to the low velocity and low energy of the particles in the vessel, their impact on the screen is only slight, and the screen remains clean and does not clog.

An improved and more effective kinetic filtration method is the hydrocyclone (Fig. 27-29). This device consists of a body that is cylindrical at the top and tapers down, funnel-like, at the bottom. Water enters the cylindrical area tangentially at the top and flows around the walls of the cylinder and the funnel. After reaching the bottom of the funnel, the jet of water rises in a vertical column through the center of the body, exiting via an outlet on the top. The heavy particles in the water stream are hurled by the centrifugal action to the extreme periphery of the body. They slide along the walls of the funnel into a closed container at the bottom, from where they are periodically removed. The tapered shape of the hydrocyclone body enhances the water velocity along its helical path downward. This increases the centrifugal forces and the efficiency of separating heavier particles. The maximum efficiency of the hydrocyclone is obtained at a pressure differential of about 8 to 10 psi between the inlet and the outlet. The ratio of the vessel size to the flow rate should be chosen accordingly.

Centrifugal sand separators are used to remove sand and other particles that are heavier than water.

Figure 27-29. Cross-sectional view of hydrocyclone sand separator.

Figure 27-30. Slotted pipe intake screen system.

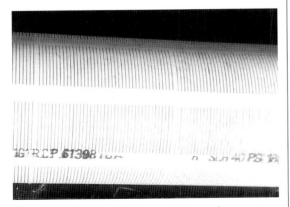

Figure 27-31. Slotted pipe used for surface water intake screen.

Sand separators may be installed on the suction side of pumps to reduce pump wear. It is important to design sand separators correctly for a particular installation. The operation of the separator depends on centrifugal forces within a vortex created by incoming flow. In addition, there is usually a significant head loss (8 to 12 psi) across the sand separator. Therefore, the separator must be matched carefully to the pump to ensure the proper flow and pressure through the separator. Centrifugal sand separators are normally self-cleaning and require minimal maintenance. Generally, under normal operation sand separators will remove particles down to about 75 microns (200 mesh). However, they cannot remove organic or other lightweight materials from the water. The hydrocyclone is often used for separating sand from water in deep-well pumping stations.

Intake Screens

If water is pumped from a surface source (lake, pond, stream, river, irrigation canal, or pit), aquatic plants, fish, and animals should be kept away from the intake pipes. This can be accomplished with an intake screen (Fig. 27-30). The efficiency of pumping will also be reduced by debris in the water entering the suction pipe. The problem can be solved by using proper selection of intake screens. The design of the screen must allow for the separation of the debris from the water without pulling the debris against the screen.

Slotted pipe

Many citrus irrigation pump stations utilizing surface water use slotted pipe for the intake mechanism (Fig. 27-31). These systems are relatively low in cost and can be efficient to use if designed and maintained properly. Intakes should be designed to limit entrance velocity into the screen to 0.5 fps or less. Low entrance velocities will allow sediment to settle out and not be taken into the system. As entrance velocities increase, the amount of sediment, algae, aquatic weeds, etc. that moves toward the screen increases. When these particles clog a portion of the intake screen, the velocity through the remaining open slots must increases to maintain the system flow rate. The increased water velocity permits more and larger suspended materials to be pulled toward the screen. As a result, the amount of material needed to be filtered out increases rapidly and the filter clogging rate is greatly accelerated. Table 27-7 lists the open area per lineal foot of screen for one manufacturer.

Example:
Determine the minimum intake length for a 2500 gpm system if a 10-inch, 20-slot PVC screen is to be used and entrance velocity is to be limited to 0.5 fps.

From Table 27-7, the 10-inch, 20-slot screen has 72 in.² open area per foot of screen

3.1 x 0.5 ft/sec x 72 in.²/ft = 112 gpm/ft

2500 gpm / 112 gpm/ft = 23.3 ft
Therefore, use a minimum of 23 feet of 10-inch, 20-slot pipe

The minimum acceptable length of 10-inch, 20-slot screen would be 23 feet. Under ideal conditions with no clogging, the entrance velocity would be 0.5 fps. If the water source has potential for significant suspended solids or organic materials, the screen length should be increased to reduce the potential for clogging.

Self-Cleaning Intake Screens
Rotating screen intake filters (Figs. 27-32 and 27-33) are a type of self-cleaning intake screen. These intakes have a screen that is continuously cleaned by a jet of water from the pump. These types of self-cleaning intake screens can be used where there is a need to remove foreign materials that could block water flow, damage or plug water distribution equipment.

The self-propelled rotating screen has an internal backwash cleaning system that keeps the water intake area free of floating and suspended materials. The cleaning jets are positioned at an angle to the screen so that the force of the jetted water both rotates the intake screen and blows accumulated debris off the screen. Thus, a smaller intake screen can be used since it is continually cleaned.

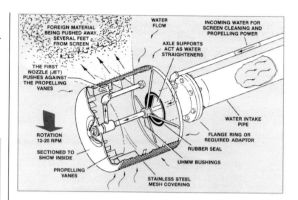

Figure 27-32. Components of self-propelled rotating screen intake filter.

Figure 27-33. Dual rotating intake screen.

Backwash cleaning is provided by a water line from the pump discharge to a stationary spray bar with a series of nozzles that are located inside the rotating screen. These nozzles spray water onto the inside of the screen, thus pushing away any foreign matter. Oftentimes an auxiliary pump is need to supply the 60 to 80 psi that is needed to effectively clean the screens.

The water spray in the self-cleaning rotating intake screen also provides the power for the screen rotation. The spray pushes against built-in propelling vanes, causing it to rotate. An additional benefit is that the rotation of screen breaks up vortex formation. As a result, very little submergence is required (typically only 6 to 12 inches). Rotating intake screens should be designed so that the intake velocity is less than 0.5 fps.

Table 27-7. Typical open area for slotted PVC intake screens. The maximum transmitting capacity (gpm) per foot of screen under ideal conditions can be calculated as 3.1 times the table value (open area in.^{2}) multiplied by the desired entrance velocity (fps).

Screen Diameter (inches)	Intake area (square inches per lineal foot of screen)					
	10-slot (0.010 in.)	20-slot (0.020 in.)	40-slot (0.040 in.)	60-slot (0.060 in.)	80-slot (0.080 in.)	100-slot (0.100 in.)
1 1/4	9	16	26	33	38	41
1 1/2	13	22	36	45	51	56
2	14	25	40	51	57	64
2 1/2	17	31	49	61	70	77
3	20	34	55	71	81	89
4	25	44	72	90	102	112
5	30	53	85	106	100	112
6	25	45	77	100	118	131
8	33	59	100	130	153	171
10	41	72	122	135	162	186
12	37	69	99	135	164	189
14	41	76	107	148	180	207
16	36	68	124	169	206	238
18	40	76	137	187	223	263
20	54	101	119	165	206	240
24	65	123	144	199	248	290

Notes:

There are no standards for open area per foot of screen. Therefore, various manufacturers may have different values. This table should be used as a guide, and specific values should be obtained from the pipe manufacturer if possible. The values may differ slightly from these figures when heavier pipe is used for extra strong construction.

Under ideal conditions the entrance velocity should be 0.5 fps or less. If the water source has potential for significant suspended solids or organic materials, the entrance velocity should be lowered to 0.1 to 0.2 fps.

Slot size refers to width of slot openings (in thousandths of an inch).

by Brian Boman

Valves are used for on-off control, regulation of the flow rate through the system, and prevention of backflow. They can also be used for pressure relief or as safety devices. In general, valves vary from simple manual on-off devices to sophisticated control mechanisms that act as metering instruments and deliver predetermined amounts of water to the system. Many types of valves have significantly greater head loss within the valve than through equivalent lengths of straight pipe. Most manufacturers specify a head loss coefficient (K) for their valves. The magnitude of the dimensionless K coefficient depends on the type of valve and the way it is constructed, and not so much on the size of the valve. Valves with smaller K coefficients will have less head loss.

On-Off Control

For normal on-off control, the best choices are gate or ball valves. The on-off service valves function by sliding or turning a flat, cylindrical or spherical flow control element over an orifice. In the fully open position, the passageway through a gate or ball valve is unrestricted. As a result, there is low pressure loss through these types of valves.

Ball Valves

Ball valves (Fig. 28-1) are often used on small diameter pipe and tubing in irrigation systems. They have a ball-shape flow control element that allows full flow to closure regulation in a 90% turn. Ball valves are full-flow units and create minimum pressure loss through the valve when fully open. Ball valves do not have linear flow characteristics (i.e., 30% flow at 30% open) when partially closed. As a result, they are difficult to use for flow regulation. Partial closing of a ball valve can cause rapid wear of internal parts. The biggest advantages of ball valves are their simple construction and compactness. Their major disadvantage is that the rapid shutoff can cause water hammer problems in irrigation systems. Thus, gate valves are typically recommended for large sizes and flow rate. Ball valves are more suitable for small size pipe and low flow rates.

Gate Valves

Gate valves are the most common type of on-off service valve (Fig. 28-2). The flow control element in a gate valve is a disc or wedge attached to the valve stem. There are various designs of these wedges, with a solid wedge being the most common. A solid wedge has the advantage of positive contact with the wedge guides, which reduces chatter when the valve is in a partially closed position. When in the fully open position, the wedge completely clears the flow path, resulting in minimum pressure loss through the valve. Other types of wedges can be used, such as split wedge, double-disc parallel wedge, or a combination of discs joined by a ball and socket that are self-aligning to each of the inclined resting faces.

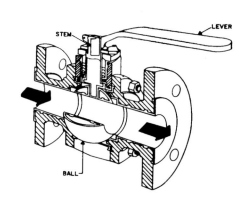

Figure 28-1. Cross section of ball valve.

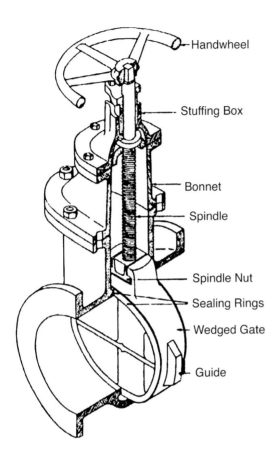

Figure 28-2. Wedged-gate, nonrising spindle gate valve.

- Handwheel
- Stuffing Box
- Bonnet
- Spindle
- Spindle Nut
- Sealing Rings
- Wedged Gate
- Guide

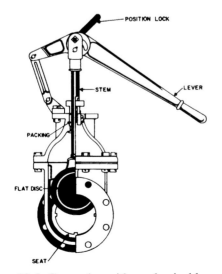

POSITION LOCK

LEVER

STEM

PACKING

FLAT DISC

SEAT

Figure 28-3. Gate valve with mechanical lever.

Gate valves are classified on the basis of the stem movement. The stem is a shaft used to control the position of the wedge. Stems can have either a rising stem or a nonrising stem. The rising-stem valve allows easy determination of how far the valve is open at a given time, since the threaded portion of the stem is exposed. However, sufficient space must be provided to allow for the rise of the stem when the valve is in the fully open position. Gate valves are normally slow acting due to the turning required for changing the position of the control element. For a quick-acting valve, the stem can be constructed in such a way that it slides up or down through the action of a mechanical lever (Fig. 28-3).

Gate valves are probably the oldest and most common valves used in irrigation systems. Gate valves have been fabricated, for the most part, in their present form since the nineteenth century. The valve consists of three main parts: housing, gate, and actuating mechanism. The outer body of the valve (housing) typically has three parts: the body, the bonnet, and the stuffing box assembly. In some models, the stuffing box forms an integral part of the bonnet; in others it is separate and joined to the bonnet by flanges or studs. Gray iron is the most common material used for casting the housing parts. Ductile iron or cast steel are used for valves designed for special working conditions, and the smaller valve sizes are made of brass or bronze. Generally, valves larger than 3 inches have flanged connections while smaller gate valves have pipe threads.

The valve body is symmetrical. Therefore, gate valves can be operated in both flow directions without any difference based on which side is the inlet or the outlet. When the valve is fully open, the head loss coefficient (K) is very low (typically 0.08 to 0.15). This is only slightly greater than the head loss through an equal length of pipe.

Gate valves are generally designed for upright operation. In special cases (such as standpipes), they can also be installed with the spindle horizontal. They are sometimes installed upside down in overhead pipelines. In such positions, friction on the opening mechanism increases, accelerating the wear on the valve and increasing the operating torque requirement. Therefore, installation of gate valves in other than an upright position should be limited to small valve sizes only.

When installing gate valves in pipelines, the large dimensions of the valve and the comparatively thin body of the housing must be taken into account. The pipe on both sides of the valve must be in exact alignment with the valve. Stresses can occur in the valve housing from any inaccuracy in installation, such as unequal tightening of the bolts or distortion of gaskets. Unlike most other valves, gate valves should be installed with the value fully closed. When closed, the tight contact between the sealing surfaces of the gate and the body prevents distortion. Installing an open valve can distort the sealing rings in the body. Any such distortion can result in permanent leakage, which no subsequent tightening of the gate can repair.

The big advantage of the gate valve is that it allows free throughflow of water without constriction and without change of flow direction. The area of water passage is approximately equal to the cross-sectional area of a pipe of the same nominal diameter as the valve. As a result, there is little loss of pressure head. There is also little risk that particulate matter in the water will be caught in the valve body and obstruct the flow. The construction of gate valves allows easy cleaning of the adjoining pipes without dismantling the valves. The main disadvantage of the gate valve is the effort necessary to open and close large-sized valves. Gate valves are also heavier and require more vertical clearance than other types of valves. Another disadvantage of the most common type of gate valve (nonspindle rising type) is that it lacks an effective indicator of gate position.

Gate valves should not be used for throttling or controlling liquid streams. Most of the shutoff using this valve takes place when the valve is almost closed. As a result, the flow control profile is not linear and is difficult to control. At the same time,

the decrease of flow area in partially closed positions significantly increases the liquid velocity. This increase can result in quick erosion of the wedge and destruction of the lower seating surface.

Diaphragm Valves

The most common diaphragm valves are the globe valve, angle valve, and Y-valve. Water flowing through diaphragm valves typically changes direction from 45° to 90° within the body. A sealing element moves parallel to the flow to open or close the orifice located in the region where the change in flow direction occurs. The sealing device is generally a resilient plate on a metal seat. The three basic designs of the diaphragm valve differ in the relative location of the inlet and outlet, and in the angle of flow direction. Although the three designs have distinct differences in appearance, direction of water flow and hydraulic properties, they have identical mechanical performance and operating mechanisms. Diaphragm valves are manufactured in nominal sizes up to 10 inches.

Globe Valve

Globe valves are widely used for flow control. A tapered plug or diaphragm closes onto a seating surface acts as a flow control element in this type of valve (Fig. 28-4). The position of the diaphragm and seating surface results in an approximately linear flow response. The globe valve is always controlled by a rising stem, so sufficient space must be provided for opening the valve. Globe valves have the inlet and outlet along the same center line. Water passing through the valve is deflected twice, by 90° each time. The changes in direction occur as the water enters the orifice and again as it leaves. The design of the valve does not provide an unrestricted flow passage when the valve is fully open, since even at this position the liquid must make two 90° turns when passing through the valve. Therefore, pressure drop in a globe valve is much greater than in a gate valve when fully open (Table 28-1). Globe valves are not recommended for on-off service in irrigation systems. However, if they are used for this purpose, the loss of pressure through the valve must be considered in the design of the system.

The sealing mechanism of the valve moves perpendicular to the axis of the pipe. Globe valves are mostly used in underground or overhead applications where space and access are limited. The two 90° changes in flow direction within the valve result in relatively high head losses. The high pressure drop is the main disadvantage of this type of valve.

Angle Valve

The inlet and outlet are at right angles to each other in an angle valve (Fig. 28-5). The water flow enters the orifice in a straight line and after leaving the orifice, it is deflected 90°. The sealing diaphragm moves parallel to the inlet axis, and at right angles to the outlet. Usually, angle valves are installed with the inlet pipe vertical (water flowing upwards) and the operating mechanism is therefore located on top of the valve. Angle valves are typically used in irrigation systems when right-angle changes in flow direction occur. Since water flowing through

the valve only changes direction 90° once, head losses in angle valves are only about half those of globe valves. Head losses are, however, considerably higher than in gate valves of equivalent nominal size.

Y-Valve

The head loss through a Y-valve (Fig. 28-6) is considerably less than for globe or angle valve designs. Like globe valves, Y-valves have both the inlet and outlet on the same center line. Unlike the globe valve, water flowing through a Y-valve is deflected only 45° to 60°. The sealing mechanism moves at the same angle of deflection to the axis of the pipe. The Y-valve with control solenoids is commonly used in automated irrigation systems. It does require ample space for installation and easy access for operation and maintenance.

The loss of head factor in globe valves, in which the flow is deflected 90° twice, is generally K = 4-5. In a Y-valve with flow deflection of 45°, the loss of head factor can be as low as K = 2.5. A 60° Y-valve normally has a K factor of 3.0 to 3.5.

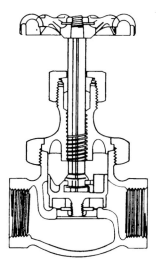

Figure 28-4. Cross section of globe valve.

Table 28-1. Pressure loss in valves (losses given in equivalent length (in feet) of standard steel pipe).

Nominal pipe size (inches)	Globe valve	Angle valve	Gate valve
1/2	17	9	0.4
3/4	22	12	0.5
1	27	15	0.6
1 1/4	38	18	0.8
1 1/2	45	22	1.0
2	58	28	1.2
2 1/2	70	35	1.4
3	90	45	1.8
4	120	60	2.3
6	170	85	3.3

The K factor for angle valves is somewhat intermediate, with values depending largely on the construction of the body and the clearance around the diaphragm and orifice.

The hydrodynamic properties of diaphragm valves make them adaptable for a variety of automatic applications (Fig. 28-7). The pressure differential on the diaphragm and the resulting hydrodynamic force tend to lift the diaphragm along the entire stroke. The area of the diaphragm exposed to the inlet and outlet pressures is constant; therefore, the force on the diaphragm is not directly due to the lift of the diaphragm, but rather depends mostly on the changes in pressure and velocity head. In manually operated diaphragm valves, these lifting forces are absorbed by the threads of the actuating mechanism, which prevent axial movement of the spindle and an uncontrolled lift of the diaphragm.

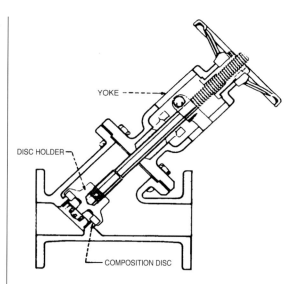

Figure 28-6. Cross section of Y-valve.

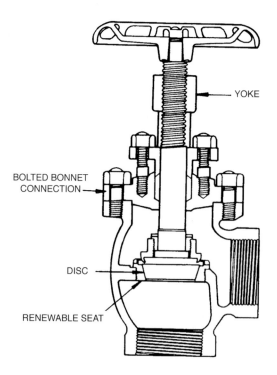

Figure 28-5. Cross section of angle valve.

Figure 28-7. Electric remote control Y-valve (courtesy of Bermad Control Valves, Inc.).

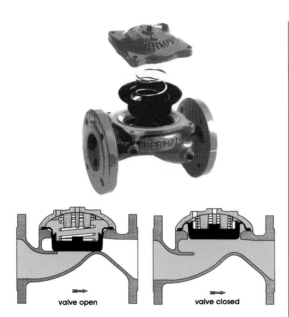

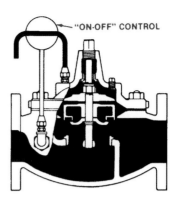

"ON-OFF" CONTROL

Figure 28-10. Hydraulically actuated valve in open position.

Figure 28-8. Hydraulically actuated diaphragm control valve (top) and schematic of operation (bottom).

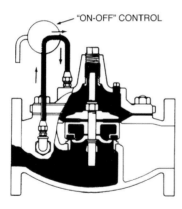

"ON-OFF" CONTROL

Figure 28-9. Hydraulically actuated valve in closed position.

In automatic valves, the smooth spindle allows the diaphragm to lift freely with the movement being counterbalanced by a spring, a weight, or a pressure-controlled piston or diaphragm (Fig. 28-8). Thus, the flow through the valve can be controlled either directly, or by a pilot-operated solenoid system. For automatic operation, a pressure chamber is placed on the body cover in the upper part of the valve. It contains the actuating element (piston or diaphragm) connected to the diaphragm

by a shaft. The chamber space above the piston (or diaphragm) is pressurized, and the space below is open to the inside of the valve or vented to the atmosphere (Fig. 28-9 and 28-10). Pipeline pressure supplies the actuation force. The area of the actuating element is larger than the effective area of the diaphragm. Therefore, the pressure in the chamber exerts a resulting downward force on the moving parts, closing the valve. When the chamber is not pressurized, water in the valve acts on the diaphragm and on the underside of the piston or diaphragm, holding the valve in an open position.

Thus, the valve can be hydraulically operated by pressurizing and depressurizing the chamber with a solenoid system. The hydraulic system consists of small-diameter tubing and valves that enable both direct and remote operation of the valve. Direct actuation can be made with a manual valve or a timer. Remote control actuation is mainly by a solenoid or small hydraulic valves installed in the solenoid system and operated either with electrical current or by pressure changes in the control tubing. Most hydraulically operated valves can also function as pressure or flow regulators.

Butterfly Valves
The rotary butterfly valve (Fig. 28-11) has been used for a long time as a regulating device in water lines. In its simplest form, a butterfly valve

313

consists of a vane inserted into the pipeline and attached to a shaft. The shaft protruding out of the conduit enables the vane to be rotated, which increases or decreases the flow area. For a long time, butterfly valves were used only in large diameter pipe. In these cases, they were used mainly for throttling and flow regulation, not for shutoff purposes. The very long perimeter of a vane makes watertight sealing nearly impossible with metal-to-metal seats. When butterfly valves were used to shut off flow, a certain amount of leakage through the valve was always expected. Sometimes special devices were incorporated into the valve body to drain the anticipated leakage.

The many advantages of the butterfly valve, including the rotary principle and the relative simplicity of construction, led to the development of tight-sealing butterfly valves for water supply in the 1960s. Several variations of the basic model have been developed with a variety of seals. The relatively light and simple construction of the body permits designing butterfly valves of very large diameters. The typical butterfly valve for irrigation systems is rubber lined, the liner being either of the interchangeable spool type, or cast and vulcanized on the inside of the body.

Normally, the direction of installation is of little importance for butterfly valves in irrigation systems. They may be installed with the flow axis in a vertical, horizontal, or inclined position and with the main shaft either vertical or horizontal. The installation with the shaft vertical is, in most cases, more convenient since the operator has easier access to the actuator, handwheel, or lever. In practice, however, the positioning of the valve depends mainly on local habits and tradition.

The direction of vane rotation is significant only if the valve is installed with the shaft in a horizontal position. The valve should be installed so that, at opening stroke, the lower part of the vane moves in the direction of flow. This prevents the edge of the vane from striking against residues that may accumulate at the bottom of the valve body. In a properly installed valve, the initial flow will flush

Figure 28-11. Butterfly valve.

residues in the first stages of opening, thus preventing interference with the rotation of the vane. If the vane rotates in an opposite direction, the debris and residues collected at the valve could obstruct the opening movement, and the vane or lining could be damaged.

In most butterfly valve models, the diameter of the vane is greater than the length of the valve body between flanges. Therefore, the vane protrudes outside of the body when the valve is in the open position. Butterfly valves should always be installed between pipes of the same diameter as the valve. Installing a butterfly valve adjacent to a check valve, water meter, or a reducing flange can result in immediate damage to both the valve and the adjoining device when the valve is opened.

The flow control element in a butterfly valve is a disc pivoted on either a horizontal or vertical axis within the valve body. This disc is placed in a position parallel to the flowing liquid when the valve is fully open, resulting in small friction losses. Butterfly valves are simple and compact and have good throttling characteristics. The flow response is approximately linear to the closing of the flow control element in these valves. Care must be taken to not open or close butterfly valves too rapidly in order to prevent surges and water hammer problems. Gear actuators, which require several turns to open or close the valve (Fig. 28-12), can reduce chances of water hammer occurring compared to the lever-type actuators.

Radial Valves

Radial valves (Fig. 28-13) are hydraulically actuated valves that are controlled by line pressure and cannot be operated by mechanical means. Therefore, they were initially used only in systems actuated either automatically or by remote control. Recently, they have been put to use in manually operated systems, actuated by a three-way service valve installed in the hydraulic solenoid system of the valve. This fairly new concept in valve design is based on the pinch-valve principle, commonly used in laboratories and in some industrial systems. In the pinch valve, an elastic tube is mechanically flattened to stop the flow. In the radial valve, an elastic sleeve shrinks radially on a centrally placed body, preventing the flow through the inside of the sleeve. The line pressure applied to the inside expands the sleeve and opens the valve. Outside pressure applied to the sleeve either throttles the flow or closes the valve.

Radial valves are simple in their construction, lightweight and of relatively small overall size. Except for a resilient sleeve, they have no moving parts such as shafts, bearings and seals, or other elements that wear easily. Valves consist of a body, central element, elastic sleeve, and a solenoid system. The body is in the shape of a short, flanged pipe with straight or slightly convex walls. It is either cast of iron, or drawn or welded from a steel sheet. The top of the body has small inlets for connecting the solenoid system. Nominal sizes of radial valves range from 1/2 to 12 inches.

The central element is generally shaped like a disc. It is located in the center of the body and held in place by struts fixed to flanges on both sides of the body. The ovoid is usually made of a plastic material. The disc is positioned at right angles to the flow and made of stainless steel or copper alloy. The elastic sleeve is spool formed, with flanges held firmly in place between the body and the connecting flanges of the valve on both its sides. It is

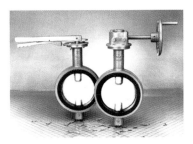

Figure 28-12. Typical lever and gear actuators for butterfly valves.

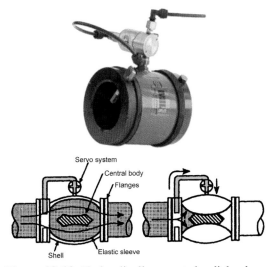

Figure 28-13. Hydraulically actuated radial valve (top) and schematic of operation (bottom).

made of natural or synthetic rubber, usually reinforced by layers of woven synthetic fibers such as nylon. In its free, unpressurized state, the central part of the sleeve clings firmly to the inner element and forms a toroidal pressure chamber between the body and outside surface of the sleeve, thus preventing low-pressure flow from passing through the valve.

The solenoid system consists primarily of a three- or multi-port valve, usually of the ball or plug type, and is either manually or power operated. It can be installed directly in one of the service connections on the body, or connected by a length of small diameter tubing. The solenoid system allows the pressure chamber to be connected either to line pressure or to drain to the atmosphere. The source of the actuating pressure can be located any place upstream. However, it is usually tapped from a threaded connection in the upstream flange of the valve. The solenoid system may also include a strainer, a needle valve for controlling the closure rate, and a remote-controlled stop valve.

When unpressurized, the sleeve is clasped tightly to the central element, closing the valve. Applying 8 to 10 psi of pressure on the inside of the valve stretches the sleeve and allows the valve to partially open. Full opening of the valve normally requires an inlet pressure of 18 to 20 psi. If the pressure chamber between the body and the sleeve is unpressurized, the valve remains open. Introducing water under pressure into the toroidal pressure chamber counteracts the inside pressure, causing the sleeve to contract again and stop the flow. Thus, the valve can be operated easily by turning a service valve that connects the pressure chamber to either the pressure source (closing) or the atmosphere (opening). By balancing the forces on the inside and outside of the elastic element, the valve can be set in any intermediate position, thus throttling the flow.

The radial valve provides a positive, watertight shutoff and a steady flow in either an open or throttled position. The principal advantages of the radial valve are that it has a very compact, light construction and only one moving part (the elastic sleeve) acting simultaneously as the actuating and sealing element. There are also no parts protruding outside the flow and actuation areas, so seals or stuffing boxes are not needed. The valve also tends to prevent cavitation. The disadvantages of the valve are the relatively high pressure demand for the initial and the full opening, lack of indication of the opening position, and a high loss of head factor (K = 4-6).

Maintenance
Valves should be placed in clean surroundings that are clear of soil and weeds. The valve locations should be marked to minimize chances of mechanical damage. To troubleshoot valves, ensure that pressure and flow conditions are appropriate for the size of valve, type of pilot, weight of the spring, thickness of diaphragm, and the application

in question. Ensure that 3-way valves, hydraulic tubing, filters, and pilots are not plugged or kinked. Verify that the diaphragm is not punctured, and that it has been installed in the correct orientation (diaphragm seam matches valve seam). Verify that the valve's accessories (3-way valve, solenoids, pilots) are connected to one another and to the valve properly (consulting operating instructions).

To troubleshoot solenoids, verify the horizontal position of the manual override. Switch the power on and off and listen for clicks. Verify venting of the solenoid when the valve opens, and verify that venting stops when the valve has closed. Verify operation of the manual override by turning it to a vertical position. If necessary, dismantle the solenoid, clean the plunger and nozzles, and check the flexibility of the spring. Verify that the plunger tips are uniform and aren't overly worn or decayed. During inactive seasons, solenoids should be energized at least once a month.

Maintenance of the Stuffing Box
In most valves, the area of flow must be separated from the part of the actuator that serves for direct operation of the valve. The most common sealing device is the stuffing box (Fig. 28-14). Its purpose is to prevent water leakage from inside the valve and along the spindle. It consists of two parts, a cup and a gland. The cup, which is the bottom part, is shaped like a round box open at the top and with a hole in the bottom, through which the spindle passes. The cup is connected to the valve cover by studs or bolts, or it may be formed as part of the cover itself.

The gland, which is the upper part of the stuffing box, is shaped like a round or elliptical flange with a ring-shaped projection fitting into the cup. The space in the cup around the spindle is filled with packing (typically hemp yarn) that has a square cross section. The packing is usually lubricated with graphite, Teflon, or other low-friction material. The gland is typically pressed into the cup by tightening two or more nuts. Tightening the nuts forces the gland into the cup, compressing the packing around the spindle to prevent leakage.

The stuffing box is widely used for sealing spindles in gate valves and most types of diaphragm valves. It has the rare property of sealing the helical (combined axial and rotary) movement of the spindle, which can rarely be achieved by a rubber seal. The stuffing box requires careful periodic maintenance, If it is not cared for properly, it becomes a source of leakage and often severe water loss.

The correct packing size is required for the seal to work properly. Adjusting the width of the yarn by flattening it with a hammer removes the lubricant and loosens the densely plaited fibers, destroying the lubricating and sealing properties of the packing. Another improper custom is wrapping the packing in a continuous spiral around the spindle instead of inserting separate rings into the cup.

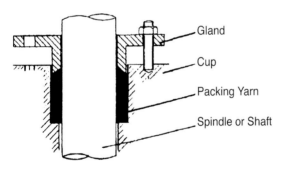

Figure 28-14. Valve stuffing box.

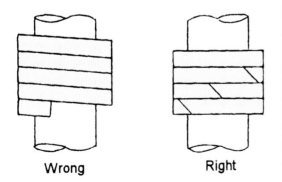

Figure 28-15. Correct and incorrect packing in stuffing box.

Such spiral-wrapped packing forms a helical surface that allows water to flow along and spill out. Also, both ends of the helix remain uneven, thus preventing formation of good sealing surfaces at the top and bottom of the packing. The proper way to fill the stuffing box is to cut the packing into rings that fit the measurements of the box, and then to insert one ring on top of the other (Fig. 28-15). Each ring should be inserted separately and set in place by pushing with the gland. Tight sealing may be achieved by cutting the edges of the rings at 45°, which allows them to overlap, and placing them so that the joining of each successive ring lies at 90° to the previous one.

Before inserting new packing, the cup and the gland should be thoroughly cleaned by removing all remnants of previous packing, sand, dust, metal chips, and rust. If not removed, debris can set in the yarn and scratch and damage the surface of the spindle. Properly inserted packing should be compressed by turning the nuts of the gland, pressing it into the cup and tightening the packing to the spindle. Careful and accurate tightening is imperative because excessive compression will squeeze the lubricant from the packing, resulting in greater friction, higher torque demand, and abrasion of the spindle.

After the initial compression of the packing, the valve is filled with water at working pressure, and the nuts are tightened again until the leakage along the spindle ceases. The spindle is then rotated, and if any further leakage occurs, the nuts are tightened slightly. After leaking stops, no additional tightening of nuts should be done. In valves installed in open spaces and subjected to heating by sun rays, the lubricant can melt and drain from the packing. Valves so placed must be checked periodically and the packing replaced if necessary. Frequent operation of a valve with dry, nonlubricated packing can seriously damage the spindle and result in permanent, irreparable leakage through the stuffing box.

A good practice is to change the packing in valves once a year. If there is leakage from the stuffing box, further compressing of the packing should not be attempted and the packing should be replaced. A leakage indicates accumulation of dirt, dry packing, or an abraded or bent spindle. Tightening the packing can only cause more damage, and stoppage of the leakage, achieved by tightening the gland, will be only temporary.

Other Valves
Needle Valves

Needle valves (Fig. 28-16) are usually intended for small flows, and their use with irrigation systems is generally limited to chemical injection. The valve's flow control element is a long-taper or needlelike plug that fits into the seat and allows for very close control of the flow. Applications of this valve are usually limited to throttling.

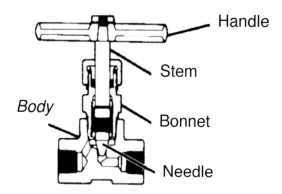

Figure 28-16. Needle valve.

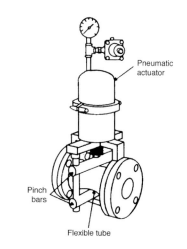

Figure 28-17. Pinch valve.

Pinch Valves

Another valve that may have use in some chemical injection applications is a pinch valve (Fig. 28-17). This valve consists of a flexible tube and a set of pinch bars, one movable and the other stationary. It has a nearly linear flow response and can be used for throttling. In addition, it has the advantage of simple design. The only component in contact with the liquid is the flexible tube, which tends to eliminate corrosion problems.

Check Valves

Check valves prevent flow in one direction. The control element may be in the form of ball, disc lift, tilting disc, flapper, or a swinging disc. This element is lifted by the pressure of liquid flowing in the normal direction. It returns to the closed position because of gravity or gravity combined with spring action when the flow stops. The pressure caused by backflow or the weight of a water column in the line increases the force, which presses the control element against the seat, further preventing flow in the reverse direction.

The design of a check valve is based on an assumption that the oncoming flow is symmetrical and that water flows evenly across the disc of the valve. This is the condition in most positions of installation. However, in some cases the velocity of flow is distributed unevenly and impinges on one side of the flap with greater velocity and force than on the other. This can result in lateral stress on the hinge and consequently result in uneven, rapid abrasion and wear of the bushings, shaft, and other moving parts. This condition may occur if a check valve is installed at a short distance downstream from a bend in a pipe. The uneven distribution of flow in the bend will affect the forces in the valve, and cause premature wear, seizing, and failure. Therefore, it is recommended to avoid installation of check valves close to a bend, regulating valve, or any flow constriction. A sufficient length of piping, inserted upstream from the check valve, will straighten and balance the flow and prevent uneven flow distribution. In system designs where such arrangement is not practicable, valves of types less affected by the flow conditions (such as globe check valves) should be used.

Ball Check Valves

Ball check valves have a control element that is a freely moving ball that fits into a seat and creates a seal (Fig. 28-18). Because the ball remains in the liquid flow path, friction loss in these valves is relatively high.

Disc and Piston-Lift Check Valves

Disc and piston-lift check valves are globe-type valves. The flow control element is a disc or piston that travels along a vertical axis (Fig. 28-19). The alignment with the valve seat is provided by the plug or disc guides. Pressure drops in these valves are less than in regular globe valves.

Flapper or Swing Check Valves

The flow control element in flapper or swing check valves is a flap or disc that pivots at a point above the main flow path (Fig. 28-20). When liquid flows in the desired direction, a disc or flap swings away. Reversal of flow brings the disc back quickly to the closed position. For better contact, a disc can be equipped with a spring that requires some pressure buildup before it opens and helps bring the disc back to the closed position.

A tilting-disc check valve (Fig. 28-21) is a modification of a standard swing check valve that provides very rapid closing due to the position of the disc. The advantage of this valve is that it doesn't chatter or flutter like swing and ball check valves.

Swing check valves should be installed so the hinge is exactly horizontal. Any inclination has a detrimental effect on the performance. Installation with a hinge even slightly inclined will result in erratic performance of the valve and in rapid wear of the moving parts. Check valves should not directly adjoin any other device in the piping. The moving parts, which can protrude outside of the flanges, can cause damage to adjoining devices. The design of a check valve absolutely must ensure free movement of the working parts, such as the hinge, bushings, and flap. A periodic check of the valve and removal of any accumulated sediments,

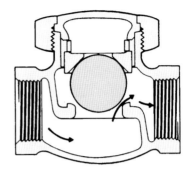

Figure 28-18. Ball check valve.

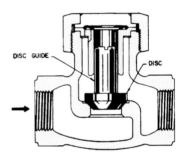

Figure 28-19. Disc check valve.

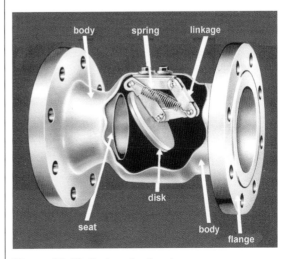

Figure 28-20. Swing check valve.

incrustations, or sand will ensure smooth, trouble-free, long-life operation.

Diaphragm Check Valves

The diaphragm check valve (Fig. 28-22) consists of a flexible sleeve that is flattened on one end. This sleeve opens on forward flow but closes against reverse flow. The sleeve is usually made with material similar to an automobile tire. Diaphragm check valves are particularly suitable for fluids containing solids, since the inside of the sleeve is soft and capable of passing suspended solids.

Foot Valves

Foot valves are most often used to maintain a pump's prime. They are installed at the end of the suction pipe in order to prevent water from leaving a suction pipe and pump when the pump is not in operation (Fig. 28-23).

Pressure Relief Valves

Pressure relief valves are used in a system to protect against excessive pressure (Fig. 28- 24). They are designed to open slowly and discharge water to relieve excess pressure in the system. They are normally held in the closed position by means of a spring-loaded disc. The spring pressure can be adjusted to provide a predetermined pressure limit. Excess pressure levels open the valve and release water. The threshold pressure is set by a screw in the top of the bonnet which adjusts the spring compression. Some models are equipped with a pilot valve (Fig. 28-25) to provide regulated opening and closing without pressure surges.

Valve Selection Summary

The main advantage of a gate valve is the straight and full flow cross section and, consequently, low energy loss. This valve, which has a small, face-to-face length, is suitable for underground installation, but in larger sizes, its overall bulk and weight can be a disadvantage. When underground, it can be operated by an elongated key attached directly to the head of the spindle. However, underground installation hinders its maintenance and repair.

Figure 28-21. Tilting disc check valve.

Figure 28 -22. Diaphragm check valve.

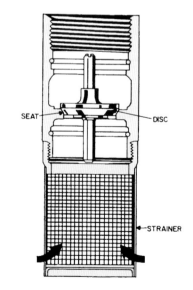

Figure 28-23. Foot valve.

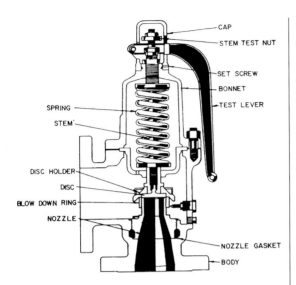

Figure 28-24. Pressure relief valve.

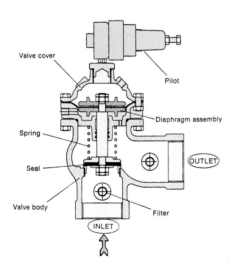

Figure 28-25. Pilot-controlled pressure relief valve.

The metal-to-metal seat can result in leakage after a period of service, and the renovation of the metal seats is troublesome and costly. Elastomer-coated gates greatly improve the performance of this type of valve. Operating torque, which depends on the

working pressure, can be very high in the last stages of closure and in the first stages of opening. A gate valve is unsuitable for throttling, and must always be held either in a fully closed or fully open position. Any throttling, however slight, can result in vibrations of the gate and quick wear of the mechanism and the gate seals. Power operation of a gate valve requires large and complicated actuators, which usually make it impractical. Danger of water hammer resulting from quick manual closure of the valve is generally not a problem due to the slow closure rate of the valve.

Diaphragm valves, because of their bulk, are limited to sizes up to about 10 inches. A resilient seal ensures watertight closure and, in case of damage, can be easily replaced. In the last stages of the closing stroke, high torque is required, especially under rising upstream pressure. The large face-to-face length of the valve, especially in the Y-type, makes it inconvenient for underground installation. The deflection of flow results in high energy losses, greater in the angle- and the globe-type, smaller in the Y-type.

Maintenance requirements of the diaphragm valve depend mostly on the type of spindle seal. Resilient-sealed spindles can operate for long periods with only periodic visual checks, while stuffing-box-sealed spindles must be cared for often. The valve can be used for throttling, but for permanent service in throttled conditions it should be equipped with appropriately designed diaphragms and spindles. The diaphragm valve is suitable for hydraulic operation and can be easily converted from hand to power actuation. Most of the hydraulically actuated devices work on the diaphragm parallel closure principle. Operation of the valve by hand is slow and difficult, especially in larger sizes. Hydraulic operation of diaphragm valves may cause severe surges.

Butterfly valves are very compact, fairly lightweight, and have a small face-to-face length that makes them convenient for underground installation. The different sealing modes with resilient seals provide a watertight shutoff, but in cases of damage to the seat, maintenance can be trouble-some because the valve parts must be removed from the pipeline and dismantled. Torque requirements are relatively low and fairly steady for most of the stroke, because of the hydraulically balanced construction. This, however, can result in a very quick closure and cause water hammer. The energy loss is low, and the valve can be efficiently used for throttling and flow regulation. The size of butterfly valves is practically unlimited, but large sizes are designed chiefly for flow regulation and not shutoff. Although butterfly valves can be adapted for actuation by external power, they require bulky, bidirectional actuators. The construction of the valve with the vane and shaft in the center of the flow area hinders maintenance, cleaning, and renovation of the adjoining piping.

Radial valves can only be hydraulically actuated. Their main advantage is that there are no moving parts either in the valve body or the actuating mechanism. As a result, there is little mechanical wear, and maintenance is minimal. Radial valves are fairly light and can be installed above or below ground. The valve is adaptable for throttling flows and for flow regulation. Radial valves generally seal well, but if the sleeve is damaged, the valve must be taken out and reconditioned with special equipment. The main disadvantages of the radial valves are the requirement for relatively high upstream pressures to open the valve and high head losses through the valve. The speed of opening or closure is not constant, often resulting in surges in the pipeline.

by Brian Boman

Throttling and Control Valves

Throttling and control valves are used when the flow rate through the pipe must be controlled at a specific rate. An ideal throttling or control valve exhibits a linear flow response to partial closing. This is accomplished by the special design of the flow path through the valve.

A combination of a valve and an actuator is usually referred to as a control valve. The system that provides automatic control of flow consists of a valve, actuator, and sensing device. Electrical hydraulic controllers and actuators are commonly used in irrigation systems. Globe and angle valves are generally adapted for automatic control becuase of the tight shutoff of these valves. Control valves allow for remote operation of irrigation systems by turning the system on and off from a distance.

A piston or a flexible diaphragm is the basic flow control element used in these valves. The diaphragm operates by pressure differentials within the valve controlled by an electric solenoid or hydraulic actuator. Power to the actuator is transmitted as electrical current through wires or hydraulic pressure through small-diameter tubing. In addition to the type of power transmitted (hydraulic or electric), control valves can be classified by the position that the valve assumes when the power is off. Some valves are normally open when the power source is off, while others are normally closed.

Electric Control Valves

Valves that use electric actuators (solenoids) are called electric control valves (Fig. 29-1). The flow control element can be in the form of a plug, disc, piston, or other similar device that controls the flow of water by opening or closing the flow path. The solenoid valves commonly used in irrigation systems rely on an electromagnetic force to either move the disc directly, or initiate the piloting action that allows water in the system to open the valve. Electric control valves can also be manually opened or closed.

Hydraulic Control Valves

The construction of hydraulic control valves is similar to the construction of electrical control valves (Fig. 29-2). The only difference is that power is transmitted by a change of pressure above the actuator (diaphragm or piston). This pressure change is transmitted through hydraulic control tubing or by diverting a portion of water flowing through the valve. Most of the hydraulic control valves are normally open, and power must be transmitted to close these valves.

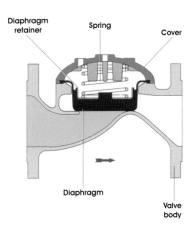

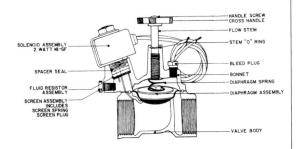

Figure 29-1. Solenoid-controlled diaphragm valve.

Figure 29-2. Hydraulic-controlled diaphragm valve.

Regulating Valves

Regulating valves control pressure and flow in an irrigation system. These valves typically operate automatically, with preset operating parameters. Regulating valves can also be used for pump output control and as stop or check valves. All regulating valves work on the principle of flow throttling. Pressure-regulating valves can be either direct acting or pilot operated. In most cases, the valves are globe, angle, Y, or radial types. Other types are generally unsuitable for automatic regulation.

Pressure-regulating valves can be distinguished as either pressure reducing or pressure sustaining. Pressure-sustaining valves will either maintain upstream pressure at its maximum permissible level by releasing excess flow, or they will maintain the minimum allowable pressure by throttling the flow through the valve. Upstream pressure governs the pressure-sustaining valve. A pressure-reducing valve will throttle flow and maintain downstream pressure at the required level if the upstream pressure is higher than the preset level. If upstream pressure is lower than the target pressure, no throttling occurs.

Direct-Acting Regulators

Pressure regulators reduce high inlet pressure to a preset outlet pressure. Fixed-pressure direct-acting regulators are not adjustable (Fig 29-3). However, they are typically less expensive and tamperproof. Direct-acting pressure regulators are controlled by the outlet pressure acting against a piston. If the pressure rises, the piston moves upwards against a spring, thus throttling the flow and reducing downstream flow. Direct-acting pressure regulators are usually preset for a fixed outlet pressure. They can be adjusted by replacing the spring or by adding inserts above the piston, thus changing the initial compression of the spring. In some models, the outlet pressure settings can be adjusted by screwing down a cap on the valve head to compress the spring (Fig. 29-4).

Direct-acting pressure-sustaining valves are constructed in a similar manner, except that the flow enters the orifice from above. An increase of the upstream pressure causes downward movement of the piston, opening the orifice. At low upstream pressure, the spring tends to close the orifice and either throttles or stops the flow. The required pressure level is set by initial compression of the spring in exactly the same way as in the reducing valve. Some types of direct-acting control valves feature an additional piston on which the control pressure acts.

Pressure regulators are preferred over pressure-regulating valves in several applications. They are used when accurate downstream pressure is required. When there are numerous submains in undulating terrain, pressure regulators are typically placed at the inlet of each riser so that each lateral will receive the same pressure. Pressure regulators are often used in drip irrigation systems to connect from the manifold to the lateral lines. A pressure regulator is typically used at the inlet to connection

Figure 29-3. Diaphragm and spring chamber of fixed-pressure regulator.

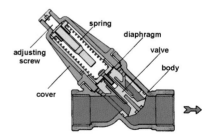

Figure 29-4. Adjustable direct-acting pressure reducing vlave.

fittings. There are several pressure regulators on the market. Most are designed to function in a non-serviceable housing that requires disposal if failure occurs or if the downstream pressure needs to be changed.

Direct pressure regulation, in spite of its simplicity, has several disadvantages. Therefore, direct-acting pressure regulators are seldom used to control irrigation systems. The main problem with the direct regulators is in their total dependence on spring action to regulate, which can cause a serious bias in the regulated parameters. The accuracy of the spring action depends on the ratio of its length to the degree of compression. Valves of even relatively small diameter must be designed with a large spring housing and are inconvenient for installation. Larger sizes are very heavy and bulky, making them impractical. Therefore, direct-action pressure regulators are seldom used with pipe diameters larger than 2 inches. However, they are often used as pilot valves in diameters up to 1/2 inch in solenoid systems of large-diameter regulators.

Pressure-Regulating Valves

For accurate regulation of flow parameters in large diameter piping, pilot-controlled valves are used (Fig. 29-5). The pressure is governed by throttling the flow in the valve and by balancing the forces acting upon the operating elements in the valve body and pressure chamber. To throttle the flow, the pressure in the chamber must be increased. The disc will then move downward, increasing the head loss in the valve. If the pressure in the chamber is lowered, the upstream pressure acts on the underside of the diaphragm. The force of the water raises the diaphragm and gradually opens the valve until the upward and downward forces balance, and the pressure stabilizes at the preset level.

The pressure in the chamber above the piston or diaphragm is controlled by a solenoid system that consists of a tubing with several devices designed to adjust the required pressure and ensure smooth operation of the valve (Fig. 29-6). Small-diameter tubing is connected by stop valves to the upstream

side of the body near the inlet flange, and to the downstream side near the outlet flange. Additional components consist of a filter, needle valve, drain tap, pilot valve, pressure gauges, and sometimes a stop valve for manual or remote actuation.

Pressure-Reducing Valve

Pressure-reducing valves are used in irrigation systems where a predetermined lower pressure is necessary for the proper functioning of certain components, such as emitters in microirrigation systems.

Figure 29-5. Hydraulically actuated pressure-reducing valve.

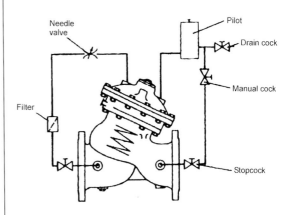

Figure 29-6. Components of pilot-controlled pressure-regulating valve.

| low downstream pressure | high downstream pressure | required downstream pressure |

Figure 29-7. Pressure-reducing valve with low, high, and target downstream pressure.

Pressure-reducing valves are used primarily for submain or manifold pressure control.

Pressure-reducing valves consist of the basic valve and a pressure reducing pilot. The valve is throttled by the action of an adjusted spring on the top of the diaphragm and the pressure of the fluid on the underside of the diaphragm (Fig. 29-7). Water from the reduced pressure side of the valve is diverted into the chamber above the diaphragm to compensate for the compression of the spring as the upstream pressure changes. This action throttles the controlling valve and allows the pressure to remain at the set level.

The downstream pressure is controlled by a pilot valve. A 3-way pilot can transmit upstream pressure to the valve chamber to reduce downstream pressure; it can vent water from the valve to increase downstream pressure, and it can block all water movements within the pilot and valve chamber to keep downstream pressure constant. The advantages of using a 3-way pilot include:

• fast reaction to downstream pressure variation;

• relatively small risk of pilot clogging because water flow is not continuous;

• relatively small head loss because the valve chamber can be fully drained by the pilot, causing the valve to stay fully open.

Pressure-Sustaining Valves

Pressure-sustaining valves consist of the basic valve and a 3-way pressure-sustaining pilot.

Pressure is maintained upstream from the valve to a preset level, while the valve outlet drains excessive pressure in order to maintain the preset inlet pressure. The control port of the pilot is connected to the upstream pressure.

When the upstream pressure is lower than the preset level, the pilot allows more water into the chamber, forcing the diaphragm down and increasing upstream pressure (Fig. 29-8). When the upstream pressure is higher than the preset level, the pilot vents water from the valve chamber, forcing the diaphragm to rise, reducing upstream pressure. When upstream pressure reaches the preset level, the water passages within the pilot are blocked, and the valve position does not change.

Applications for pressure-sustaining valves include filter stations to maintain adequate backflush pressure on hilly terrain. If such stations are used, a main system pressure must be monitored.

For applications that require a valve to both sustain an upstream pressure and regulate a downstream pressure, combination valves are available. The regulated downstream pressure should be at least 10 psi higher than the upstream-sustained pressure setting.

In some instances (especially with electric pumps), a pump control valve (Fig 29-9) may be required. They are designed for installation at the pump outlet to prevent starting and stopping surges. These valves open slowly to allow gradual filling of the irrigation mainlines. The pump start valves are

low downstream pressure   **high downstream pressure**   **required downstream pressure**

Figure 29-8. Pressure-sustaining valve operation with low, high, and target upstream pressure.

synchronized to start and stop the pump while the valve is closed. When power failures occur, the valves immediately shut off. The valves have quick-check valve action without causing pressure surges in the system.

Choosing a Regulating Valve

Not all types of valves and pilots can suit every regulating requirement. Each has its advantages and disadvantages, which can be especially significant, depending upon conditions in which the valve operates. No single type is fully adaptable to all conditions of a system, but thorough consideration of the various features of regulating valves and system requirements can help ensure the best possible results.

Piston-actuated regulating valves usually maintain very stable preset pressures. This is due to the rigidity of the piston and a constant area along its entire stroke, plus the friction between the seals and cylinder that restrains the moving parts. The piston can withstand high working pressures, and its features allow the pressure chamber to be a size that is not much larger than the valve's nominal diameter. Therefore, the whole device is compact and easily installed in confined areas.

The disadvantage of a piston-actuated regulating valve is its sensitivity to the quality of water. Silt, organic matter, and other debris in the water can deposit on the seals and cylinder walls, impairing the movement of the piston. The result can be leakage through the seals. Operation with clean, high-quality water should pose no problems. However,

good filtration is essential in areas using surface water or poor quality groundwater.

Diaphragm-actuated valves are less sensitive to water quality and present no problems in sealing the pressure chamber. The movement of the diaphragm is due to its elasticity, and no sliding of seals on the cylinder wall takes place. One disadvantage of the diaphragm-actuated valve is that the whole mechanism can move freely. Since the effective area of the diaphragm constantly changes during both the upward and downward strokes, the water force on the diaphragm is constantly changing. On its downward stroke, the disc can easily pass the point of balance and travel backward in order to correct the bias. This can take place

Figure 29-9. Pump control valve.

whenever a change of disc position is necessary. In a system with strongly fluctuating pressures, these factors can result in instability of regulation and "hunting" as the valve tries to adjust.

The stroke of the diaphragm-actuated mechanism is limited by the size of the diaphragm. In order to obtain a full opening, the diaphragm's diameter should be at least twice or three times the nominal diameter of the valve. This requires a comparatively large housing, making the valve heavy and difficult to install in limited spaces. Diaphragm-actuated regulating valves for irrigation use generally have diaphragms of reinforced rubber, either natural or synthetic.

Radial valves are often preferred to piston- or diaphragm-actuated valves. They are compact, fairly good for use in water of poor quality, and the only moving part is the sleevelike diaphragm that, because of its small mass, has practically no inertia. A very important feature of radial valves is their outstanding resistance to cavitation erosion, mainly because of the lining of the flow area with a resilient material. Therefore, this valve is suitable for high pressure differentials that may occur in regulating systems with large flows and high velocities.

Limitations of Pressure Regulation

In order to assure the proper functioning of a regulating valve, its hydraulic and mechanical properties must be considered. There are several limitations of pressure regulators that may restrict their effectiveness or harm other parts of the system. One of the problems that cannot be solved by a pressure-regulating valve is the regulation of pressure when water is not flowing in the system (static pressure). If no water is flowing in the downstream section of a pipeline, a pressure-reducing valve cannot ensure that the pressure remains below the set value. If there is leakage through the seals of the regulating valve or through the pilot (which is not designed for leakproof sealing), water can enter into the downstream piping and cause an increase of pressure. Even when no leakage occurs through the regulating valve, water hammer or surges

upstream can lift the valve's diaphragm for an instant and allow some water to pass downstream. Due to its very low compressibility, even small amounts of water can cause a substantial increase of pressure in the sealed part of the system.

Theoretically, the pressure raised by water hammer should reach the disc and the pressure chamber at the same moment and prevent any lifting of the disc. But in reality, the resilience of the PVC pipe or the diaphragm will absorb part of the energy, and the pressure reaching the chamber will be delayed and lowered. This happens when the valve is controlled by a positioning pilot, which keeps the pressure chamber sealed off and will not open with a transient rise of upstream pressure. The same can happen with a pressure-sustaining valve installed to prevent a rise of upstream pressure. A clogged filter, plugged tubing, or any obstacle in the pilot can prevent the opening of the valve at increased pressure, and it can cause damage to the system. Such cases are rare and will not occur in well-maintained valves.

Therefore, a pressure-regulating valve should not be used to protect a water system against static overpressure. It should serve only to stabilize and control flow parameters while the system performs. In exceptional cases, when a pressure-regulating valve is used for protection against dangerous static pressures, an auxiliary release device, such as a safety valve, should be added to the system.

Adjustment

The performance of pilot-controlled valves is regulated by two parameters: the preset control pressure and the speed of reaction (sensitivity). The control pressure (downstream in a pressure-reducing valve and upstream in a pressure-sustaining valve) is regulated by adjusting the compression of the spring in the pilot valve. In all pilot valves, an adjusting screw fitted with a counter nut is located at the top of the housing cover. To adjust the pressure, the nut is loosened and the screw rotated. Clockwise turning of the screw compresses the spring and increases the force exerted on the diaphragm, thus requiring a higher control pressure to move the diaphragm and the plunger upwards. Consequently the set pressure rises. Counterclockwise turning has the reverse effect.

The adjustment of the preset pressure is made while the regulating valve operates in normal conditions, and it is constantly monitored by the pressure gauge. Usually the pilot should be adjusted in small increments, allowing a short stabilizing period of 3 to 5 seconds after each step. In well-designed pilots, the adjusting screw has a fine enough thread so that each adjustment step can be made by a full turning of the screw. In pilots with coarse-threaded screws, the adjustments should be made in approximately quarter turns or somewhat greater, depending on the personal experience of the operator. When the required pressure is reached and stabilized, the lock nut should be tightened again to prevent accidental alteration of the set pressure.

After adjusting the required pressure setting, the regulating valve should be manually operated several times and the reaction time closely observed. Any additional adjustments or "fine tuning" can then be made. A stabilizing period of 3 to 5 seconds maximum is optimal for floating-type pilots. In positioning-type pilots, where the speed of regulation depends on filling and draining the pressure chamber, the reaction time can be longer, particularly if the working pressure is low.

Another control parameter is the speed of movement (sensitivity) of the regulating valve. This adjustment is made by the proper setting of the needle valve that regulates the service flow. When the speed of movement is too high, "hunting" or water hammer can result. If the reaction time is too slow, periods of uncontrolled pressure can occur, which can result in dangerous stresses to the system. The adjustment of flow velocity in the solenoid system is made by removing the protective cap or key from the head of the needle valve, and then by turning the needle clockwise to reduce the flow, or counterclockwise to increase it. Reducing the flow will result in slower filling of the pressure chamber, and consequently a slower response of

the valve. After adjusting of the flow, the protective cap should be carefully replaced to prevent accidental setting changes.

Adjusting the flow velocity can often produce a change in the control pressure. Therefore, the pressure should be checked after each change and reset if necessary. Usually, changing the set pressure will not affect the sensitivity of the regulating valve, but this may happen in extreme conditions. Therefore, after adjusting or changing any control parameter, the valve should be manually operated several times while closely observing the pressure gauges and position indicator.

The performance of regulating valves can be impaired greatly by the presence of air in the pressure chamber. The time necessary for compressing an air bubble can greatly prolong the reaction period. Therefore, air should be completely released from the pressure chamber when adjusting a new valve.

A stable, nonfluctuating pressure in the system is the main purpose of a regulating valve. This can be achieved by carefully choosing the suitable type and size of both the valve and pilot by proper adjustment of the control parameters. It is also essential to periodically check valve performance and perform maintenance in a timely manner.

Maintenance

Regulating valves perform an important function in the irrigation system and contribute greatly to smooth and trouble-free performance. Therefore, they should be cared for and inspected periodically. Frequent superficial inspections should include reading and recording the pressures shown by the gauges, visually examining the overall state of the valves for leakage and dirt, and removing any accumulated debris. At least once a year, regulating valves should be given more thorough maintenance, consisting of manually operating the valve several cycles through the full open and close positions, flushing the solenoid system, and checking the accuracy of pressure gauges. All valves should be actuated at least twice to remove any sediments. Filters should be flushed and the screens thoroughly cleaned. The sensitivity of the valves should be checked. If not satisfactory, the needle valves should be dismantled and cleaned. Any incrustations on the cone of a needle valve should be cleaned, since they may cause irregular valve performance.

Most problems with regulating valves can be remedied by cleaning the filter and needle valve, or by manually actuating the valve several times along its full stroke. The range of movement of the valve parts in regular performance is relatively short and can be impaired by sediments and incrustations on the stem or on the cylinder. Full stroke movement will usually clean the parts and restore correct the valve action.

The pilot of a pressure-sustaining valve can be flushed by back pressure from the outlet side, with the inlet valve closed and the drain and outlet valve open. This, however, cannot be done with the pilot of a pressure-reducing valve because the back pressure will close the orifice. If it is necessary to clean such a pilot, it should be removed from the system and flushed separately. Alternatively, the pilot spring should be compressed to a pressure higher than the outlet pressure before back-flushing the system. An internal inspection of a regulating valve is advisable every few years (3 to 5 years, according to the working conditions, flow and pressure, water quality, etc.). It should include disconnecting the solenoid system, disassembling the valve, cleaning out all the accumulated dirt, sediments and incrustations, checking the body and the moving parts, and examining the inner surface of the housing and all the metal parts for signs of corrosion and cavitation. Corrosion often can be found on the cover of the pressure chamber and in the body near the valve orifice, especially if the seat is made of a copper alloy fitted into a cast iron or steel body.

Cavitation erosion can be found in valves that work under high pressure differential. Cavitation attacks occur chiefly on the outlet side, in the seat ring, and at the inside surface of the body around the orifice. In valves incorrectly installed, cavitation can cause permanent leakage through the valve in its closed position, or even perforate the body of the valve. Cavitation erosion usually results in a distinct form consisting of a depression with a rugged inner surface and rim that are characteristic of the problem.

All corrosion debris should be removed and the metal surfaces examined. Eroded or corroded seats should be replaced. The elastic sealing elements should be inspected for possible cracks and changes in resilience and replaced if necessary. The elasticity of the diaphragm should be checked. Cracks around the perimeter of the diaphragm can often form as a result of material fatigue. They can be made visible by bending the material in suspected places. Cracked diaphragms can easily bust and disable the valve, so they should be replaced.

A very effective device for monitoring the performance of a regulating valve is a pressure recorder. The recorder connected to the control side (or to both sides of a valve) will trace the action of the valve over a period of time, and record problems that can be difficult to observe when the gauges are checked randomly. Thorough analysis of a recording can show whether or not the valve performs as designed, is suitable for its specific purpose, is properly located, and if the right type of pilot valve is installed and properly set.

Air and Vacuum Relief Valves
Air/vacuum relief valves are used to release accumulated air from pipelines, both during the initial filling of the pipeline and during pressurized operation. These valves are used to allow air to enter into pipelines to prevent undesirable air vacuums from occurring. Air accumulation occurs naturally within pipelines because air is dissolved (entrained) in water. As the pressure in the system is reduced, the air is released from the water. Air accumulation in pipeline systems is undesirable because it reduces the effective cross-sectional area of flow, causes pressure reductions, reduces the efficiency of pump stations, and corrodes metal

pipes and valves. If uncontrolled, air accumulation can become a safety problem.

In general, air tends to accumulate at the tops of pipelines during filling operations. When the pipe fills and water travels at velocities over 2 to 3 feet per second, air accumulation is normally minimal. If large air bubbles accumulate, they tend to become compressed at low points in the system. As these bubbles travel to the high points, they may expand and cause surges to occur. Air significantly reduces the efficiencies at pumping stations. If there is more than 10% air by volume, it may cause the pump to stop completely. Air cavitation can also corrode metal surfaces. Perhaps most importantly, the high compressibility of air causes it to act as a spring and accumulate energy. This may cause damage to life and equipment should pipeline failure occur.

Water measurements may be severely distorted by the flow of an air-water mixture or by the flow of free air in meters. The presence of air causes higher readings, potentially resulting in substantial variations from actual water quantities. Also, the high velocity of air flow and the absence of water, which usually lubricates the rotating elements, may cause serious deterioration of water meters and influence the accuracy of measurements. Appropriately sized and located air valves can prevent air flow or air accumulation in water-measuring devices and the subsequent wear and bias of measurements.

A vacuum is created when water evacuates the pipeline and is not immediately replaced with water or air. It typically occurs after sudden valve closure, after pump shutoff, and during drainage of the pipeline from a high elevation to a low elevation. A vacuum in the system may result in the collapse of mains and submains, soil ingestion into lateral lines, and clogged emitters. Sand may also be ingested through the gasketed fittings in the mainline.

Air Valves

Air valves (Fig. 29-10) ensure the undisturbed flow of water in an irrigation system by regulating the

amounts of free air entrapped in the pipeline. An air valve is an automatic device, operating by itself and affected only by the system conditions. The size, type, and location of air valves must be considered specifically for each irrigation system. An air valve can protect the piping system from various hydraulic disturbances. However, if improperly chosen or located in a wrong place, they can cause severe functional problems and interfere with the proper performance of the system.

In properly designed irrigation systems, hydraulic disturbances becuase of air accumulation or movement are negligible. However, many hydraulic problems arise from the uncontrolled release of air from pipes through orifices of air valves that are too large. In many cases, the presence of free air in pipelines may reduce the damaging effects of hydraulic disturbances, such as water hammer and cavitation. On the other hand, installation of air valves in water supply systems is often necessary to ensure the smooth operation of the system and to prevent flow stoppage or pipe failure. However, indiscriminate inclusion of air valves in system designs may also lead to severe operational problems.

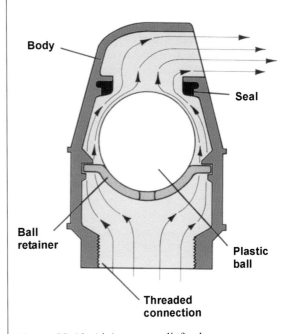

Figure 29-10. Air/vacuum relief valve.

Therefore, the designer must be aware of flow conditions in the proposed system and anticipate difficulties that could arise from air flow in the system and from the uncontrolled intake or draining of air through the various inlets and outlets.

Choosing an air valve with the suitable orifice size for a given irrigation system can prevent damaging pressure surges or vacuum while the pipeline is filled with water or drained. Collapse of light-weight pipes can be prevented by introducing air into the pipe, thus preventing formation of partial vacuum when a pipeline is drained. This can be effected by means of large-capacity, low-pressure air valves. The design of these, however, must be carefully considered to prevent sudden, uncontrolled air release through the same valves.

When filling an empty pipeline with water, the air must be released from the system. Such air release must be carefully controlled in order to prevent surges resulting from the remainder of air leaving the pipe. Air escapes through the orifice at a very high velocity, even at low pressure differentials. Escape through a too-large orifice will result in a high flow rate of the air and water. As the air rapidly escapes from the line, the water follows. When the air is exhausted and the air relief valve closes, the sudden stoppage of the water column may result in a disastrous pressure surge. Therefore, the size of the air release valve orifice must always be carefully calculated by a designer knowledgeable in the principles of air release.

Types of Air/Vacuum Relief Valves
Small-orifice air relief valves provide air relief while the pipeline is pressurized, automatically venting small pockets of air as they accumulate at high points in the system. They are hydro-mechanical devices that operate with floats and orifice sizes ranging from 1/16 inch to 3/8 inch in diameter. After the pipeline is pressurized, the float seats against the orifice and prevents air or water from escaping. As air bubbles accumulate at the top of the valve, the water level drops and causes the float to drop becuase it is no longer buoyant. As the float drops and unseats from the orifice, air

escapes from the top of the valve. The water level then rises again, seating the float against the orifice. This cycle repeats itself continuously during the course of the irrigation.

Small-orifice air relief valves will neither provide vacuum protection, nor will they release the large amounts of air while a large pipeline diameter is filling. Small-orifice valves are designed so that they do not slam shut and cause pipeline surges. They can release air at working pressures normal for the pipeline. Small-orifice air valves are commonly called "automatic air valves," even though the operation of both types is automatic. For pipes of 2 inches in diameter and smaller, some small-orifice air relief valves are able to provide both vacuum relief and air relief.

Large-orifice (kinetic) air valves are designed to intake or release large volumes of air at high flow rates and at very low pressures. Large-orifice air/vacuum relief valves operate similarly to small-orifice valves, but they are designed with larger orifices to release or accept much larger quantities of air. The intake of air occurs when the pressure in the pipeline falls below the atmospheric pressure. The release pressure is usually limited to 2 psi or less.

Large-orifice valves release and admit large quantities of air when the pipeline is unpressurized during draining or filling. During the filling, air that occupies the empty pipeline must be evacuated ahead of the incoming water, and this must be achieved in a controlled manner to prevent surge and water hammer. As water evacuates a pipeline during the draining procedure, air must be allowed to re-enter to prevent a vacuum from occurring and possible backflow or pipe collapse from negative pressure. Kinetic valves should be designed with floats that resist premature closure, and with orifices large enough to release and admit the largest quantity of air possible.

Combination air valves (Fig. 29-11) incorporate the features of both small- and large-orifice valves, providing a combination of air/vacuum relief while the pipeline is pressurized and unpressurized. They release large quantities of air while the pipeline is filling, release small quantities of air while the pipeline is pressurized, and admit large quantities of air while the pipeline is draining. Combination valves should be designed to provide maximum air/vacuum relief before the floats seat.

The nominal size of an air valve is designated by the size of its connection to the pipeline, usually by a standard NPT size. Small-orifice air valves are available in sizes from 1/2 to 2 inches with threaded connections. Large-orifice air valves come in sizes from 1 to 8 inches.

Operation

During the filling operation, an air/vacuum relief valve remains in its open position (Fig. 29-12). The float lies at the bottom, allowing air to flow freely between the pipeline and the orifice. In small-orifice air valves, where the very high flow velocity of air entering from the pipeline may lift the float and close the orifice, the float rests in a crib that allows air to pass around.

Water entering the air valve from the pipeline expels the air from the valve body, and the float, which is lighter than water, rises with the water surface. When water fills the air valve body, the top of the float touches the orifice, blocking it and stopping the flow of air. With the pressure rising in the pipeline, the force acting on the float increases, and the orifice is sealed off completely. As a result of the curvature of the float top, some air remains in the upper part of the body near the orifice. The air compresses with rising pressure, allowing the water level to rise more and forcing the float further into the orifice.

Free air flowing with water in the pipeline enters the air valve from the bottom, allowing water to drain from the valve body. The buoyancy of the float decreases as the water surface recedes, but the pressure in the air valve does not change. In this stage, the position of the float depends on the equilibrium of forces, resulting from the inside pressure, the areas on which this pressure acts, and the weight of the float. In large-orifice air valves, the

large-orifice size results in a large lift force. The float remains at the orifice and does not permit exchange of air between the pipeline and the atmosphere, even when the valve is filled with air.

In small-orifice air valves (Fig. 29-13), the small size of orifice results in small lift force, and the weight of the float forces it down, thus opening the orifice and releasing air from the air valve. After the

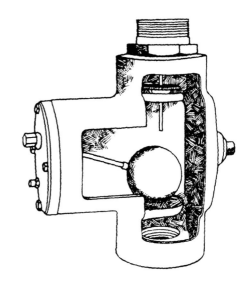

Figure 29-11. Continuous-acting air/vacuum relief valve.

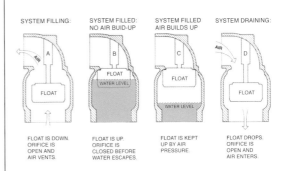

Figure 29-12. Operation of large-orifice air release valves.

release of air, water again enters the air valve body, raising the float and closing the orifice.

When the pipeline is depressurized as a result of either shutting down or disconnecting the water source and draining the pipeline, the pressure in the system falls to atmospheric pressure and a vacuum may develop. The float drops along with the water level in the air valve, opening the orifice and allowing inflow of air into the system.

In the continuous-acting combination air vent/vacuum relief valve (Fig. 29-14), the large orifice is open during the filling operation. As a result, it is able to vent large quantities of air to prevent the formation of air blocks. While the system is operating (pressurized), the large orifice is closed to prevent the escape of water. The small orifice automatically vents air that accumulates. Even small quantities of air can be vented at maximum operating pressure, thus preventing buildup of air pockets in the system. When the system drains, the large orifice opens when the pressure drops, thus relieving the back pressure and preventing the formation of a vacuum that may collapse the pipe.

Sizing of Air Valves

The design of water systems seldom takes into account dangers that can be caused by inappropriate

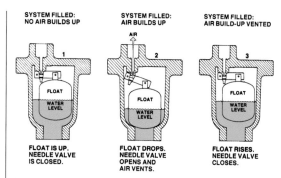

Figure 29-13. Operation of continuous-acting, small-orifice air vent valves.

sizing of air valves. Attention should be paid to prevention of water hammer caused by the sudden closure of an air valve in the final stage of filling a pipe with water. Another important consideration is to prevent excessive vacuum, which can cause collapse of the pipe when the system is drained. For safety purposes, the system should be designed so that even in case of faulty control equipment or misjudged manual regulation, the system will not suffer damage from hydraulic occurrences.

The low-pressure air valve serves mainly for draining and filling pipelines. When a pipe lying on a slope is drained through an open valve, air must be introduced at the upper end of the line. The open

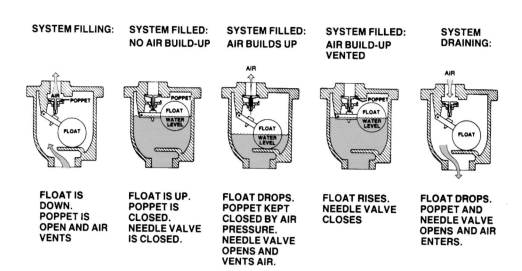

Figure 29-14. Operation of combination air vent/vacuum relief and continuous-acting air vent valves.

valve will prevent a vacuum from developing in the line, which can damage system components and cause leaks, cavitation, and pipe collapsing. The magnitude of vacuum will depend on flow parameters and pipe conditions and on the size of the air valve orifice. As a rule of thumb, vacuum should not exceed 16 feet of head.

When a pipeline is filled with water, an air valve placed at the upper end will release air, thus allowing the water to flow along freely. Flow velocity then depends on the hydraulic parameters (slope, head at inlet, hydraulic conductivity of the pipe), on the size of the air valve orifice, and on air pressure at the orifice. When water reaches the air valve and all the air has escaped, the valve closes and the flow stops. The closure is sudden and the rapid stoppage of flow results in pressure surge, called water hammer. The magnitude of the water hammer depends on the velocity of the water when the valve closes. In irrigation systems, water hammer surges should be less than 50% of the working pressure of the system. Therefore, the air valve orifice should be sized so as to control the discharge of escaping air and to prevent excessively high flow velocity of water.

Air valves serve the dual purposes of preventing vacuum when the line is draining and restraining flow velocity when the line is filled. Normally, these purposes are incompatible. To reduce the vacuum, a large orifice is required. To reduce the flow, the orifice should be smaller. The ideal solution is to use two or more one-way valves, and to size them accordingly—one for draining, and another for air release during filling. However, most system designs use two-way valves for both the filling and draining, and this requires serious consideration of the limiting parameters.

Large-orifice valves may be sized using one of two methods. The 4:1 ratio method is often used where the vent opening is 1/4 the pipeline size. For instance, a valve with a 1-inch diameter orifice or more should be specified to protect a pipe of 4-inch diameter. Valves may also be sized by calculating the volume of air to be released or admitted, and

comparing the amount to manufacturer-supplied charts.

Example:
Calculate the volume of air discharge in cubic feet per minute in a 12-inch pipeline that is 1000 feet long and flowing at a rate of 1000 gpm at 5 fps (feet per second).

Pipeline volume =
$\pi D^2/4$ x length = 3.1416 x $12^2/4$ x 1000 = 785 ft³

Calculate time to drain/fill.

Time = 1000 ft/5 fps = 200 seconds = 3.3 minutes

Calculate air discharge in cubic feet per minute.

Air Discharge = 785 ft³ / 3.3 minutes = 238 cubic feet per minute

Solution:
An air release valve with a capacity of 238 cubic feet per minute would be suitable for this application.

Small-orifice valves are sized by calculating the anticipated volume of air to be discharged while the pipeline is pressurized and comparing it to manufacturer-provided charts.

Example:
Select an automatic air release valve for a system with a flow of 2000 gpm at 60 psi (assume water has 2% air by volume).

Volume of water = 2000 gpm/7.48 gal/ft³ = 267 ft³/min

Volume of air to be discharged = 267 ft³ /min x 2% = 5.34 ft³/min

Solution:
An automatic air release valve with a capacity of 5.5 ft³/min would be acceptable.

When specifying a combination valve, the size of the kinetic relief should be given first consideration, and the size of the automatic relief given second. Automatic air relief is normally required anywhere kinetic air relief is required.

Installation of Air Valves

Air valves should be installed aboveground, either on a special riser or as part of the installation of a control device such as a shutoff valve or water meter. Installation of an air valve in a pit below ground level makes inspection and maintenance difficult. When the location of the air valve is near a water outlet such as a valve, it can sometimes be installed on a common riser. In such a case, the air valve must be placed in a straight vertical line with the riser in order to enable the air to flow freely.

Air valves installed at high points of a pipeline should be designed to account for entrainment of air along the downstream portion of the pipeline. Air pockets can accumulate in high points in a pipeline when water is flowing at low velocities or when there is no water in the line. When there is no flow, the air can be released through an air valve placed at high points. However, under low-flow conditions, the air may become entrained along the downstream portion of the pipeline, preventing its release through a valve at the peak. Therefore, air vents should be provided on the downward slope of the pipeline in order to remove air efficiently.

Air vent/vacuum relief valves should be placed in several locations of the pipeline system (Fig. 29-15). Unfortunately, most irrigation systems are lacking in proper air/vacuum relief. Valves should be considered for the following locations:

1. Near pumps, to release and introduce air to and from the line between the pump and water source

2. At high points relative to the horizontal grade

3. At high points relative to the hydraulic grade line

4. Near changes in pipe direction (especially before and after sharp changes)

5. In long, moderately inclined pipe sections with conspicuous high points

6. Along every 1500 feet of straight pipe

Automatic valves are used whenever there is a change in pipe slope in comparison with the hydraulic gradient, whenever there is a reduction in pipe size or downstream of pressure-regulating or pressure-sustaining valves, and after any location where air entrainment will occur. Kinetic valves are used after any in-line valve, at all high points in

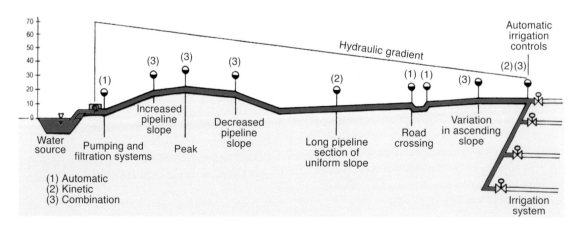

Figure 29-15. Typical locations for air/vacuum relief valves.

the system, at all turns of 90° or more, 100 feet prior to the end of pipelines, upstream of check valves, downstream of booster pumps, and every 1/2 mile of pipeline. Combination valves should be used wherever requirements for both air release and vacuum relief exist, such as near long pipeline lengths of uniform slope (every 1500 feet), prior to water meters, on filtration system manifolds, and on pumping plants.

Maintenance of Air Valves

Regular maintenance is the most important condition for proper, efficient operation of air valves. Since air valves are often at locations away from valves and other control devices, they often remain uninspected for long periods of time. Therefore, strict maintenance schedules should be established.

All air valves in the irrigation system should be inspected and serviced at least twice annually. While servicing, the air valve cover should be removed and the body thoroughly cleaned. Sediments, silt, or mud accumulated in the valve cup and on the float should be removed. The float should be tested for distortion. If there are any irregularities or surface roughness, it should be replaced. Corrosion or accumulations on the inside of the valve wall should be removed and the inside surface coated, if necessary. Special attention should be given to areas at the inlet and outlet of the air valve that are often partially clogged. Foreign bodies and incrustations should removed from the passages in order to permit an entirely free flow of air and water.

Chapter 30. Control and Automation

by Brian Boman, Steven W. Smith, and Bill Tullos

Introduction

Microirrigation (drip and microsprinkler) is the predominant method of irrigation for citrus in Florida. Typically, microirrigation systems are managed to apply relatively small amounts of water on a more frequent basis than for other types of irrigation systems. With chemigation, microirrigation systems are even more versatile as they can provide an economical method of applying fertilizer and agricultural chemicals on a timely basis. Typically, microirrigation systems require a higher level of management expertise than other irrigation methods. Microirrigation systems are more complex, require greater filtration and water treatment, and typically have high maintenance costs compared to other types of irrigation.

Irrigations generally must be scheduled more frequently with microirrigation systems because they reach only a fraction of the root zone as compared to other types of systems. One way of managing the higher demands of microirrigation is the use of automation and central control systems. These methodologies control filter backflushing, provide extensive records of water use, and allow both remote checks of system performance and efficient control of water flows to various zones. These methodologies also permit injection of water conditioners, fertilizers, and agricultural chemicals. More sophisticated systems allow irrigations to be scheduled based on evapotranspiration (ET) calculations from nearby weather stations.

In many citrus microirrigation systems, a controller is an important and integral part of the irrigation system. Controllers can help to achieve labor savings in addition to applying water in the necessary quantity and at the right time to achieve high levels of efficiency in water, energy, and chemical uses. Irrigation controllers have been available for many years in the form of mechanical and electromechanical irrigation timers. These devices have evolved into complex computer-based systems that allow accurate control of water, energy and chemicals while responding to environmental changes and crop demands.

Control Strategies

Two general types of controllers are used to control irrigation systems: open control loop systems and closed control loop systems. Closed control loops have feedback from sensors, make decisions, and apply the results of these decisions to the irrigation system. Open control loop systems apply a preset action, such as is done with simple irrigation timers.

Open Control Loop Systems

In an open control loop system, the operator makes the decision on the amount of water that will be applied and when the irrigation event will occur. This information is programmed into the controller, and the water is applied according the desired schedule. Open loop control systems use either the irrigation duration or a specified applied volume for control purposes. Open loop controllers normally have a clock that is used to start irrigations. Termination of the irrigation can be based on a preset time or may be based on a predefined volume of water passing through a flowmeter.

Open loop control systems are typically low in cost and readily available from a variety of vendors. There are many variations in design and complexity, with flexibility related to the number of stations and how schedules are specified. The drawback of open loop systems is their inability to respond automatically to changing conditions in the environment. In addition, they may require frequent resetting to achieve high levels of irrigation efficiency.

Closed Control Loop Systems

In closed control loop systems, the operator develops a general control strategy. Once the general strategy is defined, the control system takes over and makes detailed decisions of when to apply water and how much water to apply. This type of system requires feedback by one or more sensors. Depending on the feedback of the sensors, the irrigation decisions are made and actions are carried out. In this type of system, the feedback and control of the system are done continuously.

Closed-loop controllers require data acquisition of environmental parameters (such as soil moisture, temperature, radiation, windspeed, etc.) as well as system parameters (pressure, flow, etc.). The state of the system is compared to a predefined desired state, and a decision whether or not to initiate an action is based on this comparison. Closed-loop controllers typically base their irrigation decisions on sensors that measure soil moisture status using sensors or use climatic data to estimate water use by plants. In some systems, both soil moisture sensors and climatic measurements are used.

The simplest form of a closed-loop control system is that of an irrigation controller that is interrupted by a moisture sensor (Fig. 30-1). The sensor is wired into the line that supplies power from the controller to the electric solenoid valve. The sensor operates as a switch that responds to soil moisture. When sufficient soil moisture is available in the soil, the sensor maintains an open circuit. When soil moisture drops below a certain threshold, the sensing device closes the circuit, allowing the controller to power the electrical valve. When the controller attempts to irrigate, irrigation will occur only if the soil moisture sensor allows it, which, in turn, occurs only when soil moisture has dropped below acceptable levels.

Independent Controllers

The term "independent irrigation controller" is applied to those controllers that are completely separate (and independent) from other controllers. In other words, there is no feedback or communication link between controllers as there can be with a centralized irrigation control. The controller is an electronic device that, in its most basic form, is an electronic calendar and clock housed in a suitable enclosure for protection from the elements. The controller provides a low-voltage output (typically 12 or 24 volts DC or 24 volts AC) to the valves and control devices for specific zones. As long as the voltage is applied, valves stay open and irrigation water is applied.

Most remote control valves are "normally closed," meaning that the valve is closed until the solenoid is actuated by the controller. A "normally open" control valve remains open until such time as the solenoid is actuated. "Normally open valves" are sometimes used as master valves in systems when it is desirable to have a continuously pressurized mainline but still have a primary valve that can be closed in the event of high flow or another alarm condition.

Electromechanical Controllers

Electromechanical controllers (Fig. 30-2) use an electrically driven clock and mechanical switching (gear arrays) to activate the irrigation stations. These types of controllers are generally very reliable and not too sensitive to the quality of the power available. They generally are not affected by spikes in the power, and unless surges and

Figure 30-1. Switching tensiometer used to control irrigation system in microirrigated citrus grove.

brownouts are of such magnitude that they will damage the motor, they will continue to operate. Even if there is a power outage, the programmed schedule will not be lost and is generally delayed only for the duration of the power outage. However, because of the mechanically based components, they are limited in the features they provide.

Electronic Controllers

Electronic controllers (Fig. 30-3) rely on solid state and integrated circuits to provide the clock/timer, memory, and control functions. These types of systems are more sensitive to powerline quality than electromechanical controllers and may be affected by spikes, surges, and brownouts. Spikes and surges are common in many areas of Florida where lightning tends to be frequent and intense.

These types of systems may require electrical suppression devices in order to operate reliably. Because of the inherent flexibility of electronic devices, these controllers tend to be very flexible and provide a large number of features at a relatively low cost.

Features

The basic minimum features of any independent irrigation controller are to provide the time of day, a day-of-the-week calendar, the ability to change the time setting on each station, and a means of physically connecting stations to valve wiring. Some

Figure 30-2. Electromechanical controller.

Figure 30-3. Electronic controller.

models offer features that make changes in programming relatively simple. For example, a desirable feature is percent scaling. This feature allows a multiplier to be applied to the time setting on every station. Older model controllers without such a feature require that every station's time be set individually, which is time consuming and frustrating to the user. Some controllers allow for different percent scaling on each program or valve group within the controller.

With percent scaling, the system operator can go to the controller, key in a new percent scaling factor, and know that the time settings on all zones have been automatically reset for the time setting on that station multiplied by the percentage. For example, assume a certain zone has a time setting of 4 hours. If the percent scaling factor is programmed to be 75%, then the valve will open for 4 hr x 75% = 3.0 hours. It is possible to make frequent program changes easily in response to tree, soil, cultural, and climatic factors if the percent scaling feature is available on the controller. For this reason, this simple feature can be very important in maintaining efficient irrigations.

Several designs of controllers are commercially available with many different features and over a wide range of costs. Most irrigation timers provide several of the following functions:

- A clock/timer that provides the basic time measurements by which schedules are executed

- A calendar selector that allows definition of which days the system is to operate

- A station time setting that defines the start time and duration for each station

- Manual start functions that allow the operator to start the automatic cycle without disturbing the preset starting time

- Manual operation of each station so the operator can manually start the irrigation cycle without making changes to the preset starting time

- A master switch that prevents activation of any station connected to the timer

- Station omission features that allow the operator to omit any specified number of stations from the next irrigation cycle

- A master valve control feature that provides control to a master system valve (a function that is used with certain types of backflow prevention equipment and also prevents flow to the system in case of failure in the system)

- Pump start features to allow a pump start solenoid to be activated whenever a station is activated, thus tying pump control with irrigation control

Some controllers allow every station to be programmed independently of other stations, and some irrigation managers consider this feature to be quite important. Programming such a controller can be more complex and time-consuming, but the flexibility may be worth it. Another feature provides for a single irrigation event to be broken up for brief periods of operation followed by brief periods of rest. In heavy soils, irrigating in this manner allows the irrigation application rate to match the soil's intake characteristics more closely. This type of control can also be important for drip systems in flatwoods areas where root zones are limited. For example, if a system needs to operate for 10 hours per day, the 10-hour duration might be broken into five 2-hour irrigation cycles separated by 2-hour nonirrigated periods. The water applied in this scenario will typically have more lateral movement and less vertical movement than if the water was applied in one setting. As a result, water will be used more efficiently, with less opportunity for leaching below the root zone of the trees.

Other desirable features that are available on many of the more advanced controllers include provisions for:

- multiple programs so that zones with young trees can be on completely separate operating schedules;

- extended and flexible calendars that adapt to imposed restrictions that are mandated by water management districts such as every-third-day schedules or night-time-only watering;

- nonvolatile memory that holds the time settings and program in the event of power loss;

- easy adaption to respond to rain and soil moisture ssensors.

Selection Criteria

Some controllers allow the addition of a handheld remote to facilitate repairs. For example, if a maintenance person completes a repair and wants to check the valve or the system performance, a handheld remote device can be used to start the valve without going back to the controller. The time saving aspects of such a device can often show a direct payback.

When evaluating controllers to pick an appropriate system for a particular project, the following factors should be considered:

- Cost, quality, and warranty

- Programming features

- Station capacity (maximum as well as station increments)

- Enclosure (suitability to outdoor or indoor installation)

- Repair alternatives

Electrical Considerations

Schedules are maintained within the controller's logic by programming the controller. When a given station or valve is to be opened, a voltage source is provided between the controller's "common" position on the terminal strip and the valve station. A volt-ohm meter can be used to verify in practice that the voltage is available and that the controller is functioning as intended.

A terminal strip with a screw for each station provides the easiest approach to wire connections inside the controller (Fig. 30-4). Some controllers have a labeled wire bundle and valve wires are attached to the appropriate controller wire using a wire nut. When voltage is applied to the station, the solenoid on the valve is actuated and allows a small water passage to open. The water pressure upstream from the valve is utilized to hydraulically open the valve. The voltage continues to be applied from the controller, and the solenoid stays active or holds for the full-time increment set on the controller.

Grounding and Surge Protection

In general, a controller should be grounded and should be protected from electrical surge (Fig. 30-5). The manufacturers should be consulted for to specific recommendations for their equipment. Solid-state electronics are more susceptible to lightning and power fluctuations than the older electromechanical designs. However, these factors should not deter one from considering solid state controllers because of their flexibility. Controller manufacturers will generally recommend:

• 8-foot, copper-clad ground rods installed next to the controller with the controller wired to ground, using a heavy gauge wire and suitable connectors. (Three ground rods in a triangular grid or stacked one on top of another can be used in order to gain an earth ground of 5 ohms or less when tested.)

• A metal oxide varistor (MOV) on each station to protect against electrical surge

Figure 30-4. Multicolored low-voltage wiring used to connect controller to valves and other control devices.

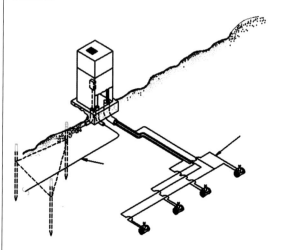

Figure 30-5. Typical grounding for 110 VAC controller.

• A MOV on the primary input side to protect the controller's electronics

Low-Voltage Wire Characteristics

Wiring used between the controller and the electric valve must be suitable for direct burial and rated as such. The term UF (underground feeder) describes wire that is suitable for direct burial and accepted for such by the National Electric Code. Wire comes in American Wire Gauge (AWG) sizes. Typical sizes in irrigation are 18 to 8 AWG, with 18 being the smallest (approximately 0.04 inch diameter) and 8 being the largest (approximately 0.13 inch diameter).

For smaller projects having short wire runs, it may be practical to consider multistrand cable, which is 18 AWG wire bundled together into a single cable. Multistrand cable is commonly available with 4-, 6-, 8-, 10-, and 12-wire cables. Each wire in a single cable is color coded differently from all other wires. The common wire is often white while the control wire is typically red or some other unique color. Each control wire should be a different color/stripe to identify it within the wire bundle. Doing so and maintaining a color scheme throughout the entire system can enhance maintenance efforts greatly when there are electrical control problems with a system. A uniquely colored wire (as opposed to a whole bundle of wires having the same color) can be quickly located anywhere in the system. When there are many wires of the same color, wires must be checked one at a time to find the desired wire. An applied voltage from a transformer, a battery or the irrigation controller, along with a volt-ohm meter, can be used to identify wiring.

Wires coming into the controller should be labeled to indicate the valve to which they are attached. Sequentially numbered wire labels are available for this purpose, which allow uniquely colored wires for individual stations.

Wire Sizing

The allowable voltage drop is the controller output voltage (typically 12 or 24 VDC or 24 VAC) minus the minimum solenoid operating voltage (manufacturer specific). The inrush current is the current necessary to initially open the solenoid valve, and this current increases as the water pressure increases because the solenoid works against the pressure. Wire resistance increases as the cross-sectional area of the wire decreases and the length of run increases. Wiring used in low-voltage control circuits is sized based on the following electrical properties:

• Allowable voltage drop

• Inrush current

• Wire resistance

The major valve manufacturers have developed wire sizing procedures for their valves to assist irrigation system designers. It is best to utilize their resources when available because some of the procedure is based on testing of their valves and using their performance criteria. Sometimes valve wire is sized for reasons other than electrical properties. Many maintenance personnel consider 14 AWG wire to be the minimum acceptable size for a purely subjective reason: they believe a heavier (14 AWG or 0.06-inch diameter) wire is less likely to be damaged or cut when mainline or wire repairs are made.

Wire Installation

Low-voltage wiring should be installed below the mainline pipe in the irrigation system. The mainline pipe can protect the wiring from cutting or nicking. When wire is not protected by the mainline, it should be installed in conduit. A warning tape installed about 6 inches deep in the trench can provide further protection and an alert to excavators. Wire should also be installed in conduit when there is good chance of damage from excavations or rodents. Wiring installed above grade should generally be in electrical conduit. High-voltage and low-voltage wire should always be installed in separate electrical conduits.

Control and common wire should be looped at 45 and 90 degree turns in the trench to provide for expansion and contraction of the wire as the ground temperature changes. The wire bundle should be taped at 6- to 10-foot intervals to keep the wire together as a bundle. A nice installation technique at the valve is to produce an expansion coil by wrapping 2 to 4 feet of wire around a shovel handle or 1-inch pipe to produce a coil that has the appearance of a spring (Fig. 30-6). A further benefit of this technique is to allow the valve top to be removed without disconnection of the wires.

All underground wire connections should be made with waterproof connectors. There are many styles of waterproof connectors available. Select one that allows for a firm connection of the two (or three) wire ends and seals the connection in silicon rubber or a similar durable and waterproof material. In larger systems with long wire runs, wire splices may be necessary at places along the mainline where there are no valves. In this case, wire splices should always be grouped together and installed in a valve box. These locations should be recorded on the as-built irrigation drawings at the completion of the installation.

Understanding the electrical characteristics of control systems facilitates troubleshooting of problems. Necessary tools and supplies include a volt-ohm meter, wire cutters and strippers, waterproof wire connectors, and wire in various gauges and colors. Some irrigation controllers provide hints or even station lights that indicate shorts to ground in the control wire for a particular valve. Broken wires can often be tracked with equipment designed to find faults and shorts to ground. This equipment can also assist in finding valves or components that have been buried belowground.

Sensors

A sensor is a device placed in the system that produces an electrical signal directly related to the parameter that is to be measured. In general, there are two types of sensors: continuous and discrete. Continuous sensors produce a continuous electrical signal, such as a voltage, current, conductivity, capacitance, or any other measurable electrical property. Continuous sensors are used when just knowing the on/off state of a sensor is not sufficient. For example, measuring pressure drop across a filter (Fig. 30-7) or determining tension in the soil with a tensiometer fitted with a pressure transducer (Fig. 30-8) requires continuous-type sensors.

Discrete sensors are basically switches (mechanical or electronic) that indicate whether an on or off condition exists. Discrete sensors are useful for indicating thresholds, such as the opening and closure of devices such as valves, alarms, etc. They can also be used to determine if a threshold of an important state variable has been reached. Some examples of discrete sensors are a float switch that detects if the level in a canal is below a minimum

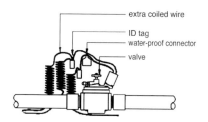

Figure 30-6. Coiled control wire and waterproof connectors at remotely activated electric valve.

Figure 30-7. Control panel for automatically flushing filters based on pressure differential between inlet and outlet of filter.

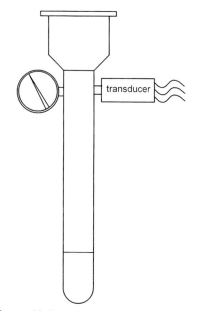

Figure 30-8. Tensiometer fitted with pressure transducer to provide continuous feedback of soil tension status.

desirable level (Fig. 30-9) or a switching tensiometer (Fig. 30-10) that detects if soil moisture is above a desired threshold. When measured over a known time period, pulses from switches can be used to measure rates such as the volume of fuel, water, or chemical solution passing through a totalizing flowmeter with a magnetically activated switch.

Sensors are an extremely important component of the control loop because they provide the basic data that drive an automatic control system. Understanding the operating principle of a sensor is very important. Sensors many times do not react directly to the variable being measured. The ideal sensor responds only to the sensed variable without responding to any other change in the environment. It is also important to understand that sensors always have a degree of inaccuracy associated with them, and they may be affected by other parameters besides the "sensed" variable. The time response of a sensor is also important. A sensor must deliver a signal that reflects the state of the system within the time frame required by the application. For example, a soil moisture sensor must be able to keep up with the changes in soil moisture that are caused by irrigation or evapotranspiration.

Some of the variables that are often measured in computer-based control systems are: flow rate, pressure, soil moisture, air temperature, windspeed, solar radiation, relative humidity, conductivity (total salts) in irrigation water, and pH of irrigation water.

The measurement of flow rate in the mainline (Fig. 30-11) is one of the most important measurements in an irrigation system. Flows that are out of range, either high or low, can be reported and acted upon. Flow sensors can be read remotely by the control system hardware directly or by adding interface hardware. Typically, flow sensors utilize a paddle or propeller inserted into the water stream that turns with the rpm (revolutions per minute) directly related to the flow velocity. Electrical pulses are generated by the sensor relative to the rpm. Remotely read flowmeters of this type are often

Figure 30-9. Float controls for drainage pump.

Figure 30-10. Tensiometer fitted with a switch to provide a discrete on/off switch.

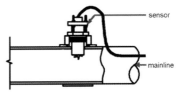

Figure 30-11. Paddle wheel flow sensor installed in mainline, allowing remote monitoring of flow rate in system.

added even if existing manually read flowmeters are already included in the system (sometimes required by the water management districts). High-flow or low-flow alarms are possible when the flowmeter is integrated with the control system.

Software used with the control system can continually check on the flow rate in the system and compare it to predefined acceptable levels. High flow conditions indicate pipe failures or stuck valves. Since the results of broken mainlines (erosion, washouts, etc.) can be disasterous, systems are often programmed to shut down when high flows are detected. If the central controller recognizes a high flow condition, it can close a master valve or shut the pump down to prevent further flow. Such action, when coupled with an alarm report issued to the central computer operator, can be quite effective in responding to a high flow and subsequently effecting a timely repair.

Wind sensors can prevent or terminate irrigation if a specified wind develops and is sustained. Rain sensors can prevent irrigation during or after significant rain. Soil moisture sensors can prevent irrigation when adequate soil moisture is already present. Sensors can be used to detect pressure and shut the system down if the pump is not primed, or to initiate flush cycles in filters.

Figure 30-12 shows a simple and low-cost rain sensor. Rain causes the porous discs in the device to swell and open a microswitch. The switch remains open as long as the discs are swollen. When the rain has passed and the ET rate is back

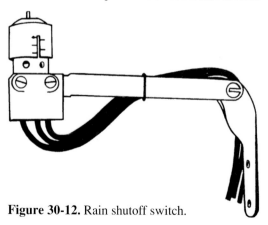

Figure 30-12. Rain shutoff switch.

up, the discs dry out and the switch closes again. This device can be implemented in two ways. With central control, the switch closure can be read by the central system, which in turn can be programmed to affect a rain shutdown at one or more sites. With an independent controller, the device can be installed at or near the controller and as a switch on the common wire. In this way, irrigation is prevented because the circuit is not completed when the switch is open.

A/D Interface

Since computer systems work internally with numbers (digits), the electrical signals resulting from the sensors must be converted to digital data. This is done through specialized hardware referred to as the Analog-to-Digital (A/D) interface. Discrete signals resulting from switch closures and threshold measurements are converted to 0s and 1s. Continuous electrical (analog) signals produced by the sensors signals are converted to a number related to the level of the sensed variable. The accuracy of the conversion is affected by the resolution of the conversion equipment. In general, the higher the resolution, the better the accuracy. An 8-bit resolution A/D board is capable of dividing the maximum input voltage in $2^8 = 256$ increments. For example, if a pressure sensor produces a voltage signal ranging from 0 to 5 volts for a range of pressure of 50 psi, an 8-bit resolution A/D board will be able to detect a change in voltage of about 5/256 volts, which will result in measurable increments of 50/256 = 0.2 psi. If the resolution of the A/D board was 12 bit, the board would be able to detect a change in voltage of about $5/2^{12}$ volts or a measurable increment of 50/4096 = 0.01 psi.

Computer-based Irrigation Control Systems

A computer-based control system consists of a combination of hardware and software that acts as a supervisor with the purpose of managing irrigation and other related practices, such as fertigation and maintenance. Generally, the computer-based control systems used to manage microirrigation systems can be divided into two categories:

- Interactive systems that collect and process information from various points in the system and allow manual control of the system from a central point by remote operation of valves or other control devices

- Fully automatic systems that control the performance of the system by automatically actuating pumps, valves, etc. in response to feedback received from the monitoring system. These systems use closed control loops which include:

  - Monitoring the state variables (pressure, flow, etc.) within the system

  - Comparing the state variables with their desired or target state

  - Deciding what actions are necessary to change the state of the system

  - Carrying out the necessary actions

Performing these functions requires a combination of hardware and software that must be implemented for each specific application.

Interactive Systems

Interactive systems are usually built around a microcomputer, either a standard PC or a specially designed unit. The information is transferred into a central unit, either directly from sensors in the pipeline or from intermediate units that collect the data from a number of sensors and then process and store them temporarily for further transfer to the central computer. These systems have features that enable the operator to transmit commands back to the various control units of the irrigation system. The field devices, such as valves (Fig. 31-13), regulators, pumps, etc., are fitted with electrically operated servo-devices that enable actuation of the pumps, closing and opening of valves, and adjusting pilots of flow regulators. This type of system permits the operator to govern the flow from the central computer by controlling flow parameters such as pressure and flow rate according to specific needs at the given time, and to receive immediate feedback on the response of the system.

Figure 30-13. Hydraulically actuated control valve with vacuum and pressure relief valves downstream.

Automatic Systems

In fully automated systems (Fig. 30-14), the human factor is eliminated and replaced by a computer specifically programmed to react appropriately to any changes in the parameters monitored by sensors. The automatic functions are activated by feedback from field units and corrections in the flow parameters by control of devices in the irrigation system until the desired performance level is attained. Automatic systems can also perform auxiliary functions such as stopping irrigation in case of rain, injecting acid to control pH, sounding alarms, etc. Most control systems include protection in emergencies, such as loss of the handled liquid due to pipe burst. They close the main valve of the whole system or of a branching when an unusually high flow rate or an unusual pressure drop is reported by the sensors.

Selection of a Central Control System

In determining the best control system for a specific project and management group, the first thing to recognize is that an informed decision will be time-consuming. If the end user already has preconceived notions about the control system, the manufacturer, or the distributor, then the user should probably act on his preconceptions. If the end user is open and unbiased and wishes to make a sound decision based on an objective evaluation of capabilities, costs, and overall effectiveness, then he or she should be prepared to put appropriate time into the effort.

Centralized Irrigation Control

Water shortages and rising power costs demand increased attention to sound water management. Large groves functioning under one management group should be particularly alert to their water management strategy. Centralized irrigation control is not only an appropriate tool for improving water management but other objectives can be accomplished at the same time.

There are many central control systems to choose from and the user base is very large and geographically diverse, with hundreds of systems throughout the country. Basic technical capabilities, reliability, and cost effectiveness have been demonstrated repeatedly to the satisfaction of even the most skeptical.

An irrigation central control system can be simply defined as a computer system that operates multiple controllers, sensors, and other irrigation devices from one central location. Today's central control system can monitor conditions in the system and surrounding areas, then control the equipment to properly respond to the conditions. This "monitor and control scenario" allows for complete system automation wherever parameters can be defined for system operation. The system can operate without personal intervention.

The monitor function of a central control system may consist of many different sensors: wind sensors, weather stations, and rain sensors are just a few of the options available. These sensors monitor their respective areas and report current conditions. The system can respond if any of the conditions are outside predefined limits. An example of sensor operations is the ability of the system to monitor rainfall. If rain occurs in a given area, the system

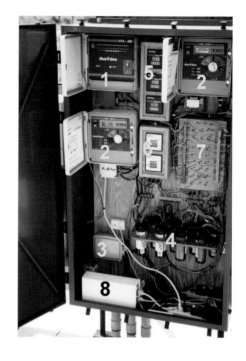

Figure 30-14. In-field controls for a centrally controlled citrus irrigation system including: 1) on-site communication, data storage, and control device; 2) valve, pump, and accessory controllers; 3) phone connection; 4) sensor decoders; 5) flow monitors; 6) pressure and water level monitors; 7) manual overrides; 8) power conditioning and surge protection.

can automatically turn off the irrigation in that area and report its actions to the central control system.

Controlling the system from the central location allows all system operations to be programmed and monitored easily and efficiently. Control actions such as adjusting watering times at all sites for seasonal fluctuations can be accomplished easily by one person from each location.

Central Control Components

Central control systems consist of a central computer, communications equipment, field controllers, and sensors. The central computer is usually located in the irrigation manager's office. The communications equipment is located both at the computer and at the field devices. Communications equipment can consist of telephone modems, radio modems, or fiber-optic modems. A middle manager device called the cluster control unit, which receives information from the central computer, is located on site to monitor and control the system equipment. These field devices are connected to irrigation valves, sensors, and other field equipment.

Weather stations can be monitored by the central computer to gather weather information and automatically calculate irrigation watering times. By gathering weather data and automatically adjusting the system, large amounts of water and money can be saved.

Most central control systems run on PC-compatible computers. In general, a faster computer with more RAM is required than in the past. Most new users purchase a high-speed computer that has plenty of hard disk capacity, additional RAM, CD drives, and a suitable backup system.

Certain minimum capabilities can be assumed from most of the central control systems available now. Most systems provide reliable radio and telephone

communication with remote sites, percent scaling to quickly accomplish day-to-day and systemwide changes in scheduling, and greatly expanded instrumentation possibilities that are limited only by imagination and budget. Percent scaling is usually possible at multiple levels. For example, a percent scaling factor may be applied globally to the entire system and another percent scaling factor may be applied to individual sites.

Figure 30-15 represents a central control system coupled with an on-site weather station or an accessible local weather station network. Together, the control system and the management system offer the potential of scheduling irrigations more closely and reactively than with any other approach. In the representation, climatic data are gathered from a weather station, and the daily reference evapotranspiration rate (ETo) is calculated and utilized in making irrigation decisions. Irrigations can be rescheduled daily or even during the course of a single day to accommodate rainfall, changes in ET, or even subjective judgments on the part of the system operator or irrigation manager.

In fact, many central control systems can be configured to automatically use ETo data to calculate crop water use and make adjustments to the run times of each zone. Whether or not a given project should be configured to perform in this way depends on people and management philosophies. There is good reason to keep the people in the loop and not allow a circumstance where the control system is left alone, without overview and oversight by knowledgeable operators.

New Capabilities

Users have numerous new capabilities to consider in central control. The look and feel of systems has been constantly improved—some dramatically improved. Pull-down menus, point-and-click (mouse) applications, and icon-driven menus now are common. New capabilities come quickly in response to wish lists from current or potential users.

Specialized consultants are available to help with initial data input, long-term modifications, and

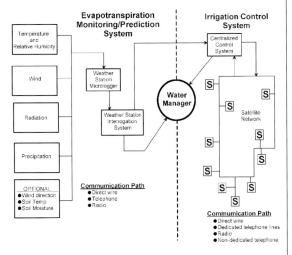

Figure 30-15. Schematic of central control system with weather station input to control irrigation events.

advice. Their services include determination or measurement of lateral precipitation rate, soil infiltration rates, lateral flow rates, and system hydraulic limitations. The initial control program can be developed and entered based on design drawings, field work, or a combination of both. Most consultants are available to provide follow-up support, training, and troubleshooting.

Some manufacturers provide an accreditation program to train and expand the knowledge base of consultants actively working with central control systems. Certification by the manufacturer is a means of promoting the service and an indication of the competence and proficiency of the consultant.

Differing Philosophies

A look at the overall philosophy of system operation can be enlightening and helpful in evaluating different systems. The control or management hierarchy built into the system and components is indicative of the system's basic philosophy. As a potential user, it is important to understand this philosophy to assure that it can be used and applied or adapted to current irrigation practices. A change in management style to match the imposed style of the system is probably not desirable.

The philosophy of some control systems includes a system with middle management. The central system communicates with a site having a cluster control unit (or CCU) at the site (Fig. 30-16). The CCU is, in turn, linked by hard wire or radio to a satellite controller. The satellite controller contains the terminal strip where the control and common wires are connected to valves. The CCU contains a computer (making it programmable and smart) and performs the middle management role in placement and hierarchy within the communication link. The CCU contains all the programs or schedules for a site. Certain actions, such as closing a valve or valves in response to a high-flow condition without the necessity of communication with the central, are possible with a CCU. One sensor at a site can affect all irrigation within the CCU.

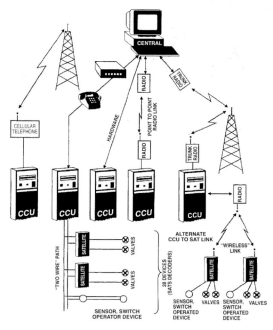

Figure 30-16. Communication components of centralized computer-controlled system.

Other systems use a direct central-to-satellite management and communication philosophy. The central control system communicates by radio or telephone directly to satellite controllers. The satellite controller may see an alarm condition (perhaps a high-flow condition at a flowmeter) and the satellite reacts, reporting actions taken when polled by the central. The central itself is programmed automatically to contact the site following each irrigation cycle. This communication process is full two-way communication between the central and the satellite controller.

Distinguishing Capabilites

Exciting new features have become available including:

- a communication approach configured by site (telephone to farm, radio to another farm, etc.);

- alarms or alphanumeric messages sent directly to pagers worn by maintenance personnel;

- automated interrogation of weather stations and automatic ETo data utilization;

- prediction of soil moisture storage based on the checkbook or water balance method of irrigation scheduling;

- user-developed condition statements (if, then, else commands) used to create specific alarms;

- multitasking environments that set the stage for dramatically expanded features in the future.

Operation of Control Systems

The information supplied from most irrigation systems is usually in the form of pressure and flow rate data. In some systems, additional parameters, such as the level in a canal or pond or an indication of a booster, are used. The sensors supplying the information are mainly pressure gauges and flow-meters with transducers adapted for telemetry and microswitches for level indication. Where transfer distances are short, the data can be supplied to the central computer as an analog output by variation of the electric potential. Such data transfer requires an individual connection for every sensor or switch, making the installation rather expensive and complicated. Therefore, in control systems involving long transfer distances, the data are encoded near the transducer units for transfer by a single channel.

In direct transfer systems, the central unit collects the instantaneous data by scanning the sensors, one after another. Depending on the number of various sensors, scanning may last several seconds or even more, so that the data received from a specific sensor reach the central computer intermittently. The interval between readouts depends on the scanning speed. The central unit receiving the data from a sensor averages four or five consecutive readings to avoid recording accidental transient data peaks or dips that may occur throughout the scanning period and bias the output. The average is then processed by the computer and usually compared to the time parameter for calculation of flow rates or for totaling the flow volumes. The processed data can be displayed digitally or graphically, stored on magnetic media for further reference, and printed as necessary.

Larger systems use intermediate field units (Fig. 30-14, No.1) that can collect data, each from a small number of sensors, thus reducing the scan period. The partially processed data are then transferred into the central unit by the same scanning method. Often the intermediate units can also act as self-contained computers, which can be specially programmed for either data display or for automatic control of the piping sector assigned to them. Such operation is independent from their main function of transferring data from sensors to the central computer, or conveying commands from the central computer to the various control devices.

The control systems are usually powered through the central unit, which is sometimes plugged directly into the main electric utility line, but more often it is connected through a rechargeable battery with a trickle charger. In order to ensure continuous operation of the control system and to avoid loss of data in case of power failure, an uninterrupted power supply unit and a lightning protector are included in the power units (Fig. 30-14, No.8).

Where utility supply power is not available, power for the control system can be provided by solar panels or from generators on diesel units. Intermediate control units, which are typically located far from standard power lines, are frequently powered this way. Wiring of an irrigation control system should be done by a competent professional to ensure that safety requirements are met and that the system meets the necessary codes. Most problems with irrigation controllers can be traced to poor electrical installation, particularly a lack of adequate grounding. Wherever electronic components are used, it is important that attention be paid to signal and powerline protection.

Communication between various units of the control system can be by wire or wireless. Data encoded by pulse requires two- or three-wire telephone cable in which one strand usually serves for energy supply from the source. For radio transmission, the units are provided with radio transceivers that operate on the usual personal phone frequencies. One common practice is to connect the various sensors by wires to the intermediate unit, which then encodes the data and forwards them to the central computer by radio or by wire. Where possible, the central computer and each field unit are provided with a modem, and the data are transmitted by ordinary telephone utility cables.

Economics

A computer model allows for analysis and comparison of the management models and control alternatives. Any project issues can be broken down into six categories:

• Control system factors

• Communication factors

• Existing irrigation system factors

• Management factors

• Water factors

• Economic or annual cost-of-money factors

The primary control system factor, from the economic perspective, is cost. What is the initial installed cost of the control system? The same is true of communication in that the initial cost and the annual cost of communication must be determined to understand the respective cost implications. Telephone tends to have a lower initial cost than radio, but radio tends to have a lower annual cost. Only a thorough analysis indicates which is most cost-effective. Further, if a particular control system is suitable only with radio, or conversely with telephone, then the system must be analyzed with this limitation in mind.

An inventory of the existing irrigation system is necessary. The number, location, and relative size of existing independent controllers must be known before a replacement concept can be developed. Historic applications and practices must be known in order to compare past practice with future practice. Water factors include the current availability and unit cost of pumping water, in addition to the projected future availability and rate increases.

The historic ET rate for the area must be known to ascertain management alternatives and develop a water management strategy. The annual costs that may be effected by central control implementation must be determined and economic factors such as rate of inflation, cost of money, and economic life must be estimated.

All of these factors can best be analyzed with a spreadsheet program. A sensitivity analysis on the data can often provide much insight into the decision process. Oftentimes, other factors such as reliability, ease of use, dealer support, and maintenance requirements are more important than overall cost.

Chapter 31. Materials and Installation

by Brian Boman

Materials

The main considerations for the choice of materials in an irrigation system are the ability of the materials to withstand mechanical stresses, extreme temperatures, solar radiation, and any chemical conditions to which they may be subjected. Mechanical stress may be due to internal conditions such as water pressure, water acidity or alkalinity, water hammer, or vacuum. Components may also be affected by such external forces as earth load, thermal expansion and contraction, and mechanical blows to the component.

There are numerous combinations of metals, composites, plastics, ceramics, and elastomers that can be used in components. Not only should materials be able to withstand mechanical and environmental stresses, but they should be compatible with each other in order to avoid internal corrosion or galvanic effects. The main considerations in component selection are the pressure rating, temperature range, chemical affinity, and cost.

The design working pressure is the most important factor in selecting materials for the distribution system. Typically, pipelines and control devices should be able to withstand twice the normal operating pressure. For instance, if the maximum normal operating pressure is 50 psi, all components should be able to withstand a surge of 100 psi.

The ambient and water temperatures should receive some consideration in the selection of construction materials. High temperatures may decrease the strength of plastics. In cases where components may be exposed to high temperatures (such as around pump stations), special high-temperature plastics or steel should be used instead of PVC. In areas subject to freezing, consideration should be given to materials that will not fracture as easily as cast iron. Ultraviolet (UV) radiation is a common problem for applications exposed to sunlight. UV radiation may cause degradation of most plastics; therefore, additives such as carbon black are used to inhibit this process. Elastomers are normally less influenced by UV radiation, and metals are usually unaffected.

Chemical resistance is often an important consideration because microirrigation systems are usually designed with the potential to inject many types of agricultural chemicals, fertilizers, and line cleaners. Consequently, unpredictable combinations of often corrosive compounds may travel through the system. Acidic (low pH) conditions will result in corrosion of most metals. The most sensitive metals to acid are aluminum and zinc. Cast-iron, carbon steel, and copper alloys are less sensitive, but for prolonged exposure to acidic conditions, a resistant material such as 316 stainless steel should be considered. High pH (alkaline) conditions may be harmful to aluminum, zinc, and titanium. Most other metals as well as most plastics and elastomers are not sensitive to alkaline conditions.

The effect of dissolved solids in water is to increase the electrical conductivity and acceleration of the corrosion processes. Common cations such as calcium (Ca^{++}), potassium (K^+), and sodium (Na^+) have almost no effect on most metals and plastics, while ammonium (NH_4) may attack some plastics and elastomers. Anions such as sulfides and sulfates are much more harmful to both metals and plastics.

Solid particles such as sand in the water stream can result in severe erosion of metal and plastic components. Erosion of surfaces can also result from cavitation. Cast-iron, mild steel, and copper alloys can be coated by very hard metals or ceramics to make them more resistant to cavitation. Plastics and elastomers are more resilient and resistant to cavitation erosion. However, they are more susceptible to abrasion damage from high-velocity solids.

Ferrous materials

Cast-iron pipe is generally used for the bodies of most meters and valves that are larger than 3 inches. Cast iron is also used for other large components that require economical pricing. For higher pressure applications and severe working conditions, ductile cast iron, cast alloy, or other special metals may be used.

Cast iron normally has a very good resistance to corrosive conditions without any special protection. Coating the bodies with protective paints such as epoxy resins is a common practice. However, under normal operating conditions, unprotected cast-iron parts normally do not corrode to a degree that impairs their performance. When exposed to cathodic materials such as copper, cast-iron components may develop corroded surfaces. These corroded areas should be periodically removed to prevent impaired function.

Components constructed from steel plate normally have low corrosion resistance. Therefore, these components must be specially coated or galvanized on both the inside and outside. In spite of this protection, serious corrosion may occur to the component if there are copper or copper alloy parts in contact with the steel. Periodic inspection and recoating are necessary to ensure long-term operation. Particular attention should be given to welds as they are very sensitive to corrosion.

Nonferrous metals

Bodies of valves smaller than 2 inches and other small components are often constructed from cast bronze. Cast bronze is also used for bodies of larger valves designed for high pressure and for various parts of valve and meter mechanisms. Brass is used for machined parts such as shafts and spindles. Brass inserts are used for bearings and threaded connections in cast-iron bodies to reduce friction and avoid corrosion.

Special copper alloys are used for extremely corrosive and erosive conditions. They are also used for applications that have high velocity. Cast aluminum is typically only used in components that are portable or where weight is a serious consideration.

Plastics

There is a wide array of synthetic resins in the forms of polymers, copolymers, composites, laminates, and coatings, each of which may have different properties. Plastic materials are less resistant to stress than metals and tend to lose strength under pressure and elevated temperatures. They are used for smaller valves and as parts for many devices such as flowmeter impellers, low-friction bushings, and components not subjected to stress.

Bodies of smaller valves are typically made of glass-reinforced polyester, polyacetal, and polycarbonate. Small mechanisms may be made of an acetal resin such as Delrin, and where low-friction components are required, Teflon linings may be applied. Nylon, Teflon, and polyacrilate glass-fiber-reinforced materials are often used for flexible small-diameter tubing and joints. Filtering elements often use acetal, polycarbonate resins, thermoplastic polyester, or high-density polyethylene. For the protection of metal bodies and components against corrosion, plastic linings and epoxy resins are often used. Valve vanes are often coated with nylon or Teflon to reduce friction.

Elastomers

Elastomer materials used for gaskets and seals must be able to withstand high stress together with both high and low temperatures. Natural or synthetic rubber has proven to be the most suitable material for gaskets and seals in irrigation systems. They give good performance for resilience, flexibility, and wear. However, they are prone to degradation by microorganisms, especially when in long contact with wet soil. Neoprene reinforced with nylon is generally suitable for flange gaskets. When corrosive materials are present, other hydrocarbon acrylonitrile elastomers such as Buena N, neoprene, Viton, or EPDM rubber should be used.

PVC Pipe

Polyvinyl chloride (PVC) pipe is the result of the chemical development of a synthetic base material and is manufactured under computer-controlled conditions. It is continuously extruded into seamless lengths that are strong, chemically resistant, have low friction loss, and are lightweight for ease of handling. There are both Types I and II and Grades I and II available. Type and grade refer to the hydrostatic design stress pressure capabilities of the pipe. Type I, Grade I, is the most commonly used for microirrigation systems.

PVC pipe may be designated as either low head, which is rated for less than 50 feet of head (22 psi), or high head, which is rated for more than 50 feet of head. Pressure ratings may be given in feet of head, class or psi rating, schedule, or SDR (Standard Dimension Ratio). Class or psi designation refers to a pressure rating in pounds per square inch (Table 31-1). The operating pressure on the job, including surges, must not exceed the class rating. PVC pipe with a schedule designation has the same outside diameter and wall thickness as iron or steel pipe of the same nominal size (Table 31-2). SDR is the ratio of the outside pipe diameter to the wall thickness. There are pressure rating differences between Types I and II and Grades I and II. This

Table 31-1. Maximum operating pressure for class and SDR-rated PVC pipe.

Pipe rating designation	Maximum operating pressure including surges (psi)
Class 80	80
Class 100	100
Class 125	125
Class 160	160
Class 200	200
Class 250	250
Class 315	315
SDR 81	50
SDR 51	75
SDR 41	100
SDR 32.5	125
SDR 26	160
SDR 21	200
SDR 17	250
SDR 13.5	315

information should be stamped on the pipe. Temperature is also important when using PVC because as the temperature increases, the safe operating pressure decreases (Table 31-3). To obtain the pressure rating, multiply the pressure rating at 73° F by the corresponding service factor.

Either an IPS or PIP size may be used (Table 31-4). IPS refers to plastic pipe which has the same outside diameter as iron pipe of the same nominal size. PIP is an industry size designation for plastic irrigation pipe. When obtaining cost estimates, be sure that all estimates are in the same size designation. Normally pipe size should be large enough so that the water velocity does not exceed 5 feet per second in order to avoid excessive friction losses and surge or water hammer problems. Pipe sections are available in 20-foot, 30-foot, 35-foot, and 40-foot lengths. Diameters available are 4 to 15 inches for low-head pipe and 1/2 to 12 inches for high-head pipe.

PVC Failure

Most PVC pipe and fittings used in the irrigation industry are manufactured from Type I, Grade I PVC compounds. These compounds have a minimum tensile strength of 7000 psi before being stressed, and a modulus of elasticity in tension of 400,000 psi at 73° F. Even after years of service, PVC pipe normally maintains its ability to withstand occasional high pressure surges. However, if PVC pipe is subjected to frequent pressure variations of a cyclic nature it can fail, even though the peak pressure never exceeds the design pressure of the pipe. The number of cycles to failure depends on the magnitude of the pressure variation. It appears that the ability of PVC pipe to withstand cyclic pressure conditions is independent of its ability to withstand constant static pressure.

Burst Failure

Burst failure in PVC pipe and fittings is usually rather dramatic (Fig. 31-1). It may begin at a point of stress concentration or weakness and continue by splitting through fittings and pipe for some distance. Sometimes, the failures will completely

Table 31-2. Operating pressures for Schedule 40 and Schedule 80 PVC pipe (Type I, Grade I at 73.4 °F).

Diameter (inches)	Maximum operating pressure (psi)	
	Schedule 40	Schedule 80
3	840	1200
4	710	1040
6	560	890
8	500	790
10	450	750
12	420	730

Table 31-3. Pressure rating service factors for temperatures from 73° to 140° F for PVC and PE pipe.

Temperature (°F)	PVC factor	PE factor
73	1.00	1.00
80	0.88	0.92
90	0.75	0.81
100	0.62	0.70
110	0.50	
120	0.40	
130	0.30	
140	0.22	

shatter a fitting and the adjacent pipe. Burst failures usually occur during hydraulic transient conditions that create large pressure variations in the system. These include rapid valve closure, pumps starting or stopping, rapidly escaping entrapped air, or an air pocket shifting within a pipeline. Burst failure will sometimes occur in a pipe or fitting that was damaged during installation or that is subject to external loads. In these cases, the failure may occur at pressures well below the expected burst limit of the product.

Long-term Pressure Failure

Long-term pressure failure (Fig. 31-2) occurs when the system operates continually at a pressure that will eventually cause failure. The failures can occur within a short time after system installation or after many years. The failures will usually appear as slits or small cracks in the pipe or fitting along the minimum wall thickness, or in an area of stress concentration. Some yielding of material will usually be evident.

Cyclic Surge Failure

Cyclic surge failure (Fig. 31-3) can occur in systems that are subject to frequent changes in flow and/or pressure. It is difficult to distinguish between cyclic failure and long-term static failure in fittings. Even though their burst strength may be equal to that of the same class of pipe, fittings will not withstand as many cycles as PVC pipe because of the stress concentrations and extra forces placed on them. There may also be a marked reduction in burst strength after subjection to a period of cyclic

Figure 31-4. Standard rigid PVC pipe dimensions (all dimensions in inches).

Size	Outside diameter	CL 100 SDR 41		CL 125 SDR 32.5		CL 160 SDR 26		CL 200 SDR 21		CL 315 SDR 13.5		Schedule 40		Schedule 80	
		ID	Wall	ID	Wall	ID	Wall	ID	Wall	ID	Wall	ID	Wall	ID	Wall
1/2	0.840									0.716	0.062	0.622	0.109	0.547	0.147
3/4	1.050							0.930	0.060	0.894	0.078	0.824	0.113	0.742	0.154
1	1.315					1.195	0.060	1.189	0.063	1.121	0.097	1.049	0.133	0.957	0.179
1 1/4	1.660					1.532	0.064	1.502	0.079	1.414	0.123	1.380	0.140	1.278	0.191
1 1/2	1.900					1.754	0.073	1.720	0.090	1.618	0.141	1.610	0.145	1.500	0.200
2	2.375					2.193	0.091	2.149	0.113	2.023	0.17	2.067	0.154	1.939	0.218
2 1/2	2.875					2.655	0.110	2.601	0.137	2.449	0.213	2.469	0.203	2.323	0.276
3	3.500			3.284	0.108	3.230	0.135	3.166	0.167	2.982	0.259	3.068	0.216	2.900	0.300
4	4.500	4.280	0.110	4.224	0.138	4.154	0.173	4.072	0.214	3.834	0.333	4.026	0.237	3.826	0.337
6	6.625	6.301	0.162	6.217	0.204	6.115	0.255	5.993	0.316	5.643	0.491	6.065	0.280	5.761	0.432
8	8.625	8.205	0.210	8.095	0.265	7.961	0.332	7.805	0.410						
10	10.750	10.226	0.262	10.088	0.331	9.924	0.413	9.728	0.511						
12	12.750	12.128	0.311	11.966	0.392	11.770	0.490	11.538	0.606						

pressure conditions. Of all the operating conditions that can create problems, cyclic pressure appears to be the most critical because fittings are the weakest system components.

Mechanical Failure

Mechanical failure covers a multitude of piping failures that are unrelated to, but may interact with, the hydraulics of the system. One of the most common types of failure is the overtightening of female threaded fittings, which results in splitting of the fitting (Fig. 31-4). Since PVC is viscoelastic, it yields easily upon threading fittings together. The threads are smooth and create little friction, and it is very easy to overtighten PVC fittings. It is possible, with very little effort, to create circumferential stress beyond the failure limit when assembling threaded fittings. This is even more pronounced when using some thread lubricants, dopes, or sealants. The failure usually appears as a split perpendicular to the threads, beginning at the leading edge and extending into the body of the fitting. Occasionally, a split at the base of the female threads will appear parallel to the thread direction. This will usually occur in a fitting with a shoulder or thickened place near the base of the threads and is more common when the male part bottoms against a shoulder.

Another type of mechanical failure occurs from improper solvent welding or improper fitting of the components of an assembly (Fig. 31-5). Improper penetration of pipe into socketed fittings significantly reduces the strength of the fitting. Improper solvent welding techniques can cause failures in the bonding, creating leaks or separation. Fittings should have full socket penetration in order to not significantly weaken the assembly.

Mechanical failure of PVC components can also occur from temperature expansion. If sufficient expansion or contraction allowances are not made by providing expansion loops, offsets, or slip joints, severe stress can be placed on the pipe and fittings. Mechanical failure occurs when inadequate thrust blocking is provided. This allows excessive pressure to be placed on a fitting as the line pressure

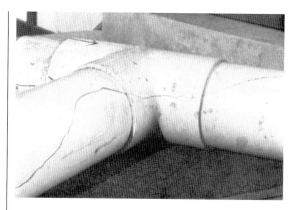

Figure 31-1. Typical burst failure in PVC pipe (reprinted from Bliesner, 1987).

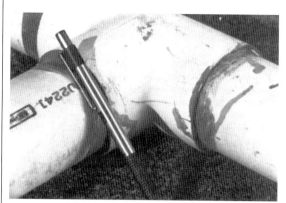

Figure 31-2. Example of long-term pressure failure in PVC pipe (reprinted from Bliesner, 1987).

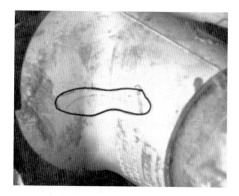

Figure 31-3. Example of typical cyclic pressure failure in PVC pipe (reprinted from Bliesner, 1987).

tries to displace it while the fitting is restrained by the pipe to which it is attached.

Underground Installation, Bedding, and Backfilling

Both solvent weld and rubber gasket joints are commonly used in irrigation systems (gasket joints are typically used only in pipe sizes of 10 inches and larger). Rubber gasket joints may be the bell-and-spigot type, or a separate twin gasket coupler may be used with plain-end pipe. Solvent welding requires extreme cleanliness, and the material has temperature and humidity precautions that must be observed to ensure that it is properly installed.

A minimum depth of 30 inches of cover over the pipe is required for protection against heavy machinery, etc. when PVC pipe is buried (Table 31-5). A trench must be wide enough to ensure proper compaction when a pipeline is buried. This is especially important for large pipe sizes. For pipe diameters greater than 4 inches, a trench at least 12 inches wider than the diameter of the pipe should be specified (ASAE Standard). Where possible, assemble the pipe aboveground and then place it in the trench. If the trench curves, be sure to assemble the pipe on the outside of the curve to eliminate the possibility of it being too short. Spray a small amount of paint on each joint of gasket pipe before dropping it in the trench so that you can see if any joints have slipped apart.

With a smooth, uniform, trench bottom, the pipe will be supported over its entire length on firm, stable material. Blocking should never be used to

Figure 31-4. Example of threaded PVC fitting failure due to overtightening (reprinted form Bliesner, 1987).

change pipe grade or to provide intermittent support over low sections in the trench. Because subsoil conditions can vary greatly, different pipe bedding problems will be encountered in various localities. In general, however, subsoil should be stable and should provide physical protection for the pipe.

The pipe should be surrounded with backfill materials having a particle size of 1/2 inch or less. Backfilling should be done in layers, and each layer should be compacted sufficiently so that lateral passive soil forces are developed uniformly.

Hand tamping or water packing may be used. Use water packing only on rapidly draining soils. Use only fine-grained material that is free of rocks and

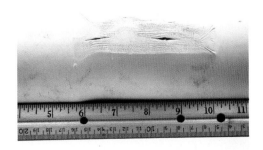

Figure 31-5. Failure in PVC pipe due to improper solvent welding of joint. Excess cement used in the joint (left) pooled and softened the pipe, resulting in failure (right).

clods for the initial coverage of the pipe. Hand tamping involves tamping the initial fill in 2-inch layers to at least 6 inches above the pipe to a soil density of at least 85% of the undisturbed sidewalls.

If the water packing method is used, the pipeline must be filled with water before beginning the backfilling operation. Backfill approximately 8 inches over the pipe, and then add water to thoroughly saturate the initial backfill without overwatering. Close the valves so the pipe remains full and allow the wetted fill to dry until firm before completing the backfill. The pipeline must be full of water to avoid floating the pipe. Always fill low-head pipe with water prior to backfilling to avoid damage to the pipe by crushing or flow reduction from flattening. When compacting sand or gravels, vibratory methods are recommended. If water packing is used, the initial backfill should be sufficient to ensure complete coverage of the pipe, and additional backfill should not be added until the water-flooded backfill is firm enough to walk on.

Sand and gravel containing a significant proportion of fine-grained material (silt, clay, etc.) are preferably compacted by mechanical tamper. If none is available, sand and gravel should be compacted by hand. In all instances, the trench should be filled completely. All backfill should be placed and spread in fairly uniform layers to eliminate voids or unfilled spaces. Large rocks, clods, and other debris larger than 3 inches in diameter should be removed. Rolling equipment or heavy tampers should be used only to consolidate the final backfill.

Thrust Blocking

Water under pressure in pipelines exerts thrust forces at tees, elbows, valves, and at any change in pipe size or direction. The size, shape, and type of thrust blocking required depends on the maximum system pressure, pipe size, type of fitting, and soil type. The design of the thrust block also requires information on the magnitude of the thrust generated in the pipeline (Table 31-6) and the bearing strength of the soil in which the pipeline is located (Table 31-7).

Concrete thrust blocks (Fig. 31-6) should be installed at changes in direction (tees, elbows), changes in pipe sizes, and at stops or ends to prevent the pipe from coming apart during operation. Each thrust block should be large enough to adequately bear the thrust of water in the pipe, sometimes as much as 5000 to 6000 pounds of pressure. The exact size of thrust block must be calculated for each pipe size and soil type. Typically, the thrust block is constructed on the outside edge of an elbow and the downstream side of a tee.

Example:
Determine appropriate thrust blocking for a 10-inch 90° elbow on a system with a pressure of 50 psi installed in a sandy soil.

From Table 31-6, the thrust that must be supported by the soil is 11,200 lb at 100 psi. At 50 psi, the thrust would be 11,200 lb ÷ 2 = 5600 lb.

From Table 31-7, the bearing strength of a sand is 1000 lb/ft^2. The contact area required

Table 31-5. Minimum depth of cover for buried pipelines.

Pipe size (inches)	Depth of cover (inches)		Minimum trench width (inches)
	Traffic areas	Nontraffic areas	
1/2 - 1 1/2	18	6	6
2 - 2 1/2	18	12	12
3 - 5	24	18	16
6 and larger	30	24	18

to achieve adequate thrust blocking would be:

contact area = 5600 lb ÷ 1000 lb/feet^2
= 5.6 feet^2

The pipeline should be thoroughly tested for leakage before backfilling operations are undertaken. With gasket joint pipe, it is necessary to partially backfill before testing to hold the line in place. Only cover the body of the pipe and leave all joints exposed. During the initial pressurization of the pipeline, fill the line very slowly and increase to design pressure over a 15- to 20-minute period.

Joining Pipes
Make sure the fittings, valves, and pipe are of the same type PVC. For example, it is unwise to use Type I and Type II PVC in the same installation. The expansion and contraction features, pressures, etc., are different, and use of mixed materials could cause failure. Make sure that the proper cement is used with the proper PVC pipe and fittings. Never use CPVC cement on Type I PVC pipe or, conversely, never use PVC cement on CPVC pipe and fittings.

Normally, PVC pipe and fittings are manufactured to produce a snug fit when assembled. This condition, however, can vary because of the minimum and maximum tolerances for which the pipe

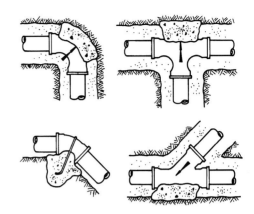

Figure 31-6. Concrete thrust block configurations used to stabilize pipeline corners above and below grade.

is produced. In the case of a fitting with the maximum diameter and a pipe with the minimum diameter, a loose fit could result. This can be remedied by interchanging fittings. Applying two coats of solvent cement under these conditions will also help ensure a good joint.

Conversely, if the pipe diameter is on the maximum side and the fitting on the minimum side, the interference may be too great, and it may be necessary to sand interchanging fittings on the pipe O.D. to permit entrance. For these specific reasons, it is important to check dry fits prior to making a

Table 31-6. Thrust developed for various fittings per 100 psi of line pressure.

Pipe size (in.)	90° elbow (lb force)	45° elbow (lb force)	Valves, Tees (lb force)
1 1/2	300	200	200
2	500	300	400
3	1000	600	800
4	1800	1100	1300
6	4000	2300	2900
8	7200	4100	5100
10	11200	6300	7900
12	16000	9100	11300

Table 31-7. Estimated bearing strength of soils.

Soil Type	Bearing Strength
Muck	0
Soft clay	500
Silt loam	750
Sand	1000
Sand and gravel	1500
Sand and gravel with clay	2000
Sand and gravel cemented with clay	4000
Hard pan	5000

solvent-welded joint. The amount of interference and taper on the fittings is greater for Schedule 40 type fittings. The Schedule 40 and lighter-wall SDR pipe have a tendency to round themselves within the Schedule 40 fittings, thus permitting a greater degree of interference. However, in the case of Schedule 80 fittings, the heavy wall on the pipe causes the pipe to be nonroundable. Usually the interference is less on Schedule 80 fittings, which in many cases will allow the pipe to bottom dry with very little interference. It is under these conditions that it may be necessary to apply more than one coat of solvent cement to the pipe and fitting if the "dry fit" seems loose.

The cement system used for PVC bonds is a solvent-based type. The solvent dissolves the mating surfaces when properly applied to each surface. The PVC resin filler contained in the cement assists in filling the gaps between pipe and fitting

surfaces. The evaporation retardant, usually cyclohexanone, slows the rate of evaporation of the prime solvent. Some clear cements are available, while most others contain pigments to match the pipe color. The most common color is gray, made from titanium dioxide and carbon black, which are considered inert pigments.

To eliminate dry spots that will not bond, it is essential to join wet mating surfaces in one minute or less after the cementing starts. The bond interface consists of a mixture of cement resin and dissolved PVC from the pipe and fitting surfaces. As the solvent evaporates, the interface becomes homogeneous with the pipe and fitting surfaces, except for residual solvent that dissipates over a period of a year or longer. The resultant homogeneous bonded area has led to the term "solvent welded," although no heat is applied to melt and fuse the bonded areas as in metal welding.

For smaller diameters and thin-wall schedules with interference fits, use Schedule 40, quick-dry, lightweight cements, also known as light-body cements. These cements are not designed to fill as much of a gap. They tend to dry and cure faster and do not dissolve into the pipe and fittings as much. Recommended curing times for various temperature ranges are given in Table 31-8.

Heavyweight, heavy-body, or Schedule 80 cements are used for larger diameters and the heavier-walled Schedule 80 pipes, which are not as roundable. These cements are designed to fill more gap, dry slower, dissolve into the pipe and fitting more, and cure somewhat slower.

Table 31-8. Recommended joint curing chart for regular- and medium-bodied cement. Note: heavy-bodied cements will require longer curing periods.

Temperature range	1/2- to 1 1/4-inch		1 1/2- to 3-inch		3 1/2- to 8-inch	
	<180 psi	180-370 psi	<180 psi	180-315 psi	<180 psi	180-315 psi
60° - 100° F	1 hr	6 hr	2 hr	12 hr	6 hr	24 hr
40° - 60° F	2 hr	12 hr	4 hr	24 hr	12 hr	48 hr
10° - 40° F	8 hr	48 hr	16 hr	96 hr	48 hr	8 days

Heavyweight cements can be, and are successfully used in place of the lighter cements. It is extremely difficult to get a satisfactory bond using lighter and quicker-drying cements with larger and heavier-wall pipe. Care should be used to avoid cement spill into the pipe I.D. Cure times required will be longer with the heavyweight cements.

The cement should still be wet when the surfaces are mated. A check should be made with the cement applied to ensure it will provide a surface that remains wet for at least one full minute with a normal full coat under the actual field conditions. This can be done by preparing a scrap piece of pipe with the primer, then applying a full, even coating stroke with the brush, and checking to see if the cement is still wet after one minute.

PVC solvent-cemented joints that are correctly assembled with good cement under reasonable field conditions never blow apart when tested after the suggested cure period under recommended test pressures. Good PVC solvent joints exhibit a completely dull surface on both surfaces when cut in half and pried apart. Leaky joints will show a continuous or an almost continuous series of shiny spots or channels from the socket bottom to the outer lip of the fitting. No bond occurs at these shiny spots. This condition can increase to the point where almost the entire cemented area is shiny, and fittings can blow off at this point. Shiny areas can be attributed to one or a combination of the following causes:

- Use of a cement that has partially or completely dried prior to installation of the fitting

- Use of a jelled cement that will not dissolve into the pipe and fitting surfaces because of loss of the prime solvent

- Insufficient cement or cement applied to only one surface

- Excess gap that cannot be satisfactorily filled

- Excess time taken to make the joint after cement was applied (in which cases it is often impossible to bottom the fitting since the lubrication effect of the cement has dissipated)

- Cementing with pipe surfaces above 110° F, resulting in evaporation of too much prime solvent

- Cementing when the pipe surfaces are wet or there is very high humidity coupled with low temperature

- Joints that have been disturbed and the bond broken prior to a firm set, or readjusted for alignment after bottoming

- Cementing surface not properly primed and dissolved prior to applying solvent cement

If the container of cement is subjected to prolonged exposure to air, the cement becomes thick and viscous, or gel-like. Chances are that this condition has been brought about by the evaporation of the solvent. If this occurs, the cement is useless. Do not try to restore the cement by stirring in a thinner. For this reason, it is suggested that smaller containers of cement be used, especially in warm or hot weather.

Prior to using an unopened can of cement, it should be shaken vigorously to ensure proper dispersion of the resin and solvents. Keep in mind that the solvents contained in PVC cements are highly flammable and should not be used near an open flame. The area in which the cement is being used should be well ventilated, and prolonged breathing of the fumes and contact with skin or eyes should be avoided. All PVC cement should be handled in the same manner as a very fast-drying lacquer.

Large Diameter Pipe

The basic solvent cement instructions apply to all sizes of pipe, but when making joints of 4-inch diameter and above, it is recommended that two workers apply the solvent cement simultaneously to pipe and fitting. Additional workers should be in a position to help push the pipe into the fitting socket while the cemented surfaces are still wet and ready for insertion. Alignment of large-diameter pipe and fittings is much more critical than when working with small-diameter pipe. As the pipe diameters increase, the range of tolerances also increases, which can result in "out-of-round" and "gap" conditions. Speed is important in making the joint and applications of heavy coats of solvent cement in these cases. When working with larger pipe diameters (8 inches and larger), checking the dry fit of pipe and fittings is more critical. In many cases where fabricated fittings are used, interference fits may not be present. Therefore, it will be necessary to apply more than one coat of cement to the pipe and fitting. It is essential to use a heavy-bodied, slow-drying cement on these large-diameter sizes. The heavy cement provides thicker layers and a higher capacity to fill gaps properly than regular cement. Heavy cements also allow slightly more open time before assembly.

In installations where bell-ended pipe is used to eliminate couplings, it is suggested that the interior surface of the bell be penetrated exceptionally well with the primer. In the process of belling pipe, some manufacturers use a silicone release agent on the belling plug, and a residue of this agent can remain inside the bell. This must be removed in the cleaning process.

Chapter 32. Emitter Selection Considerations

by Brian Boman

Microirrigation is by far the most common method of citrus irrigation in Florida. Microirrigation systems have the advantage of lower initial costs than permanent solid set sprinkler systems for widely spaced tree crops. Microsprinklers are often preferred because they provide a greater degree of freeze protection than drip and conventional overhead irrigation methods. With chemigation, microirrigation also provides an economical method of applying fertilizer and other agricultural chemicals on a timely basis. One of the drawbacks of microsprinkler systems is that they require more maintenance, including filtration and water treatment, than conventional overhead systems. Also, irrigations must be scheduled more frequently since the microsprinkler systems wet only a fraction of the root zone as compared to overhead sprinkler systems. Overall, microirrigation systems require a higher level of management expertise than other irrigation methods.

Irrigation systems for Florida citrus should be designed to provide all of the tree's water needs during extended drought periods. Most Florida citrus is planted on sandy soils that have low water-holding capacity. In flatwoods areas, high water tables generally limit citrus root zones to 18 inches or less. Even in the deep sands of Central Florida, usually only about 2 to 3 feet of root zone are managed with microirrigation. Rainfall in the Florida citrus belt can be highly variable, ranging from adequate to excess to drought within a few weeks. Experience has shown that there are several important factors to consider when designing, installing, operating, and maintaining a microirrigation system in Florida. Some of the more important factors are:

- application uniformity;

- clogging;

- insect problems;

- wear;

- wetting patterns;

- emitter maintenance;

- scheduling when and how much to irrigate.

Emitter Hydraulic Characteristics

Microirrigation emitter flow rates have a different response to pressure variations. The response of a specific emitter depends on its design and construction. The relationship between the emitter operating pressure and flow rate is given by the following equation:

$$Q = K_d \times P^x \qquad \text{Eq. 32-1}$$

where,

Q = flow rate (gph),
P = hydraulic head or pressure (psi) at the emitter,
K_d = the emitter discharge coefficient, which is a constant dependent on units,
x = the pressure discharge exponent.

The emitter exponent is a measure of flow rate sensitivity to pressure changes. The value of 'x' is usually between 0 and 1. The larger the 'x' value, the more sensitive the emitter is to pressure variation. A value of 1.0 means that for each 10% change in pressure, there is a corresponding 10% change in flow rate. In contrast, an 'x' value of 0 means that the emitter flow rate does not change as pressure changes. Normally, the 'x' coefficient is about 0.5 for labyrinth-type and orifice control emitters, less than 0.5 for pressure-compensating emitters, and greater than 0.5 for long flow path or spaghetti tube emitters.

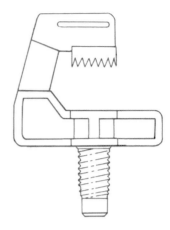

Figure 32-2. Turbulent flow (orifice control) microsprinkler emitter.

Example:
Determine the emitter flow rate if lateral line pressure is 15 psi and the emitter K_d and x values are 2.6 and 0.5, respectively.

Solution:
$Q = K_d \times P^X = 2.6 \times 15^{0.5} = 10.1$ gph.

Laminar Flow Emitters

Laminar flow devices usually have long and narrow flow paths (Fig 32-1). These devices (spaghetti or capillary tubing and spiral path emitters) are generally simple and relatively inexpensive. Water in laminar flow emitters moves at low velocities. Flow is regulated by dissipating energy by friction of the water against the walls of the passages. The smaller the tubing diameter or the longer the passage pathway, the more resistance there is from friction. Flow rate is controlled by tubing diameter and length.

One of the negative aspects of laminar flow devices is that their discharges are relatively pressure sensitive (the flow coefficient 'x' is often 0.7 to 1.0). The emitter discharge rates may change considerably with pressure changes in the system. They are more susceptible to clogging because their flow passages are small. The discharge rates of laminar flow emitters will change with water temperature changes.

Turbulent Flow Emitters

Flow velocity in turbulent flow emitters is higher than in laminar flow devices. The discharge rate of turbulent flow emitters is regulated by dissipating energy by friction of water against the walls of the passages and between the fluid particles themselves. Drippers that use orifice control and most microsprinklers are turbulent flow devices

Figure 32-1. Laminar flow drip emitter.

(Fig. 32-2). Discharge rate at any given pressure is governed primarily by the orifice diameter. Turbulent flow emitters have shorter flow paths and larger diameter passages than laminar flow devices. Thus, flow velocities are greater, and the potential for clogging is less than for laminar flow devices. Flow rates for turbulent flow emitters are less sensitive to pressure (flow coefficient 'x' is about 0.5) and less sensitive to water temperature than in laminar flow devices.

The most commonly used orifice control microsprinkler emitters used in Florida citrus are turbulent flow devices. With orifice control emitters, the base of the emitter determines the discharge rate. The top of the emitter determines the pattern and diameter of spread. Numerous types are available from various manufacturers. As pressure increases, the water is thrown farther from the emitter (Table 32-1). Therefore, there is an increase in the effective coverage area. In general, the diameter increases more than the flow, so the precipitation rate as expressed in inches per hour is decreased.

Vortex Control Emitters

Vortex control emitters (Fig 32-3) are less sensitive to pressure variations than laminar or turbulent

Table 32-1. Effects of pressure on discharge rate and average application depth for typical orifice control emitters with 360° patterns.

Emitter base color	Orifice diameter (inches)	Operating pressure (psi)	Discharge rate (gph)	Coverage diameter (ft)	Coverage area (ft²)	Application rate (inches/hr)
Blue Base	0.04	10	8.0	8.2	59	0.24
		15	9.0	9.5	71	0.20
		20	10.0	11.2	99	0.16
		25	11.3	13.3	139	0.13
Green Base	0.05	10	10.5	11.0	95	0.18
		15	13.0	12.6	125	0.17
		20	14.8	14.5	165	0.14
		25	16.5	16.2	206	0.13
Red Base	0.06	10	15.9	12.3	119	0.22
		15	19.5	14.9	174	0.18
		20	22.9	17.9	252	0.15
		25	25.9	19.3	292	0.14

Note: Different manufacturers have different flow, diameter, and coverage characteristics for the same color emitters. Therefore, it is important to verify the emitter performance when replacing emitters with a different brand.

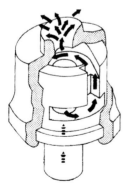

Figure 32-3. Vortex control microsprinkler emitter.

flow emitters (flow coefficient 'x' is about 0.4). In vortex emitters, water is forced to circulate, forming a vortex or whirlpool at the center of the emitter. As the water rotates, centrifugal force pushes it towards the outer edge of the vortex. This action causes a low-pressure area to be formed in the center, where the orifice is located. The result is a reduction in the energy of water at the discharge

point, and a controlled discharge rate. Emitter discharge rate is controlled by vortex design and orifice diameter.

Pressure-compensating Emitters
Pressure-compensating emitters (Fig 32-4) use excess inlet pressure to modify the shape, length, or diameter of the flow path to control the discharge rate. Generally, there is a diaphragm made of an elastic material that deforms to control the discharge rate. As the pressure increases, the diaphragm restricts the passage diameter. Pressure-compensating emitters are designed to discharge at a fairly constant rate over a wide range of pressures (flow coefficient 'x' is normally less than 0.1).

One of the drawbacks of pressure-compensating emitters is that the elasticity of the diaphragm may change over time. As a result, the discharge and pressure compensation characteristics may change over time as well. In addition, diaphragms will often retain some moisture when the pressure is off. The moisture may allow bacterial growth within the emitter. As a result, emitter clogging may occur. Another problem can result from the invasion of ants seeking a food source. The ants may feed on the bacteria and destroy the diaphragm or clog the emitter with body parts.

Emitter Manufacturing Variation

In addition to variation in flow due to pressure, variation in flow between emitters due to manufacturing also occurs. No two emitters can be identically manufactured; some variation will exist from emitter to emitter. The manufacturing coefficient of variation C_{vm} is defined as follows:

$$C_{vm} = \frac{S_{dm}}{X_m} \qquad \textbf{Eq. 32-2}$$

where,

    C_{vm} = manufacturing coefficient of variation
    S_{dm} = standard deviation of emitter flow rate for new emitters operated at the same pressure
    X_m = mean flow rate

A C_{vm} of 0.10 implies that 68% of all flow rates would be within plus or minus 10% of the mean flow rate. The design of the emitter, the material used in its construction, and the precision with which it is manufactured determine the variation for any particular emitter. With recent improvements in manufacturing processes, most emitters have C_{vm} values less than 0.10. Pressure-compensating drip emitters generally tend to have somewhat higher C_{vm} than labyrinth path emitters. C_{vm} values of 0.05 or less are considered excellent, 0.05 to 0.10 are good, 0.10 to 0.15 are marginal, and greater than 0.15 are unacceptable. Independent laboratory tests are performed for emitters of most manufacturers.

Stake Assemblies

Microsprinkler field installations typically use 10- to 20-gph emitters. Spray emitters usually have slotted caps or deflector plates that distribute water in distinct streams. Spinners use a moving part that rotates to disperse the water stream over the wetted diameter. Both types of microsprinklers are often

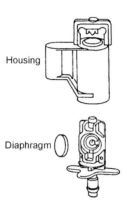

Figure 32-4. Components of pressure-compensating microsprinkler emitter.

preferred over drip systems since they have larger diameter wetting patterns that are especially desirable in areas with coarse-textured soils where lateral movement of soil water is limited.

Emitters are normally attached to stake assemblies that raise the emitter 8 to 10 inches above the ground (Fig. 32-5). The elevated position of the emitters provides a larger wetting pattern, and allows water to be dispersed over low-lying weeds and grass. The stake assemblies usually have 2- to 3-foot lengths of 4-mm spaghetti tubing and are connected to the polyethylene lateral tubing with a barbed or threaded connector (Fig. 32-6). Experience has shown that the barbed-by-barbed connections generally are more durable in the field than other types of connections. Threaded connectors typically have thinner walls and are, therefore, more prone to breakage. Failure of the connectors is most commonly due to impact on the stake assembly or tubing during harvesting operations.

Microirrigation water applications are generally frequent and at low application rates. Therefore, variations in emitter discharge rates, even though small in magnitude, may represent relatively large variations when viewed as a percentage of the total application. There are several factors that may affect the discharge uniformity of emitters, such as pressure differences within the system, elevation changes, design characteristics, stake assemblies, emitter wear, and the unit-to-unit variation of emission devices during manufacture. However, the importance of emitter performance characteristics is often overlooked in the design and operation of microsprinkler systems. To prevent problems associated with pressure loss in the spaghetti tubing, care should be taken to recognize its potential importance during the design process.

Field checks of emitter discharge rates normally yield application rates that are less than the manufacturer's specified rates for a particular pressure in the lateral lines. Oftentimes, the reduced discharge is a result of pressure losses in the stake assembly. The standard 4-mm-diameter spaghetti tubing used by most manufacturers can significantly decrease discharge rates from the manufacturer's specified emitter discharge rates. One study showed that the discharge rates in 1.22-mm orifice diameter (green base) emitters (17.5-gph manufacturer-specified flow rate) had a 12% average decrease when 4-mm-diameter spaghetti tubing was used as compared to 6-mm tubing. Even greater differences were noted with emitters having 0.080 in. orifices (32 gph manufacturer-specified discharge rate). The 4-mm tubing restricted the average discharge

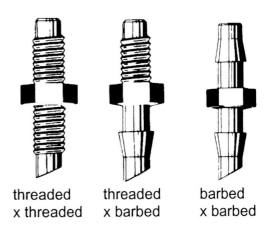

threaded x threaded   **threaded x barbed**   **barbed x barbed**

Figure 32-6. Common types of spaghetti tubing to lateral tubing connectors.

to 19.9 gph compared to 32 gph with 10-mm tubing, a nearly 40% reduction in emitter discharge resulting from friction losses in the 4-mm spaghetti tubing/stake assembly.

Effects of spaghetti tubing on emitter discharge rates were evaluated in a study that used 0.030 in. (black base), 0.040 in. (blue base), 0.050 in. (green base), 0.060 in. (red base), and 0.070 in. orifice (orange base) emitters on both 4- and 6-mm spaghetti tubing and also plugged directly into the lateral with no spaghetti tubing. The emitters were coupled to stake assemblies with various lengths of spaghetti tubing and operated at 15, 25, and 35 psi pressure. Emitters on 6-mm spaghetti tubing assemblies had higher flow rates than those with 4-mm tubing. The advantage of the 6-mm tubing was especially apparent in the red and orange base emitters (with discharge rates greater than 20 gph). The 4-mm tubing reduced discharge rates of orange base emitters to only 75% of the no-tubing rates. Even the black blase and blue base emitters had 5% to 10% reductions from specified flow rates when 4-mm spaghetti tubing was used. Pressure drops in the stake/spaghetti tubing assembly can rob a significant portion of the operating pressure (Fig. 32-7). Even at discharge rates less than 10 gph with a system pressure of 25 psi, pressure losses in the 4-mm stake assemblies were nearly 3 psi. Once the discharge rates reached 32 gph,

Figure 32-5. Microsprinkler stake assembly.

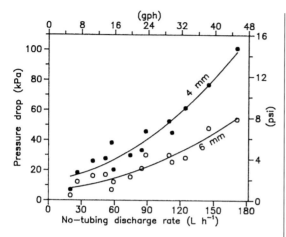

Figure 32-7. Calculated pressure drop in stake assembly for emitters with 4-mm spaghetti tubing as compared to 6-mm tubing.

over 50% of the system pressure was lost in the stake/spaghetti tubing assembly.

Microsprinkler system designs should be based on actual installed emitter discharges, which may be different from those in the manufacturer's literature. Specifications given for emitters often don't account for pressure losses in the stake assemblies. Improper design discharge rates may result in inefficient system designs in terms of both cost and performance. Systems designed to require continuous operation in all zones in order to satisfy peak demands may result in unwanted crop stress if design discharge rates cannot be delivered. Emitters that discharge less than the design discharge rates will require increased run times, increasing operating and maintenance costs. If necessary, most systems can tolerate small increases in zone watering duration such as that which may be required for the smaller orifice sizes. However, the 15% to 25% increase in zone watering times that may be required with the larger orifice bases may be too great for many systems. In addition, decreased pressure and discharge rates generally result in poor wetting patterns.

Wetting Patterns

Wetting patterns of microsprinklers can be an important consideration in sandy soils. They are

especially important when root zones are shallow, a condition that is typical in flatwoods areas. In one study, several microirrigation spinner and spray emitters were evaluated to determine their distribution patterns under no-wind conditions. Emitters were operated in a greenhouse for 6 hours with the applied water caught in a grid of catch cans on a 1-foot spacing. Spinner-type emitters had much higher application uniformities than the spray-type emitters. Spinners had most of the wetted area receiving near-average application depths, with nearly continuous wetting throughout the pattern (Fig 32-8). Spray-type emitters were characterized by wetted spokes radiating from the emitter (Fig. 32-9). The spray-type emitters typically had 50% to 75% of the area within the coverage diameter receiving little or no wetting. Both types of emitters had higher uniformity when the pressure was 20 psi or higher compared to 15 psi.

Most of the spray emitters have very little wetting in the area between the streams, with some spray emitters having little or no wetting in a 8- to 12-foot diameter area centered at the emitter (Fig. 32-9). Application rates at the end of the radials for spray emitters can be quite high (the solid black areas in Fig. 32-9 represent catch cans receiving more than 6 times the average application). For example, a green base emitter with a 15-gph discharge and 15-foot diameter pattern will have an average application rate of 0.14 inch per hour. If the pattern for the emitter was similar to that in Fig. 32-9, the darkest areas would receive over 3.4 inches during a 4-hour irrigation, while the areas with dots would remain essentially unwetted.

In sandy soils where lateral redistribution of applied water is limited, these types of emitters would be unacceptable. They are also a poor choice on some of the lighter soils where the root zone is shallow. Under these conditions, consideration should be given to selecting emitters that have more uniform wetting in order to minimize nutrient-leaching losses and to provide a sufficiently large volume of soil that is wetted so moisture stress does not occur.

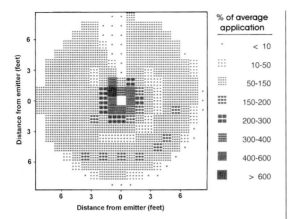

Figure 32-8. Catch distribution pattern for a 0.040-inch orifice diameter spinner-type emitter operated at 20 psi for 6 hours.

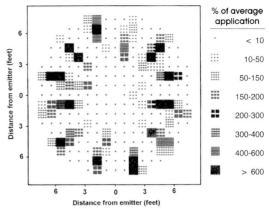

Figure 32-9. Catch distribution pattern for a 0.040-inch orifice diameter spray-type emitter operated at 20 psi for 6 hours.

Young Trees

During the first year or two, citrus growers typically use either microsprinklers with top hats or units that have 90°, 120°, or 180° deflectors that confine the discharge to a small area (typically 3 to 4 feet in diameter) around the tree (Fig. 32-10). The small area wetted by the emitter requires special attention. Application rates per unit area from these assemblies can be quite high (Table 32-2). For example, a 15-gph emitter with a young tree deflector confining the discharge to an area 3 feet in diameter will have an application rate of 3.4 inches per hour. The time required to apply the equivalent of a 3/4-inch depth would only be 13 minutes. Since many large irrigation systems require 20 to 30 minutes for water to travel from the pump to the farthest emitter, this can pose operational problems. When the application rates are high, the number of fertilization applications needs to be increased and the run times may need to be reduced to minimize nutrient leaching.

In a young tree nutrient leaching study, it was demonstrated that the small wetted diameter of these types of microsprinklers could affect quantity of leached nitrogen on sandy soils. The irrigation system in the study was capable of completing a fertigation and flush cycle in less than 30 minutes. Even so, leaching occurred on many occasions due to percolation below the shallow root zone. In a

typical grove situation, materials injected at the pump station may take 30 minutes or more to reach the most distant trees. Therefore, the fertigation/flush cycle times in large systems need to typically be 1 hour or more. The possibility of leaching soluble nutrients is greatest on sandy soils with small-diameter wetting patterns when application rates exceed 0.75 inch per hour.

Mature Trees

In a study that compared drip, seepage tubing, and microsprinkler irrigation, it was found that increased irrigation coverage in a high-density 'Hamlin' orange grove resulted in greater fruit

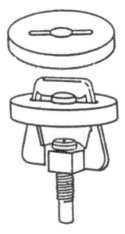

Figure 32-10. 360° emitter with "top hat" down-spray for young trees.

Table 32-2. Typical application rates and time required to apply the equivalents of 3/4-inch and 1-inch depths for microsprinklers with young-tree downsprays.

Wetted diameter (ft)	Application rate and minutes required to apply depths of 3/4 inch or 1 inch											
	10 gph			12 gph			15 gph			17 gph		
	in./hr	3/4 in.	1 in.	in./hr	3/4 in.	1 in.	in./hr	3/4 in.	1 in.	in./hr	3/4 in.	1 in.
3	2.27	20	26	2.84	16	21	3.40	13	18	3.97	11	15
4	1.28	35	47	1.60	28	38	1.91	24	31	2.23	20	27
5	0.82	55	73	1.02	44	59	1.23	37	49	1.43	31	42
6	0.57	79	106	0.71	63	85	0.85	53	70	0.99	45	60

yield in the deep sandy soils of Central Florida. In a study that compared drip, microsprinkler, and overhead sprinkler irrigation, increased irrigation coverage of the orchard floor increased fruit yield for large grapefruit trees. When microsprinkler and drip systems were compared on 'Valencia' orange grown on deep, sandy soils, it was reported that about 50% coverage of the root zone was necessary for adequate growth and yield.

In a flatwoods study, microsprinklers with coverage diameters of 11, 15, 19, and 23 feet were compared on mature 'Ruby Red' grapefruit trees planted at a 25-foot across-row by 20-foot within-row spacing on 50-foot wide beds (tree density of 89 trees per acre). The areas wetted by the microsprinklers represented 19%, 35%, 57%, and 83% of the orchard floor. All emitters had discharge rates of about 18 gph. The area that was wetted by the microsprinklers did not affect fruit size or juice, acid, and soluble solids percentages. However, the trees irrigated with the largest area of coverage (23-foot diameter) had less cumulative fruit yield and total soluble solids per unit area over the 4 seasons than trees irrigated with smaller coverage areas. The very low application rate (less than 0.05 inch per hour) of the 23-foot diameter emitters made managing the water applications difficult.

In general, larger wetting patterns are considered more satisfactory for citrus. However, larger pattern emitters need a corresponding discharge rate to make it possible to manage them effectively. When the average discharge rate per unit of wetted

area decreases to less than 0.08 inch per hour, it requires very long run times to move the water into the mid and lower root zone with the microsprinklers. Typically there is more potential for wind drift, evaporation, and wetting of non-productive areas (especially on bedded citrus) as the diameter of wetting increases. In high-density plantings, smaller wetting patterns that provide intertree wetting may be preferred over larger wetting patterns that may become distorted due to "shadows" in the coverage caused by interference by tree trunks and low branches.

The average application rate of a microsprinkler can be estimated using the following equation for flow rates in gph and the radius of application (r) in feet:

$$\frac{\text{inches}}{\text{hour}} \approx 0.5 \text{ x } \frac{\text{gph}}{r^2} \qquad \textbf{Eq. 32-3}$$

Example:
Estimate the average application depth for a 10-gph emitter with a 12-foot diameter wetting pattern.

$$\frac{\text{inches}}{\text{hour}} \approx 0.5 \text{ x } \frac{\text{gph}}{r^2} = 0.5 \text{ x } \frac{10 \text{ gph}}{6 \text{ ft}^2}$$

$$= 0.14 \text{ inch per hour}$$

Effects of Wear
Wear on emitter orifices over long periods of operation can significantly increase microsprinkler discharge rates and wetting patterns, especially if the irrigation water contains sand. In one study,

Table 32-3. Average application depths (inches per hour) for various microsprinkler coverage diameters and discharge rates (shaded discharge/diameter combinations of 0.08 inch per hour and less may be difficult to manage effectively).

Wetted diameter (ft)	Emitter application rate (inches per hour)					
	7.5 gph	10 gph	12.5 gph	15.0 gph	17.5 gph	20 gph
6	0.42	0.57	0.71	0.85	0.99	1.13
8	0.24	0.32	0.40	0.48	0.56	0.64
10	0.15	0.20	0.25	0.31	0.36	0.41
12	0.11	0.14	0.18	0.21	0.25	0.28
14	0.08	0.10	0.30	0.16	0.18	0.21
16	0.06	0.08	0.10	0.12	0.14	0.16
18	0.05	0.06	0.08	0.09	0.11	0.13

several combinations of flow control method, base size, and operating pressures were examined to determine their effects on long-term microsprinkler discharge rates. Flow control methods were by orifice control (OC), by vortex control (VC), and by pressure compensation (PC). Blue base emitters with nominal discharges of 10 gph and green base emitters with nominal discharges of 15 gph with the three flow control methods were operated at 30 psi for 2000 hours.

Initially, emitter discharges were individually measured from 20 emitters, and then repeated after 500, 1000, and 2000 hours. OC and VC emitters operated at 30 psi had increases in discharge rates of 8% to 11% after 2000 hours of operation (Fig. 32-11). After 500 hours of operation, green base PC emitters had reductions in discharge of about 7% when operated at 20 psi and 18% when operated at 30 psi. These reductions in flow rate resulted from changes in the properties of the diaphragm. After 2000 hours of operation, the discharges of the PC emitters were about 8% less than initial values.

Microsprinkler systems need to be designed with consideration given to the effects of wear on system components. Pumps, filters, and other system components should be designed with enough extra capacity to accommodate increases in discharge rates as much as 10% to 15% due to wear on emitter orifices. The amount of increase in emitter discharge rates over time depends on several factors such as operating pressure, materials used in emitter manufacturing, water source, filtration, and hours of operation. If the water contains sand or abrasive grit, the wear problem can be greatly accelerated.

Emitter Clogging

One of the biggest challenges of using micro-irrigation systems is clogging of emitters, resulting in lowered system performance and water stress to the nonirrigated plants. Both surface and groundwater can experience rapid changes when drawn into an irrigation system. Changes can occur in pressure, temperature, velocity, and composition, which then cause changes in chemical reaction balances and bioactivity. These can result in reactions that cause the formation of scales and slimes that impair the irrigation system.

The main problem with surface water systems is slime and bacterial contamination. Even slight mineral loads in the water are able to support

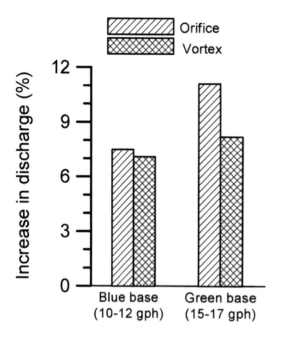

Figure 32-11. Increase in discharge rate after 2000 hours of operation at 30 psi for blue and green orifice and vortex-control emitters.

significant bacterial populations. This is especially true with surface waters that have an abundant source of carbon dissolved in them to support the metabolic processes of the microorganisms. Scale formation is the primary problem in systems served by groundwater. The appearance and development rate of a scale layer are dependent on the composition and relative concentrations of the minerals in the water.

Prevention of clogging by scale can be best accomplished by frequent water treatment. The chemicals involved are generally composed of a variety of organic and inorganic compounds that have been found to prevent and dissolve scales. Product stability, injection rates, biological control capabilities, and required dosages vary widely. Shock treatments are directed towards existing problems. In order for shock treatments to be effective, the system must be completely flushed after each shock treatment. Preventative treatments usually require less manual labor since little or no line flushing is required and can be accomplished with normal

irrigation practices. The two major shock treatments used are chlorination (used to kill bacteria and algae) and acidification (used to dissolve or break up existing scale from the water lines). Acid injection is also used to reduce pH in order to make chlorination more effective.

Effects of Emitter Design on Clogging

Several types of microirrigation spinner and spray emitters were evaluated over a 3-year period to determine long-term maintenance characteristics. Emitters were located in a block of mature 'Temple' orange trees on single beds with a 30-foot between-row times 22.5-foot within-row spacing. The microsprinkler system used surface water with 50-mesh screen filtration. Emitters were all in the same irrigation zone and received water of the same quality at the same time and for the same duration.

All emitters were thoroughly examined twice a year. At the time of examination, the conditions of all emitters were recorded and any that were malfunctioning were replaced. The average clogging rate over the study period ranged from 1.6% to 34.4%, depending on the emitter model. The design that used an enclosed cap to disperse water (Cap, Fig. 32-12) had the highest clogging rates. The least amount of clogging occured in a model that had a plug that sealed the orifice (Spinner w/plug, Fig. 32-12) when there was no water pressure. The emitter design with the cap had nearly 20 times the total clogging over the 3-year study as the design with an orifice plug (Fig. 32-13). Therefore, consideration should be given to long-term maintenance requirements and costs, resulting from emitter design as well as discharge rate, pattern diameter, and the initial cost of the emitter.

Orifice Diameter

Orifice diameter can be an important consideration in emitter clogging, and was evaluated in a field study in a flatwoods grove. The clogging rates of microsprinkler emitters with orifice diameters of 0.030 in. (black base), 0.040 in. (blue base), 0.050 in. (green base), and 0.060 in. (red base) were evaluated in a citrus grove over a 3-year period.

The experiment used a randomized complete block design with 8 emitters per plot and 5 replications of each treatment. The irrigation water was drawn from a pond and filtered through a sand media system. Every 3 months, each emitter was inspected, and those that were clogged were examined to determine the cause of clogging. All effected emitters were thoroughly cleaned and/or replaced. Overall, 46% of the clogging was due to algae, 34% was from ants and spiders, 16% was from snails, and 4% was from physical particles such as sand and bits of PVC. Most of the clogging from physical particles occurred during the first year after the system was installed. The clogging rate was directly related to the orifice diameter (Fig. 32-14). About 22% of the black base emitters required cleaning or replacement during each quarter compared to about 14% of the blue base, 7% of the green base, and 5% of the red base emitters. If the water source permits, emitter orifice diameters of 0.05 in.(green base) or larger are recommended to minimize the maintenance costs associated with clogging of emitters.

Selenisa Sueroides

Larvae of *Selenisa sueroides* are known to use the spaghetti tubing on microsprinkler assemblies in citrus groves as an alternate host to native plant species with hollow stems. The *S. sueroides* caterpillars chew holes in the spaghetti tubing in order to enter and pupate. The life cycle of *S. sueroides* is approximately 56 days from oviposition on foliage to adult emergence from the pupal chamber. *S. sueroides* larvae require a hollow stem in which to pupate, and they use the spaghetti tubing as an alternate host to native plant species with hollow stems. They bore holes about 5 mm in diameter in the tubing and enter the hollow interior (Fig. 32-15). The larvae then seal the hole with silk and pupate.

One study reported more severe damage by *S. sueroides* to stake assemblies constructed with black tubing than those with colored spaghetti tubing. The study also showed that tubing material could affect the severity of damage, with higher incidence of damage on poly tubing compared to

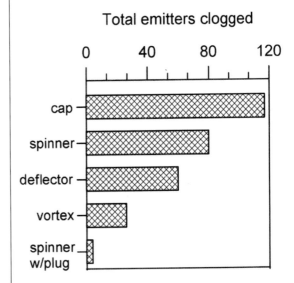

Figure 32-13. Total number of emitters clogged by type of water dispersion mechanism during 6 semiannual inspections.

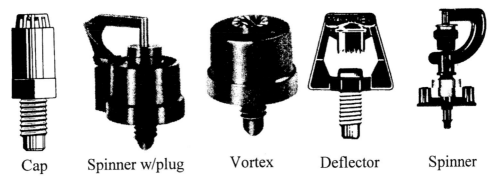

Figure 32-12. Types of emitters evaluated for clogging potential in mature citrus grove.

vinyl spaghetti tubing, and even less damage on assemblies made with colored tubing.

Another study evaluated the effectiveness of spraying emitter assemblies with Sevin and Teflon sprays to prevent damage from *S. sueroides* caterpillars. Sevin showed little effect, but stake assemblies sprayed with Teflon had about 40% less damage than the untreated assemblies. In addition, units constructed with polyethylene tubing suffered more damage from *S. sueroides* than those made with vinyl tubing. The Teflon spray coating provided slightly better protection than Sevin or the nontreated assemblies. However, none of the combinations of spaghetti tubing material and spray treatment were effective in preventing damage from *S. sueroides*. All of the nontreated and the Sevin-treated units had damage, along with about 80% of those treated with Teflon. About 20% of the units had holes bored completely through the spaghetti tubing. The larvae of *S. sueroides* may try unsuccessfully to make several holes before succeeding. Therefore, it was assumed that given enough time, the number of assemblies needing replacement would be directly related to the incidence of unsuccessful attempts to penetrate the tubing.

It may be more practical to take care of the maintenance problems caused by the *S. sueroides* caterpillars through methods other than treating the emitters. A program to eliminate known host plants through herbicide application and mowing is probably more cost-effective. These host plants include the several *Aeschynomene* species found in South Florida: Sesbania, coffeewood, poinciana, and senna. Feeding specimens of *S. sueroides* have been found on *Triadica sebifera*, Para grass, wild locust, Chinese tallow, *Pithecellobium dulce* (cat claw), and *Daubentonia punicea*. If infestations of *S. sueroides* caterpillars persist after mid-September, spray applications of liquid Teflon to the emitter assemblies should provide some protection from *S. sueroides* larvae.

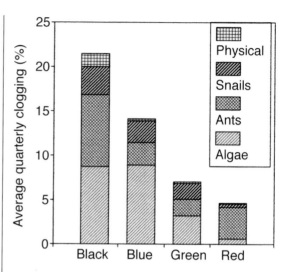

Figure 32-14. Average quarterly clogging rates by emitter base.

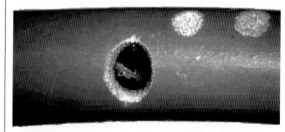

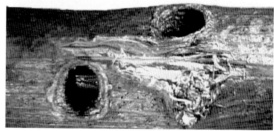

Figure 32-15. Holes bored in native *Aeschynomene* (bottom) by *S. sueroides* and completed and attempted holes in 4-mm spaghetti tubing.

Ant Damage

Ants are a frequent cause of plugging in emitters, resulting in nonuniform watering and trees not receiving adequate water. Typically, a *S. invicta* colony is started by a single queen who excavates a burrow in the soil after she alights from the mating flight. Her initial brood of 10 to 20 individuals will reach maturity within 24 to 30 days. These workers form the nucleus of the colony that can become a nest of 11,000 workers within a year.

The foraging activity of the ants is regulated to a great extent by the prevailing soil temperature. Little significant foraging occurs when the temperature at the 5-cm depth is below 60° F or above 100° F. Maximum foraging occurs between 70° and 95° F. Ants are highly omnivorous and opportunistic, so they may feed upon whatever plant or animal material they encounter. Their diet primarily consists of arthropods (insects, spiders, and myri-opods), earthworms, and other small invertebrates, although some feeding on plants does occur.

Micropulsators

Ants can also physically damage certain types of emitters. Reports have shown damage caused by *S. invicta* on the silicone diaphragms of micropulsators. Micropulsators (Fig. 32-16) have become an alternative for irrigation systems with limited water supplies, or to retrofit microsprinklers into existing drip systems. Micropulsators provide low discharge rates, about 3 gph, and have a distribution pattern about 9 feet in diameter. In operation, water entering the interior chamber displaces the silicon diaphragm and as the diaphragm stretches, the seal at the top of the chamber is broken and the water is expelled. The diaphragm retracts and the cycle begins again.

In one study, ants were found to have severely damaged silicon diaphragms of micropulsators. The units, which had only been in the field for about 16 months, had various degrees of damage. Most of the emitters had portions of the silicon diaphragm removed (Fig. 32-17). In several units, most of the silicon diaphragm had been removed. The destruction of the components was attributed to action by *S. invicta.*

In many cases, rupture of the diaphragm occurred from the weakening of the diaphragm sides. The ants were thought to have been attracted by moisture and/or bacteria in the diaphragm and probably entered the expansion chamber of the

micropulsator through the vent holes in the sides of the case.

The study showed that only minor damage occurred to the silicon components when the vent passageway widths were limited to a width of 0.02 inch. To ensure that ants do not enter passageways, vent openings need to be less than the 0.02 inch used in the ant-bar style micropulsators. Although the 0.02-inch openings did not completely exclude ants, the ant-bar design was effective at preventing moderate and severe damage to the silicon diaphragms. As a result, none of the ant-bar units were in danger of failure.

Pressure-Compensating Drip Tubing

Damage to the diaphragms in pressure-compensating drip and microsprinkler emitters has

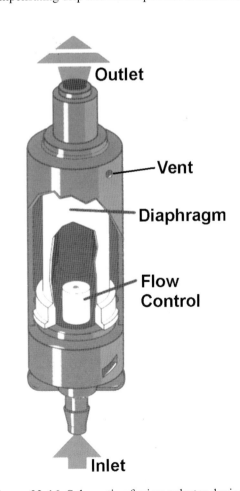

Figure 32-16. Schematic of micropulsator device.

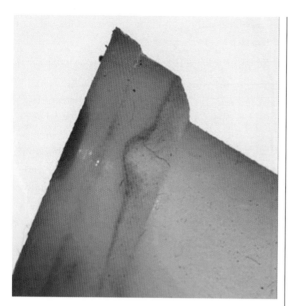

Figure 32-17. Damage to micropulsator diaphragm base from *S. invicta* with silicon removal areas highlighted.

Figure 32-18. Ant damage to ends of used emitter diaphragms (left and bottom) compared to new diaphragm (upper right).

been reported to be caused by ants. Typically, damage is most apparent when systems are started following periods of nonuse due to adequate rainfall.

In one study, emitters were randomly cut out of several thousand feet of tubing that was replaced because of ant damage. It was obvious why many of the emitters were discharging streams of water instead of having a drip action. The diaphragms in many of the emitters had significant portions missing (Fig. 32-18) or had holes (Fig. 32-19) that destroyed the pressure and flow-reducing abilities of the emitters.

Most of the used emitters also had portions of the molded body missing. Magnification showed the scratch marks typical of ant damage on other plastic components. When several of the emitters were taken apart, ant body parts were found inside (Fig 32-20). The combination of enlarged orifice and lack of pressure compensation (due to diaphragm failure) has the potential to drastically alter the hydraulics of the system and result in overall unacceptable system performance.

Maintenance

In a survey, Florida citrus production managers were asked questions concerning irrigation water source, filtration, emitters, clogging problems, water treatment, and estimated microirrigation system maintenance costs. The 61 respondents represented about 130,000 acres of citrus located throughout the state. Slightly less than half of the respondents used wells as the water source, while the rest used surface water. About 25% of the respondents, accounting for 22,000 acres (17% of total), used drip systems, while most (75%) used microsprinklers.

Most of the responses indicated clogging was caused by two or more factors. The percentage of plugged emitters attributed to insects and to physical particles was about the same, each being about 22% of the total. Clogging due to the formation of precipitates only accounted for about 12% of the total plugging. Biological plugging of emitters was by far the most severe problem reported in the survey. The respondents attributed about 45% of the emitter clogging to biological factors such as bacteria, slimes, and algae.

The average annual cost of emitter maintenance ranged from a few dollars to more than $60 per acre. Most of the reported costs were in the $10 to $40 per acre range, with an overall average of $25 per acre (Fig. 32-21). Drip systems averaged $17

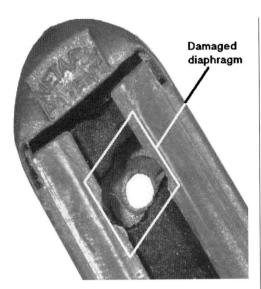

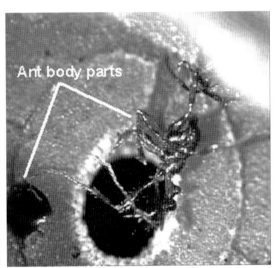

Figure 32-19. Emitter with cover removed, showing ant-caused destruction of diaphragm in area of orifice.

Figure 32-20. Enlarged orifice with ant body parts found inside drip emitter.

per acre in annual maintenance costs, while microsprinkler in systems averaged $30 per acre. Assuming the survey is representative of the entire acreage that is irrigated with microsprinklers in Florida, the maintenance of systems is a $13+ million per year problem to Florida citrus growers.

Less than half of those surveyed were chemically treating the irrigation water. Of the 45% that were using chemicals to treat the irrigation water, about 25% were using commercial water conditioners, 20% used acid, 35% injected chlorine, and about 20% were using a combination of two or more materials. Water treatment was continuous for 27% of the respondents with another 21% reporting injections at 1- to 3-week intervals. Monthly injections were indicated by 27%, 15% reported 2- to 3-month intervals, 3% used 4- to 6-month injections, and 6% reported treating the system at intervals greater than 6 months.

Considerations

Microirrigation systems should be treated like any other piece of expensive equipment. With installed costs of $800 to $1,200 per acre, most large systems represent a considerable investment. During the design stage, consideration should be given to long-term maintenance costs in the economic analysis. For a variety of cultural, economic, and water supply reasons, microsprinklers with orifice diameters of about 0.040 inch are the most commonly used emitter sizes in Florida citrus groves. These blue base microsprinklers have been shown to be similar to the green base (0.050 inch orifice) emitters in terms of insect, physical, and snail clogging. However, the blue base emitters are more prone to clogging from algae, the most common type of plugging. If the end user of the microsprinkler system has concerns about future clogging problems, system designers should consider emitters with larger orifice diameters, because they have been demonstrated to decrease clogging rates.

Larger emitters, however, require water sources capable of supplying the required rate. Table 32-3 lists the gpm required per acre required to meet various combinations of emitter discharge rates and tree planting densities.

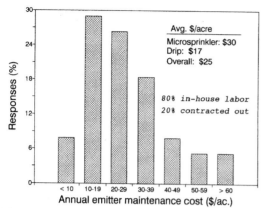

Figure 32-21. Distribution of annual micro-irrigation system maintenance costs from survey responses.

> Example:
> Compare the pumping requirements for 10-gph emitters versus 16-gph emitters for a block that has 76 net acres with a tree planting density of 140 trees/ac.
>
> From Table 32-3:
>
> 10 gph @ 140 trees/ac = 10 gal/hr x 60 min/hr x 140 trees/ac = 23 gpm/ac
> 23 gpm/ac x 76 ac = 1748 gpm
> 16 gph @ 140 trees/ac = 16 gal/hr x 60 min/hr x 140 trees/ac = 37 gpm/ac
> 37 gpm/ac x 76 acres = 2812 gpm
>
> If 1748 gpm are available, how many acres could be irrigated with 16-gph emitters?
>
> 1748 gpm/37 gpm/ac = 47 acres

The emitter selection process should consider uniformity as well as other factors, such as cost, wind effects, system constraints, maintenance, and soil type, so that the best emitter for a particular field condition is selected. Designers should take time to ensure that the wetting pattern of emitters is compatible with the soils, tree spacing, and rooting pattern of the trees. Emitters with distinct spoke patterns may not be as efficient as more uniform patterns where shallow root systems and very sandy soils are encountered. Consideration also must be given to the water requirements of the

mature tree. Higher density plantings with smaller trees will require less water per tree than more widely spaced trees. Larger wetting patterns may be more desirable for more widely spaced trees. In all cases, designers should ensure that tree water requirements can be met with reasonable run times, and not result in percolation of nutrients and water below the root zone.

When selecting the microsprinklers to irrigate a particular block, consideration should be given to tree water requirements, soils, and the operating philosophy of the grower. The emitter that is selected must be capable of applying the maximum tree water replacement quantity required during an extended drought period. The emitter must have a coverage diameter to allow this volume of water to be stored in the wetted area without causing deep percolation below the root zone. The emitter should also have a sufficient application rate in the wetted area, so that the grower-required run times are compatible with the power unit and labor availability. For example, since on-peak electric power costs can be several times off-peak costs, emitters must be able to apply the required volume of water during off-peak hours.

Emitter design and operating conditions can influence maintenance requirements. Shielding the orifice when the emitter is not operating is an effective method of reducing clogging problems. Larger-diameter orifices are more effective in flushing out particles, insects, or algae than smaller orifices. Designs that incorporate these features and also provide secure attachment of the emitter to the stake assembly should reduce maintenance requirements. Most importantly, designers should verify actual system performance by field testing to ensure that actual operation meets design standards. When repairing broken or missing emitters, operators should take care to replace the original equipment with emitters that have similar discharge characteristics to maintain distribution uniformity within the system. Whatever design is employed, it is essential that proper system

Table 32-4. Required flow rate (gpm/acre) for various emitter discharge rates (gph/tree) and tree planting densities.

gph/tree	Trees per acre						
	80	100	120	140	160	180	200
6	8	10	12	14	16	18	20
8	11	13	16	19	21	24	2
10	13	17	20	23	27	30	33
12	16	20	24	28	32	36	40
14	19	23	28	33	37	42	47
16	21	27	32	37	43	48	53
18	24	30	36	42	48	54	60
20	27	33	40	47	53	60	67
22	29	37	44	51	59	66	73
24	32	40	48	56	64	72	80

filtration, flushing, chlorination, and operation procedures be followed to minimize maintenance problems.

A high application rate can lead to leaching of nutrients and pesticides. Application rate can be increased by increasing emitter output or decreasing spray pattern size. With small diameter spray patterns and higher application rates, relatively short irrigation durations can drive water below the main root zone. By adjusting spray diameter, irrigation duration, and emitter output, microirrigation systems can be managed to meet tree water needs while reducing overirrigation and chemical leaching.

by Brian Boman and Dorota Haman

Chemical application through irrigation systems is called chemigation. Chemigation has been practiced for many years, especially for fertilizer application (fertigation). However, other chemicals are also being applied through irrigation systems with increasing frequency. The primary reason for chemigation is economy. It is normally less expensive to apply chemicals with irrigation water than by other methods. The other major advantage is the ability to apply chemicals only when needed, and in required amounts. This "prescription" application not only emulates plant needs closer than traditional methods, but also minimizes the possibility of environmental pollution. Through chemigation, chemicals can be applied in amounts needed, and thus large quantities are not subject to leaching losses if heavy rainfalls follow applications. This reduces adverse environmental impacts in addition to saving the time and money needed to reapply the materials.

Backflow Prevention

Currently, Florida state law requires that backflow prevention equipment be installed and maintained on irrigation systems in which chemicals are injected for agricultural purposes (Fig. 33-1). The possible dangers in chemigation include backflow of chemicals to the water source, causing contamination and water backflow into the chemical storage tank. Backflow to the storage tank can rupture the tank or cause overflow, contaminating the area around the tank, and possibly contaminating the water source. Safety equipment is available that, when properly used, will protect both the water supply and the purity of the chemical in the storage tank.

Once the problems of contamination with chemicals are solved, the risk of liability in chemigation is not much greater than the risk from the field use of chemicals applied by other means. For technical reasons, such as reduced wind drift, rapid movement into the soil, and high dilution rates, chemigation could result in less risk of liability than the traditional methods of chemical application, if proper backflow preventors and other safety devices are used.

The rules governing the installation of backflow prevention devices are found in Section 487.055 of the Florida Statute (Appendix 6). The rules relating to backflow protection were designed to protect the surface and groundwater resources of the state.

An antisyphon device is a safety device used to prevent backflow of a mixture of water and chemicals into the water supply. Chemicals can be any substances that are intentionally added to water for agricultural purposes, while toxic chemicals are pesticides with labels bearing the signal words "Danger" or "Poison."

Safety Equipment

Any irrigation system designed or used for the application of chemicals shall be equipped with the following components:

Figure 33-1. Typical backflow prevention device with vacuum breaker, check valve, and low-pressure drain.

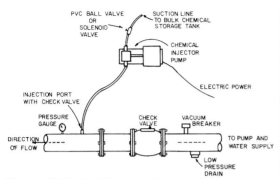

Figure 33-2. Backflow requirements for systems where nontoxic chemicals will be injected.

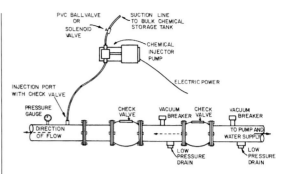

Figure 33-3. Backflow requirements for systems where toxic chemicals will be injected.

a) A functional check valve located in the irrigation supply line between the irrigation pump and the point of injection of chemicals (required). The check should be installed so that it is no more than 10 degrees from the horizontal. The check valve will prevent water from flowing from a higher elevation or pressure in the irrigation system back into the well or surface water supply. It will also prevent water from being syphoned back to the water source. Thus, water contaminated with chemicals cannot flow back into the water supply. A single antisyphon device assembly (Fig. 33-2) can be used for those systems where nontoxic chemicals, such as fertilizers, will be injected. A double antisyphon device assembly (Fig. 33-3) is required for systems where toxic chemicals will be injected. The functioning of each device in the double assembly system must be capable of being checked independently of other devices to ensure the effectiveness of the system.

b) A low-pressure drain with an orifice size of at least 3/4 inch in diameter (required). State law requires that the drain be located on the bottom of the horizontal pipe between the check valve and the water source. It must be located so that the water flow does not drain back to the water source. It must be level, must not extend beyond the inside surface of the pipe, and the outside opening of the drain must be above grade. A clearance of two inches between the drain and ground surface is required to ensure that the drain will operate freely.

c) A vacuum breaker, installed on the top of the horizontal pipe between the check valve and the irrigation pump, and opposite to the low-pressure drain. The vacuum breaker needs to have an orifice size of at least 3/4 inch in diameter, and must be located upright and above the irrigation pipe so that it functions effectively. The vacuum breaker will allow air to enter the pipe when pumping stops so that water flowing back to the pump will not create a suction, pulling additional water and chemicals from the irrigation system with it.

d) A functional check valve on the chemical injection line. If injector pumps are used, they need to be installed so that when water flow ceases, the injector pumps will not operate. In addition, a method should be provided for positive shutoff of the chemical supply when the injection system is not in use. When chemical injector pumps are used, power supplies must be interconnected so that the injector pump cannot operate unless the irrigation pump is also operating. If the injector pump is mechanically driven (from a drive belt with an engine-driven pump (Fig. 33-4) or by water flow in the irrigation system), the power supply interconnection is not needed. In these cases, when the engine stops, the injector pump will also stop.

When the chemical injector pump is electrically driven (Fig. 33-5), its electrical circuit must be interconnected with the irrigation pump's electrical circuit to ensure that it stops when the irrigation pump stops. The injector pump can also be controlled, using a pressure switch or flow switch that automatically disconnects power to the injection pump when pressure or flow is discontinued in the irrigation system. These precautions ensure that the chemical injector pump does not continue to inject into an empty irrigation pipeline, or worse, backwards into the water supply.

Spring-loaded electrical switches can be installed for the testing and calibration of the chemical injection pump when the irrigation pump is not operating. Only a spring-loaded switch, which requires the presence of an operator to engage the switch, is permissible. A multiposition switch with automatic and manual operation positions is not permissible because it would be possible for the operator to accidentally leave the switch in a manual operation position and override its safety function.

If chemicals are injected by means other than electric injector pumps, interconnected power supplies are not required. However, all the other backflow prevention devices are required.

Storage Tank Lines

A check valve on the chemical injection line must be used to prevent backwards water flow from the irrigation system to the chemical storage tank. This precaution prevents dilution of the chemical by the irrigation water. It also prevents possible rupture or overflow of the chemical storage tank and pollution of the surrounding area.

If chemical injection pumps are used, chemical injection line check valves are typically spring-loaded and require a relatively high pressure to allow fluid to flow through them. These valves only permit flow that is a result of the high pressure generated by the pump. When the injector pump is not operating, chemicals will not leak because of the small static pressure created by the chemical level in the storage tank.

A valve must be provided for positive shutoff of the chemical supply when the injection system is not in use. This item can be a manual gate valve, ball valve, "normally closed" automatic valve, or

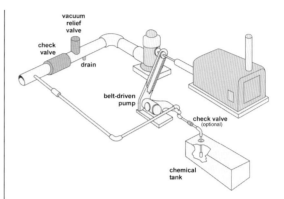

Figure 33-4. Engine-driven injection pump not requiring interconnection shutoff (guards removed for illustration).

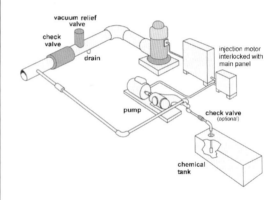

Figure 33-5. Requirements for electric injection pump. Note that for toxic chemicals, double backflow prevention is required.

other positive shutoff valve. The valve must be installed near the bulk chemical storage tank, on the suction side of the injection pump if an injection pump is used. It must be open only when the injector pump is operating.

An advantage of using an automatic valve is that it will shut off the chemical supply automatically when the injector pump shuts off. A disadvantage is that corrosive chemicals may cause the valve to fail to operate after a period of time. A PVC ball valve or gate valve will be less affected by corrosion; however, it will require manual operation. A good practice is to install both the manual and automatic valves. A manual valve located at the chemical tank will provide positive shutoff of chemicals when the irrigation system is not in use. All check valves, low-pressure drains, and vacuum breakers should be kept free of corrosion or other buildup and operative at all times during operation of the system.

Summary of Safety Requirements

The safety equipment components are located and function in such a way as to prevent contamination of ground and surface water by the applied chemicals. The devices incorporate common sense to minimize spills and operator hazards. Table 33-1 lists the commonly required devices, their proper location, and the purpose they seek to achieve.

Chemical Storage Tanks

Chemical storage tanks must be located in an area that is remote from the well site or surface water supply. Tanks should also be sloped so that contamination of the water supply will not occur if the tank ruptures or if a spill occurs.

The chemical supply tank should be constructed of material that withstands the corrosive chemicals that may be stored in it. Some chemicals and tank materials are subject to degradation by sunlight; therefore, chemical tanks are often painted to exclude sunlight. In some cases, the chemical tank will need to be diked to contain the chemical in the event of a tank failure. State law requires that chemical tanks must be placed in containment structures (or dikes) if used to store chemicals that are hazardous, such as pesticides (Figs. 33-6 and 33-7). Containment can be achieved by construction of a watertight concrete pad with cinder block walls sufficiently large to hold 1.5 times the

Table 33-1. Descriptions of required safety devices for chemical injection.

Device	Description/Location	Purpose
Irrigation check valve	Between well and injection points	Prevents chemicals from flowing backwards and entering the water source
Injection line check valve	At the injection point; it is a one-way valve with a 10 psi spring that closes when not under pressure	Prevents water from flowing backwards into the chemical tanks, which would cause the tank to overflow and spill
Vacuum relief valve	Between check valve and well	Prevent vacuum when pump shuts off; reduces chance of backflow
Low-pressure cutoff	On irrigation pipeline	Turns off injector power when irrigation water pressure is low
Low-pressure drain	Between well and irrigation line check valve	Discharges any water that might leak through the check valve after irrigation pump is working
Normally closed solenoid valve	Between injection pump and chemical tank	Prevents tank from emptying unless injector is working
Interlock	Between injection pump and irrigation pump control panel	Prevents injection if irrigation pump stops

Note: If there is no possibility of water source contamination (e.g., the injection point is downstream of an air gap), some of these devices may not be required.

Figure 33-6. Polyethylene containment tank (adapted from Burt et al., 1995).

Figure 33-7. Concrete containment area (adapted from Burt et al., 1995).

capacity of the chemical tank in the event of tank failure. Soil liners can be used under the tanks in permeable soil areas or where toxic chemicals are being used.

The size of the supply tank should be at least large enough to contain all the chemical for one injection, for the entire area irrigated. The volume of the tank can be determined by:

$$V = \frac{r \times A \times n}{c \times d} \qquad \text{Eq. 33-1}$$

where,

  V = volume (gallons)
  r = rate of application (lb/ac)
  A = area to be fertigated (acres)
  c = concentration of fertilizer source (N-P-K, decimal)
  n = number of applications between tank fillings
  d = density of fertilizer material (lb/gal)

Example:
Determine the chemical tank size required for a 100-acre citrus block given the following criteria: sufficient storage for two fertigations, application rate per fertigation is 6 lb N per acre. Fertilizer material is a 9-2-9 solution made from NH_4NO_3, KCl, and phosphoric acid with a density = 10.6 lb/gal.

Convert the N concentration of the 9-2-9 solution to a fraction → 9% N = 0.09 N

$$V = \frac{r \times A \times n}{c \times d} = \frac{6 \text{ lb/ac} \times 100 \text{ ac}}{10.6 \text{ lb/gal} \times 0.09 \text{ N}}$$

x 2 fertigations

= 1257 gallons

Ten percent additional storage should be added to accommodate dead storage at the bottom of the tank. Therefore, a 1400 gallon chemical storage tank would be the minimum size.

Other Backflow Requirements
Some counties and municipalities have backflow prevention regulations that may be more restrictive than state law. All public water supply systems have more restrictive requirements. The Florida Department of Environmental Protection (FDEP) has regulations concerning the usage of chemical storage tanks in Florida.

Compliance with the state law governing backflow prevention from irrigation systems does not alleviate the need to comply with other regulations that may apply. Rather, the state law should be considered to stipulate only the minimum backflow prevention requirements for irrigation systems in Florida.

Equipment Installation and Maintenance
To be serviceable, all equipment must be properly installed. Electrical installations should be in accordance with state and local codes. Only UL-approved equipment and materials developed for outdoor conditions should be used. Water and electricity are a potentially dangerous mixture.

All valve and pipe components must be pressure-rated to be able to withstand the high pressures of chemical injection. Chemicals and their concentrations must be compatible with the irrigation system materials. Storage tanks must be designed for the chemicals being used, and they must be properly located, installed, and maintained to guard against spillage.

Chemigation safety is more than the right equipment properly installed; equipment requires regular maintenance. Many chemicals are highly corrosive. Corrosion-resistant components should be used and maintained by flushing with clean water between uses. All components should be checked before use and replaced before they become inoperable.

Chemical Injection Methods
There are several methods of chemical injection into an irrigation system. The choice of appropriate methods and equipment depends on several factors. If solid materials will be injected, they need agitation and mixing at the pump site. Liquid fertilizers and agricultural chemicals, on the other hand, can be injected directly from their storage tanks. Injection of most fertilizer materials normally can be accomplished without great concern for workers getting exposure to the materials. However, when handling and injecting acids and toxic pesticides, worker safety is of great concern.

Some installations may require more than one injector because of vastly different flow rate requirements for the materials used. For instance, fertilizer injections are normally at a rate of at least 0.1% of the system flow rate. If the irrigation system delivers 1000 gpm, the injection rate should be at least 1 gpm. The injection rate for acids, water conditioners, and some pesticides may be less than 10% of

that for fertilizers, making it impossible to use the same injection device for both applications.

Sometimes, it is desirable to limit the amount of chemical that can be applied during an irrigation event. For instance, it may be advantageous to limit the amount of fertilizer so that a large over-application will not seriously damage or kill young trees. Other applicators may want to inject a specific volume each time, even though the run times or pressure may vary. Oftentimes, the surest way of limiting the applied quantity is to use a larger storage or nurse tank to fill a smaller injection tank. With only a limited volume of chemical in the tank, it will be impossible to inject too much material even if other safeguards fail.

On electric pumps, controllers or timers can be used to limit the duration that injection pumps can operate. On water-powered pumps, volumetric controls can be used to shut the injection system off once a specified volume has been injected.

When installing a chemical injection system, it should be designed so that one can easily flush clean water through the injectors and fittings. The life of most injectors will be extended if they are flushed after use. Frequent flushing helps maintain gaskets and metal components and may prevent encrustations from developing within the injector.

Normally, it is desirable to inject materials upstream from filters. The filters should trap any contaminants or precipitates that occur as a result of the injections. However, because of their corrosive effect, acids should normally be injected downstream from the filters. It is also necessary to discontinue injections during filter backwash cycles. On filter systems with automatic backwash controls, a controller should be installed to control both the backwash cycles and the injectors.

Injectors
Injection methods can be classified according to the method of operation. These methods include the use of centrifugal pumps, positive displacement pumps, pressure differential methods, and the

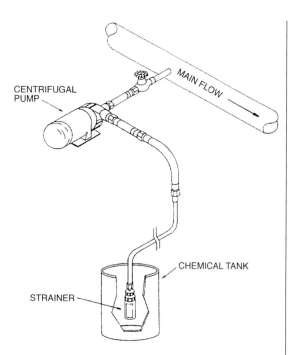

Figure 33-8. Centrifugal pump used to inject into the irrigation line.

Figure 33-9. Piston injection pump.

venturi principle. Some injectors use a combination of these methods.

Centrifugal Pumps

Small, radial flow centrifugal pumps (booster pumps) can be used to inject chemicals into irrigation systems (Fig. 33-8). For a centrifugal pump to operate as an injector, it is necessary that the pressure produced by the pump be higher than the pressure in the irrigation line. However, the flow rate of the chemical from the pump depends on the pressure in the irrigation mainline. The higher the

line pressure, the smaller the flow rate from the injection pump. Therefore, centrifugal pumps require calibration while operating. It is not recommended that this type of pump be used when the injection rate must be very precisely controlled.

Positive Displacement Pumps

Positive displacement pumps are frequently used for injection of chemicals into a pressurized irrigation system. Generally, the volume of fluid pumped is independent of the pressure encountered at the discharge point. However, if the internal parts of the pump deform due to increased pressure (as in a mechanically driven diaphragm pump), the displacement volume of the pump will change, and the injection rate will not be constant. Excessive pressure at the discharge may also result in some backflow through the clearances of the pump parts (for example, between the gears and the housing in the gear pump).

Reciprocating pumps have a piston or a diaphragm that displaces a specific amount of solution with each stroke. The change in internal volume of the pump creates high pressure, which forces the solution into the discharge pipe. Piston, fluid-filled diaphragm, and piston/diaphragm pumps generally provide a constant flow rate independent of the discharge pressure. However, even with these pumps, excessive discharge pressure should be avoided (i.e., a closed valve in a discharge line) since it may result in pump damage.

The operation of a piston pump (Fig. 33-9) is similar to the operation of the cylinder of an automobile engine. On an intake stroke, the solution enters the cylinder through the suction check valve. On a compression stroke, the solution is forced into the discharge line through the discharge check valve.

The operation of a diaphragm pump (Fig. 33-10) is similar to that of a piston pump. The pulsating motion is transmitted to the diaphragm through a fluid or a mechanical drive, and then through the diaphragm to the solution being injected.

Combination pumps usually contain a piston that forces oil or other fluid against a diaphragm that

Figure 33-10. Diaphragm-metering pump.

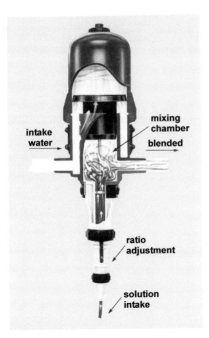

Figure 33-11. Action of water-driven proportional injector.

displaces the concentrated solution. The advantage of these pumps is that they combine the high precision of a piston pump with the resistance to chemicals characteristic of diaphragm pumps.

Reciprocating pumps are often electrically driven. The solution injection rate from an electrically driven pump is approximately constant regardless of the water flow rate. Thus, the injection rate must be adjusted between zones if the flow rate is not constant to all zones.

To ensure constant concentration of chemical in the irrigation line, an electrically driven injector can be equipped with a water flow sensor to detect changes in flow rate and automatically adjust the speed of the injector or injection time. The other possibility is to measure the conductivity of the irrigation water (if fertilizers are being injected) and use this information for automatic adjustment. Sensors that measure the conductivity must be recalibrated for different chemicals.

Proportional Injectors

Some piston and diaphragm pumps are driven by a water motor. As water flows through the injector, it causes a cam to turn and push the piston back and forth. In a diaphragm pump, the piston or cam motion is transmitted to the diaphragm. Consequently, since the revolution of the cam depends on the flow rate of water in the irrigation system, oscillation of the piston and/or diaphragm also varies with water flow rate. In this case, the solution flow is proportional to the flow rate in the irrigation system.

Proportional injectors utilize water flowing in the system to operate the injector (Fig. 33-11). A volumetric hydraulic motor drives a volumetric dosing pump. The hydraulic motor is composed of a piston, the upper and lower faces of which are connected alternately to the inlet and outlet of the water supply via a four-way valve. The four-way valve is connected to an overcentering device actuated by two rods located on the piston stem. Therefore, the hydraulic motor moves up and down once every time the cylinder is filled with a known volume. The dosing pump, driven by the piston, sucks up and injects the required volume of stock solution. The amount injected is adjusted by altering the free stroke of the dosing piston, using the adjusting nut on the outside of the piston.

Piston and diaphragm pumps inject solutions in concentrated pulses separated in time. Some pumps are equipped with double-acting pistons or diaphragms to minimize variations in the concentration of chemicals in the irrigation system. If the length of pipe between the injection port and the first point of application is short, a blending tank should follow the injection to ensure adequate mixing of water and fertilizer.

Rotary Pumps

Rotary pumps transfer solution from suction to discharge through the action of rotating gears, lobes or other similar mechanisms. Both gear- and lobe-type rotary pumps are sometimes used for chemical injection into irrigation systems. The operation of a gear or lobe pump is based on the partial vacuum that is created by the unmeshing of the rotating gears (Fig. 33-12) or lobes (Fig. 33-13). This vacuum causes the solution to flow into the pump. Then, it is carried between the gears or lobes and the casing to the discharge side of the pump. Gear and lobe pumps produce approximately constant flow for a given rotor speed, and the injection rate does not change with flow rate in the irrigation system. Flow sensors can be used to ensure a constant injection rate.

Peristaltic Pumps

Peristaltic pumps (Fig. 33-14) are used mostly in chemical laboratories, but they can be used for solution injection into small irrigation systems. Their capacity is limited, and most of them produce a pressure of only 30 to 40 psi. A flexible tube is pressed by a set of rollers, and an even flow is produced by this squeezing action. The pump is suitable for pumping corrosive chemicals because the pumped liquid is isolated from all moving parts of the pump.

Pressure Differential Methods

The pressure differential concept for injection is

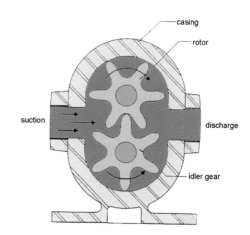

Figure 33-12. Gear injection pump.

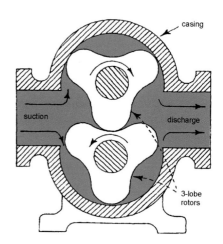

Figure 33-13. Lobe injection pump.

quite simple: if the pressure at the point of injection is lower than at the point of intake of the solution, the solution will flow into the line. There are several injection techniques that use this principle. They can be separated into two distinctive groups based on which side of the pump injection takes place: the suction side or discharge side.

Suction Line Injection

The suction line injection technique can be used in irrigation systems using centrifugal pumps that are pumping water from the surface source, such as a pond, lake, or canal. It is approved only for injection of fertilizer. Suction line injection is not permitted for irrigation systems pumping from wells.

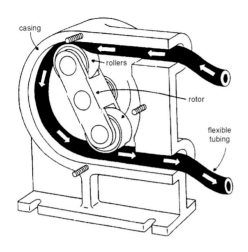

Figure 33-14. Peristaltic pump.

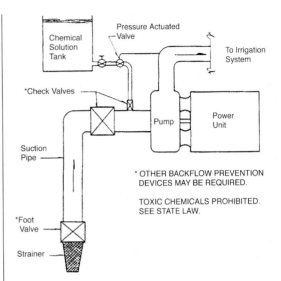

Figure 33-15. Injection on the suction line (legal only for fertilizer with surface water).

This method requires only a minimum investment. The equipment necessary for this type of injection is a pipe or a hose, a few fittings, and an open container to hold the fertilizer solution (Fig. 33-15). The rate of solution flow depends on the suction produced by the irrigation pump, the length and size of the suction line, and the level of solution in the supply tank. The injection rate can not be easily adjusted.

Discharge Line Injection
Discharge line injection requires a differential pressure to be created downstream from the pump. This is usually done by redirecting a portion of the mainline flow through a chemical tank while providing a pressure drop in the irrigation line between the point where the water is taken and the point where the solution enters the irrigation line. The pressure drop is accomplished by using some kind of restriction in the line, such as a valve, orifice, pressure regulator, or other device that would create a pressure drop. The use of valves allows for adjustment of the pressure drop, which also allows for some adjustment of the injection rate.

Pressurized Mixing Tanks
A mixing tank injector operates at the discharge line on a pressure differential concept. The water is diverted from the main flow, mixed with fertilizer, and injected or drawn back into the main stream of the system (Fig. 33-16). A measured amount of fertilizer required for one injection is placed in the cylinder. The flow back into the main line is often controlled by a metering device installed on the inlet side of the injector. The concentration of the injection changes as the solution becomes diluted as the water enters the tank during injection. To operate, there must be a pressure differential in the irrigation line between the inlet and outlet of the injector.

Proportional mixers are modifications of pressurized mixing tanks. They operate on the displacement principle. The chemical is placed in a collapsible bag that separates the solution from the water. Water pressure from the high-pressure side forces the solution from the bag, through the regulating valve and into the mainline. As the solution flows out, the bag contracts, and water on the outside of the bag displaces the volume. As long as the pressure and the flow rate in the system do not vary significantly, the injection rate will remain fairly constant. In systems where flow fluctuations can be expected, a proportioning control valve should be used. The proportioning valve responds to the changes of flow, not to pressure changes.

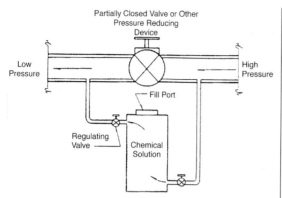

Figure 33-16. Pressurized mixing tank using a pressure-reducing device to create pressure differential.

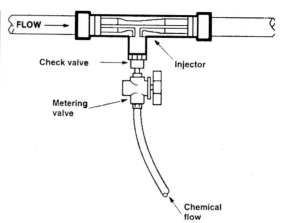

Figure 33-17. Venturi with metering valve suitable for chemical injection.

Venturi Injectors

Chemicals can be injected into a pressurized pipe using the venturi principle. A venturi injector is a tapered constriction (Fig. 33-17) that operates on the principle that a pressure drop results from the change in velocity of the water as it passes through the constriction. The pressure drop through a venturi must be sufficient to create a negative pressure (vacuum) relative to atmospheric pressure in order for the solution to flow from a tank into the injector.

A venturi injector does not require external power to operate. There are no moving parts, which increases its life and decreases probability of failure. The injector is usually constructed of plastic, which makes it resistant to most chemicals. It requires minimal operator attention and maintenance. The device is very simple, and it is low cost compared to other equipment of similar function and capability. It is easy to adapt to most new or existing systems, provided that a sufficient pressure differential can be created.

Venturi injectors come in various sizes and can be operated under different pressure conditions. Suction capacity (injection rate), head loss required, and the range of working pressures will depend on the model. It is important to note that as the level in the supply tank drops, the injection rate decreases. To avoid this problem, some manufacturers utilize a small float-controlled injection tank located near the supply tank. A float valve in the line connected to the supply tank maintains the level in the injection tank; thus, a fairly constant injection rate can be acheived.

A small venturi can be used to inject small solution flow rates into a relatively large main line by shunting a portion of the flow through the injector (Fig. 33-18). To ensure that the water will flow through the shunt, a pressure drop must occur in the main line. For this reason, the injector is used around a point of restriction, such as a valve, orifice, pressure regulator, or other device that creates a differential pressure. A centrifugal pump, used to provide additional pressure in the shunt (Fig. 33-19), can also be used.

Most venturi injectors require at least a 20% differential pressure to initiate a vacuum. If there is only a small pressure differential in the irrigation pipeline, a large venturi can be used to create a pressure drop (Fig. 33-20). The large venturi can either be installed in the main line or in a bypass line. The pressure difference between the inlet and the throat of the large venturi can be used to power the chemical injection in the smaller venturi.

Combination Methods

There are some injectors on the market that employ combinations of the different principles of injection at the same time. The most common combination is a pressure differential combined with a

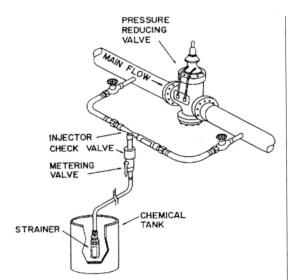

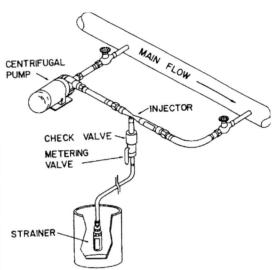

Figure 33-18. A small venturi in a bypass line used in conjunction with a pressure-reducing valve to inject agricultural chemicals.

Figure 33-19. The use of a booster pump to create adequate pressure differential to operate a venturi for chemical injection.

venturi meter or some measuring device, which operates on the venturi principle.

Direct use of pressure differential in combination with a venturi can be found in some systems where the pressure drop required for a venturi may be difficult to provide because of design restrictions of the existing irrigation system. The combination of a venturi device with a pressurized chemical tank may be used in this case (Fig. 33-21). The chemicals are placed in the tank. Since the water flowing through the tank is under pressure, a sealed airtight pressure supply tank constructed to withstand the maximum operating pressure is required. In this case, the injection rate will change gradually because of the change of solution concentration in the tank as the water enters during injection.

Various metering valves that are used with mixing and proportioning tanks operate on pressure or flow changes in the irrigation system. There are many designs of these valves. Frequently, they involve some application of the venturi meter or changing the orifice diameter. The manufacturer should be contacted for descriptions and operation instructions for various metering and proportioning valves.

Chemical injection on the suction side of a centrifugal pump is generally not permitted in Florida. The exception is a system that uses a surface water supply with only fertilizers being injected into the system. Florida backflow prevention law requires that a double protection of a check valve and a foot valve be used upstream of injection port in this case.

According to the Environmental Protection Agency (EPA), only piston and diaphragm injection pumps can be used for pesticides and other toxic chemicals. Other methods can be used for injection of fertilizers or cleaning agents, such as chlorine or acids. Table 33-2 lists some of the advantages and disadvantages of the various types of injection devices.

Calibration of Fertilizer Injection Systems
Each method of fertilizer injection must be calibrated by the user. Calibration procedures vary, depending upon the injection method used and the specific design of the injection equipment. In all cases, the user must verify that the manufacturer's calibration or the method being used is correct by using a chemical flowmeter, which is accurate in the flow range of gallons per hour (or other rate

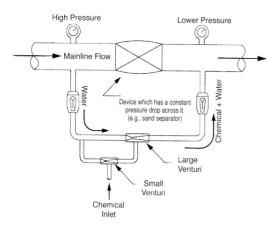

Figure 33-20. The use of a large venturi to create adequate pressure differential to operate a smaller venturi for chemical injection.

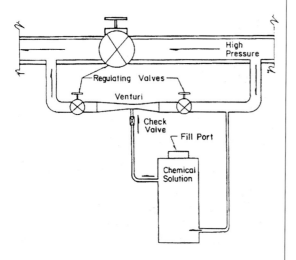

Figure 33-21. Combination of pressurized tank and venturi injector.

being injected), or by measuring the injection rate volumetrically.

Chemical Flowmeters

Flowmeters (Fig. 33-22) are available that can be used to directly measure the solution flow rate while the injection system is operating under field conditions. Meters can often be mounted on the low-pressure (suction) side of injection pumps. If a chemical flowmeter is used on the high-pressure

side of an injector, be certain that the flowmeter is rated for the pressure being used before installing it in that position. Failure to use a properly installed, adequately pressure-rated meter may cause it to be damaged, and it may be hazardous to individuals working in the area.

Volumetric Flow Rate Measurement

To measure flow rates volumetrically, a container of known volume (such as a graduated cylinder) and a stopwatch or other accurate timer are needed. Measure the time required to fill the container. Then calculate the flow rate as the volume per time, typically in units of gallons per hour (gal/hr or gph).

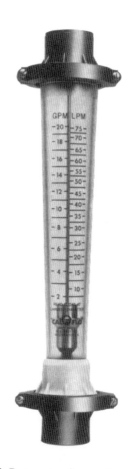

Figure 33-22. Rotameter flowmeter.

Table 33-2. Comparison of various chemical injection methods.

Injector	Advantages	Disadvantages
Centrifugal pump	Low cost Can be adjusted while running	Calibration depends on system pressure
Piston pump	Very high pressure High precision Linear calibration Calibration independent of pressure	High cost May need to stop to adjust calibration Chemical flow not continuous
Diaphragm pump	Can adjust calibration while injecting High chemical resistance	Nonlinear calibration Calibration depends on system pressure Chemical flow not continuous Medium to high cost
Piston/diaphragm pump	High precision Linear calibration Very high pressure Calibration independent of pressure High precision	High cost May need to stop to adust calibration
Gear and lobe pumps	Injection rate can be adjusted when running	Fluid pumped cannot be abrasive Injection rate is dependent on system pressure Continuity of chemical flow depends on number of lobes in lobe pump
Peristaltic pump	High chemical resistance Major adjustments can be made by changing tubing size Injection rate can be adjusted while running	Short tube life expectancy Injection rate dependent on system pressure Low to medium injection pressure
Suction line port	Very low cost Injection rate can be adjusted while running	Permitted only for surface water source with injection of fertilizer Injection rate depends on main pump operation
Proportional mixers	Low to medium cost Calibrate while operating Injection rates accurately controlled	Pressure differential required Volume to be injected is limited by size of injector
Pressurized mixing tanks	Medium cost Easy operation Total chemical volume accurately controlled	Pressure differential required Variable chemical concentration Cannot be calibrated for constant injection rate Frequent refills required
Venturi	Low cost Water powered Simple to use Calibrate while operating No moving parts	Pressure drop created in system Calibration depends on solution level in tank
Combination proportion mixers and venturi injectors	Greater precision than proportional mixer or venturi	Higher cost than proportional mixer or venturi alone

Example:
Assume that a 100-ml graduated cylinder and stopwatch were used to measure injection rates, with 90 ml of fertilizer solution collected in 4 minutes and 3 seconds. Determine the injection rate.

To convert ml to gal, divide by 3785 ml/gal.
90 ml = 90/3785 = 0.0238 gal

To convert seconds to hours, divide by 3600 sec/hr.
4 min, 30 sec = 270 sec
270/3600 = 0.075 hr.
0.0238 gal/0.075 hr = 0.32 gal/hr

For many injection methods, the injection rate will decrease as system pressure increases. Therefore, the calibration procedure should be done on each zone while the system is operating at typical pressure and flow rates. It is always a good idea to measure the rate of fertilizer removal from the storage tank to provide a check on calibration. Measure the drop in the tank level over a specific time period (typically 1 hour) to verify the injection rate.

Example:
The level in a 12-ft diameter vertical supply tank drops 10 inches during a 1-hour injection period. Determine the injection rate.

Calculate the volume (ft^3) of liquid removed from the tank.

Convert depth of 10 inches to feet: 10 inches/ 12 inches/ft = 0.83 ft

$$Area = \frac{\pi d^2}{4} = \frac{3.14 \times 12^2}{4} = 113 \text{ ft}^2$$

Volume = area x depth = 113 ft^2 x 0.83 ft = 93.8 ft^3

Convert to gallons (7.5 gallons per ft^3)
Volume = 93.8 ft^3 x 7.5 gal/ft^3 = 704 gal

Example:
The initial level in a 5-ft diameter x 8-ft long horizontal supply tank is 38 inches from the bottom. After 1 hour, the level has fallen to 28 inches. Determine the injection rate.

Calculate the total volume of tank:
Volume = area x length = 19.6 ft^2 x 8 ft = 157 ft^3

$$Area = \frac{\pi d^2}{4} = \frac{3.14 \times 5^2}{4} = 113 \text{ ft}^2$$

Convert to gallons (7.5 gallons per ft^3)
Volume = 157 ft^3 x 7.5 gal/ft^3 = 1178 gal

Refer to Appendix 7: Approximate Gallons Contained in a Partially Filled Horizontal Cylindrical Tank. Total depth = 5 ft x 12 inches/ft = 60 inches

Initial TD (percentage of total tank diameter)
TD = 38 inches/60 inches = 0.63 = 63%
From Table A7-1 with TD = 63% → TC = 66.39
Initial volume = 1178 gal x 66.39% = 782 gal

Final TD
TD = 28 inches/60 inches = 0.47 = 47%
From Table A7-1 with TD = 47% → TC = 46.19
Final volume = 1178 gal x 46.19% = 544 gal

The change in volume of the tank is 782 - 544 = 238 gal
The injection rate is 238 gal/1 hour = 238 gph or about 4.0 gpm

It is a good idea to inject fertilizers from a small, graduated supply tank rather than to pump directly from a large bulk storage tank. The small tank should be sized to contain the fertilizer solution for one application, and only that amount should be placed in the small tank before irrigation. This procedure can improve the effectiveness of fertilizer injection because only the amount of fertilizer in the small supply tank can be injected during irrigation, thus preventing excess applications from accidentally being made. The amount of fertilizer injected can easily and accurately be read if the supply tank is relatively small and has graduations permanently marked on it. Another benefit is that only the fertilizer in the small tank will be diluted if backflow from the irrigation system occurs from failure of the injection pump and backflow prevention system.

For injection methods that use a suction tubing between the injection pump and the supply tank, the injection rate can be measured with a solution flowmeter or by connecting the tubing to a graduated cylinder. Measurements should be made while the injector is operating under normal conditions, including normal injection rates and normal irrigation system operating pressures. Then, adjustments in the injection rate can be made as the injection system operates.

Calculating Fertilizer Injection Rates
For all methods of injection, the required fertilizer injection rate must be known. The required injection rate can be calculated from the following equations for microsprinkler systems.

The fertilizer injection rate in gallons per hour (gph) is calculated from:

Rate = (100 x A x F)/(P x H x W)   **Eq. 33-2**

where,

Rate = fertilizer injection rate (gph)
A = area to be irrigated (ac)
F = fertilizer amount to be applied per acre (lb/ac)
P = fertilizer fraction, percent of fertilizer per gallon of fluid injected (%)
H = fertilizer injection time (hr)
W = weight of fertilizer solution (lb/gal)

Example:
Assume that 8 lb per acre of nitrogen is applied to a 75-acre citrus block using a microsprinkler system. The fertilizer to be used is a 10-0-10 solution that weighs 10.5 lb/gal. The irrigation cycle is 4 hours, and fertilizer injection begins 1 hour after the system has reached normal operating pressure. Fertilizer will be injected for 2 hours, leaving 1 hour to flush the fertilizer from the irrigation system.

$$Rate = \frac{(100 \times 75 \text{ ac} \times 8 \text{ lb/ac})}{(10\% \times 2.0 \text{ hr} \times 10.5 \text{ lb/gal})} = 286 \text{ gph}$$

The required 8 lb/ac of N can be applied by injecting 286 gal of 10-0-10 fertilizer per hour for the 2.0 hr injection time. The total volume to be injected would be 286 gal/hr times 2.0 hrs = 572 gallons.

It is important to note that microsprinkler irrigation systems do not irrigate the entire soil surface, and that the fertilizers applied, using these systems will be delivered only to the irrigated portion of the soil surface. For example, if only 50% of the soil surface is irrigated with the spray system, the N application rate in the irrigated zone will be 16 lb/ac, and in the nonirrigated zone will be 0 lb/ac. Likewise, if only 20% of the soil surface is irrigated, the application rate in the irrigated area would be 5 times the average on a gross acre basis. Because water and fertilizers are not applied to the entire soil surface when microirrigation systems are used, fertilizer applications to microirrigated crops are often made on the basis of individual plants, rather than on a gross acre basis. In this case, the following equation can be used:

Rate = (100 x A x Fp x NP)/(P x H x W)   **Eq. 33-3**

where,

Fp = amount of fertilizer to be applied per plant (lb/plant)
NP = number of plants per acre
A = area to be irrigated (ac)
P = fertilizer fraction, percent of fertilizer per gallon of fluid injected (%)
H = fertilizer injection time (hr)
W = weight of fertilizer solution (lb/gal)
Rate = fertilizer injection rate (gph)

Example:
Assume that 0.05 lb of N (from an 8-0-8 solution weighing 10.4 lb per gal.) is to be applied to each tree in a 40-acre grove of young citrus

trees with 151 trees per acre. The irrigation system is operated for a total of 3 hr per irrigation. After start-up of the irrigation system, fertilizer is injected for 2 hr, followed by almost 1 hour of irrigation to flush the fertilizer from the system.

$$Rate = \frac{(100 \times 40 \text{ ac} \times 0.05 \text{ lb/tree} \times 151 \text{ tree/ac})}{(8\% \times 2\text{hr} \times 10.4 \text{ lb/gal})}$$

$= 181$ gph

Thus, the required 0.05 lb of N per tree can be applied to 40 acres by injecting 181 gph for the 2 hr of fertilizer injection time. The total volume to be injected would be 181 gal/hr times 2 hrs = 362 gal.

Chapter 34. Efficiency, Uniformity, and System Evaluation

by Brian Boman

Properly designed and operated irrigation systems can shorten the time from planting to economical harvest, increase the quality and quantity of fruit harvested, and increase the efficiency of water use. Citrus yields are dependent on many conditions related to the interaction between plant, soil, water, and atmosphere. An irrigation system allows water to be available to the plant in sufficient quantities and quality to ensure that a lack of water is not a limiting factor in crop production. In addition, irrigation may serve other purposes, such as modifying the microclimate and the chemical and biological properties of the soil. Water is available to plants only if it is within the root zone. Therefore, the uniformity and efficiency of water applications are important factors in irrigation system design and operation.

No irrigation system will apply water without some waste or losses because the cost to prevent all losses is prohibitive. Thus, some water losses are expected and accepted in proper irrigation system design, installation, and management. However, excessive waste may be caused by poor irrigation system design, improper installation, poor management, and equipment failures. Waste may occur as nonuniform water applications, excessive applications, evaporation or wind drift during application, surface runoff or subsurface (lateral) flow from the irrigated area, canal seepage, percolation below the root zone, evaporation from the irrigation distribution system, leakage from defective pipe connections, or other losses.

It is not possible to apply the exact amount of irrigation water required with perfect uniformity because of variations in soil properties, variations in irrigation system components, pressure losses in systems due to friction and elevation changes, or other causes. When the correct average amount of water is applied, nonuniform water applications

waste water due to excess applications in some areas, while crop yields may be reduced due to inadequate applications in other areas.

Irrigation efficiencies vary with the type of irrigation system and with many other factors such as soil, crop, and climate characteristics, as well as with the level of maintenance and management of the irrigation system. Water may be lost because of evaporation or wind drift during application, especially for sprinkler and spray types of irrigation systems. Surface runoff, subsurface flow from the irrigated area, canal seepage, percolation below the crop root zone, and evaporation from a water distribution system during application will also reduce irrigation efficiencies. Conversely, recovery and reuse of surface runoff and subsurface water will increase irrigation efficiencies.

Irrigation Efficiency

One definition of water application efficiency is the ratio of the quantity of water effectively put into the crop root zone to the quantity delivered to the field, the efficiency being expressed as a percentage. The concept of efficiency is related to the distribution of water within the root zone. Efficiency is always less than 100%, mainly because of the limited control over the way in which water is applied and how it will distribute itself in the soil. Low efficiency in microsprinkler irrigation is usually due to a combination of the following factors:

- Poor moisture distribution pattern in relation to root distribution

- Losses due to deep percolation

- Losses due to evaporation from the soil surface

- Wind drift and evaporation of emitter spray

A system with high uniformity is not necessarily efficient. Consider a highly uniform system where the discharges are nearly identical for all emitters. If the distribution pattern allows water to move out of the root zone, the resulting efficiency will be low because a significant part of the water applied is beyond the root system. However, this does not imply that uniformity and efficiency are unrelated. To attain high levels of efficiency, a uniform discharge is required. In other words, it is not possible for a system to have high efficiency with a low degree of uniformity while supplying the crop water demand.

Regardless of the design, there will be pressure and emitter variations causing the application rates of each emitter to vary slightly. Normally, these emitter-to-emitter variations are maintained within reasonable limits. However, when uniformity is low, part of the irrigated block will be overwatered and another part of the block will be underwatered. In order to compensate for the underirrigated region, a larger volume of water is applied by running the system for longer periods of time. This results in a decrease in water application efficiency because a significant portion of the crop becomes overirrigated.

There are many meaningful definitions of efficiency that relate to irrigation and crop water use. In general, the term irrigation efficiency refers to the ratio of the volume of water delivered by an irrigation system to the volume that is input to the system. Irrigation efficiencies can be defined for components of irrigation systems, for entire field- or farm-scale irrigation systems, and for multifarm or regional irrigation projects.

Reservoir Storage Efficiency

Reservoir storage efficiency (E_s) is the ratio of the volume of irrigation water available from an irrigation reservoir to the volume of water delivered to the reservoir. This ratio is normally less than 1.0 because of seepage and evaporation losses. The amount of seepage loss will strongly depend on the properties of the materials from which the reservoir is constructed. Seepage losses may be reduced by

lining reservoirs with impermeable soils (typically clays) or with plastic liners. In some flatwoods locations, seepage flows may be both into and out of reservoirs, depending on the water table elevation.

Evaporation losses from reservoirs can be reduced by designing reservoirs with smaller surface areas and greater depths. Shallow water depths heat up and evaporate at higher rates than in deep areas. Since evaporation is a surface process, less surface area will result in lower evaporation losses.

Transpiration losses occurring as a result of vegetative growth in and around a reservoir can be reduced by preventing or minimizing this growth. Vegetative growth along the shoreline can also be reduced by minimizing shallow water areas, although some vegetation will normally be required to stabilize the soil embankments.

Water Conveyance Efficiency

Water conveyance efficiency (E_c) is the ratio of the volume of water delivered for irrigation to the volume of water placed in the conveyance system. This ratio is normally less than 1.0 for open channel conveyance systems, but it may be approximately 1.0 for pipeline conveyance systems.

Losses from open channel conveyance systems occur due to seepage, evaporation, and transpiration. These losses can be reduced by using lined channels and controlling vegetative growth. Some evaporation losses will be unavoidable. Open channels are used in flatwoods areas where existing high water tables and restrictive soil layers minimize seepage losses. However, even under these conditions, E_c is very site-specific and must be determined by measurements taken at the site or estimated by persons experienced with these systems.

Irrigation Application Efficiency

Application efficiencies (E_a) are affected by irrigation system management practices. Since it is not possible to measure and apply the exact amount of water required in the crop root zone at precisely the time that available soil water is depleted, excess water applications will sometimes occur. As a result, E_a will be reduced. Irrigation application efficiency is the ratio of the volume of irrigation water stored in the root zone and available for crop use (evapotranspiration) to the volume delivered from the irrigation system. This ratio is always less than 1.0 due to losses from evaporation, wind drift, deep percolation, lateral seepage, and runoff which may occur during irrigation. Irrigation application efficiency can be calculated by:

$$E_a = \frac{V_b}{V_f} \times 100 \qquad \textbf{Eq. 34-1}$$

where,

E_a = irrigation system efficiency (%)

V_b = volume of water beneficially used (stored in root zone)

V_f = volume of water delivered to the field

Example:
Determine the irrigation efficiency of an 80-acre citrus microirrigation system given the following conditions:

Annual crop ET: 45 inches
Annual effective rainfall: 33 inches
Pumping rate: 1800 gpm
Annual hours of operation: 320 hours

Solution: V_b = ET - effective rainfall
= (45-33) in. x 80 ac = 960 acre-inches

V_f = pumping rate (gpm) x hours operated
x 60 min/hour / 27,154 gal/ac-in.
= 1800 gpm x 320 hr x 60 min/hr / 27,154 gal / ac-in.
= 1272 acre-inches
E_a = 960 / 1280 = 75 percent

Overall irrigation efficiency (E_o) is calculated by multiplying the efficiencies of the components. For a system which includes reservoir storage, water conveyance, and water application, the overall irrigation efficiency is defined as:

$$E_o = E_s \times E_c \times E_a \qquad \textbf{Eq. 34-2}$$

Example:
Determine the overall irrigation efficiency for a flatwoods citrus microirrigation system that is using water from an open reservoir with a 60% storage efficiency, conveying it using an open channel from which 20% is lost in transit (80% conveyance efficiency), and using microsprinklers with an application efficiency of 85%.

E_o = 0.60 x 0.80 x 0.85 = 41%.

Effective Irrigation Efficiency
Effective irrigation efficiency (E_e) is the overall irrigation efficiency corrected for water that is either reused or restored to the water source without a reduction in water quality. Tailwater recovery systems allow runoff from one block to be recycled or used on another, resulting in an increase of E_e above E_o. A citrus seepage/flood system, where water is drained from one block but used to irrigate the next block, is an example of a production system where E_e is greater than E_o. In addition, if seepage from conveyance ditches flows into the block being irrigated, this flow will not be lost from the irrigation system, and E_e will be greater than E_o.

If irrigation water moves from the root zone due to lateral flow or deep percolation, its quality may be degraded by salts, fertilizers, or pesticides. If this water cannot be intercepted for reuse in the same production system, it is considered to be lost and E_a and E_o are reduced. Thus, lateral flow and deep percolation will reduce irrigation efficiencies unless interceptor drains or ditches are installed to recover this water for reuse. The effective irrigation efficiency is defined as:

$$E_e = E_o + FR \times (1.0 - E_o) \qquad \textbf{Eq. 34-3}$$

where,

FR is the fraction of the water lost that is recovered. Some of the water that leaves an irrigated field due to runoff, seepage, or deep percolation might be recovered in some cases. Losses due to evaporation, wind drift, and transpiration cannot be recovered.

Crop Water Use Efficiency

Crop water use efficiency (WUE) is defined as the ratio of the mass of marketable yield or biomass produced per unit of water used. For this definition, crop use efficiency has units of production unit per water volume unit. Units typically used are tons per acre-inch. This definition is not a true efficiency because it does not express a dimensionless ratio. It does, however, have the advantage of comparing both yield and water used; thus, it is often used in economic comparisons of alternative crops.

Irrigation Water Use Efficiency

There is no general agreement on a single definition of irrigation water use efficiency (E_u). However, E_u can be defined as the ratio of the volume of water beneficially used to the volume delivered from the irrigation system. Water that is beneficially used includes that which is applied for leaching of salts from the crop root zone, crop cooling, freeze protection, and other such uses, in addition to that which is stored in the crop root zone for evapotranspiration. This ratio expresses the fraction of each unit of water delivered that is beneficially used.

For example, excess water beyond that which can be stored in the crop root zone may be required to leach salts from the crop root zone if poor quality water is being used for irrigation. Since this would be a beneficial use, the irrigation water use efficiency would remain high, although the irrigation application efficiency would be reduced because all of the water applied was not stored in the crop root zone.

E_u can also be defined as the ratio of the increase in production of the marketable crop component to the volume of water applied by irrigation for irrigated as compared to than nonirrigated production.

$$E_u = \frac{Y_i - Y_o}{V_i}$$

Eq. 34-4

where,

Y_i = mass of marketable crop produced with irrigation

Y_o = mass of marketable crop produced without irrigation

V_i = volume of irrigation water applied

Although this definition of E_u is not a true efficiency, it expresses the potential increase in crop production attributable to irrigation, which can be used for economic evaluations of the profitability of proposed irrigation projects. It is not meaningful if the crop cannot be produced without irrigation. Because both of these definitions of irrigation water use efficiency are sometimes used, it is always necessary to clearly define E_u and to give its units when this efficiency term is used.

Factors Affecting Irrigation Efficiencies

Pressurized irrigation systems include overhead sprinklers, portable guns, microsprinklers, and drip systems. Because networks of pressurized pipelines rather than soil hydraulic properties are used to distribute water, the field-scale uniformity of water application and the associated irrigation application efficiency are more strongly dependent on the hydraulic properties of the pipe network designed than site-specific soil hydraulic properties. Thus, application efficiencies of well-designed and well-managed pressurized irrigation systems are much less variable than application efficiencies of seepage/flood irrigation systems, which depend heavily on soil hydraulic characteristics.

Sprinkler Systems

During water applications, sprinkler irrigation systems lose water due to evaporation and wind drift. More water is lost during windy conditions than calm conditions. When water is discharged at high angles over great distances and at great heights above the ground surface, more water is lost because of greater opportunity time for evaporation. More water is also lost during high evaporative demand periods (hot, dry days) than during low demand periods (cool, cloudy, humid days). Thus, sprinkler irrigation systems usually apply water more efficiently at night (and early mornings and late evenings) than during the day. In addition, greater water losses occur from systems that operate at high pressures that result in small droplet sizes. Small droplets are more readily carried by wind, and they expose more surface area to the atmosphere for evaporation.

Sprinkler irrigation application efficiencies are reduced by nonuniform application when some areas are overirrigated, causing water and nutrients to be lost to deep percolation, while other areas are underirrigated, resulting in reduced crop yields. Thus, system design affects application efficiency. Nonuniformity also occurs if pressure losses within the irrigation system are excessive either because of friction losses or elevation changes. Other causes of nonuniformity, such as clogged nozzles or enlarged nozzles from abrasion by pumping sand, also reduce application efficiencies.

Overhead systems have sprinklers permanently installed at spacings that result in optimum uniformity. However, wind, incorrect operating pressure, and component wear or failure can still distort application patterns and reduce uniformity and application efficiency. Sprinkler water application patterns must overlap sufficiently (typically about 50%) to apply water uniformly. Because of this need for overlap, nonuniformity occurs at the edges of fields where overlap is not possible. Part-circle sprinklers can be used at the edges of fields to improve uniformity, but they are more mechanically complicated and more expensive than full-circle sprinklers.

Big gun sprinklers discharge high flow rates at high pressures and irrigate circular land areas. When they are moved from location to location, there is usually some overlap of the previously irrigated area. Because of nonuniform water applications where patterns overlap, and because of the greater wind drift and evaporation losses, portable guns typically have lower application efficiencies than solid set systems.

Microirrigation Systems

Application efficiencies of microirrigation systems are typically high becuase these systems distribute water near, or directly into, the tree's root zone. Pressure loss from friction or elevation changes is the main factor affecting efficiency of microirrigation systems. Management problems such as overirrigation or clogged emitters can also be important factors.

The number of emitters per tree and the placement of emitters with respect to the tree's root zone can influence application efficiencies. On sandy soils, drip systems only wet the soil within 1 to 2 feet of the emitter. Evaporation losses are typically small because of the limited surface area wetted and the rapid surface drying of sandy soils. Wind usually does not affect the efficiency of these systems. Application efficiencies of drip systems are primarily dependent on the hydraulic properties of the systems and on their maintenance and management. However, soil hydraulic properties influence water movement in the soil away from drip emitters, thereby influencing application efficiencies. This can be an important consideration for young trees with limited root zones.

Microsprinkler system efficiencies are typically less than those of drip systems. These systems distribute water over a significant fraction of the crop root zone. But because water is sprayed in very small droplets, evaporation and wind drift losses may occur. Wind distortion of spray patterns may also occur. Therefore, wind drift and evaporation losses can be high if microsprinkler systems are operated under windy conditions on hot, dry days. Thus, management to avoid these losses is important to achieving high application efficiencies with these systems.

Seepage/Flood Systems

With seepage/flood irrigation systems, water is distributed by flow through the soil profile or over the soil surface. Because water is distributed by gravity flow, the uniformity of water application and the associated irrigation application efficiency are strongly dependent on the soil topography and hydraulic properties. In addition, losses of irrigation water due to lateral flow are highly dependent on the soil water status of surrounding land areas. As a result of these site-specific factors, water application efficiencies of seepage systems may vary widely.

In order to irrigate, the water table above a restrictive soil layer must be raised by pumping water into open ditches or underground conduits. Not all of the water pumped is available for tree use, and depending on the depth to the natural water table, large quantities may be required to build and maintain the water table. Substantial reductions in efficiency may result from losses due to deep percolation below the crop tree root zone and to subsurface lateral flow to surrounding areas. The magnitudes of losses by both of these mechanisms are site- and time-specific as they depend on the permeability of restrictive soil layers and the management practices occurring in surrounding fields. Thus, irrigation application efficiencies for these types of systems can vary widely, depending upon site-specific conditions.

Water distribution from seepage/flood irrigation systems occurs below the soil surface. Therefore, wind and other climatic factors do not affect the uniformity of water application. Also, evaporation losses from both surface ditch systems and underground pipes are about equal because the soil surface is wet and evaporation occurs at near potential rates with both systems. Conveyance losses in open ditches depend on the hydraulic properties of the soil in the ditch, the depth of the water table, and the location of the water source with respect to the irrigated field. Losses may be very significant when water is conveyed long distances.

Application efficiencies of seepage/flood systems greatly depend on the soil hydraulic characteristics, permeability of restrictive layers, water table levels, and the characteristics of surrounding land areas. Application efficiencies are significantly increased if the excess irrigation water drained from the citrus groves is reused. If drainage water is reused, the application efficiency has been observed to be high (75%), while if the excess water is lost, then the application efficiency has been observed to be low (25%).

Potential System Application Efficiencies
Potential irrigation system application efficiencies are application efficiencies that can be achieved with well-designed and well-managed irrigation systems. Application efficiencies can vary widely, depending upon how well a system is designed and managed, time of year, and climatic conditions. However, average seasonal application efficiencies of well-designed systems that are scheduled to maintain adequate soil moisture levels to meet crop water requirements for evapotranspiration (ET) will be much less variable. The application efficiencies listed in Table 34-1 are for typical Florida conditions when irrigations are scheduled to meet crop water requirements (ET). The values given are seasonal values that represent average production conditions throughout the growing season. Irrigation application efficiencies from specific events will vary more widely than table values due to the stage of tree development, time of year, climatic conditions, and other factors.

Application efficiencies will be reduced if water is applied for leaching of salts, freeze protection, establishment of young plants, crop cooling, or other uses. Because uniformity of water application of gravity irrigation systems is strongly influenced by soil hydraulic properties, application efficiencies of gravity flow irrigation systems range much more widely than those of pressurized irrigation systems.

Uniformity
Microirrigation systems have the potential to be a very efficient way to irrigate. However, to be efficient, irrigation water must be uniformly applied. This means that at each irrigation, approximately the same amount of water must be applied to all of the trees. If irrigation is not uniformly applied, some areas will get too much water and others will get too little. As a result, tree growth will also be nonuniform, and water will be wasted where too

Table 34-1. Average seasonal irrigation system application efficiencies (E_a) for well-designed and well-managed citrus irrigation systems in Florida. Application efficiencies are site-specific and range widely among systems (adapted from Smajstrla et al., 1991).

System type	Range	Average
Portable guns	60-70	65
Overhead sprinklers	70-80	75
Drip	70-90	85
Microsprinkler	70-85	80
Seepage (flood)	25-75	50

much is applied. Uniformity is especially important when the irrigation system is used to apply chemicals along with the irrigation water because the chemicals will only be applied as uniformly as the irrigation water.

Microirrigation systems are hydraulic networks used to deliver water to the trees. The systems rely on emitters to distribute water. The coefficient of uniformity is a measure of the hydrodynamic behavior of the system. Uniformity is an indicator of how equal (or unequal) the application rates are from each individual emitter. A low coefficient of uniformity indicates that the application rates from the emitters are significantly different, while a high coefficient of uniformity indicates that the application rates from the delivery devices are very similar. The coefficient of uniformity by itself is not a measure of how well the system is distributing water within the root zone. Several factors affect uniformity in microirrigation systems:

- Inadequate selection of pipe diameters and/or emitter type and discharge

- Emitter clogging

- Changes in properties of emitters with time

- Changes in system components such as pump efficiency or pressure regulation

- Significant effects of wind conditions on system operation (microsprinklers)

Uniformity is particularly important for micro-irrigation systems because of the repetitive nature of nonuniformity. Nonuniform applications are repeated during each irrigation event and tend to accumulate. Uniform water application is essential to achieve high irrigation efficiency and to minimize nutrient leaching. A highly uniform application does not ensure high efficiency because water can be uniformly overapplied. However, a highly efficient irrigation system and good crop yield require uniform application. Therefore, the primary goal in the design and management of microirrigation systems is to achieve and maintain uniform water application.

The uniformity of water application from a micro-irrigation system is affected both by the water pressure distribution in the pipe network and by the hydraulic properties of the emitters used. The effects of pressure variations in the pipe network (hydraulic uniformity) and variations due to the emitter characteristics (emitter performance variation) both affect uniformity of water application. The emitter hydraulic properties include the effects of emitter design, water quality, water temperature, and other factors on emitter flow rate. Factors such as emitter plugging and wear of emitter components will affect water distribution as emitters age.

Field Tests for Uniformity

Three separate tests are useful in evaluating the performance of a microirrigation system: overall water application uniformity, hydraulic uniformity or pressure variation, and emitter performance variation. If the overall water application uniformity is high, there is no need to perform further tests. If the water application uniformity is low, then hydraulic uniformity tests should be conducted in order to determine the cause of the low uniformity. The hydraulic uniformity test will indicate whether the cause of the low water application uniformity is excessive pressure variation in the system or emitter performance problems such as plugging.

Before actually performing the catch-can evaluation, a visual inspection of the irrigation system should be made. This includes recording the type and model of emitters, distribution pattern, height and location of emitter, length and diameter of laterals, type of filter, and location of regulators. Also included are tree in-row and across-row spacing and the number of emitters per tree.

If there is significant topographic change within the area irrigated by the subunit, a means of determining elevation, such as a hand level, is recommended. A sketch of the subunit layout with dimensions and elevations can be useful in interpreting the results of the field evaluation. After the preliminary information has been gathered, pressurize the irrigation system. Make sure that it is operating under normal conditions. Record inlet and outlet filter pressures. If a system flowmeter exists, record the flowrate.

Flushing the laterals to determine which materials may be accumulating and becoming a potential source to plug emitters is suggested. Use a nylon sock over the end of the lateral to catch the flushed-out materials for inspection.

Field Measurements

The first step in the evaluation procedure is to determine the field location of the emitters to be tested. Primary consideration should be given to elevation changes and to the distance from the pump station. Field uniformity is measured by a random sample of the flow volumes from the individual emitters. The catch-can measurements should be collected randomly throughout the zone and the measurement points should characterize the entire subunit: some measurements near the inlet, some at the end of the laterals and manifold. If there are elevation differences, also take them into account when selecting measurement points. The more measurements, the more accurate the results will be. The older the system, the greater the number of measurements required to get accurate results.

Line pressure can be determined using the pressure gauge. Pressure should be taken at the inlet and the end of several laterals when the terrain is level. On hilly terrain, pressures should be determined at the high and low points along laterals as well as the

inlet and the end of laterals. Discharge should be taken from the same locations as pressure. The greater the number of observations, the greater the accuracy of the data.

The discharge from emitters is collected for a specified period (usually 2 to 8 minutes, depending on discharge rate). With microsprinklers, the emitter can be inverted into an appropriate catch cylinder. For drip systems, the catch can be collected in a low, flat pan that is placed under the tubing. If the tubing is elevated to receive the catch sample, errors will be introduced. The catch quantity is divided by the number of minutes to obtain discharge in milliliters per minute. Divide the milliliter-per-minute reading by 63 to calculate gallons per hour.

The hydraulic uniformity can be estimated by measuring pressures at points distributed throughout the subunit. Pressures should be measured to the nearest 1 psi using a portable pressure gauge connected to a flexible tube. Gauges are also available with a fitting for direct insertion into the lateral line. The pressure gauge can be equipped with a device that slips over the emitter allowing the pressure to be measured with the emitter in place. If the lateral has more than 10 emitters, pressure will not be significantly impacted by stopping flow from one emitter.

Microsprinklers are typically connected to the lateral line by a connecting 'spaghetti' tube. Because of pressure losses in the connecting tube, pressure should be measured at the end of the tube near the emitter while the emitter is operating. In many cases, pressure is checked with a pressure gauge screwed into a T-fitting in the lateral line. While this does provide pressure, it is more of a static pressure reading (i.e., not involving moving water) and can be 10% to 20% above dynamic (water-moving) pressure. A more realistic reading should be taken at the emitter itself. This can be accomplished by attaching a pressure gauge on a "T" with the straight-through portion of the "T" on the stake assembly, and with the emitter in the other end (Fig. 34-1). This dynamic pressure gauge provides a realistic measure of pressure losses throughout the field as well as pressure losses through distribution tubing of the microsprinkler assembly.

Distribution Uniformity

One of the keys to efficient irrigation is uniform application of water. The more uniformly water is applied, the higher the potential irrigation efficiency. Variation in emitter discharge rates may result from pressure (hydraulic) variation within the piping network, variations in flow from one emitter to the next because of manufacturing differences, or emitter plugging. When uniformity is poor, some trees may have to be overirrigated in order to get enough water to trees where the application rates are lower. If every emitter discharged water at the same rate, uniformity would be 100 percent. However, this is not economically achievable. Typically, the minimum acceptable uniformity for microirrigation systems is about 85 percent.

Uniformity is often expressed as a statistical representation of discharges measured in the field from several emitters. A common measure of irrigation application uniformity is distribution uniformity (DU), which is defined as:

$$D U = \frac{L_q}{X_m} \times 100 \qquad \textbf{Eq. 34-5}$$

where,

L_q = average of the lower quarter of catch-can observations

X_m = average of all catch-can observations

DU calculations require measuring a minimum of 16 flow volumes from the individual emitters. Since drip emitters may need to run for a longer time than microsprinklers to get a representative sample, it is expedient to sample drippers in groups of close proximity so that more than one emitter at a time can be observed.

Example:
A microirrigation system uniformity test was performed by taking random catch-can measurements of emitter discharges. A

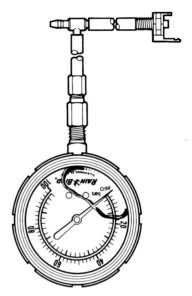

Figure 34-1. Pressure gauge with "T" connection suitable for testing pressure at emitter.

graduated cylinder was used to catch the discharge from 16 blue base emitters that had a specified discharge rate of 10.5 gph. The following were the volumes (ml) of water measured for each of the emitters in the 2-minute catch period: 1260, 1320, 1385, 1330, 1425, 1390, 1345, 1410, 1418, 1285, 1316, 1435, 1495, 1248, 1504, and 1288 (1 gal = 3785 ml).

Solution: Sum of all catches = 21,854 ml

X_m = 1366 ml = 0.36 gal

Avg. discharge rate = 0.36 gal/2 min x 60 min/hr = 10.8 gph

L_q = (1248+1260+1285+1288)/4 = 1270

DU = (1270/1366) x 100 = 92%

When the variation in emitter flow rates increases, the uniformity of water application decreases. The relationship between uniformity and efficiency can be expressed as the potential application of efficiency based on the lower quarter of flow rates (E_{plq}). If the water applied to the lower quarter application rates just meets the soil water deficit, then E_{plq} is equal to DU. Irrigating based on the mean of the lower quarter means that 12.5% of the field will be underirrigated, while the remainder will be overirrigated. If uniformity is high, only a small amount of water is lost to percolation below the root zone. If uniformity is poor, the volume of water leached from the root zone is relatively large.

Coefficient of Variation

The uniformity of water application can also be calculated from the statistical distribution of emitter flow rates that are measured in the field. The terms used include statistical uniformity (SU) and coefficient of variation (Cv), which is the standard deviation divided by the mean. The standard deviation (S_d) is defined as:

$$S_d = \sqrt{\frac{Sum\ (Xi - Xm)^2}{N - 1}} \qquad \textbf{Eq. 34-6}$$

where,

X_i = observed depth or volume
X_m = mean of observed flows
N = number of observations
Sum = the sum of all of the squares of all of the observations minus the mean

where,

Cv = coefficient of variation
S_d = standard deviation
X_m = mean flow rate

Example:
Using the catch data from the previous example, determine the Cv of the system using the 16 catch values (X_m = 1366 ml)

$(1260 - 1366)^2 = -106^2 = 11,236$
$(1418 - 1366)^2 = 52^2 = 2704$
$(1320 - 1366)^2 = -46^2 = 2116$
$(1285 - 1366)^2 = -81^2 = 6561$
$(1285 - 1366)^2 = -81^2 = 6561$
$(1316 - 1366)^2 = -50^2 = 2500$
$(1330 - 1366)^2 = -36^2 = 1296$
$(1435 - 1366)^2 = 69^2 = 476$
$(1425 - 1366)^2 = 59^2 = 3481$
$(1495 - 1366)^2 = 129^2 = 16,641$
$(1390 - 1366)^2 = 24^2 = 576$

$(1248 - 1366)^2 = -118^2 = 13,924$
$(1345 - 1366)^2 = -21^2 = 441$
$(1504 - 1366)^2 = 138^2 = 19,04$
$(1410 - 1366)^2 = 44^2 = 1936$
$(1288 - 1366)^2 = -78^2 = 6084$

Sum $(X_i - X_m)^2 = 11,236 + 2116 + 6564 + 1296 + 3481 + 576 + 441 + 1936 + 2704 + 6561 + 2500 + 4761 + 16,641 + 13,924 + 19,044 + 6084 = 99,862$

$$S_d = \sqrt{\frac{Sum\ (Xi - Xm)^2}{N - 1}} = \sqrt{\frac{99,862}{15}} = 81.6$$

$Cv = S_d / X_m = 81.6 / 1366 = 0.06 = 6\%$

The significance of S_d is that about 2/3 of all flow rates will fall within plus or minus one standard deviation of the mean (based on the properties of a 'normal' distribution). The assumption of a normal distribution is reasonable if emitter plugging is not occurring. A Cv of 0.1 means that 67% of all flow rates would be within plus or minus 10% of the mean flow rate. The 6% Cv in the above example shows an excellent flow distribution with 67% of the emitters having discharges within 6% of the mean.

When emitter flow rates are measured in the field, Cv includes the effects of variability in emitter flow rate from all causes, including water pressure distributions, emitter hydraulic properties, and emitter plugging.

Statistical Uniformity

The Statistical Uniformity (SU) coefficient is another measure of uniformity and is given as:

$$SU = \left(1 - \frac{S_d}{X_m}\right) \times 100 \qquad \textbf{Eq.34-7}$$

Example:
Using the catch data from the previous examples, determine the SU of the irrigation system.

Solution: X_m = 1366

$$S_d = 81.6$$

$$SU = [1 - (81.6/1366)] \times 100 = 94\%$$

SU can also be calculated using a statistical uniformity nomograph (Fig. 34-2). This nomograph is used to determine both water application uniformity from emitter flow rate data and hydraulic variation from pressure distribution data. To accurately determine uniformity, measurements should be made at a minimum of 18 points located throughout each irrigated zone. More may be required for greater accuracy. Computations will be simplified if the number of points measured is a multiple of 6. The statistical coefficient of variation is then calculated from these data points.

Care should be taken to distribute the measurement points throughout the irrigated zone. Some points should be located near the inlet, some near the center, and some at the distant end. Also, some should be located at points of high elevation and some at points of low elevation. However, the specific emitters to be tested should be randomly selected at each location. Do not visually inspect the emitters to select those with certain flow characteristics before making measurements. In fact, to avoid being influenced by the appearance of an emitter as it operates, emitters could be selected and flagged before the irrigation system is turned on.

The following steps are required to estimate application uniformity from Fig. 34-2:

1. Calculate 1/6 of the number of data points.

2. Find the sum of the lowest 1/6 of the volumes measured. For 18 data points this will be the sum of the 3 lowest volumes measured.

3. Find the sum of the highest 1/6 of the volumes measured.

4. Locate the sum of the highest 1/6 of the volumes on the vertical axis in Fig. 34-2. Draw a horizontal line across the graph from that point. If this sum does not fit on the scale, the sums calculated in Steps 2 and 3 can both be multiplied or divided by a common factor. The absolute values are not important; only the

relative differences between the high and low values are critical.

5. Locate the sum of the lowest 1/6 volumes on the horizontal axis in Fig. 34-2. Draw a vertical line up the graph from that point.

6. Read the water application uniformity at the intersection of the two lines drawn.

Example:
Assume that water from 18 emitters was collected at random locations throughout a zone. The volume of water caught in a graduated cylinder in 2 minutes was measured. The following volumes were recorded (ml):

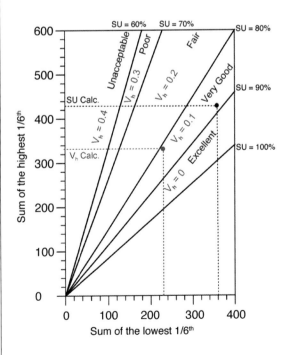

Figure 34-2. Statistical uniformity nomograph or the determination of hydraulic and water application uniformity coefficients.

Location	Volume
1	605 ▼
2	660
3	725 ▲
4	690
5	710 ▲
6	680
7	640
8	605 ▼
9	675
10	715 ▲
11	630
12	685
13	665
14	645
15	650
16	590 ▼
17	660
18	685

1. 1/6 of 18 data points = 3

2. Sum of lowest 3 values = 590 + 605 + 605
   = 1800

3. Sum of highest 3 values = 725 + 715 + 710
   = 2150

4. Divide both highest and lowest values by 5
   to obtain numbers in the range of Fig. 34-2.
   1800/5 = 360 and 2150/5 = 430. Locate 430
   on the vertical axis and draw a horizontal
   line across the graph from that point.

5. Locate 360 on the horizontal axis in Fig.
   34-2 and draw a vertical line from that
   point.

6. The intersection of these two lines occurs
   between the 80% and 90% lines. This indicates a "Very Good" water application uniformity coefficient of about 88%.

It can be concluded that the irrigation system
was designed and constructed to achieve a
high degree of uniformity of water application
throughout the zone. Since there was a high
water application uniformity, it can be also be
concluded that there is no significant emitter
plugging.

The American Society of Agricultural Engineers
(ASAE) recommends that microirrigation system
uniformity be expressed in terms of SU. Typically,
SU values should be greater than 80% to be considered adequate (Table 34-2). The major advantage to using the statistical uniformity measure is
that sources of nonuniformity can be separated and
thus identified. Knowing the contribution of various factors to nonuniformity is of great value in determining the corrective action that is needed to
improve water application uniformity. However,
most mobile irrigation laboratories use the more
traditional DU. Assuming a normal distribution,
DU is related to statistical parameters and SU as:

$$ DU = \left[1 - \left(1.27 \times \frac{S_d}{X_m} \right) \right] \times 100 \qquad \textbf{Eq.34-8} $$

Example:
Using the data from the Cv example, calculate
the DU.

$$ DU = \left[1 - \left(1.27 \times \frac{81.6}{1366} \right) \right] \times 100 = 91\% $$

(Compared to SU = 94%)

Hydraulic Uniformity

When water flows through a pipe network, pressure losses occur because of friction losses in the
pipes and fittings. Pressure changes also occur as
water flows uphill (pressure loss) or downhill
(pressure gain) in a pipe network. If a microirrigation system is poorly designed or improperly
installed, pressure losses may be excessive because
components are too small for design flow rates or
slopes are too steep for the components selected.
For these reasons, water application uniformity
may be greatly affected by the design of the pipe
network.

Hydraulic uniformity refers to the effects of pressure variations on the uniformity of water application from a microirrigation system. The definition
of hydraulic uniformity (U_h) is similar to that of
water application uniformity, except that the

Table 34-2. Criteria for rating microirrigation system uniformity on emitter discharge rates (adapted from EP-405.1, ASAE, 1996).

Rating	SU	DU
Excellent	> 90%	> 87%
Good	80-90%	75 - 87%
Fair	70-80%	62 - 75%
Poor	< 70%	< 62%

emitter discharge exponent (defined in Chapter 32) must also be considered. The emitter exponent is based on the relationship between emitter operating pressure and flow rate. Because x is different for different types of emitters, the allowable pressure variation is also different for each emitter type.

$$U_h = 100 (1 - x \cdot V_h) \qquad \textbf{Eq. 34-9}$$

where,

U_h = hydraulic uniformity based on pressure distributions

x = emitter discharge exponent

V_h = hydraulic variation, which is the statistical coefficient of variation of pressures

A low value of U_h is most often due to improper design. However, improper installation of components or the installation of the wrong components can also reduce U_h. Low values of U_h may be due to pipe sizes that are too small, laterals that are too long, laterals that are incorrectly oriented with respect to slope, improper emitter selection, or other causes.

The hydraulic uniformity of a microirrigation system is estimated by measuring pressures at points distributed throughout each irrigated zone. Although it is not necessary to measure pressures at the same emitters where flows were measured, it is normally convenient to do while flow rates are being measured. The pressure distribution data are analyzed similar to flow data using Fig. 34-2. The only difference is that the hydraulic uniformity must be calculated using V_h and the emitter discharge exponent.

Example:
Assume that pressure was measured at 18 emitters at random locations throughout a zone. The emitters used were typical orifice control microsprinklers with an emitter exponent (x) of 0.5. The following pressures were recorded (psi):

Location	Pressure
1	19
2	15 ▼
3	19
4	18
5	21 ▲
6	20
7	20
8	19
9	21
10	23 ▲
11	18
12	23 ▲
13	16 ▼
14	17
15	15 ▼
16	18
17	19
18	21

1. 1/6 of 18 data points = 3

2. Sum of lowest 3 values = 15 + 15 + 16 = 46

3. Sum of highest 3 values = 21 + 23 + 23 = 67

4. Multiply both highest and lowest values by 5 to obtain numbers in the range of Fig. 34-2.
   46 x 5 = 230 and 67 x 5 = 335.
   Locate the value for the highest 1/6th (335) on the vertical axis and draw a horizontal line across the graph from that point.

5. Locate 230 on the horizontal axis of Fig. 34-2 and draw a vertical line from that point.

6. The intersection of these two lines occurs in the section where $V_h = 0.1$, in the "Very Good" category.

7. Calculate the hydraulic uniformity (U_h) from Eq. 34-9 (using x = 0.5) as:

Uh - 100 (1 - 0.5(0.1)) = 95%

This hydraulic uniformity would be classified in the "Excellent" category (from Table 34-2). If a low water application uniformity was calculated from the volume measurements, the cause of that low uniformity would not be from poor hydraulic uniformity. Rather, the cause would be poor emitter performance, probably emitter plugging.

Effective Uniformity

If there is more than one emitter per tree (typically the case with drip systems), the effective uniformity is greater than the field-measured uniformity. To estimate the effective uniformity (SU_e), use the following relationship (this equation assumes randomness and independence):

$$SU_e = \left[1 - \frac{Cv_t}{\sqrt{n}} \right] \times 100 \qquad \textbf{Eq. 34-10}$$

where,

Cv_t = field-measured coefficient of variation

n = number of emitters per plant.

Example:
A citrus drip irrigation system was evaluated to have a field-measured coefficient of variation of 0.20, and there were 5 drippers per tree. What is the effective uniformity SU_e?

$$SU_e = \left[1 - \frac{Cv_t}{\sqrt{n}} \right] \times 100 = \left[1 - \frac{0.2}{\sqrt{5}} \right] \times 100 = 91\%$$

Emitter Manufacturing Variation

In addition to flow variations due to pressure, variations between emitters of the same type also occur due to manufacturing variations in the tiny plastic components. Because their orifice diameters are very small, microirrigation emitters are easily plugged or partially plugged from particulate

matter, chemical precipitates, and organic growths. For these reasons, water application uniformity may be greatly affected by the emitter performance.

The manufacturing coefficient of variation (C_{vm}) is defined as the statistical coefficient of variation (standard deviation divided by the mean discharge rate) in emitter discharge rates when new emitters of the same type are operated at identical pressures and water temperatures (see Chapter 32). Under these identical operating conditions, differences in flow rates observed are assumed to be due to variations in emitter components. When comparing emitters with similar flow properties, the highest uniformity will be obtained by selecting the emitter with the smallest manufacturing variation.

A C_{vm} of 0.10 implies that 68 percent of all flow rates would be within plus or minus 10 percent of the mean flow rate. The design of the emitter, the material used in its construction, and the precision with which it is manufactured determine the variation for any particular emitter. With recent improvements in manufacturing processes, most emitters have C_{vm} values < 0.10. Pressure-compensating drip emitters generally tend to have somewhat higher C_{vm} than labyrinth path emitters. C_{vm} values of 0.05 or less are considered excellent, 0.05 to 0.10 are good, 0.10 to 0.15 are marginal, and greater than 0.15 are unacceptable. Independent laboratory tests are performed for emitters of most manufacturers. For information, contact the Center for Irrigation Technology, California State University Fresno, 5370 N. Chestnut Ave., Fresno, CA 93740-0018. Phone: (559) 278-2066. Web address: http://cati.csufresno.edu/cit

Emitter Performance Variation Evaluation

Emitter performance variation, V_{pf}, refers to nonuniformity in water application that is caused by the emitters. If the emitter performance variation is high, it is normally due to emitter plugging or from manufacturing variation among emitters. Emitter performance variation can be evaluated by measuring emitter flow rates at known pressures. This can be done by removing the emitters and testing them in a laboratory. However, the nomograph in Fig. 34-3 simplifies the procedure for determining the emitter performance variation from the hydraulic and water application uniformities. The emitter performance variation can be estimated by:

1. locating the calculated hydraulic (pressure) uniformity coefficient (U_h) on the upper bar of the nomograph;

2. locating the calculated water application uniformity coefficient (SU) on the center bar of the nomograph; and

3. drawing a straight line from the measured hydraulic uniformity (upper bar) through the water application uniformity (center bar), and extending it down to the lower bar. Read the emitter performance variation (V_{pf}) on the lower bar.

Example:
Assume that the U_h of a new microirrigation system measured immediately after installation was 95%. The SU from emitter flow rate measurements was 93%. From Fig. 34-3, a straight line drawn through $U_h = 95\%$ and $U_s = 93\%$ intersects the lower bar at an emitter performance variation (V_{pf}) of 5%. This value would be expected to be approximately the coefficient of manufacturing variation for the emitter because this system is newly installed and no emitter plugging has occurred.

Assume that the same irrigation system was again evaluated after operating for 1 year. At this time, U_h was again found to be 95%, but the SU was now 88%. From Fig. 34-3, a straight line drawn through these two points shows that V_{pf} increased to 11%. Because the U_h remained unchanged, the hydraulics of the system was not the cause of the lower application uniformity measured. The cause of the lower SU is a change in emitter performance, probably emitter plugging.

Wetting Patterns
Evaluating pattern uniformity for spinner-type emitters is much easier than for spray- or fan-pattern emitters due to the distribution of water relative to the emitter. To evaluate spinners, catch cans should be placed at one foot intervals in two or more radial legs from the spinner-type emitter. The cans can be of any size, but all should be of the same height and diameter. The system should be run for one hour or more, at which time the volume of water caught in each can should be measured with a graduated cylinder. As a general rule of thumb, a satisfactory distribution pattern should result in 50% or more of the water caught in 50% or more of the cans.

Evaluating distribution uniformity for spray-type emitters with catch cans is more difficult due to the elongated, alternating, wet and dry surface patterns around the head. This petal pattern makes it difficult to place the cans directly under the spray jet or in the petal pattern itself. In addition, interpretation

Figure 34-3. Nomograph for determining emitter performance variation.

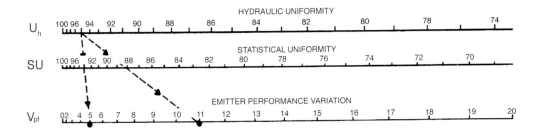

of any catch-can data may not truly reflect subsoil water redistribution. Visual observations of soil wetting patterns from each radial leg should be used for evaluation of spray-type emitters. The ideal distribution should resemble a petal pattern along each radial. The more uniform each radial leg is in creating a petal pattern, the more uniform will be the wetting in the soil.

The most representative approach to distribution uniformity is to use a soil probe. The soil probe will indicate the horizontal and vertical distribution of applied water relative to the tree's drip line. In sandy soils, the soil should be probed between 2 to 4 hours following the irrigation. In most situations, citrus trees should have 33% to 50% of their potential rooting zone wetted (i.e., with 12 x 24-ft tree spacing (288 ft^2), at least 95 ft^2 should be watered). The soil probe will also provide information on how deep the water is moving. Excessive water movement below the crop's root zone may result in leaching of fertilizers, waterlogged conditions, and/ or waste of water.

Chapter 35. Fertigation and Nutrient Management

by Brian Boman and Thomas Obreza

Fertigation is the application of liquid fertilizer through an irrigation system. Microirrigation and fertigation offer the potential for precise control of nutrients and water, which are the main grower-controlled inputs to plant growth. A major benefit of fertigation is that it provides greater flexibility and control of applied nutrients than conventional broadcast applications. Fertilizers are applied when needed and in small doses, so water-soluble nutrients are less subject to leaching by excess rainfall or overirrigation.

Care must be exercised to avoid emitter plugging problems resulting from reactions of the fertilizer with the irrigation water. The fertilizer source must be water-soluble. Chemical reactions between fertilizer materials can result in the formation of precipitates, which can plug the irrigation system. The uniformity of the fertilizer application depends on the uniformity of the water application. Therefore, high water application uniformity is very important for fertigation.

Nitrogen

Nitrogen (N) is the plant nutrient most often injected as fertilizer into microirrigation systems. One of the major benefits of small, frequent nitrogen applications is a potential reduction in leaching of nitrate into the groundwater. Only small amounts of N are applied at any one time; therefore, excess nitrate is not present to be leached in the event of heavy rainfall. Nitrogen can be applied using a number of different compounds, but urea and ammonium nitrate are the most desirable sources because they have a low plugging risk. Anhydrous or aqua ammonia are not recommended for use in microirrigation systems because they will increase the pH of the irrigation water. Consequently, calcium, magnesium, and phosphorus may precipitate in the line and increase the plugging

potential. Ammonium sulfate and calcium nitrate can be dissolved in water, but they also may cause plugging problems. If calcium or magnesium levels are high in the irrigation water, ammonium phosphate may cause precipitates to form, which can plug emitters. Nitrogen can contribute to microbial growth if it is applied continuously and remains in the irrigation line after the system has been shut off.

Nitrogen movement in the soil depends on the type of nitrogen fertilizer. The ammonium cation is less mobile in the soil than nitrate. The depth of movement depends on the cation exchange capacity (CEC) of the soil, and the rate of fertilizer application. Application of ammonium fertilizer to the soil surface may result in loss to the atmosphere by ammonium volatilization, especially if soil pH is greater than 7. Most ammonium will be transformed biologically to nitrate within two to three weeks at soil temperatures in the 75° to 90° F range. Nitrate will move with the irrigation water to the wetted front. Thus, with subsequent irrigations, nitrate may be leached beyond the root zone or may be pushed to the periphery of the wetted soil volume and only part of the root zone will have access to it. Urea is very soluble in irrigation water, and it is not adsorbed by soil. Thus, it will move deeper below the soil surface than ammonium, but will not leach as easily as nitrate. A balance between ammonium and nitrate in the nitrogen fertilizer is usually recommended.

Some water sources (such as recycled wastewater) may contain a significant amount of nitrate. This nitrogen should be taken into account when determining tree fertilizer requirements. The nitrogen added to the crop due to nitrate in the irrigation source water can be determined as follows:

$$N = C_n \times I_n \times D_i \qquad \textbf{Eq. 35-1}$$

where,

N = nitrogen (lb/ac),
C_n = a constant for unit conversion (0.226)
I_n = NO_3-N concentration in the irrigation water (mg/L)
D_i = depth of irrigation water applied (inches).

> Example:
> Determine the nitrogen supplied by the irrigation water if 14 inches of water are applied annually and the NO_3-N concentration is 10 mg/L.
>
> N = 0.226 x 10 x 14 = 32 lb N/ac

Nitrogen Cycle

Compounds containing nitrogen are of great importance in the life processes of all plants and animals. The chemistry of nitrogen is complex because of the numerous oxidation states that it can assume, and because of the fact that changes in the oxidation state can be brought about by living organisms.

Because of environmental concerns, nitrate (NO_3^-) is of particular interest. It is very mobile and easily transported by water. In surface water systems, NO_3^- is a nutrient source and can contribute to the overproduction of algae or other aquatic life, resulting in eutrophication of surface water bodies. Nitrate in groundwater is of even greater concern because groundwater is the principal domestic water because in many areas. The EPA has established a drinking water maximum concentration level (MCL) of 10 mg/L as N in NO_3^-, which is equivalent to 45 mg/L NO_3^-. Nitrogen is a very complex nutrient and it exists in the environment in many forms. It is continually transformed due to biological and chemical influences. Nitrogen can be divided into two categories:

1. Organic N contains carbon in the compound and exists in plant residues, animal waste, biosolids, septic effluent, and food processing waste.

2. Inorganic N contains no carbon in the compound and exists as ammonium (NH_4^+), nitrite (NO_2^-), nitrate (NO_3^-), and nitrogen gas (N_2).

Understanding the behavior of N in the soil is essential for good fertilizer management. Many N sources are available for use in supplying N to crops. In addition to inorganic (commercial) fertilizer N, animal manures and waste products are significant sources of organic N. Nitrogen fixation by legume crops can also supply significant amounts of N.

Sources of NO_3^- are both man-made and natural. The principal man-made sources of nitrate are commercial fertilizer, and septic and sewage systems. The ultimate source of N used by plants is N_2 gas, which constitutes 78% of the earth's atmosphere. Nitrogen gas is converted to plant-available N by one of the following methods:

- Fixation by microorganisms that live symbiotically on the roots of legumes (also certain nonlegumes)

- Fixation by free-living or nonsymbiotic soil microorganisms

- Fixation as oxides of N by atmospheric electrical discharges

- Fixation by the manufacture of synthetic N fertilizer (Haber-Bosch process).

The virtually unlimited supply of nitrogen in the atmosphere is in dynamic equilibrium with the various fixed forms in the soil-plant-water system. The N cycle can be divided into N inputs and outputs (Fig. 35-1). Understanding this process can influence how nitrogen is managed to minimize its negative effects on the environment, while maximizing the beneficial value of N for plant growth.

Animals and higher plants are incapable of utilizing nitrogen directly from the atmosphere. The nitrogen cycle with inputs, outputs, and cycling is complex. N derived from plant and animal residues and from the atmosphere through electric, combustion, and industrial processes is added to the soil. N in these residues is mobilized as ammonium (NH_4^+) by soil organisms as an end product of decomposition. Plant roots absorb a portion of the NH_4^+, but much of the NH_4^+ is converted to nitrate (NO_3^-) by nitrifying bacteria, in a process called nitrification. The NO_3^- is taken up by plant roots and is used to produce the protein in crops that are eaten by humans and fed to livestock. NO_3^- is lost to groundwater or surface water as a result of downward movement of percolated water through the soil. NO_3^- is also converted by denitrifying bacteria into N_2 and nitrogen oxides that escape into the atmosphere. The major processes of the nitrogen cycle (Fig. 35-1) are: N-mineralization, nitrification, NO_3^- mobility, denitrification, and volitilization.

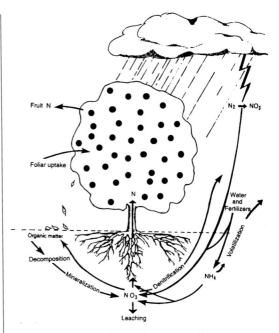

Figure 35-1. The citrus nitrogen cycle.

N-mineralization

The conversion of organic N to NH_4^+ is called mineralization. Mineralization occurs through the activity of heterotrophic microorganisms, which are organisms that require organic carbon compounds (organic matter) for their energy source. The NH_4^+ produced by mineralization is subject to several fates:

- Converted to NO_2^- and then to NO_3^- by the process of nitrification

- Absorbed directly by higher plants

- Utilized by heterotrophic organisms to further decompose organic residues

- Fixed in a biologically unavailable form in the lattices of certain clay minerals

- Released to the atmosphere as N_2

The quantity of N mineralized during the growing season can be estimated. Soil organic matter contains about 5% N by weight; during a single growing season, 1% to 4% of the organic N is mineralized to inorganic N.

Example:
If a soil contained 3% organic matter (OM) in the top 6 inches and 2% mineralization occurred, calculate the N made available (lb/acre). (Assume an acre of soil 6 inches deep weighs 2 million lb). 2,000,000 lb x 3% OM x 2% mineralization x 5% N = 60 lb N

N-immobilization

The conversion of inorganic N (NH_4^+ and NO^{-2}) to organic N is termed immobilization, and is basically the reverse of N mineralization. If decomposing OM contains low N relative to C, the microorganisms will immobilize NH_4^+ or NO_3^- in the soil. Since soil organisms need N in a C:N ratio of about 8:1 or less, they will utilize inorganic N. They are very effective at competing with plants for available N. Thus, N fertilizer is often applied to compensate for N immobilization. After the decomposition of the low residuals, the N in the microorganisms is mineralized back to NH_4^+.

Nitrification

The biological conversion of NH_4^+ to NO_3^- is called nitrification. This is a two-step process in which NH_4^+ first changes to NO_2^-, then to NO_3^-. This process is performed by autotrophic bacteria that obtain their energy from the oxidation of N, and their

C from CO_2. Nitrate leaching from field soil must be carefully controlled because of the serious impact that it can have on the environment. Since NO_3^- is very mobile and subject to leaching in the soil, understanding the factors that affect nitrification will provide insight into best management practices to minimize nitrate losses by leaching. Factors that affect nitrification in the soil are:

- Supply of NH_4^+: If conditions do not favor mineralization of NH_4^+ from organic matter, nitrification does not occur. (If organic residue with a high C:N ratio is plowed into the soil before planting, microorganisms will tie up available N while trying to decompose the residue.)

- Population of nitrifying organisms: Soils differ in their ability to nitrify NH_4^+, even under similar conditions of temperature, moisture, and level of added NH_4^+. One factor that may be responsible for this is the variation in the number of nitrifying organisms present.

- Soil pH: Nitrification takes place over a wide range of pH (4.5 to 10) conditions, but the optimum is thought to be about 8.5. The nitrifying bacteria also need an adequate supply of Ca. Thus, liming of low pH soil helps nitrification influence by both driving the soil pH closer to optimum, and also by providing more available Ca for the activity of the nitrifying organisms.

- Soil aeration: Aerobic nitrobacteria will not produce NO_3^- in the absence of O_2.

- Soil moisture: Nitrobacterial activity is sensitive to soil moisture. Nitrification is greatest under moist (but not saturated) soil conditions.

- Temperature: Nitrification increases between 40° and 95° F. Activity decreases at temperatures over 95° F.

NO_3^- mobility

The nitrate anion is very soluble in water, so leaching is a major cause of N loss from soils in humid climates or under irrigated conditions. Under irrigated conditions, the factors that most affect nitrate leaching from irrigation are:

- timing of irrigation;
- duration of irrigation;
- uniformity of irrigation;
- the amount of NO_3^- available for leaching.

Denitrification

Crop removal and leaching constitute the major N losses from the soil; however, under certain conditions, inorganic N ions can be converted to gases and lost to the atmosphere. When soil becomes waterlogged, O_2 is excluded and anaerobic decomposition takes place. Some anaerobic organisms have the ability to obtain their O_2 from sulfate and nitrate, with the accompanying release of N, and N_2. This is a common occurrence under Florida flatwoods growing conditions.

Volatilization

Volatilization is the loss of N from the soil to the atmosphere. Ammonia volatization can be significant, especially from calcareous (pH>7) soils under warm conditions.

Ammonium Process

The major source of nitrogen in synthetic fertilizers is ammonia, and it is produced by the Haber-Bosh process. Purified nitrogen gas (N_2) reacts with hydrogen gas (H_2) at high temperature and pressure to form ammonia:

$$N_2 + 3H_2 \xrightarrow{\text{heat/pressure}} = 2NH_3$$

Ammonia is the starting point from which nearly all other nitrogen fertilizers are made (Table 35-1). It contains 82% nitrogen and is the cheapest source of nitrogen compared to other nitrogen materials. It is used extensively in many areas of the U.S., either as anhydrous ammonia or aqua ammonia, accounting for nearly half of the total nitrogen fertilizer consumed. Ammonia is used in the U.S. primarily as a direct-application material. A direct-application material is a fertilizer material that is purchased as the pure material, rather than as a blend with other fertilizer materials in mixed fertilizers.

Losses of nitrogen from direct application of anhydrous ammonia to soils can be severe, particularly

Table 35-1. Synthetic nitrogen fertilizer materials that are derived from ammonia.

Reaction	Material Grade	Commercial N-P_2O_5-K_2O
$NH_3 + 2O_2 \rightarrow HNO_3 + H_2O$	Nitric acid	———
$NH_3 + HNO_3 \rightarrow NH_4NO_3$	Ammonium nitrate	33.5-0-0
$2NH_3 + H_2SO_4 \rightarrow (NH_4)_2SO_4$	Ammonium sulfate	20.5-0-0
$NH_3 + H_3PO_4 \rightarrow NH_4H_2PO_4$	Monoammonium phosphate	11-48-0
$NH_3 + H_3PO_4 \rightarrow (NH_4)_2HPO_4$	Diammonium phosphate	18-46-0
$2NH_3 + CO_2 \rightarrow (NH_2)_2CO + H_2O$	Urea	45-0-0

when applied to sandy soils. Application of anhydrous ammonia to sandy soils in Florida is not recommended because of excessive volatilization losses.

Urea and ammonium nitrate are mixed together with water and are often sold as a solution fertilizer containing up to 32% nitrogen. This material is quite stable and can be applied with relatively inexpensive, nonpressurized equipment. Nitrogen solutions are quite popular for direct applications to soil and are a major source of nitrogen for liquid fertilizer.

Nitrate-nitrogen (NO_3-N) is very mobile in soil and will move freely in irrigation and rain water. Since water percolates through sandy soils quite rapidly, application of large quantities of nitrate-nitrogen may increase nitrogen losses because of potential leaching.

Ammonium-nitrogen (NH_4-N) is adsorbed on cation exchange sites within the soil, and is retained, to some extent, against leaching. However, it is rapidly converted to nitrate-nitrogen by soil microorganisms and may be leached readily in this form. Nitrate-nitrogen is also lost rapidly from flooded soils because as soon as the oxygen in the soil is depleted, certain microorganisms can immediately begin to utilize the oxygen present in nitrate-nitrogen. This process, called denitrification, converts nitrate-nitrogen back into nitrogen gas (N_2), which escapes to the atmosphere and is unavailable to plants. Recovery of fertilizer nitrogen by crops rarely exceeds 70% and often is less than 50%.

Other chemical processes can occur to cause nitrogen losses from soil. For example, ammonium sources of nitrogen should never be surface-applied to soils recently limed or containing free calcium carbonate. In the alkaline pH environment surrounding the lime, ammonium will revert to ammonia gas and escape into the atmosphere:

$$NH_4 + OH^- \longrightarrow NH_3 + H_2O$$

The application of urea directly to the soil surface should also be avoided. Urea is quickly broken down into ammonia and carbon dioxide by the enzyme urease, which is normally abundant in cultivated soils:

$$NH_2 + CO \xrightarrow[H_2O]{urase} NH_3 + CO_2$$

It is always a good idea to mix urea thoroughly with the soil to minimize gaseous losses of nitrogen. All nitrogen fertilizers containing ammonium nitrogen leave an acid residue in soil as a result of the nitrification process:

$$2NH_4^+ + 4O_2 \longrightarrow 2NO_3^- + 4H^+ + 2H_2O$$

The use of high rates of nitrogen on sandy soils low in calcium needs careful attention. These soils are poorly buffered against changes in pH caused by the acidity released during nitrification. Florida has many soils in agricultural production that are sandy, low in organic matter (the principle component that aids in buffering) and contain less than

Table 35-2. Pounds of lime required to neutralize the acidity produced by one pound of nitrogen when applied as a particular fertilizer.

Material	N (%)	Lb of lime per lb N
Ammonium sulfate	20.5	5.35
Urea	45.0	1.80
Ammonium nitrate	33.5	1.80
Anhydrous ammonia	82.0	1.80
Nitrogen solutions	21-31	1.08

200 pounds of calcium per acre. These soils tend to be only mildly acid initially, but should be monitored carefully when large amounts of ammonium-nitrogen fertilizer are being applied (Table 35-2).

Ammonium sulfate is by far the most acid-forming source of nitrogen. This is because all the nitrogen is present as ammonium-nitrogen. Ammonium nitrate requires much less lime per pound of nitrogen since only half of the nitrogen is present as ammonium. If ammonium sulfate is used instead of ammonium nitrate, more than twice as much lime is needed to neutralize the acidity produced. The initial reactions involving urea and anhydrous ammonia in soil are quite basic, which tends to neutralize, to some extent, the acidity produced during subsequent nitrification.

Fertilizer Solubility

Several dry fertilizer products (Table 35-3) used for making fertilizer solutions are marketed with or without a protective conditioner. Whenever possible, the "solution grade" form of these products should be purchased to avoid having to deal with the conditioners and the potential plugging problems they can cause. Most dry-solid fertilizers are manufactured with a coating (commonly clay, diatomaceous earth, or hydrated silica) to keep the moisture from being absorbed by the fertilizer pellets. To avoid having these materials create plugging problems, it is best to prepare a small amount of the mix to observe what happens to the coating. If the coating settles to the bottom of the container, the clear transparent liquid can be taken from the top portion without disturbing the bottom sediment. If a scum forms on the surface, conditioners may need to be added to facilitate the removal of the conditioner by skimming.

When urea, ammonium nitrate, calcium nitrate, and potassium nitrate are dissolved, heat is absorbed from the water, and a very cold solution results. Consequently, it may not be possible to dissolve as much fertilizer as needed to achieve the desired concentration. Oftentimes, it is necessary to let the mixture stand for several hours and warm to a temperature that will allow all the mixture to dissolve.

Before injecting fertilizer solutions, a "jar test" should be conducted to determine the clogging potential of the solution. Some of the fertilizer solution should be mixed with irrigation water in a jar to determine if any precipitate or milkiness occurs within one to two hours. If cloudiness does occur, there is a chance that injection of the chemical will cause line or emitter plugging. If different fertilizer solutions are to be injected simultaneously into the irrigation system, they all should be mixed in the jar. The jar test should be conducted at about the same dilution rate that is used in the irrigation system.

Nitrogen

Urea, ammonium nitrate, calcium nitrate, potassium nitrate, and ammonium sulfate are very soluble in water. These nitrogen fertilizer materials are readily available on the market and are used extensively in the preparation of single nutrient or multinutrient fertilizer solutions.

Phosphorus

Commercial fertilizers contain the guaranteed percentage of P_2O_5 on the label as water-soluble

Table 35-3. Solubility rates for various fertilizer materials used to prepare fertigation solutions (adapted from California Fertilizer Association, 1980).

	Grade	Form	Temp °F	Solubility gm/100 ml	Solubility lb/gal
Nitrogen Fertilizers					
Ammonium Nitrate	34-0-0	NH_4NO_3	32	18.3	9.87
Ammonium Polysulfide	20-0-0	NH_4S_x		high	high
Ammonium Sulfate	21-0-0	$(NH_4)_2SO_4$	32	70.6	5.89
Ammonium Thiosulfate	12-0-0	$(NH_4)_2S_2O_3$		v. high	v. high
Calcium Nitrate	15.5-0-0	$Ca(NO_3)_2$	62	121.2	10.11
Urea	46-0-0	$CO(NH_2)_2$		100.0	8.34
Urea Sulfuric Acid	28-0-0	$CO(NH_2)_2 \cdot 9H_2SO_4$		high	high
Urea Ammonium Nitrate	32-0-0	$CO(NH_2)_2 \cdot NH_4NO_3$		high	high
Phosphate Fertilizers					
Ammonium Phosphate	8-24-0	$NH_4H_2PO_4$		moderate	moderate
Ammonium Polyphosphate	10-34-0	$(NH_4)_5P_3O_{10}$ & others		high	high
Ammonium Polyphosphate	16-37-0	$(NH_4)_7P_5O_{16}$ & others		high	high
Phosphoric Acid, green	0-52-0	H_3PO_4		45.7	high
Phosphoric Acid, white	0-54-0	H_3PO_4		45.7	high
Potash Fertilizer					
Potassium Chloride	0-0-60	KCl	68	34.7	2.89
Potassium Nitrate	13-0-44	KNO_3	32	13.3	1.10
Potassium Sulfate	0-0-50	K_2SO_4	77	12	1.00
Potassium Thiosulfate	0-0-25-17S	$K_2S_2O_3$		150	12.5
Monopotassium Sulfate	0-52-34	KH_2PO_4		33	2.75
Micronutrients					
Borax	11 % B	$N_2B_4O_7 \cdot 10H_2O$	32	2.10	0.17
Boric Acid	17.5% B	H_3BO_3	86	6.35	0.53
Solubor	20% B	$Na_2B_8O_{13} \cdot 4H_2O$	86	22	1.84
Copper Sulfate (acidified)	25% Cu	$CuSO_4 \cdot 5H_2O$	32	31.6	2.63
Cupric Chloride (acidified)		$CuCl_2$	32	71	5.93
Gypsum	23% Ca	$CaSO_4 \cdot 2H_2O$		0.241	0.02
Iron Sulfate (acidified)	20% Fe	$FeSO_4 \cdot 7H_2O$		15.65	1.31
Magnesium Sulfate	9.67% Mg	$MgSO_4 \cdot 7H_2O$	68	71	5.93
Manganese Sulfate (acidified)	27% Mn	$MnSO_4 \cdot 4H_2O$	32	105.3	8.79
Ammonium Molybdate	54% Mo	$(NH_4)_6Mo_7O_{24} \cdot 4H_2O$		43	3.59
Sodium Molybdate	39% Mo	Na_2MoO_4	44.3		3.70
Zinc Sulfate	36% Zn	$ZnSO_4 \cdot 7H_2O$	68	96.5	8.05
Zinc Chelate	5% -14% Zn	DTPA & EDTA		v. Sol.	v. Sol.
Manganese Chelate	5% -12% Mn	DTPA & EDTA		v. Sol.	v. Sol.
Iron Chelate	4% -14% Fe	DTPA & EDDHA		v. Sol.	v. Sol.
Copper Chelate	5% -14% Cu	DTPA & EDTA		v. Sol.	v. Sol.
Sulfuric Acid	95%	H_2SO_4		v. high	v. high

and citrate-soluble phosphate. Phosphorus is not very mobile in many soils and is much less likely than nitrogen to be lost when applied conventionally. Plants generally need phosphorus early in their life cycle, so it is important that this element, if deficient in the soil, be applied during or before planting. If the plant shows phosphorus deficiency symptoms during the growing season, injection of phosphorus into the irrigation water allows for later stage correction.

Phosphorus fertilizer injection may cause emitter plugging. Solid precipitation in the line occurs most often due to interaction between the fertilizer and the irrigation water. Most dry phosphorus fertilizers (including ammonium phosphate and superphosphates) cannot be injected into irrigation water because they have low solubility. Monoammonium phosphate (MAP), diammonium phosphate (DAP), monobasic potassium phosphate, phosphoric acid, urea phosphate, liquid ammonium polyphosphate, and long chain linear polyphosphates are watersoluble. However, they can still have precipitation problems when injected into water with high calcium concentration. Problems occur when the polyphosphate injection rates are too low to offset the buffering effects of the calcium and magnesium concentrations in the irrigation water.

The application of ammonium polyphosphate fertilizers to water that is high in calcium will almost always result in the formation of precipitants that can plug the emitters. These precipitants are very stable and not easily dissolved. Phosphorus and calcium, when in solution together, may form di- and tricalcium phosphates, which are relatively insoluble compounds. Similarly, phosphorus and magnesium can form magnesium phosphates that are also insoluble and plug emitters. Of considerable concern in South Florida is the formation of iron phosphates, which are very stable. Given the high levels of calcium, iron, and bicarbonate in Florida irrigation water, phosphorus should not be injected unless significant precautions are taken.

Phosphoric acid is sometimes injected into microirrigation systems. It not only provides phosphorus, but also lowers the pH of the water, which can prevent the precipitation problems previously mentioned. This practice will be effective as long as the pH of the fertilizer-irrigation water mixture remains low. As the pH rises due to dilution, phosphates precipitate. One approach that is sometimes successful is to supplement the phosphoric acid injections with sulfuric or urea sulfuric acid to ensure that the irrigation water pH will remain low (pH < 4.0). Continuous use of phosphoric acid at levels in excess of 25 mg/L, however, can produce zinc deficiencies. Phosphoric acid injection will be effective only as long as the pH of the fertigated water remains very low. Combined Ca and Mg should remain below 50 ppm and bicarbonate should remain less than 150 ppm.

Potassium

Potassium fertilizers are all water-soluble, and injection of K through microirrigation systems has been very successful. The problem most often associated with potassium injection is that solid precipitants form in the mixing tank when potassium is mixed with other fertilizers. The potassium sources most often used in microirrigation systems are potassium chloride (KCl) and potassium nitrate (KNO_3). Potassium phosphates should not be injected into microirrigation systems. Potassium sulfate is not very soluble and may not dissolve in the irrigation water. Potassium thiosulfate (KTS) is compatible with urea and ammonium polyphosphate solutions. However, it should not be mixed with acids or acidified fertilizers. When KTS is blended with urea ammonium nitrate solutions, a jar test is recommended before mixing large quantities. Under certain mixing proportions, particularly when an insufficient amount of water is used in the mix, potassium in KTS can combine with nitrates in the mix to form potassium nitrate crystals. If this happens, adding more water and/or heating the solution should bring the crystals back into solution.

Calcium

Fertilizers containing calcium should be flushed from all tanks, pumps, filters, and tubing prior to injecting any phosphorus, urea ammonium nitrate, or urea sulfuric fertilizer. The irrigation lines must be flushed to remove all incompatible fertilizer products before a calcium-containing fertilizer solution is injected. Calcium should not be injected with any sulfate form of fertilizer. It combines to create insoluble gypsum.

Micronutrients

Several metal micronutrient forms are relatively insoluble, and therefore not used for fertigation purposes. These include the carbonate, oxide, or hydroxide forms of zinc, manganese, copper, and iron. These relatively inexpensive materials can be broadcast and incorporated into the soil. However, they constitute a long-term source of micronutrients and will supply only a low level of nutrients for many years.

The sulfate form of copper, iron, manganese, and zinc is the most common and usually the least expensive source of micronutrients. These metal sulfates are water-soluble and are easily injected. However, using these materials for fertigation is not very successful in alleviating a micronutrient deficiency, because the metal ion has a strong electrical charge (2^+) and becomes attracted to the cation exchange sites of clay and organic matter particles, where it tends to reside near the soil surface. Consequently, the micronutrient usually does not reach the major plant root zone. If the soil pH is high, manganese, iron, and copper are changed into unavailable forms, and little or no benefit will be obtained from their use. If the metal sulfate solutions are acidified, however, the availability of the micronutrient can be prolonged in the soil.

Common Fertigation Materials

Ammonium Nitrate Solution (20-0-0)

NH_4NO_3 is ammonium nitrate fertilizer dissolved in water with a density of 10.5 pounds per gallon. It is the most widely used nitrogen source used for Florida citrus.

Urea Ammonium Nitrate Solution (32-0-0)

$(NH_2)_2CO \cdot NH_4NO_3$: Urea ammonium nitrate solution is manufactured by combining urea (46% N) and ammonium nitrate (35% N) on an equal nitrogen content basis. The combination of urea and ammonium nitrate contains the highest concentration of nitrogen of all the nitrogen solution products. When urea ammonium nitrate solutions are combined with calcium nitrate, a thick, milky-white insoluble precipitate forms, presenting a serious potential plugging problem.

Calcium Nitrate (15.5-0-0-19 Ca)

$Ca(NO_3)_2$: This fertilizer is high in nitrate-nitrogen (14.5%) with 1% ammonium-nitrogen and supplies calcium. The product can be combined with ammonium nitrate, magnesium nitrate, potassium nitrate, and muriate of potash. It should not be combined with any products containing phosphates, sulfates or thiosulfates.

Ammonium Thiosulfate (12-0-0-26)

$(NH_4)_2S_2O_3$ is used as both a fertilizer and as an acidulating agent. When applied to the soil, *Thiobacillus* bacteria oxidize the free sulfur to sulfuric acid. The acid then dissolves lime in the soil and forms gypsum. The gypsum helps to maintain a good, well-granulated, aerated, and porous soil structure. Ammonium thiosulfate is ideal for treatment of calcareous (high lime) soils. It is compatible with neutral or alkaline phosphate liquid fertilizers and nitrogen fertilizers. Ammonium thiosulfate can be applied in liquid mixes or by itself.

Ammonium thiosulfate should not be mixed with acidic compounds because it will decompose into elemental sulfur and ammonium sulfate at pH < 6. Application to neutral and acidic soils (without free lime) may result in a pronounced drop in soil pH over several weeks. The extent of the pH drop in these types of soils depends upon the total amount of this fertilizer applied, the cation exchange capacity of the soil, and the buffering capacity of the soil. The higher the clay content and the higher the lime content of the soil (i.e., the larger the buffering capacity), the slower the pH will drop with the same fertilizer application.

Phosphoric Acid (0-54-0)

H_3PO_4 has a density of approximately 14.1 pounds per gallon. The acid is a syrupy liquid that requires storage in stainless steel (No. 316) tanks. Phosphoric acid can be used in many formulations of nitrogen, phosphorus, and potassium mixes. Phosphoric acid should never be mixed with any calcium fertilizer. It will form insoluble calcium phosphate, which can plug irrigation lines.

Potassium Chloride (0-0-62)

Potassium chloride (KCl) is generally the least expensive source of potassium and is the most popular K fertilizer applied through fertigation. It may not be desirable for use on citrus if irrigation water contains high salinity levels.

Potassium Nitrate (13-0-46)

Potassium nitrate (KNO_3) is expensive, but the consumer benefits from both the nitrogen and the potassium in the product. It is an excellent choice of potassium fertilizer for areas where irrigation water salinity problems are present. It is less soluble than potassium chloride but more soluble than potassium sulfate.

Potassium Sulfate (0-0-52)

K_2SO_4 can be an alternative to KCl in high-salinity areas and provides a source of sulfur. It is fairly popular for fertigation. It is less soluble than potassium chloride and potassium nitrate.

Potassium Thiosulfate (0-0-25-17 and 0-0-22-23)

$K_2S_2O_3$ (KTS) is marketed in two grades and is a neutral to basic, chloride-free, clear liquid solution. This product can be blended with other fertilizers, but KTS blends should not be acidified below pH 6.0. The proper mixing sequence for KTS is: water, pesticide, KTS and/or other fertilizer. Always perform a jar test before injecting blends. Potassium thiosulfate provides not only potassium, but the thiosulfate is oxidized by *Thiobacillus* bacteria to produce sulfuric acid. This acid reacts with calcium carbonate in the soil, which releases additional calcium for the plant. Thus, potassium thiosulfate use on calcareous soils not only supplies potassium and sulfur but aids in increasing the availability of calcium to plants.

Sulfuric Acid

H_2SO_4 is not a fertilizer and thus, has a grade of 0-0-0. It has a density of approximately 15.3 pounds per gallon when concentrated. Sulfuric acid is a clear liquid when pure; however, much of the agricultural material may have a brown to black color. It has no odor and pours as an oily liquid. It is injected into high bicarbonate water to control the pH by reducing it to about pH 6.5 to 7.0. It is sometimes injected directly into calcareous soils (high lime) where the reaction produces gypsum. Sulfuric acid should not be injected with calcium fertilizers because calcium sulfate (gypsum) will form and create a creamy suspension very much like cottage cheese, which can easily plug the lines. O.S.H.A. requirements for safe handling preclude fertilizer dealers from storing sulfuric acid on the premises; therefore, it is difficult to find a source of sulfuric acid. Sulfuric acid is extremely corrosive and must be handled with proper equipment and clothing. Never combine urea and sulfuric acid in the field.

Urea Solid (46-0-0) and Urea Solution (23-0-0)

$CO(NH_2)_2$ (Urea) is sold as 46-0-0 dry fertilizer or as a liquid 23-0-0 urea solution. Commercial urea contains about 2.25% biuret, a byproduct that forms only during the manufacturing process. It can inhibit plant growth or damage plants. Urea with less than 0.25% biuret content should be used for foliar applications. Urea should never be mixed with sulfuric acid.

Urea Sulfuric Acid

Urea sulfuric acid ($CO(NH_2)_2 \cdot H_2SO_4$) is an acidic fertilizer that combines urea and sulfuric acid. By combining the two materials into one product, many disadvantages of using these materials individually are eliminated. The sulfuric acid decreases the potential ammonia volatilization losses from the soil surface and ammonia damage in the root zone that can occur with the use of urea alone. Urea sulfuric acid is safer to use than sulfuric acid alone. Urea sulfuric acid is well suited for fertigation and can be used for other purposes such as: acidifying the irrigation water (reducing plugging potential from carbonates and bicarbonates); using maintenance injections to keep lines and emitters clear of calcium carbonate deposits; cleaning irrigation lines once they have been plugged; and acidifying the soil.

Corrosion

Fertilizers and other injected chemicals can be corrosive to irrigation equipment. Table 36-4 lists the relative corrosion of six metals immersed in eight different fertilizer solutions for four days.

Injection Time

To determine the time required to fertigate and flush the system, the time for water to travel from the injection point to the furthest emitter must be known. The travel time for a chemical that is water-soluble can be estimated by the average velocity of the irrigation water. The travel time from the injection point to the last emitter can be calculated by summing the travel times for each pipe segment. For a pipe segment, travel time can be determined as follows:

$$T = D/V \qquad \text{Eq. 35-3}$$

where,

T = time of travel (minutes)
D = distance or length of pipe (ft)
V = velocity (ft/min)

Table 35-4. Relative corrosion of various metals (adapted from Martin, 1953). Solutions were made by dissolving 100 lb of material in 100 gallons of water. Metals were immersed for 4 days. Ratings: 0 = none, 1 = slight, 2 = moderate, 3 = considerable, 4 = severe.

Kind of Metal	Calcium Nitrate	Sodium Nitrate	Ammonium Nitrate	Ammonium Sulfate	Urea	Phosphoric Acid	Diammonium Phosphate	17-0-17 mix $(NH_4)_2SO_4$ + DAP + K_2SO_4
Galvanized Iron	2	1	4	3	1	4	1	2
Sheet Aluminum	0	2	1	1	0	2	2	1
Stainless Steel	0	0	0	0	0	1	0	0
Phospho-Bronze	1	0	3	3	0	2	4	4
Yellow Brass	1	0	2	2	0	2	4	4
Solution pH	5.6	8.6	5.9	5.0	7.6	0.4		7.3

Observations:
Ammonium nitrate, phosphoric acid, and ammonium sulfate are very corrosive.
Brass and bronze are corroded by phosphate, especially if ammonium is present.
Copper is very corrosive to aluminum, even in small doses.
While 316-grade stainless is corrosion resistant, other grades of stainless may corrode.

Table 35-5. General acid compatibility of component materials.

Material	Occasional	Continual
Buna-N	No	No
Ceramic/Graphite	Yes	Yes
EPDM	Yes	Yes
Hypalon	Yes	No
Leather	No	No
Neopreme	Yes	Yes
Teflon	Yes	Yes
Viton	Yes	Yes
Aluminum	Yes	No
PVC	Yes	Yes
Brass	Yes	No
Polypropylene	Yes	Yes
303 stainless	Yes	No
316 stainless	Yes	Yes
Galvanized	No	No

If recommended only for occasional use, components should be rinsed and neutralized with soda after use.

Example:
Determine the travel time from the injection point to the manifold based on average velocity, given 1000 ft of 6-inch ID pipe with a flow rate of 500 gpm.

$V = (500 \text{ gpm}/7.48 \text{ gal/ft}^3)/((0.5 \text{ /ft})_2 \times 3.14/4)$
$V = 344 \text{ ft/min}$

$T = D/V = (1000 \text{ ft} / 344 \text{ ft/min}) = 2.9 \text{ min.}$

The travel time for a lateral of uniform diameter with evenly spaced emitters that have equal discharge rates can be estimated as follows:

$$T = t \times (0.577 + Ln(N)) \qquad \textbf{Eq. 35-4}$$

where,

T = travel time for entire lateral (minutes)
t = travel time between the last two emitters (minutes)
N = total number of emitters on the lateral
Ln = natural logarithm

The value of t is determined by:

$$t = A \times S/q \qquad \textbf{Eq. 35-5}$$

where,

A = cross-sectional area of the pipe (ft^2)
S = emitter spacing (ft)
q = emitter discharge (ft^3/min)

Example:
Determine the travel time based on average velocity for: emitter flow = 1.0 gph, lateral diameter = 0.632 inch, emitter spacing = 3 ft, and lateral length = 630 ft.

Dia = 0.632 in. x 1 ft/12 in. = 0.0526
A = 0.0526 ft^2 x 3.14 1/4 = 0.00218 ft^2
S = 3 ft
q = 1.0 gal/hr x 1 ft^3/7.48 gal x 1 hr/60min
　　= 0.00222 ft^3/min
t = AS/q = 0.00218 ft^2 x 1 ft / 0.00222 ft^3/min
　　= 2.95 minutes
T = 2.95 [0.577 + Ln (630)] = 21 minutes

The above procedure can be applied to the submain if it has equally spaced lateral outlets. However, the manifold is often tapered (nonuniform diameter), and in that case, a step-by-step analysis must be performed. Total travel time is the sum of the all the segment travel times for the entire pipe system from the injection point to the last emitter, which would typically include the mainline, submain, manifold, and lateral line.

Travel time calculations are based on average velocity of water in the pipeline. Actually, velocity is higher at the center of the pipe than near the pipe wall. So, to ensure complete flushing, the flush time should be twice the calculated travel time.

For an existing irrigation system, chemical travel time can be easily measured in the field by injecting a dye, acid, or fertilizer salt. If a fertilizer is used, a simple electrical conductivity (EC) meter can detect when the fertilizer has arrived at the farthest outlet of the irrigation system. Sampling should be continuous until the chemical arrives, which will be indicated by an increase in the EC of the water. Similarly, a pH meter (or pH strips) may be used if the injected material is an acid.

Fertilizer Concerns

Growers should only inject water-soluble fertilizers or fertilizer suspensions that are compatible with their irrigation system and crop production system. Because they are potentially corrosive, fertilizers should be flushed from the irrigation system after each application. Fertilizer solutions should always be injected before (upstream of) the filters in microirrigation systems. The compatibility of fertilizer solutions with the irrigation water and with any other chemicals being injected should be tested to avoid the formation of chemical precipitates in the irrigation system.

Care must be taken to ensure that injected materials do not react with dissolved solids in the irrigation water in such a way as to form precipitates or deposits in the irrigation system. The chemicals must be soluble and remain in solution throughout the operating conditions of the irrigation system. The fertilizers selected to be injected into the irrigation water need to be entirely soluble in water and should not react with salts or chemicals in the water. Most nitrogen sources cause few clogging problems. The exceptions are anhydrous ammonia, aqua ammonia, and ammonium phosphate, which increase the pH of the water and cause precipitates with calcium and magnesium to form. Application of most forms of phosphorous through the system can result in extensive clogging. However, phosphoric acid can be safely injected in most waters because it acidifies the solution to a point where precipitation is prevented. All of the common potassium fertilizers are readily soluble and present no clogging problems. Fertilizers can be highly

corrosive and are a potential health hazard to skin and eyes. Therefore, all system components, including pumps, injection devices, lines, filters, and tanks, should be inspected prior to use. There should be a routine monitoring program of the fertigation process with particular emphasis on the start-up and shutdown periods. Injection rates and times should be calibrated and rechecked frequently to ensure proper operation of the system. Leaks, runoff, excess applications, and application to areas with open water should be prevented. All system components should be flushed with clean water following each use.

When injecting fertilizers, the salinity of the irrigation water with the fertilizer in it should be checked. Heavy dosages of fertilizers can cause leaf burn, even if relatively low-salinity water is used. It is generally preferable to inject small dosages of fertilizer frequently rather than making fewer applications at a high rate.

It is essential that proper and legal backflow prevention devices be used in the irrigation system to prevent fertilizers from being back-siphoned into the water supply. The injection device itself should have a screen and check valve. It is recommended that injection take place upstream from filters so that any contaminants or precipitates can be filtered out.

Fertigation rates and times should be calibrated for each area that is fertigated. The flushing time needs to be at least as long as the travel time in the system from the injection point to the furthest emitter. In many microirrigation systems, this time is often 20 to 30 minutes. Fertilizer injections need to be at least this amount of time, and flush times need to exceed this travel time so that nutrients will not remain in the lateral tubing and promote algal growth.

Chapter 36. Prevention of Emitter Clogging

by Brian Boman

Water Source and Classification

Agricultural irrigation waters are commonly divided into two categories: well water and surface water. However, when considering irrigation water quality and its effects on the functioning of a microirrigation system, classifying water as being either aerobic or anaerobic is more advantageous. Aerobic waters normally come from surface water or from shallow wells. They generally support a variety of bacterial activity and, depending on the source, have a wide range of dissolved minerals in them. Anaerobic waters normally come from deeper wells that tap below 600 feet in the Floridan Aquifer, or 300 to 800 feet in the shallower Hawthorne formation. These waters are deep enough in the geologic strata that impervious or reactive layers generally prevent oxygen from penetrating into the aquifer. These waters can be rich in minerals, especially in the Floridan Aquifer, and they are usually relatively free of bacterial contamination.

The main problems in systems served with aerobic waters tend to be slime and bacterial contamination. Even slight mineral loads in the water are able to support significant bacterial populations. This is especially true of surface waters that have an abundant dissolved carbon source to support the metabolic processes of the microorganisms. Scale formation, sometimes very strong, is nearly always associated with aerobic slime deposits.

The most prevalent problem in systems served by anaerobic waters is the development of scales. The appearance and development rate of a scale layer is dependent on the composition and relative concentrations of the minerals in the water. Bacterial problems associated with anaerobic water are normally limited to few specific strains of bacteria. For example, one quite common bacteria associated with the Floridan and upper Hawthorne

formations is *Desulfovibrio sulfuricans*. These bacteria can be recognized from the typical sulfur smell they produce. Thus, general visual observation of the water source and ability to classify it as aerobic or anaerobic allows us to do some initial predictions as to what type of problems, if any, may be expected.

Nearly all irrigation water sources in Florida can be considered relatively stable. Stability means there is little or no change in the chemical analysis and biological and physical composition over time. Even though surface waters have seasonal variations due to rainfall patterns, they tend to have similar average characteristics on an annual basis. However, once water is drawn into an irrigation system, the water experiences rapid changes. Changes in pressure, temperature, solar radiation exposure (UV and IR), velocity, and composition (filtration) can result in altered chemical reaction and bioactivity balances within the water. The dynamic state of the water can become the trigger that initiates a plugging problem.

Emitter Clogging

The physical causes of irrigation system flow restrictions or plugging can be roughly divided into four classes: biological clogging, chemical precipitates, insects, and particulate matter. Chemical precipitates are normally a result of reactions of dissolved inorganic constituents such as Ca, Mg, Mn, or Fe with oxygen, sulfates, phosphates, and silicates (Table 36-1). Precipitates can also form from reactions with organic materials such as humates, sugars, amino acids, or degraded plant and bacterial matter, which are an integral part of any surface water. Biological clogging is the result of the growth of organisms such as bacteria, algae, and fungi. The amount of particulate matter in water varies considerably depending on the water source and filtration devices. Particulate clogging is often caused by suspended clay, silt, sand, or plant material that is carried by the irrigation water (Table 36-2). The most common insect-clogging problems result from ant or spider activity in the emitter orifices. Another pest of microsprinkler systems is the snail, which may become lodged in the emitters and distort wetting patterns in addition to reducing discharge rates.

Clogging problems due to a single factor, such as calcium deposition, can develop. However, in most cases the plugging is a result of several independent or interdependent factors. For example, formation of a precipitate often is associated with bacterial activity, changes in the water pH, and fluctuations in the temperature and pressure of the water.

Chemical Precipitation and Scale Formation

Mineral precipitation and scale formation are common problems when groundwater is used in micro-irrigation systems. All groundwater contains salt in solution. Water temperature, pressure, and/or pH may change when water is removed from the aquifer by pumping. This can initiate reactions that cause salts to precipitate from the water solution.

The scale-forming reactions in irrigation systems result from the chemical response to changing conditions of the irrigation water. The formation of precipitates depends on several factors. For example, depending on pH, temperature, concentration of species in solution, time, etc., Ca and PO_4 can form seven different combinations. Three of those (hypo-, meta-, and pyrophosphate) are

Table 36-1. Major water constituents associated with microirrigation system clogging problems.

Cations (+)	Anions (-)	Anion precipitate compounds
Scale forming		
Ca calcium	CO_3 carbonate	Ca, Mg
Mg magnesium	HCO_3 bicarbonate	Ca, Mg
Fe iron	PO_4 phosphate	Ca, Mg, Fe
Mn manganese	OH hydroxide	Ca, Fe
	SO_4 sulphate	Ca, Mg, Fe, Mn
Soluble compounds		
Na sodium	NO_3 nitrate	
K potassium	Cl chloride	

Table 36-2. Major water constituents associated with microirrigation system clogging problems.

Physical	Chemical	Biological
organic debris	Ca or Mg carbonates	filaments
aquatic weeds	Ca sulfate	slimes
algae	ferric iron	microbial deposits
snails and fish	fertilizers - phosphate, ammonia, Mn, Zn, Cu	iron ochre
PVC shavings	metal hydroxides, carbonates, silicates, and sulfides	magnanese ochre
soil particles		sulfur ochre

Table 36-3. Precipitate formations and characteristic colors for various cation and anion combinations.

Cations	CO$_3$	PO$_4$	SO$_4$	OH	SiO$_3$/SiO$_4$	Oxide
_Ca_						
Precipitate Formula	CaCO$_3$	Ca$_3$(PO$_4$)$_2$	CaSO$_4$	Ca(OH)$_2$	CaOSiO	CaO
Color of precipitate	colorless white	colorless white	colorless	colorless	colorless	white
_Mg_						
Precipitate formula	MgCO$_3$	Mg$_3$(PO$_4$)$_2$	MgSO$_4$·7H$_2$O	Mg(OH)$_2$	Mg$_2$SiO$_4$	MgO
Color of precipitate	colorless white	colorless white	colorless	colorless	white	colorless
_Fe_						
Precipitate formula	FeCO$_3$	FePO$_4$·2H$_2$O	FeSO$_4$	Fe(OH)$_2$	Fe$_2$SiO$_4$	Fe$_2$O$_3$
Color of precipitate	gray amorphous	white-black yellow-white pink	yellow white black-green	green white	colorless	black red-brown
_Mn_						
Precipitate formula	MNCO$_3$	MN$_3$(PO$_4$)$_2$	MNSO$_4$	MN(OH)^2	Mn$_2$SiO$_4$	MN$_2$O$_3$
Color of precipitate	rose light brown	rose gray red	pink rose red	white-pink brown-black	red	black dark red olive

insoluble. The naturally occurring brushite (CaHPO$_4$·2H$_2$O) and whitlockite (Ca$_3$(PO$_4$)$_2$) have solubilities of 2.6 lb per 1000 gal and 0.17 lb per 1000 gal, respectively (Table 36-3). Monoorthophosphate and pyrophosphate pentahydrate have significantly higher solubilities. The solubilities of all of the above compounds decrease with increasing pH.

Phosphates not only precipitate rapidly, but they form very stable compounds. Reactions with Ca, Mg and Fe are of special concern since these elements are widely found in Florida waters. These elements react with phosphates, producing compounds that render the phosphates unavailable for the plants. Therefore, if there are high calcium, magnesium, or iron loads in the irrigation water, it is questionable whether liquid phosphate fertilizer will produce desired effects in the targeted plants.

In most irrigation systems, flows range from hundreds to thousands of gallons per minute. As a result, large amounts of dissolved chemicals are continually being moved throughout the irrigation system. For example, a system with a 2500 gpm flow rate, pumping water with 1000 ppm total dissolved solids (TDS), will have more than 20 lb of additional solids entering the system each minute. Correlating laboratory reports of water quality to real-life precipitate potential for irrigation systems with high flow rates is usually very difficult. However, information on the theoretical potential for precipitate formation for various combinations of cations and anions can be useful to determine if the potential for clogging problems may exist. Although there are usually several possible combinations that can occur, only a single cation concentration is presented for each compound (Table 36-3).

Calcium Carbonate Scale
Groundwater often contains a relatively high amount of Ca and bicarbonate. This occurs as carbon dioxide, released by bacterial action in the soil, and is absorbed into the percolating rainwater as it infiltrates through the soil. This creates a weak acid

(carbonic acid), which tends to dissolve carbonates in soil and limestone formations. These ions are absorbed into solution as the water remains in the ground. When the groundwater is pumped to the surface, the carbon dioxide escapes, and there is a resulting increase in the pH of the water. This increase may cause water to be supersaturated with calcium carbonate, which then precipitates and can form scale deposits within the microirrigation system.

Water high in Ca, HCO_3, pH, and temperature will likely precipitate $CaCO_3$, which can plug emitters and other components of the microirrigation system by the gradual buildup of precipitant scale in water passageways. A method for calculating the stability of water relative to precipitation of calcium carbonate is known as the Langelier Saturation Index (LSI). The LSI has been widely used as a means of predicting scale formation in industrial and municipal water systems. If the LSI is positive, then scale formation is likely. The higher the value, the greater the scaling potential. The LSI produces good estimates of potential scale development if the TDS is less than 4000.

<u>Iron and Manganese Scale</u>
Plugging of the microirrigation system emitters due to iron in the irrigation water is a common problem for citrus growers in many parts of Florida. Iron is frequently found in groundwater, since approximately five percent of the earth's surface is made up of iron. Water, rich in carbon dioxide, readily dissolves iron from the soil as it percolates to the underground formation. Iron concentrations in irrigation water in excess of 3 ppm are relatively common in some areas.

Iron exists in irrigation water in three basic forms: as elemental metallic iron, and in the ferrous (Fe^{++}) or ferric (Fe^{+++}) state. Iron is present in groundwater due to the solubility of ferric bicarbonate that dissolves as a result of the action of carbon dioxide on iron deposits in the ground. Iron remains in the soluble ferrous state as long as the water remains underground, where molecular oxygen is scarce.

When iron-bearing water is first brought to the surface, it is usually clear and colorless. After exposure to air, it soon turns reddish brown. This is caused by molecular oxygen entering the water as carbon dioxide escapes, which oxidizes the ferrous ions (Fe^{++}) by changing them to ferric ions (Fe^{+++}). Then the ferric ions combine with free hydroxyl ions (OH^-) to form the insoluble gelatinous compound known as ferric hydroxide [$Fe(OH)_3$]. The precipitation of ferric hydroxide can form scale, which will plug the irrigation system. This scale is notoriously difficult to dissolve. Ferrous iron can also combine with carbonate ions to form iron bicarbonate. This form of iron plugging may take several years to develop before becoming a serious problem. Growers often do not realize that a problem exists until the scale has accumulated for several irrigation seasons, and then it is often very difficult to remove.

The specific form that iron takes in water is dependent on the pH and the dissolved oxygen content of the water. In natural systems in which O_2 levels are low or absent and pH is from 6.5 to 7.5, iron occurs primarily as dissolved ferrous ions (Fe^{++}). Ferrous iron is unstable when it comes into contact with oxygen. This may occur in the pumping operation as water is removed from the aquifer and mixing between the anaerobic and aerobic zones occurs (Fig. 36-1).

Iron in Groundwater and Irrigation Systems

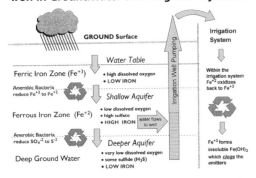

Figure 36-1. Iron status in groundwater and irrigation systems (from Pitts and Capece, 1997).

A less common problem, but as potentially troublesome, is iron sulfide precipitation. Dissolved iron in the presence of sulfides can form a black, insoluble precipitant. A combination of greater than 0.6 ppm iron and greater than 2.0 ppm total sulfides present in the water will generally create an iron sulfide sludge. Wells that draw water from two formations—one high in sulfides and the other high in iron—would be candidates for this type of problem. Treatment for iron sulfide includes a combination of aeration, acidification, and chlorination.

Manganese resembles iron in its chemical behavior and occurrence in groundwater. However, problems generally occur at lower concentration levels than iron. When exposed to O_2, manganese (Mn^{++}) is precipitated to form a black, sooty deposit.

In municipal water treatment, iron and manganese are usually removed through oxidation with the atmospheric oxygen. As air is bubbled through the water, the iron and manganese are precipitated. At pH levels of 7.0, about 15 minutes are required for 90% conversion of ferrous ions. Manganese precipitation is somewhat slower.

Both iron and manganese can be kept in solution by adding a small amount of sodium hexametaphosphate to the water. The polyphosphate stabilizes the iron and manganese compounds and delays their precipitation. The time length of the delay varies with the iron and manganese concentration and the amount of polyphosphate added. The polyphosphate must be added before the water is exposed to oxygen. If the mixing of aerobic water occurs during the pumping operation, the iron or manganese may oxidize before the stabilizing chemical can be added to the water.

Physical Components

Suspended matter consists of undissolved material such as clay, silt, sand, plant particles, and dead and living microorganisms. As they accumulate, these particles plug filters and system components. The matter is also responsible for abrasion on the soft plastic parts of the system, and the resulting microscopic scratches in the surfaces provide attractive attachment areas for scales and microbial life. For example, clay is microscopically abrasive, and it also has strongly charged surfaces that attract

Table 36-4. Guideline for interpretation of water analysis results (adapted from Nakayma and Bucks, 1986).

Water Quality Parameter	Degree of Problem		
	Slight	Moderate	Severe
Salinity			
EC (mmho.cm)	< 0.8	0.8 - 3.0	> 3.0
TDS (ppm)	< 500	500 - 2000	> 2000
Toxicity			
Sodium (SAR)	< 3.0	3.0 - 9.0	> 9.0
Chloride (ppm)	< 140	140 - 350	> 350
Boron (ppm)	< 0.5	0.5 - 2.0	> 2.0
Emitter Plugging			
Iron (ppm)	< 0.3	0.3 - 3.0	> 3.0
Manganese (ppm)	< 0.1	0.1 - 1.5	> 1.5
pH	< 7.0	7.0 - 8.0	> 8.0
Sulfides (ppm)	< 0.1	0.1 - 0.2	> 0.2
Calcium Carbonate (ppm)			
Suspended Solids (ppm)	< 50	50 - 100	> 100
Hardness ($CaCo_3$)	> 100	100 - 200	> 200
Alkalinity	< 150	150 - 300	> 300
Bacteria (#/L)	< 10,000	10,000 - 50,000	> 50,000
Bacteria (#/gal)	< 2600	2600 - 13,000	> 13,000

bacteria and scale. The result is an aggressive cycle where clay attracts bacteria, which attract more clay and other matter, which attract more clay, etc. Consequently, the mass of the slime and scale grows much faster than a laboratory water quality analysis would indicate.

Organic matter, in the form of plant parts, excretion products from cell metabolism, and dead cells, makes excellent attachment places for bacteria, minerals, and other suspended matter in water and on the tubing walls. These organic materials react with metals in solution, and also serve as a carbon source for bacterial activity. The role of the suspended organic matter as a food source for fungi, algae, and bacteria is probably more detrimental to irrigation systems than reactivity with dissolved inorganic compounds. Proper filtration and maintenance can greatly reduce the clogging problems associated with physical particles.

Suspended solids can easily disable a microirrigation system through emitter clogging. While larger sand particles can be removed through filtration, small particles (silts and clays) are not removed by standard filters. The presence of these particles may influence the selection of emitters.

In a standard laboratory analysis, suspended solids are measured by passing a volume of water through an extremely fine filter. The filter paper is then dried and weighed. The mg/L of suspended solids is the difference between the prefiltering and postfiltering weights of the filter paper. However, knowing the mg/L of the suspended solids does not tell one the size of the particles in the water. Measurement of the particle-size distribution of a water sample requires a special optical laboratory instrument.

Sand in water supplies is removed by filtration. However, silt and clay will tend to pass through the irrigation system filter. Therefore, if the silt or clay load is very high, it is often preferable to build a settling basin for preliminary treatment prior to filtration. The size of the settling basin will be determined by the system flow rate and the settling velocity of the particles to be removed. This settling velocity, in turn, is determined by the particle size, shape, and density. In most cases, silt and clay will pass through the microirrigation system and do not have to be removed. If biological growth is occurring within the piping network, the biological elements will sometimes attach with silt and clay particles to form aggregates large enough to plug the emitters.

Very fine (clay-sized) colloidal particles are too small to be economically removed by means of a settling basin because they settle so slowly that in most cases a prohibitively large settling basin would be required. In addition, water turbulence and wave action can keep the particles in suspension. Silt and clay particles that enter into the microirrigation system may settle out of the water in the lateral lines or emitters, where they may become cemented together by the action of bacteria to form large and potentially troublesome masses of slime. In order to combat this slime problem, chlorination is often practiced to curb the growth of any biological organisms. In addition, manifolds and lateral lines should be regularly flushed to remove sediments.

Biological Clogging

The biological components of water that can cause plugging problems in a microirrigation system are bacteria, algae, and larger organisms (insects). Algae is often a problem in surface water ponds. Available nutrients and sunlight can cause algae blooms. If present in sufficient quantity, these organisms can clog filter systems. The presence of algae can be determined by a chlorophyll laboratory test.

The principal groups of organisms found in surface waters are classified as protista, plants, and animals. The protista category includes bacteria, fungi, protozoa, and algae. Protista are the most important group of organisms with respect to microirrigation system water quality, especially the bacteria and algae.

Several nuisance bacteria are of concern in irrigation systems. They can discolor the water, cause

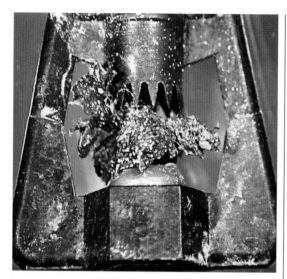

Figure 36-2. Algae growth in orifice and on bridge of microsprinkler emitter.

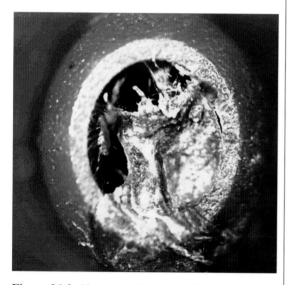

Figure 36-3. Algae growth in base of microsprinkler.

odors, and result in precipitates. Most importantly, they can form highly active slime. They are capable of mass reproduction, forming capsular structures, making extracellular excretions, and depositing nutrients as a result of metabolic processes. Bacterial plugging develops rapidly when there are mineral food sources available in addition to suspended particles with which bacteria can attach. Microscopic observation has revealed that bacteria attach to the tube surfaces either by

physical trapping by the micropores, or as a result of secondary chemical attraction. This chemical adhesion occurs when an organic or inorganic compound formed in the water bonds to the wall of the tubing and the bacteria attach to it.

Algae are a very large group of microorganisms that are capable of photosynthesis (they need sunlight), and are present virtually everywhere even just a little moisture is present. Algae contribute to plugging in several ways. Floating and suspended algae increase the turbidity of the water, resulting in filter plugging and frequent maintenance on intake screens. Dead algae, which remain in filters and the distribution lines, can be a good carbon source for other bacteria. Algae that pass through filters travel through the system and attach to surfaces near emitter orifices where light is available. As the algae grow, they restrict emitter discharges and eventually plug the emitter. It is not uncommon to see algae growth completely cover and plug drippers and microsprinklers.

Algae

Algae clogging of emitters normally occurs from growth in the emitter orifice or bridge (Fig. 36-2), or in the base of the emitters (Fig. 36-3). Algae are photosynthetic organisms that may exist as single cells, as loosely organized groups of cells (colonies), as sheetlike structures, or as intricately branched filaments. Algae is most commonly found in surface water and not typically expected to be present in groundwater due to its photosynthetic requirement. However, algae can be found nearly anywhere. Algae occur in the most severe habitats on earth. Some species grow on snow in perpetually freezing temperatures; others thrive in hot springs at temperatures of 160° F or more. A few can survive in extremely saline bodies of water; yet others withstand the pressure and low light intensity conditions of 300 feet or more below the surface of lakes or oceans.

Algae may grow profusely in surface water supply reservoirs and may become very dense, particularly if the water contains plant nutrients, especially nitrogen and/or phosphorus. Algae can be a great nuisance in surface waters because, when conditions are right, they will rapidly reproduce and cover streams, lakes, and reservoirs in large floating colonies called blooms. Algal blooms are usually characteristic of what is called "eutrophic" lake. In many cases, algae may cause difficulty with primary screening or filtration systems because of a tendency for algae to become entangled within the screen meshes.

Algae is typically not associated with groundwater since it requires sunlight for photosynthesis. However, algae has been found in significant amounts in irrigation water from shallow wells in Florida. This is likely due to large-sized macropores and very shallow wells.

Algae are unicellular or multicellular, autotrophic, and photosynthetic protists. Like other organisms, algae require nutrients to reproduce. The principal nutrients required are carbon dioxide, nitrogen, and phosphorus. Trace elements, such as iron, copper, and molybdenum, are also important. There are four broad classes of algae:

1. Green (Chlorophyta). The green algae are principally a freshwater species, and can be unicellular or multicellular. A distinctive feature of this group is that the chlorophyll and other pigments are contained in chloroplasts, membrane-surrounded structures that are the sites of photosynthesis. Common green algae are those of the *Chlorella* group.

2. Motile Green (Volvocales Euglenophyta). Colonial in nature, these algae are bright green, unicellular, and flagellated. Euglena is a member of this group of algae—as is Mastigophora—containing chlorophyll.

3. Yellow-Green or Golden-Brown (Chrysophyta). Most forms of the Chrysophyta are unicellular. They are freshwater inhabitants, and their characteristic color is due to yellowish brown pigments that conceal the chlorophyll. Of this group of algae, the most important are the diatoms. They are found in both fresh and salt waters. Diatoms have shells that are composed mainly of silica. Deposits of these shells are known as diatomaceous earth.

4. Blue-Green (Cyanophya). The blue-green algae are of very simple form and are similar to bacteria in several respects. They are unicellular, usually enclosed in a sheath, and have no flagella. They differ from other algae in that their chlorophyll is not contained in chloroplasts but is diffused throughout the cell.

Green algae can only grow in the presence of light. Algae will not grow in buried pipelines or in black polyethylene laterals or emitters. However, enough light can enter through exposed white PVC pipes or fittings to permit growth in some parts of the system. Where practical, exposed PVC pipe and fittings should be painted to reduce the possibility of algal growth within the system. Since algae can cause clogging problems when washed into laterals or emitters, chlorination is the recommended treatment to kill algae growing within the irrigation system.

Iron Bacteria

A common problem in some areas of Florida is emitter plugging due to iron bacteria. Iron bacteria frequently thrive in iron-bearing water. It is unclear whether the iron bacteria exist in groundwater before well construction and simply multiply as the amount of iron increases due to pumping, or whether the bacteria are introduced into the aquifer from the subsoil during well construction. Water well drillers should use great care to avoid the introduction of iron bacteria into the aquifer during the well-drilling process. All drilling fluid should be mixed with chlorinated water at a 10 ppm free chlorine residual.

Some of the most important nuisance microbes are iron bacteria that change soluble iron (Fe^{3+}) into insoluble ferric hydroxide ($Fe(OH)_2$). The bacteria create a slime that forms aggregates called ochre, which are usually red, yellow, or tan in color. The species *Sphaerotilus* generates sheath formations and deposits the iron around itself. The result can be a thick, slimy carpet woven from tubes and

Figure 36-4. External dripper plugged by sulfur slime.

crusted with iron deposits. These formations (globs) can then be severed from the tube walls due to changes in pressure, temperature, flows, etc., and completely plug the emitters. Another iron microbe, *Gallionella*, forms stalks or ribbons of iron that are deposited around itself by secretions. This deposit can rapidly trap suspended particles and thus increase the food source, resulting in exponential growth of the colony.

Emitter plugging from iron-oxidizing bacteria is especially difficult to control. Even very small iron concentrations (less than 0.5 ppm) are sufficient to provide healthy bacterial growth. Iron bacterial growth generally appears reddish. These bacteria oxidize iron from the irrigation water as an energy source. Precipitation of the iron and rapid growth of the bacteria create a voluminous material that can completely plug a microirrigation system in a matter of a few weeks.

It has long been recognized that bacteria play an important role in the oxidation of ferrous iron in groundwater systems. Only recently, however, has it been known that bacteria play an important role in reducing ferric iron in the soil to a ferrous state in which it is then dissolved into solution. Since ferric iron reduction by bacteria occurs where O_2 is nearly absent, and thus there is little possibility for reoxidation, this short-circuits the iron cycle and can lead to the accumulation of high concentrations of ferrous iron in groundwater. It appears that these iron-reducing bacteria occur most frequently in the area between the aerobic and anaerobic zones of an

aquifer system. Thus, it may be possible during well construction to case off these high iron zones in order to minimize the iron problem.

Sulfur Bacteria

Some Florida drip irrigation systems have ceased to function properly because of filter and emitter clogging caused by sulfur bacteria. An abundant species of bacteria causing this problem is *Thiothrix nivea*, a common inhabitant of warm mineral springs in Florida. These bacteria oxidize hydrogen sulfide to sulfur and can clog small openings within a brief period of time. *Beggiatoa*, another sulfur bacterial genus, is also often found in microirrigation systems. Continuous chlorine treatment is an effective control for the sulfur bacteria problem. Another group of bacteria that can cause problems is one that fixes sulfur or sulfur derivatives. A beneficial bacterial genus in this group is *Thiobacillus*, which is tolerant to pH levels below 1.0. These bacteria utilize elemental sulfur (S) in their metabolic processes and convert it into sulfuric acid. These bacteria are one reason powdered sulfur can be applied on the ground to control the soil pH. *Desulfovibrio sulfuricans* also tolerate low pH, but they change sulfate (SO_4) and other sulfur compounds to hydrogen sulfide (H_2S), an acidic gas. The bacterial formation (Fig. 36-4) is a whitish slime with a strong rotten egg smell, typical of H_2S. When conditions have been right, some irrigation systems have been impaired in just a couple of weeks due to this bacteria. Therefore, if the smell is evident in an irrigation system, continual careful observation of the system and a well-planned preventive maintenance program are advisable in order to avoid exponential population growth.

Ion Toxicity

Ion toxicity problems are different from salinity problems, in that they occur within the plant itself and are not caused by water stress. Toxicity can occur when certain ions (Na, Cl, and B) are taken up with the soil water and accumulate in the leaves during transpiration to an extent that results in damage to the plant. When using microsprayers, water with high salt levels can damage plants if the

Figure 36-5. Microsprinkler emitter clogged by ants.

Figure 36-6. Autoflusher on end of lateral line.

irrigation water is directly applied to the leaves. The degree of damage depends on time of day, salt concentration, crop sensitivity, and crop water use. If damage is severe enough, crop yield is reduced.

Toxicity can result from the excess of several ions, but chloride toxicity is most common. Problems result when the chloride content of leaves reaches 0.3% of leaf dry weight. Levels above 0.5% should be considered excessive. Common symptoms are burned necrotic (dead tissue or dry-appearing) edges on leaves. Other symptoms that are difficult to assess prior to leaf burn include reduced root growth, decreased flowering, smaller leaf size, and impaired shoot growth.

A small amount of boron is essential for plant growth; however, amounts only slightly above the needed levels can be toxic. Concentrations under 0.5 mg/L do not pose toxicity problems. Between 0.5 and 2.0 mg/L causes some problems. Concentrations above 2.0 mg/L cause severe problems.

Insect and Pest Damage

A number of pests can increase maintenance costs. Rats and mice have been reported to chew holes in drip tubing, larvae of Tortricidae (Lepidoptera) and pupae of *Chrysoperla externa* (Hagan) have also caused emitter clogging. Larvae of *Selenisa sueroides* have been reported to damage irrigation systems in citrus groves by boring holes into the spaghetti tubing. Imported red fire ants have destroyed silicon diaphragms in microsprinkler units

The best way to minimize insect plugging of emitters is to have an aggressive ant control program. Ants are able to enter the very small passageways of microsprinkler orifices and clog the emitter (Fig. 36-5). In fact, higher clogging rates can be expected for emitters with orifices less than 0.035 inch in diameter. Large diameter emitters can more readily flush out ants that may enter them. Emitters that utilize a plug to cover the orifice when the pressure is off have been demonstrated to have lower plugging rates. Frequent operation of the system is another way of minimizing the insect problem. Autoflush units (Fig. 36-6) on lateral lines often result in ant entry to lateral lines when the system is off. If ants are a problem, the use of autoflushers is discouraged.

Water Treatment

Most of the methods to treat irrigation system clogging problems have come from municipal or industrial water treatment industries. The types of treatment can be roughly divided into two categories: preventive treatment and treatment of an already existing problem. Prevention can be best accomplished by continuous water treatment. The chemicals involved are generally composed of a variety of organic and inorganic compounds that have been found to prevent and dissolve scale-forming elemental matter. The stability of the product, method of operation, biological control capabilities, and required dosages vary widely.

Where the continuous water treatment is intended to take care of the disease, and consequently prevent the development of the symptom, shock treatments are directed towards the symptoms alone. Another inherent difference between prevention and shock treatment is that the system must be completely flushed after each shock treatment in order to be effective. When preventive treatments are properly made, the need for flushing should be greatly diminished. The major shock treatments used are acid and chlorine injection. The materials cannot be combined since mixing them may result in explosion.

Chlorination

Chlorination is a widely used and quite effective method to kill bacteria. Active chlorine is commercially available as a solid, liquid, or gas. Gas chlorine is 100% available as Cl_2, while calcium hypochlorite (solid) is 65% available and sodium hypochlorite (liquid) is generally 10% available. Based on these factors, the following conversions are derived:

Calcium hypochlorite: 1.54 lb = 1 lb chlorine gas
Sodium hypochlorite: 1.2 lb = 1 lb chlorine gas, or
  100 gal = 83.7 lb chlorine gas

The use of calcium hypochlorite is not advisable due to its high calcium concentration and increase in pH, which can create worse problems than the injection was able to solve. The use of gaseous chlorine (Cl_2) for water treatment requires the presence of trained licensed personnel, and it subjects the facility to local, state, and federal hazardous materials reporting requirements. The remaining alternative, sodium hypochlorite, is therefore the most widely used. Sodium hypochlorite is corrosive, and limited success of the treatment is due to the fact that it is only partly effective in penetrating through bacterial layers, and it is sensitive to the pH and mineral content of the water.

Cost comparisons between materials should be based on weight of available chlorine. Chlorine gas is typically the most economical chlorine source (Table 36-5). However, ease of use, equipment cost and maintenance, safety, tanks, etc., all need to be considered before selecting a method.

C1- Reaction in Water

When chlorine is introduced into the irrigation water, hydrolysis produces hypochlorous acid ($HOCl^-$). When HOCl undergoes an ionization reaction, hypochlorite (OCl^-) is formed. Together HOCl and OCl^- are referred to as free available chlorine. HOCl is 40 to 80 times more effective in killing microorganisms than OCl^-. As water pH increases above neutral, a higher portion of the free available chlorine is in the OCl^- form, and less is present as HOCl (Fig. 36-7). Therefore, chlorination is most effective when the pH is less than about 7.2. In addition to the pH sensitivity, the chlorine is subject to a demand in relation to mineral constituents of the water. This can quickly add up so that effective treatment becomes impractical or uneconomical. It is not uncommon for combined

Table 36-5. Cost comparison for chlorine sources.

Source	Typical Cost	Availability (%)	Cost per lb of available chlorine
Chlorine gas	$0.05/lb	100	$0.50
Calcium hypochlorite	$1.50/lb	65	$2.31
Sodium hypochlorite	$1.00/gal	10	$1.20

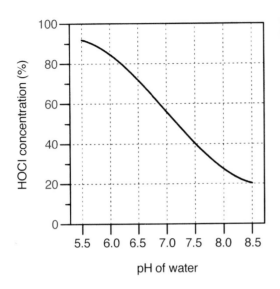

Figure 36-7. HOCl concentration at 77° F as a function of solution pH.

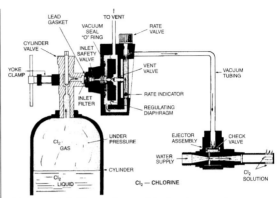

Figure 36-8. Typical gas chlorination equipment includes regulator and venturi ejector assembly.

Figure 36-9. Portable gas chlorination equipment used for citrus irrigation system.

(biological and chemical) chlorine requirements to exceed 50 to 60 ppm of available chlorine.

One of the most damaging drawbacks of chlorine treatment is its oxidizing capacity. Chlorine will readily oxidize soluble iron into insoluble Fe_2O_3, thus potentially creating a greater plugging threat. Not only does the chlorine precipitate the iron, but the fresh precipitate will then function as a food source for new bacterial growth. In this case, the biocidal effectiveness of chlorine ceases, and the process of system impairment is accelerated by the chlorination treatment. As with some pesticide combinations, a compatibility agent is needed to prevent precipitates. Presently, there are no commercially available compatible water treatment products that can be used with chlorine to prevent harmful precipitates from forming.

Due to pH sensitivity and variable chemical chlorine demand, continuous low-level chlorination is not usually a feasible method of water treatment. In addition, sodium hypochlorite degrades over time, limiting effective field storage periods to less than two months. Chlorinating during fertigation cycles is ineffective. Whenever nitrogen is present in the water, the biocidal action of chlorine greatly diminishes. Chlorine reacts with nitrogen sources

in the water and forms chloramines that are about 1/150 as active as disinfectants as chlorine alone.

Gas Chlorination

When using a vacuum-type, solution-feed gas chlorination system (Figs. 36-8 and 36-9), the pressure in the cylinder or container is reduced to a vacuum immediately as it enters the chlorinator. Under this condition, no reliquification of chlorine can occur, and pressure changes due to temperature changes of the chlorine supply are automatically compensated for. Chlorine can be withdrawn from a single 100- or 150-pound cylinder continuously at a rate of slightly over 100 pounds per 24 hours.

Where very high rates are required from 100 or 150 pound cylinders, it is necessary to manifold cylinders together in order to obtain full capacity. In this case, the connectors and piping between the cylinders and the chlorinator must be heated so that they never cool below the temperature of the chlorine cylinders. If this should occur, then liquid chlorine would re-form in the gas supply lines, causing fouling of the chlorination equipment.

The chlorinator ejector, which creates the vacuum and mixes chlorine with water before injecting it into the system, can be located near the chlorinator, but preferably should be located at or near the point of solution injection. A vacuum line then connects the chlorinator to the ejector, and any opening in this line or damage to the line would cause the chlorinator inlet valve to close tight. This not only is the simplest and most economical installation, but also the safest. The chlorinator system is easily turned on and off, merely by turning on and off the water supply to the ejector. This can also be set up to be automatic for shock treatment by using a low-cost timer and installing a solenoid valve in the water supply line.

Chlorine Injection Rate Calculation:
Gas: lb/day = 0.012 x ppm x gpm
Liquid: gal/hr = (0.006 x ppm x gpm) / % Cl
gpm = flow rate of system
ppm = ppm of Cl to be injected

Conditions:
1500 gpm system
15 ppm Cl_2 injection rate

NaOCl - 10% Cl_2 (liquid chlorine solution)
gal/hr = (0.006 x ppm x gpm) / % Cl
gal/hr = (0.006 x 15 x 1500) / 10
    = 13.5 gal/hr

NaOCl - 5.25% Cl_2 (household bleach)
gal/hr = (0.006 x 15 x 1500) / 5.25
    = 25.7 gal/hr

Gas - 100% Cl_2
lb/day = (0.012 x 15 x 1500)
    = 270 lb/day

Acidification

Acidification has been used mostly in an attempt to dissolve existing scale from the water lines. A factor of great concern when using acids is the personal safety of the people involved with the injection and cleanup operation, not to mention the effects of new hazardous materials reporting requirements. Mineral acids are extremely corrosive and can cause severe burns if accidental exposure occurs. When making acid injections, potential exists to cause damage to the plants either as acid burn or as toxicity, since the low pH water acidifies the soil.

As a rule of thumb, the pH of water in the system should be lowered to near 2.0 to achieve maximum effectiveness of the acidification. Acidification treatment lowers the pH so the compounds and precipitates that are strongly bonded and insoluble under normal conditions can dissociate. The acid supplies an excessive amount of highly reactive positive hydrogen ions to the irrigation water. These hydrogen ions react with the dissociated anions to form soluble compounds that can be flushed from the system.

Phosphoric acid (H_3PO_4) has been one of the favorites used for acid treatment of microirrigation systems. However, in many cases it is very difficult to inject sufficient phosphoric acid to lower the pH below 3.0. Many times the free dissociated phosphates react with existing precipitates in the lines, and more stable compounds are formed. These compounds are more difficult to remove. If improperly made, the result of phosphoric acid injection may be the formation of a more severe problem.

The other inorganic acids commonly used include nitric (HNO_3), hydrochloric (HCl, commonly called muriatic), and sulfuric (H_2SO_4) acids. Muriatic acid is probably the most effective in dissolving a majority of the scales. Muriatic acid generally does not produce harmful precipitation reactions, and also it is often chemically the most effective in dissolving the precipitates.

The chemistry of acid treatments requires collisions between the hydrogen ions and the precipitates or stable compounds that are trying to be flushed. If no contact is made, no dissociation will occur. If there is no dissociation, the treatment fails to solve the problem. Therefore, the water treatment problem requires ensuring that the chemical gets to where it can do the most good without losing its effectiveness to undesirable side reactions.

In order for the acid to reach precipitate attached to tubing walls near the end of the system, the acid has to travel a relatively long distance within the irrigation distribution lines. During its travel, it will encounter an enormous number of collisions with everything in the system. Should one of these collisions be with a constituent such as Ca that can potentially react with the acid, the acid unit will be unavailable for cleaning action. This is absorption of the acid into the water, called the "acid demand" of the water. Therefore, the more potentially active material there is in the water, the more acid it takes to meet the acid demand of the water.

The quantity of acid required to lower the pH of the irrigation water to the desired level (usually about pH = 2.0) should normally be determined by laboratory titration with a water sample. If this is not possible, a field test may be conducted by filling a 55-gallon drum with the irrigation water. Slowly add a small measured portion of the acid that you wish to inject and mix thoroughly. Measure the pH of the water, and repeat until the desired pH is attained. The required injection rate can be calculated from the acid: water ratio that attains the target pH.

Acid Injection Rate Example:
1500 gpm system
Water pH = 8.1
Desired pH = 6.5 for maintenance
    pH = 2.5 for shock treatment
Titration: 3 ml in 10 gal of water = pH 6.5
    12 ml in 10 gal = pH 2.5

(1 gal = 3785 ml . . . ml x 0.0002642 = gal)

Maintenance - pH 6.5
Step 1 - convert acid volume to gal
    3 ml x 0.0002642 ml/gal = 0.00079 gal
Step 2 - find acid:water ratio
    0.00079 gal acid/10 gal water = 0.000079

---

    injection rate = 1500 gpm x 0.000079
    = 0.12 gpm

    0.12 gal/min x 60 min/hour = 7.2 gal/hr

Shock - pH 2.5
    12 ml x 0.0002642 ml/gal = 0.00317 gal
    0.00317 gal / 10 gal = 0.000317
    injection rate = 1500 x 0.000317
    = 0.48 gpm

    0.48 gpm x 60 min = 28.8 gal/hr

It generally takes substantial amounts of acid to do any good in a degraded system. Many times the results have not been long lasting, or they have been unsatisfactory, especially when weighed against the cost of materials and labor involved in the flushing and cleanup process. Normally, it is less costly and safer to initiate other system cleaning treatments before acid treatments become necessary.

Water Conditioners
Presently available water conditioners can be divided into two categories: chemical treatment, which includes the scale inhibitors and bacterially active compounds, and physical devices. The scale inhibitors are the most common of the chemical alternatives. These products, which are usually mixtures of several synthetic polymers or polyphosphates, prevent formation of the scale by interfering with the reaction mechanisms of the precipitate. Often these mixtures are accompanied with a range of surfactants or penetration agents to aid the penetration and dismantling of the biological and crystalline formations that have attached to the system parts.

The scale inhibitors do not control bacteria. They merely remove some of the food source. They may also penetrate the bacterial layers that have accumulated on the irrigation system components, thereby reducing or alleviating the bacterial problem.

Generally, these scale inhibitor mixtures can be injected into the system at low rates (<10 ppm). Most of the scale inhibitors have undergone adequate testing so that fairly accurate predictions can be made concerning how much of each active ingredient is required to prevent precipitation reactions from occurring. Hence, treatment recommendations for each system can be custom tailored if an accurate water quality analysis is available.

However, the presence of some constituents in the water makes predictions of the treatment effectiveness difficult. If P, Ca, or Fe are in the water, the chemistry can become quite complex. Trying to dissolve ferrous sulfate requires a different approach than ferrous phosphate. Furthermore, it is difficult to predict how much additional material will be removed from the irrigation system components and whether there will be other synergistic reactions that will alter the demand of the treatment chemical. Therefore, initial applications of scale inhibitors in poorly maintained systems may require significantly higher rates than those published by the manufacturer for system maintenance. This increased rate requirement may be needed for several applications, until the system is adequately cleaned up to go back to maintenance rates. Always consult with the appropriate product representatives if any problems occur.

Although there are many commercially available biocides, they are generally not used with scale inhibitors. The use of biocides is classified by EPA as a pesticide application, and thus they require appropriate licensing by the agency. Unfortunately, the present market volume is not sufficient to justify the cost of the pesticide licensing process. Therefore, the only means of obtaining potential biocides has been to attempt to get licensing for biocides that have been approved for drinking water. Today in the marketplace we have two alternatives: chlorine and copper.

The bioactivity of copper is based on the fact that it is toxic to many strains of bacteria as it interferes with their respiratory functions. As a result, the bacterial population is prevented from growing and will eventually die due to copper activity. The major problem with use of copper in the past has been the short activity periods and the large quantities that were needed to reach sufficient concentrations for proper control. However, recent developments in chelation and bonding techniques have overcome this problem, and the newer materials approved by EPA for injection can be used at low injection rates (in the order of 1 ppm).

One particular benefit of these copper compounds is that it is usually not necessary to use them during the entire irrigation cycle. They can be injected into the system during the latter part of the cycle, leaving an effective residue in line to prevent bacterial growth. Another benefit of controlled bacterial activity is that the mineral deposition rate will slow down since the biological portion of the precipitate accumulation is eliminated. However, it should be noted that systems with bacterial problems may still require the use of other forms of treatment due to chemical scale formation that these copper compounds are unable to prevent.

The other devices that have been employed to control deposit formations include various magnetic treatment systems and ion exchange resins. The former normally involves an installation of either inflow or external units that magnetize water, and thus alter the energy levels of the precipitation reaction so that it may not occur. The ion exchange resins are basically chemical filters that remove the harmful particles from the water by exchanging them with sodium and chloride, for example. Both of these approaches have been proven effective in systems with low flow rates. However, the magnetic treatment has been plagued by a rapidly decreasing effectiveness as water travels away from the treated area. As a result, these systems are often economically prohibitive. The resin systems, due to size and recharging requirements, are too expensive for agricultural use.

Injection Considerations

It is essential that proper and legal backflow prevention devices be used in the irrigation system to prevent injected materials from being back-siphoned into the water supply. State regulations should be consulted for the type of water supply and injection device to be used (see Appendix). In addition, local ordinances need to be consulted as they can be more restrictive than the state law in some parts of Florida.

The injection device itself should have a screen and check valve. The screen will prevent undissolved particles from entering the system. The check valve in the supply is required in order to prevent backflow into the fertilizer tank. It is recommended that injection take place upstream of filters, so that any contaminants or precipitates can be filtered out. Injection should cease during filter backflush operations. On filters that use a portion of the supply water to continuously backflush, the injection location needs to be downstream of the filter. Injection of very acidic and corrosive materials should also take place downstream of filters that may be subject to corrosion.

Injection rates and times should be calibrated for each block that is treated. Injection times need to be at least as long as the travel time in the system from the injection point to the farthest emitter. This can be measured by starting the system and bringing it up to the normal operating pressure and flow rate. If fertigation is practiced, take a measurement of the electrical conductivity (EC) of the water prior to injecting any fertilizer. Once injection begins, relocate to the far end of the system. EC measurement should be taken at the farthest emitter to detect the point in time when the EC dramatically increases. In large systems, this time is often 20 to 30 minutes. An alternate method to determine travel time is to inject liquid soap and visually determine the time for foaming to appear at the farthest emitters. Injection times need to be at least as long as the travel time to the farthest part of the block.

Injection rates can be measured either with a chemical flowmeter or volumetrically. If a chemical flowmeter is used on the high-pressure side of an injector, make sure that the flowmeter is rated for the pressure being used. Volumetric measurements can be made by determining the time

required to inject a known quantity of material under normal operating conditions.

If the injected solution had approximately the same specific weight as water, the required injection rate can be calculated with the equation:

$$I = \frac{ppm \times Q}{AI\% \times 10,000} \qquad \textbf{Eq. 36-1}$$

where,

I = injection rate in gpm
Q = system flow rate in gpm
AI% = percentage active ingredient in the injected solution.

Example:
Assume that it is desired to inject sodium hypochlorite (with 10% AI) at a rate of 15 ppm Cl. The block has 72 grove acres with a tree density of 145 tree/ac. Each tree is irrigated with a blue base microsprinkler emitter discharging 10.9 gal/hr.

The system flow rate should be about:

72 ac x 145 tree/ac x 10.9 gal/hr x 1 hr/60 min
= 2000 gpm

The injection rate would be:

I = (15 x 2000) / (10 x 10,000)
= 0.3 gpm or 18 gph

All materials should be injected according to their label. Another factor of great concern is the personal safety of the people involved with the injection and cleanup operation. Chlorine and mineral acids are extremely corrosive and can cause severe burns should accidental exposure occur. Proper safety equipment and handling procedures are essential.

Fertigation
Surveys of irrigation waters throughout Florida have indicated a high variability of fertilizer solubility with different water sources. If fertilizers will be injected into the irrigation water, they need to be

entirely soluble in water and must not react with salts or chemicals in the water. A simple compatibility test should be conducted before any chemical is injected into an irrigation system. Take a clean bottle and fill it with water from the irrigation system supply. Add a small amount of the chemical to be injected so that the concentration is about the same as anticipated in the system. Shake the mixture and let it stand for 24 hours. If any precipitates or reactions occur, injection of the material is not recommended. Injection of more than one chemical at a time is not recommended unless chemical compatibility tests are performed.

Most nitrogen sources should cause few clogging problems. The exceptions include anhydrous ammonia, aqua ammonia, and ammonium phosphate, which increase the pH of the water and cause precipitates with calcium and magnesium to form. Application of most forms of phosphorous through the system can result in extensive clogging. The phosphorous compounds react with the calcium and magnesium found in most waters and form precipitates. However, phosphoric acid can normally be safely injected at low rates in most waters since it acidifies the solution to a point where precipitation is prevented. All of the common potassium fertilizers are readily soluble and present no clogging problems with most waters.

Other Considerations and Recommendations
The type of emitter that is selected can influence clogging susceptibility. In a 3-year study, clogging rates ranging from 1.6% to 34.4% were reported, depending on emitter model. Designs that used enclosed caps to disperse water had the highest clogging rates. Emitters that shielded the orifice when the emitter was not in use had reduced clogging problems. Larger diameter orifices are less prone to clogging since they are more effective in flushing out particles, insects, or algae than smaller orifices.

Systems that remain unused for months at a time usually experience much higher emitter plugging rates than those used regularly. Oftentimes, microsprinkler systems are not used for several months

during the summertime due to plentiful rainfall. During these times, it is recommended that the systems be run at least once a month just for maintenance purposes. Start up the system and run for 30 minutes or more at the normal operating pressure. Biocides may be injected prior to shutdown to help minimize bacterial and algal growth within the system.

In new irrigation systems, oily surfaces of the tubing inhibit bacterial and chemical attachment. Often there are no signs of scale or slime deposits for periods lasting from a few weeks to a couple of years, depending on the water quality and frequency of use. Once attachment and accumulation of precipitates or slimes has started, the process develops faster as the layer grows. At this point, system treatment must be started or the cost and effort required to reclaim system performance will greatly increase. When the slime or scale accumulation reaches the exponential growth stage, the system experiences frequent plugging that gets worse cycle by cycle. System maintenance becomes labor-intensive and parts need to be frequently replaced.

Microirrigation systems should be treated like any other piece of expensive equipment. With installed costs of $800 to $1,200 per acre or more, many of the 2000+ gpm systems that are commonly used represent investments of over $100,000. These systems require frequent inspections and maintenance in order to preserve the original design performance characteristics. Anticipate the development of clogging problems and treat before severe shock treatments are required. Be knowledgeable about the causes and cures for clogging problems. Make sure any treatment that you make is directed at your particular problem. What works for your neighbor may not work for you if your conditions are not the same. The water chemistry of irrigation systems and injected materials is very complex. Effectiveness of treatments can vary considerably from time to time. Therefore, don't be afraid to experiment with other products and techniques if what you've tried is not satisfactory.

Flushing

Injection systems should be installed so that they can be easily flushed with clean water. Most injectors should be flushed after each use. This will increase the life of gaskets and metals, and help to prevent encrustation problems. When the system is used to inject more than one chemical, complete flushing of the unit between injections will minimize the potential for chemical precipitation and other reactions. Additionally, the entire irrigation system should be flushed following the injection period to purge all the chemical from the irrigation system. If chemical is left in the line, it may drain unevenly and result in a nonuniform application. Nutrients left in the line may encourage bacterial growth.

Flushing of pipelines is needed to remove any particles that accumulate in microirrigation pipelines before they build up to sizes and amounts that cause plugging problems. Particles accumulate in pipelines from several sources. Filtration systems do not remove all suspended materials from the water. Due to the high cost of removing very small particles, agricultural filters are usually designed to only remove particle sizes larger than about 20% of the emitter orifice diameter. Therefore, filters do not remove clay, silt, and some very fine sand-size particles. Although these particles are small enough to be discharged through the emitters, they can cause plugging problems if enough are present. They may travel through the filters as individual particles, but then flocculate or become attached to organic residues, and thus become large enough to plug emitters.

Organic growths in pipelines, especially bacteria, are difficult to completely eliminate, and they provide the 'glue' that sticks small particles together. Many types of algae are too small to be filtered; thus, they can readily enter pipelines. Even if they are killed by chlorination or other chemical treatment, they can cause small particles to stick together. As the algae die, the cells rupture and the organic residue becomes an adhesive for small particles or groups of cells.

Other material can also enter irrigation pipelines and require flushing. Sand and other soil particles can travel backwards through emitters, especially if pipelines and emitters are buried. However, even on the surface, dust and debris can be blown or washed into the emitters. Small insects also enter emitters, sometimes carrying debris with them. Insects and other debris can also enter pipelines through flush valves at the ends of laterals and other pipelines, especially through automatic valves, which are normally open.

Chemical precipitates can also occur in irrigation pipelines. Minerals and salts may be dissolved in water and then precipitated because of changes in water temperature, pressure, exposure to oxygen, or injection of materials such as fertilizer. This problem must be addressed by avoiding chemical precipitation altogether because it is difficult and often impossible to flush chemical precipitates from irrigation systems, especially when they adhere to system components such as pipe walls and emitter passageways. Chemical precipitation can be avoided by injecting only water-soluble or acid-forming fertilizers, or by the injection of acid or other chemical water treatments.

Most debris in pipelines begins as small particles, which then join together to cause plugging problems. These small particles are light enough to be suspended and readily transported by the flowing water when the velocity is high. However, the velocity of water flow in a microirrigation system decreases as water is discharged along the length of the pipelines, especially along the lateral lines. At the very end of laterals, the water flow is reduced to the flow rate of the last emitters. Most debris accumulates at this point because the velocity is no longer fast enough to carry particulate matter. As a result, even small particles settle to the bottom of the laterals. Emitter plugging from this cause normally begins at the ends of the laterals.

The accumulation of loose debris in pipelines can readily be removed by flushing. To be effective, flushing must be done often enough to remove debris before it accumulates in large amounts and plugs emitters. Also, flushing velocities must be sufficient to dislodge and transport the accumulated debris.

Lateral and manifold pipelines should always be equipped with flush valves. Flush valves are required because the velocities in these multiple outlet pipes are low at the downstream end during normal irrigation operation. Conversely, the velocity in mains and submains remains high throughout the length of the pipeline; thus, low-velocity zones where debris would normally accumulate are avoided. Some designers specify flush valves at the ends of main and submain pipelines; however, flushing is less often needed at these locations.

Flushing Procedures
Irrigation lines are flushed by opening the ends of the lines during operation and allowing water to freely discharge, carrying particulate matter along. The goal of flushing is to discharge water at sufficient velocity so that any particulate matter will be suspended and removed from the system with the flush water. Pipelines must be equipped with valves or other means of allowing the pipeline to be opened quickly. Either manual or automatic valves can be used, or flexible PE tube can simply be folded over and clamped or tied. Manual valves are often used if infrequent flushing is adequate, while automatic valves are preferred when frequent flushing is needed.

When flushing pipelines, it is important to observe the type and amount of debris discharged. Use a clean white bucket or other suitable container to collect the material that is flushed as soon as the valve is opened. A cheesecloth or similar filter placed over the end of the pipe can also be used to trap the flushed sediment. If large amounts of material are discharged, then more frequent flushing may be required, while if only a small amount is flushed, it may be possible to increase the time between flushings. It is important to collect and observe the flushed material regularly because water quality may change, requiring changes in the flushing frequency throughout the year.

Figure 36-10. Collecting material flushed from irrigation lateral lines.

Observing the type and amount of material flushed (Fig. 36-10) may also suggest other changes in management practices. For example, the presence of organic growths could indicate the need for chlorination or other chemical applications. The presence of large particles might indicate the failure of a filter. The presence of chemical precipitates that have been deposited and then flaked off of the pipe walls might suggest a need to inject acid or change fertilizer injection procedures.

Because the flush water is dirty, it is normally discarded rather than reused. For this reason, flushing should not occur when chemicals other than pipeline cleaning chemicals such as chlorine are being injected. This will prevent loss of fertilizers or pesticides and avoid possible pollution problems.

Flushing Frequency
The frequency of flushing depends on the amount of debris that is removed. In some systems, the water quality is very good, and only very small amounts of debris are found. In those cases, only infrequent flushing (weekly, biweekly, or monthly) flushing is required. Pipelines should be flushed at least once a month during the irrigation season. In some systems, a large amount of debris accumulates in the pipelines each time the irrigation system operates (such as in systems that use surface water that contains large amounts of suspended algae, clay, and silt particles). In those cases, it may be necessary to flush laterals during each irrigation, using automatic flush valves.

During the summer rainy season, systems may be idle for weeks or months at a time when irrigation is not required. It is a good idea to start these systems and flush them every 4 to 6 weeks, even when irrigation is not required. The potential for system plugging is typically heightened when hot, humid climate conditions favor bacterial and algae growth in pipelines. Operating and flushing the system during these times (even though irrigation is not necessary for plant growth) can help remove plugging sources before they accumulate to the level that large chemical shock treatments are required to reclaim the system.

Flushing Duration
The duration of flushing depends on many factors, especially the water quality and the system design. It is difficult to accurately estimate the time required to adequately flush a pipeline before the system is installed. Fortunately, however, it is simple to determine in the field. Flushing should occur until the water discharged runs clean. This normally only requires a short time—a minute or two is normally sufficient because the debris mainly accumulates at the end of the pipeline near the flush valve. Depending on the system design, amount and type of debris, and the flushing velocity, longer times may be required for some systems.

Flushing Velocities
For proper flushing to occur, the discharge velocity must be high enough to dislodge and transport particulate matter from the pipelines. Typically, a minimum velocity of 1 foot per second (fps) is recommended. However, a velocity of 2 fps may be needed where larger particle sizes need to be discharged. This often occurs in microsprinkler systems where coarser filters are used than those required for drip emitters because of the large microsprinkler orifice sizes. The coarser filters allow larger particles to enter the system with the irrigation water, and therefore require higher flushing velocities. In general, the flushing velocity should be as high as possible in order to dislodge and transport as many particles as possible, but never less than 1 fps. Higher flushing velocities

will aid particle removal and shorten the flushing time needed.

Microirrigation systems must be properly designed so that adequate flushing velocities can be obtained. It is usually not possible to flush all laterals at once, because sufficient velocity will not be available for all laterals. Rather, open only a few flush valves at once, allow them to operate, and then close them before moving to the next valves. Flush the manifolds first, and then the laterals served by the manifolds that have just been cleaned.

Table 36-6 can be used to determine if adequate flushing velocity is being obtained for polyvinyl chloride (PVC) and polyethylene (PE) plastic pipes commonly used for microirrigation manifolds and laterals. Because velocities are difficult to measure in the field, this table shows the flow rates that will produce a velocity of 1 fps in commonly used pipe sizes. These flow rates should be doubled for systems or where larger particle sizes must be flushed.

Flow rates can be measured in the field by collecting the flush water in a graduated cylinder or other graduated container, and measuring the time required to collect a given volume of water. Table 36-7 shows flow rates in gallons per minute (gpm) based on the time required to catch a specified number of gallons.

Example:
When flushing a 2-inch PVC manifold pipeline in a drip irrigation system, 5 gallons of water is collected in 20 seconds. Is the flushing rate adequate?

$$Q = \frac{5 \text{ gal}}{20 \text{ sec} \times 1 \text{ min}/60 \text{ sec}} = 15 \text{ gpm}$$

From Table 36-6, the minimum required flow rate for effective flushing of a 2-inch pipeline is 12 gpm; therefore, the measured flow rate is adequate.

If the flow rate was not adequate, the flow would need to be increased by shutting off flow to other sections of the irrigation system to direct it to this

447

manifold for the flushing operation. This can be done with manual valves that are only used during flushing. For this reason, the irrigation system designer should take this need into account when the system is designed and installed.

> Examples:
> V = Q/A
> V = velocity (ft/sec)
> Q = flow rate (ft³/sec)
> A = cross-sectional area of pipe/tubing (ft²)
>
> Conversions: 7.48 gallons per ft³
> 12 inches per ft
> 60 seconds per minute

It takes 2 minutes to flush 3 gallons from a 3/4-inch lateral. Is the velocity adequate to flush out solids?

> A = πD 2/4
> 3/4 inch / 12 = 0.0625 ft
> A = 3.1416 x 0.0625 2/4 = 0.0031 ft²
> Q = 3 gal/2 min /7.48 = 0.201 ft³/min
> Q = 0.201 ft³/min / 60 sec/min
> = 0.0034 ft³/sec
>
> V = Q/A = 0.0031 ft² / 0.0034 ft³/sec
> = 0.9 ft/sec
> Velocity is NOT adequate to flush laterals.

What is the minimum flow rate required to flush with at least 2 ft/sec in 3/4-inch lateral tubing?

> Q = V x A
> Q = 2 ft/sec x 0.0031 ft² = 0.0062 ft³/sec
> 0.0062 ft³/sec x 60 sec/min x 7.48 gal/ft³
> = 2.8 gpm

What is the minimum flow rate required to flush with at least 2 ft/sec in 1-inch lateral tubing?

> Q = V x A
> A = 3.1416 x (1/12)2/4 = 0.00545 ft²
> Q = 2 ft/sec x 0.00545 ft² = 0.0109 ft³/sec

> 0.0109 ft³/sec x 60 sec/min x 7.48 gal/ft³
> = 4.9 gpm

What is the minimum flow rate required to flush with at least 2 ft/sec in a 4-inch submain?

> Q = V x A
> A = 3.1416 x (4/12) 2/4 = 0.087 ft²
>
> Q = 2 ft/sec x 0.087 ft² = 0.175 ft³/sec
> 0.175 ft³/sec x 60 sec/min x 7.48 gal/ft³
> = 78 gpm

If the measured flushing velocity is not adequate, something must be done to increase the discharge rate per lateral. This can be done by simply closing some of the flush valves and flushing fewer laterals at once. Flush valves should be closed until the required flow rate is obtained from the remaining open valves. In most cases, this trial-and-error field procedure will need to be followed by growers to calibrate their systems. The number of laterals that can be flushed at the same time depends on many factors, and the easiest way to determine it is to measure it for each irrigation zone by opening and closing flush valves until the required minimum flow rate is obtained.

Higher velocities than those shown in Table 36-6 will be more effective. Discharge rates lower than those presented should not be used. In some systems, the water quality is poor, and very frequent flushing is required. This can be done by installing an automatic flush valve on each lateral. Typically, normally open valves close at low pressures (1 to 3 psi). Thus, the first surge of water in each lateral is discharged through the valve, carrying debris with it. This approach is effective in many cases, but may not always be adequate because the valves may not provide either the required flush velocity or duration. Thus, it may still be necessary to periodically manually flush these systems. In areas where ants are a problem, autoflush devices are not recommended.

Table 36-6. Minimum flow rates (gpm) required to flush microirrigation lateral and manifold lines at a velocity of 1 fps.

Pipe/tubing size (inches)	Flow required for 1 fps (gpm)	Pipe/tubing size (inches)	Flow required for 1 fps (gpm)
1/2	1	3	26
3/4	1.7	4	42
1	2.7	6	92
1 1/4	4.7	8	155
1 1/2	6.3	10	240
2	12	12	340
2 1/2	17		

Note: Polyethylene tubing assumed for sizes below 2 inches. Class 160 PVC pipe assumed for sized of 2 inches and larger.

Table 36-7. Flow rate (gpm) based on the time required to catch a specified number of gallons.

Time (sec.)	Gallons caught					
	1	2	3	4	5	10
10	6.0	12	18	24	30	60
20	3.0	6.0	9.0	12	15	30
30	2.0	4.0	6.0	8.0	10	20
40	1.5	3.0	4.5	6.0	7.5	15
50	1.2	2.4	3.6	4.8	6.0	12
60	1.0	2.0	3.0	4.0	5.0	10
70	0.09	1.7	2.6	3.4	4.3	8.6
80	0.08	1.5	2.3	3.0	3.8	7.6
90	0.07	1.3	2.0	2.7	3.3	6.6
100	0.05	1.0	1.5	2.0	2.5	5.0

Chapter 37. Aquatic Weed Control

by Brian Boman, Chris Wilson, Vernon Vandiver, Jr., and Jack Hebb

If a Florida citrus grower has ditches and water, then likely he or she will have aquatic weed problems. In natural aquatic systems, vascular aquatic plants contribute to maintaining the balance of nature and offer food, protection, oxygen, and habitat to aquatic species. In agricultural waterways, canals, and ditches, it is usually desirable to maintain much less cover of aquatic plants than would be desirable in a natural lake or river system. One can strive to maintain a diverse habitat in their agricultural water bodies with these lower levels of aquatic plants, while sustaining crop success and avoiding loss of income due to aquatic weed problems. For example, over the long-term, the inefficiency of a clogged intake system on an irrigation pump can result in excessive fuel consumption and diminished water delivery to the trees. The consequences can be especially severe if clogging occurs during a freeze event when irrigation is being used for cold protection. Overabundant aquatic weed growth can also cause lowered drainage rates following heavy rains, resulting in severe root pruning, increased disease incidence, and fruit drop.

Cultural Control Measures
Drawdown
Draining water from ditches and allowing them to dry out can be an effective method for controlling some aquatic vegetation. Water removal from ditches is an old reliable concept, practiced for years to kill aquatic plants. However, there are some species that can withstand periods of time without water. In order to get good aquatic weed control, drawdown usually needs to be accompanied by the application of a herbicide. With organic farming, the use of conventional herbicides cancels certification standards, even in bordering ditches. Therefore, organic growers must use mechanical means and/or biological control in conjunction with drawdown to remove vegetation. This is called integrated control.

Figure 37-1. Aquatic weeds in drainage ditch.

Figure 37-2. Debris baffle installed on canal culverts.

Screening
Application of catch screens in ditches is another concept that has been effective in some cases for vegetative control. This concept is popular with pumping systems that move massive amounts of water and materials into waterways. This technique allows plant material to be screened along the waterway before entering the intake pipes. As more restrictive measures concerning aquatic weed discharges become adopted by water management districts, it may become necessary to adopt screening procedures for the growers in those districts.

Debris baffles (Fig. 37-2) on outfall structures reduce off-site discharges of floating aquatic

Figure 37-3. Ribbon barrier installed on canal.

Figure 37-4. Hyacinth barrier installation.

Figure 37-5. Mechanical aquatic weed removal from canal.

vegetation into the primary canal system. The debris baffle is attached to the outfall structure, and prevents floating debris from passing through the outfall culvert. The requirements to retrofit baffles on existing outfall structures are site specific. The baffle creates the potential for additional elevated water stages upstream of the structure if aquatic weeds are allowed to build up and restrict flow. Accumulated debris needs to be removed periodically to ensure free flow through the structure. Debris baffles should be used in conjunction with other aquatic weed control strategies.

Ribbon barriers (Fig 37-3) installed upstream of outfall structures reduce off-site discharges of floating aquatic vegetation into the primary canal system. Under typical low-flow conditions, a barrier with an 18-inch skirt is recommended. If high-flow conditions are typically experienced, a barrier with a 30-inch skirt is recommended. Ribbon barriers are most effective when attached to the bank and allowed to move vertically according to canal stage levels. Ribbon barriers should be utilized in conjunction with chemical or biological aquatic weed control programs.

Water-hyacinths are floating aquatic plant species that grow to a height ranging from several inches to several feet. The plants are characterized by smooth leaves attached to a spongy bulb-shaped stalk. Reproduction is primarily through the production of daughter plants.

Hyacinth barriers (Fig. 37-4) installed upstream of outfall structures reduce discharges of floating aquatic vegetation into the primary canal system. Hyacinth barriers are not suitable for all sites, and should only be installed in ditches or canals with low-flow potentials. Hyacinth barriers should be installed in combination with an aquatic weed removal program.

Excavation

Excavation is perhaps one of the oldest and still most preferred methods of aquatic weed control (Fig. 37-5). Usually, the process is done in one of two ways: either by a screen rake, which removes only the vegetation from the top of the water, or by earth removal, which allows for bottom weed, top weed, and ditch bank weed removal. Weeds chopped along canal banks with a disc or mower can wash back into the waterway, thus recreating the weed problem. Weeds chopped along ditch banks also can wash down the ditchbank slope with rainfall, enter the drain pipes, and reenter the waterway.

Removal is accomplished by different types of equipment such as cranes, trackhoes, backhoes, etc., depending on the location and the situation. Weeds are removed and placed on the upper ditch slope so that drainage will be directed away from the bank and slope. Oftentimes, significant amounts of sediment are removed from ditch bottoms during the process. Care should be taken to place fragmented weeds back from the slopes because of their ability to reestablish themselves within the canal or ditch within a relatively short period of time. Weeds should not be chopped along the canal bank with a disc, but rather within the grove site with a blade mower. Weeds chopped along canal banks with a chopper disc or mower have a tendency to wash back into the waterway, thus enhancing or exacerbating the weed problem. Weeds chopped along ditch banks also have the tendency to wash down the slope with rainfall, enter the drain pipes, and reenter the waterway. The same is also true for sediment or earth removal from canals. Until grass is reestablished, sediments will move (with rainfall or dredging operations) down the slope, into the drains, and back into the canal.

Harvesting

Harvesting is a method accomplished by both dry and wet operations of equipment. Usually in dig operations, the aquatic environment is allowed to dry and then specially adapted machines cut the vegetation by sickle bar or blade; the product is then removed and loaded onto platforms. These loaders remove the product off-site to other locations, such as pastures or other areas for livestock, or to be chopped.

The wet harvesting operation is similar to the dig in respect to loaded material. However, the cutting machinery is usually in the form of a paddle wheel boat or a floating combine. This type of operation can be quite intensive and expensive for control of aquatic weeds. Most of this type of work is done on large massive areas of marsh, such as with rice farming, or large bodies of water, such as lakes or rivers.

Biological Controls

One of the most important ingredients for control of localized infestations of pests is the use of natural organisms. Many of the aquatic weed problems in Florida derive their origin from exotics or nonnative species. As these nonnatives have entered the Florida habitat, they have had no native predators to keep them in check. As a result, they have proliferated in size and quantity in their aquatic environments.

Scientists have traveled to the origin (home country) of some of the exotic species to search for natural predators that limit the growth of these aquatic weeds. In some cases, this research has been successful. However, there are frequently problems with the organism's ability to adapt to the Florida environment. Some of the exotic plant species have been controlled in population and size by foreign pests. However, the majority of these exotic aquatic plant species have not been adequately controlled by biological control organisms alone.

Insects and Diseases

Some exotic plant species have been controlled by the introduction of biological control agents. The alligator weed flea beetle (*Agasicles hygrophila*) was introduced into the United States from South America in 1964. This beetle has done a remarkable job of reducing problems with alligator weed. In fact, alligator weed is not considered a major aquatic problem in most areas of the state.

Various biological control agents have been tested on water-hyacinths throughout the years. Of these predator introductions, the most effective have been two types of water-hyacinth weevil (*Neochetina eichorniae* and *Neochetina bruchi*) and the water-hyacinth mite (*Orthagalumna terebrantis*). In addition, the fungus *Cercospona rodmanic* has been imported and found to have some effect on the water-hyacinth. When infected water-hyacinths are introduced into areas with healthy hyacinths, some control results as the disease spreads and infects healthy plants.

Research continues on new species and strains of predators, pathogens, and parasites that have potential for control of aquatic weed species. As new biological control methods are introduced, citrus growers are encouraged to incorporate their

Figure 37-6. Triploid grass carp.

benefits as part of an overall aquatic weed control program.

Triploid Grass Carp

Triploid grass carp (Fig. 37-6) feed upon aquatic vegetation. Triploid grass carp (*Clenopharyngodon idella*) are exotic, hybrid fish (with three sets of chromosomes) that cannot reproduce. Their introduction into water bodies requires permitting from the Florida Fish and Wildlife Conservation Commission in Tallahassee. Usually, the permitting requires a fish barrier retention structure on outfall structures. This is to contain the grass carp and reduce the chances of introduction to primary canals, streams, or lakes.

All grass carp used in Florida under this permit system must be certified as triploid by the Florida Fish and Wildlife Conservation Commission or the U.S. Fish and Wildlife Service. Instructions on purchasing these fish and the certification process are included with the Commission's permit, which must be received prior to possession of any grass carp. The one-page application for certification and possession of grass carp for control of aquatic weeds may be obtained by contacting:

Florida Fish and Wildlife
Conservation Commission
620 South Meridian Street
Tallahassee, FL 32301
Telephone: (850) 488-4066 or (850) 488-4069

The lack of reproduction on the part of the grass carp means that under high weed pressure, a large quantity of fish must be introduced to initiate adequate weed control. Under the right conditions, these fish can eventually attain weights of up to 30 pounds or more. However, keep in mind that during the introduction period, the fish are small.

Juvenile fish can only consume a fraction of what adults can consume. Usually a high-density aquatic weed situation would best be served by lowering the weed density with herbicides or mechanical control before the fish are introduced.

Grass carp are associated with control of submerged aquatic species such as hydrilla, elodea, and certain types of algae (Table 37-1). Herbivorous fish are not a quick solution to eradicate massive aquatic weed problems. An effective biological control program requires maintaining adequate water depths and oxygen levels in waterways. In addition, there must be a long-term equilibrium between fish density and the quantity of aquatic plants required to support the fish population. Most fish are host selective in nature, and prefer to feed on only a few weed species. Therefore, under a biological control program, some groups of plants may seem to proliferate in the absence of natural controls in an aquatic environment.

Implementation Considerations
• Targeted canals and ditches should maintain a minimum year-round water depth of 2 feet.

• Fish barriers should be installed to contain grass carp within targeted ditches.

• Stocking densities should be determined by the Florida Fish and Wildlife Conservation Commission or the Florida Cooperative Extension Service.

• An on-site inspection should be conducted to determine the feasibility of stocking grass carp.

• Grass carp are effective at controlling submerged aquatic vegetation and can deter the growth of floating aquatic vegetation.

• Grass carp can withstand dissolved oxygen levels as low as 2 ppm.

• Recommended stocking size should be no less than 12 inches in length or one pound in weight.

Chemical Control
The objective of an herbicide program is to control aquatic weeds within grove drainage ditches.

Table 37-1. Preference by grass carp for common aquatic plants and their effectiveness in providing control in lakes and ditches.

Common name	Scientific name	Preference	Effectiveness
Alligator-weed	*Alternanthera philoxeroides*	Low	Low
Bladderwort	*Utriclularia* sp.	High	High
Cattail	*Typha* sp.	Low	Low
Coontail	*Ceratophyllum demersum*	High	High
Duckweed	*Lemna minor*	High	Low
Fanwort	*Cambomba caroliniana*	High	High
Filamentous algae	*Spirogyra* sp., many others	Moderate	Low
Hydrilla	*Hydrilla verticellata*	High	High
Hygrophila	*Hygrophila polysperma*	High	*Moderate
Planktonic algae	Many species	Low	Low
Pondweed	*Potamogeton* sp.	Moderate	Moderate
Sedges	*Cyperus* sp.	Low	Low
Smartweed	*Polygonum* sp.	Low	Low
Southern naiad	*Najas guadalupensis*	High	High
Spatterdock	*Nuphar luteum*	Low	Low
Spikerush	*Eleocharis* sp.	Low	Low
Water-primrose	*Ludwiga* sp.	Low	Low
Water-hyacinth	*Eichornia crassipes*	Low	Low
Watermeal	*Wolffia* sp.	High	Low
White water-lily	*Nymphaea odorata*	Low	Low
Willows	*Salix* sp.	Low	Low

*Suggested stocking rate is 20 to100 fish per acre in canals, depending on the type of weed infestation.

Citrus growers have relied extensively on chemical control for effective reduction of invasive weed species in and along waterways (Fig. 37-7). Chemical control of plant weed species is accomplished with various types of herbicides. Usually these herbicides will fall into one or two of the following categories:

- Contact - Applied materials must contact the plant surface (specifically the leaf) and desiccation follows.

- Systemic - Materials that are applied are translocated from the plant part (specifically roots or leaves) and move either upward or downward through the plant.

- Selective - These herbicides are designated for use on specific species.

- Nonselective - These herbicides will kill a wide range of species.

- Preemergent - These herbicides are specifically designed to kill weed species before or immediately after seedling emergence. No preemergent herbicides are labeled for aquatic weed control.

- Postemergent - These herbicides are designed to kill weed species after plant emergence.

One of the primary criteria for evaluating an aquatic herbicide is understanding the chemical's environmental profile. The first step is to ensure

Figure 37-7. Wiper used to apply aquatic herbicides to grove ditches.

that the product is registered for the intended use. These guidelines are specified on the product's EPA label. Beyond that, users must consider the impact of herbicide toxicity levels on humans, fish, and wildlife, and their implications in terms of water use restrictions and transportation concerns. Toxicity levels vary according to the chemistry and concentration of the aquatic herbicide. Use restrictions specified on the label may require a waiting period during which the treated water is not safe for irrigation, fish consumption, watering livestock, and/or swimming. Generally, water treated with most copper and chelated copper herbicides can be used immediately for irrigation, stock watering, swimming, fishing, and domestic use.

Other herbicides may have use restrictions of up to 30 days. Use considerations are also important when evaluating an aquatic herbicide. Proper usage is contingent on the correct identification of the plant species to be controlled. This can be done independently using a good aquatic vegetation guidebook or the services of a professional biologist. Canal and ditch water applications are primarily focused on the control of submersed, rooted plants, and filamentous algae. Speed of effectiveness should also be considered when selecting herbicides for canal applications. Slow-acting herbicides can allow continued plant growth to impede water flow.

Once the proper herbicides have been identified for the application, competitive products and brands may then be compared based on ease of application. There are several factors that facilitate this comparison. Most important are the specific directions listed on the label. Some products must be applied more frequently than others. Aquatic herbicides provide control lasting two weeks and longer. Products that offer longer control need to be applied less frequently.

Cost is a relative term when discussing aquatic herbicides, as there are a number of trade-offs. Variations in purchase price among the different chemistries are offset by factors such as product efficacy, application frequency, handling requirements, and related legal/risk costs. It is a good idea to evaluate the supplier's technical support network. Some aquatic herbicides are sold direct from the manufacturer, while others are marketed locally by knowledgeable herbicide distributors. In either case, the manufacturer or distributor may be able to assist in identifying vegetation, calculating dosage requirements, and providing application equipment recommendations.

The variety of choices for chemical control has allowed managers and operators to time applications based on the characteristics of the herbicide, and to choose the appropriate control mechanism based on plant species. Most important of all, the operator should always read the herbicide label before mixing, loading, or applying a herbicide. The label is the law, and it is unlawful to detach, deface, alter, or destroy the label. It is also unlawful to use a pesticide in a manner that is inconsistent with or not specified on the label. The herbicide label contains a great deal of information concerning the product, and should be read carefully and thoroughly before use. Make sure that from the label you can determine the following:

- Signal word for toxicity to humans (i.e., "Caution," "Warning," "Danger")

- Personal protective equipment needed (i.e., gloves, boots, coveralls, hats, aprons, eyewear, respirator)

- Environmental hazards (i.e., to fish, mollusks, etc.)

- Environmental and weather conditions that may prohibit applications

- The type of weed species controlled by the herbicide

- Application rate under various aquatic conditions (i.e., murky water, clear water, etc.)

- Where the herbicide can be applied (i.e., water, ditch bank, etc.)

- Time of year for applying the herbicide (i.e., season, plant growth stage, etc.)

- Restrictions concerning other water uses (i.e., irrigation, livestock watering, fishing, etc.)

Some herbicides can kill submerged or emergent aquatic weed species that produce oxygen in the water. In addition, degradation of plant materials will also consume oxygen within the water column. As a result, dissolved oxygen levels can fall below levels that are needed to sustain fish populations. Therefore, treat canals and ditches early in the year before plants get too dense. When possible, selectively control species that shade or crowd out other native species. Avoid treatments on cloudy days when dissolved oxygen will naturally be lower. If a large portion of a water body is extensively covered with plants, treat no more than 1/3 to 1/2 of the area at one time. This will allow time between applications for oxygen recovery. During the winter months, optimal performance of the herbicide will be achieved when water temperatures are relatively warmer.

Aquatic Herbicides

There are several herbicides that can be used for aquatic weed control. Each material has advantages and disadvantages. The selection of the most appropriate material should be based on the target species, alternative control measures, and the effects on other aquatic organisms. Table 37-2 lists the LC_{50} values for some species. These values are the concentrations at which 50% of the population would be expected to die within 96 hours after

exposure. Consult UF/IFAS Extension Publication SP-55 for the most current list of materials.

Copper Products (Copper Sulfate and Copper Chelates)

Water systems labeled for use:
Copper sulfate: impounded waters, lakes, ponds, reservoirs, and irrigation systems. Copper chelates: ornamental, fish, and fire ponds; potable water reservoirs; freshwater lakes and fish hatcheries.

Mode of action:
Contact herbicide, often used in combination with other contact herbicides.

Duration of herbicidal activity:
Copper sulfate may persist for up to seven days before the free copper is precipitated to insoluble forms that are not active. As the hardness of the water increases, the persistence of the free copper decreases. The chelated coppers can be used where hard water may precipitate uncomplexed forms of copper too rapidly.

Precautions:
Copper sulfate can be very corrosive to steel and galvanized pipe. Chelated coppers are virtually noncorrosive. Contact with skin and eyes may be irritating. As water hardness decreases, toxicity to fish increases. Copper sulfate may be toxic to fish species at recommended dosages. Generally, the chelated coppers are nontoxic to trout, tropical fish, ornamental fish, and other sensitive fish at recommended dosages.

2,4-D Products

Water systems labeled for use:
Treatment is for management of emerged and floating aquatic weeds on ditch banks, in ditches, and in other waters within Florida citrus groves.

Mode of action:
2,4-D is a selective, translocated phenoxy compound used as a postemergent herbicide.

Duration of herbicidal activity:
Treated ditches should not be used to irrigate susceptible crops. More effective herbicidal activity is obtained when 2,4-D is applied to actively growing plants. A rain-free period of at least 6

hours is required to ensure lethal dose absorption by treated foliage. Never treat more than 1/2 of a lake at a time due to oxygen depletion by dying plants.

Precautions:

Do not apply when the temperature is over 90° F. Avoid drift. There are species that are susceptible to drift or volatilization. These include tomatoes, grapes, fruit trees, and ornamentals. Low volatility esters can become volatile at temperatures of 90° F or above. When 2,4-D products are applied to aquatic plants, decaying weeds may affect the flavor of the water for short periods of time. Salts from excessive 2,4-D applications may temporarily inhibit seed germination or plant growth on some soils. Heavy rains will not leach all of the chemical out of the soil. Amine salts are less hazardous to use than highly volatile ester formulations near susceptible crops.

Diquat

Water systems labeled for use:

May be used in slowly moving bodies of water, ponds, lakes, rivers, drainage and flood control canals, ditches, and reservoirs.

Mode of action:

Contact herbicide.

Duration of herbicidal activity:

Diquat is rapidly and completely inactivated by soil.

Precautions:

Do not apply to muddy water because the diquat will be inactivated. Never treat more than 1/3 to 1/2 of a densely vegetated pond at any one time because rapidly decaying vegetation will deplete oxygen, thereby suffocating fish. Skin contact may cause irritation. Avoid drift.

Diuron

Water systems labeled for use:

Irrigation and drainage ditches that have been drained of water for a period of 72 hours. After 72 hours, diuron is fixed to the soil and the ditch may then be used.

Mode of action:

Diuron is readily absorbed through the root system, less so through foliage, and is translocated

upward toward plant foliage.

Duration of herbicidal activity:

Control duration will vary with the amount of chemical applied, soil type, rainfall, water flow, and other conditions. Usually control will last for a period of 10 to 12 months.

Precautions:

May irritate eyes, nose, throat, and skin. Avoid breathing dust. Apply before expected seasonal rainfall. Do not treat ditches that have desirable tree roots extended into the band, since the tree may be injured. Prevent drift of dry powder to desirable plants. Do not contaminate any body of water.

Endothall

Water systems labeled for use:

Irrigation and drainage canals, ponds, and lakes.

Mode of action:

Contact herbicide.

Duration of herbicidal activity:

Microbiological breakdown is fairly rapid in water and soil with a short herbicidal duration.

Precautions:

Some formulations of endothall should not be used where fish are an important resource. Fish may be killed by dosages necessary to kill weeds. See IFAS Publication Circular 707 for detailed information. Skin contact may cause irritation. May be corrosive to application equipment.

Fluridone

Water systems labeled for use:

Lakes, ponds, ditches, canals, and reservoirs.

Mode of action:

Fluridone is absorbed by the foliage and translocated into the actively growing shoots where destruction of the chlorophyll pigments occurs, resulting in white growing points.

Duration of herbicidal activity.

Depending upon application and the vegetation being controlled, control may last one year.

Precautions:

Do not use treated water for irrigation of agronomic crops or turf for 7 to 30 days following treatment. Trees or shrubs growing in treated water may be injured. Higher treatment rates will

Table 37-2. Toxicity of aquatic and ditch bank herbicides to selected aquatic organisms.

Compound	Treatment[a] rate (ppm)	Toxicity (96-hr LC_{50})[b]		
		Bluegill/sunfish (ppm)	Rainbow trout (ppm)	Invertebrates (ppm)
Copper sulfate	0.1-1.0			17.0[c]
Soft water		0.9	0.01	
Hard water		7.3	-	
Copper chelate	0.1-1.0			19.0[d]
Soft water		1.2	<0.2	
Hard water		7.5	4.0	
2,4-D Amine	negligible[e]	524	377	184[f]
2,4-D BEE	1.25-2.5[g]	0.61	2.0	7.2[c]
Diquat	0.12-0.72	>115	21	>100[h]
Diuron	negligible	8.2	1.6	0.164
Imazapyr	negligible	>100	>100	>100[e]
Endothall (Aquathol)	1.0-5.0	501	529	320[i]
Endothall (Hydrothol)	0.1-3.0	1.2	1.3	0.36[e]
Fluridone	0.01-0.15	14.3	11.7	6.3[e]
Glyphosate (Rodeo)	negligible	>1000	>1000	930[e]

[a]Estimated concentration in water after application according to label instructions.

[b]Toxicity varies according to experimental conditions. The 96-hr LC_{50} values are typical of various sources and represent the concentrations at which 50% of the population would be expected to die within 96 hours after exposure.

[c]Freshwater shrimp.

[d]Blue shrimp.

[e]Labeled only for foliar or ditch bank application. Therefore, concentrations in water are negligible.

[f]*Daphnia*.

[g]Calculated for label rates of 26.7% G.

[h]*Gammarus fasciatus*.

[i]*Daphnia*, 48 hr LC_{50}.

be required if there is a large turnover in water volume.

Glyphosate

Water systems labeled for use:
Lakes, ponds, streams, rivers, ditches, canals, reservoirs, and other freshwater bodies.

Mode of action:
Glyphosate is absorbed by foliage and translocated throughout the plant and root system, killing the entire plant.

Duration of herbicidal activity:
Only effective at the time of treatment.

Precautions:
Not to be used for submersed or preemergent vegetation. Floating mats of vegetation will require treatment. A rain-free period of 6 hours after application is required. May be corrosive to galvanized steel. Avoid drift to desirable vegetation as glyphosate is nonselective and will affect contacted vegetation.

Imazapyr

Water systems labeled for use:
Nonirrigation ditch banks and similar areas.

Mode of action:
Both foliage- and root-absorbed and translocated throughout the entire plant.

Duration of herbicidal activity:
Provides control of existing and germinating seedlings throughout growing season.

Precautions:
Do not contaminate any water supply. Do not apply on ditches used for irrigation. Do not treat in areas where desirable tree roots are visible. Prevent drift to desirable plants. Should not be mixed or stored in unlined steel containers or spray tanks.

Triclopyr

Water systems labeled for use:
Nonirrigation ditch banks, not labeled for use in water (Experimental Use Permit required).

Mode of action:
Triclopyr induces characteristic auxin-type responses in growing plants. It is absorbed by both leaves and roots, and it is readily translocated throughout the plant. Foliar applications have achieved maximum plant response to treatment when applied soon after full leaf development, when soil moisture is adequate for normal plant growth.

Duration of herbicidal activity:
Time required for 50% breakdown in soil is between 10 and 46 days, depending on environmental conditions and soil type. At label rates, phytotoxic residues in soils should cause no problems. Triclopyr has a 2- to 6-hour half-life in water.

Precautions:
Keep out of lakes, streams, or ponds. Do not contaminate water by cleaning of equipment or disposal of wastes.

Formulas for Herbicide Calculations

Definitions:
AI = active ingredient
ppm_v = parts per million by volume
ppm_w = parts per million by weight

Active ingredient (AI)

Gallons of liquid formulation = **Eq. 37-1**

$$\frac{\text{AI required (lb)}}{\text{AI concentration (lb/gal)}}$$

lb of dry formulation = **Eq. 37-2**

$$\frac{\text{AI required (lb)}}{\text{AI in formulation (\% expressed as decimal)}}$$

Herbicide applications to ponds or lakes

Volume of pond in cu ft = **Eq. 37-3**
surface area in sq ft x average depth in ft

Volume of pond in ac-ft = **Eq. 37-4**
surface area in acres x average depth in ft

Volume of pond in ac-ft = **Eq. 37-5**

$$\frac{\text{volume of pond in cu ft}}{43{,}560 \text{ ft}^2 \text{ per ac}}$$

Total gal of chemical = **Eq. 37-6**
ac ft x ppm_v x 0.33

$$ppm_w = \frac{\text{AI of chem applied (lb)}}{\text{volume (ac ft) x 2.72}} \qquad \textbf{Eq. 37-7}$$

Total lb AI required =   **Eq. 37-8**
ac-ft x 2.72 x ppm_w desired

Total gal of liquid formulation =   **Eq. 37-9**

$$\frac{\text{ac-ft x 2.72 x } ppm_w \text{ desired}}{\text{AI of concentrate (lb/gal)}}$$

Acre-feet calculation
Ac-ft = (acres) x (average depth in feet) **Eq. 37-10**

Acreage calculations
Rectangular shape:

$$\text{Acres} = \frac{\text{width (ft) x length (ft)}}{43,560 \text{ ft}^2 \text{ per ac}} \qquad \textbf{Eq. 37-11}$$

Circular shape:

$$\text{Acres} = \frac{3.14 \text{ x (radius in ft)}^2}{43,560 \text{ ft}^2 \text{ per ac}} \qquad \textbf{Eq. 37-12}$$

Herbicide application coverage

$$\text{Ac/hr} = \frac{\text{swath width (ft) x speed (mph)}}{8.25} \qquad \textbf{Eq. 37-13}$$

Volume of herbicide concentrate required
Herbicide concentrate needed (gal) =

$$\left(\frac{\text{AI in mixture (lb)}}{\text{AI of herbicide (lb / gal)}} \right) \text{x acres}$$

Examples:

1. A grove requires a herbicide application at the rate of 1 pound of AI per acre. If the herbicide being applied is an emulsifiable concentrate with a 2 EC label, what are the gallons of liquid formulation required per acre?

   Use Eq. 37-1,

   $$\text{Gal/ac} = \frac{1 \text{ lb AI}}{2 \text{ lb/gal}} = 0.5 \text{ gal/ac}$$

2. A grove requires 2 pounds of AI per acre for application purposes. A granular herbicide is selected and the container label states that it contains 20% AI of a compound. How many pounds of dry formulated herbicide are needed per acre?

   Use Eq. 37-2,

   $$\text{lb/ac} = \frac{2 \text{ lb/ac AI}}{0.20 \text{ AI}} = 10 \text{ lb/ac}$$

3. A circular pond has a diameter of 300 ft. Find the volume of the pond if the average depth is 10 feet. How much chemical (gallons) is required to treat the pond if the rate of treatment calls for 5 parts per million by volume (ppm_v)?

   Use Eq. 37-12,
   for area (radius = 300/2 = 150 ft)

   $$\text{acres} = \frac{3.14 \text{ x (150 ft)}^2}{43,560} = 1.62 \text{ ac}$$

   Use Eq. 37-4, for volume
   ac-ft = 1.62 ac x 10 ft = 16.2 ac-ft

   Use Eq. 37-6, to determine application rate
   gal of chemical required = 16.2 ac-ft x 5
   x 0.33 = 26.7 gal

4. If a chemical contains 4 lb AI /gal and 20 gal are sprayed into a ditch containing 5 ac-ft of water, find the ppm by weight of the applied material.

   Total AI = 4 lb/gal x 20 gal = 80 pounds AI

   Use Eq. 37-7, to calculate ppm_w

   $$ppm_w = \frac{80 \text{ pounds AI}}{5 \text{ ac-ft x 2.72}} = 5.9 \text{ } ppm_w$$

5. If herbicide is applied with a 6-ft boom operating at 2 mph, how many acres per minute is the grower covering?

Use. Eq. 37-13,

$$\text{Ac/hr} = \frac{6 \text{ ft x 2 mph}}{8.25} = 1.45 \text{ ac/hr}$$

6. A 50-acre block is to be sprayed with a herbicide that calls for 4 pounds active ingredient per acre. If the grower uses a 2E concentrate material, how much herbicide concentrate is required to spray the block?

Use Eq. 37-14,

$$\text{gal} = \frac{50 \text{ acres x 4 lb AI/ac}}{2 \text{ lb AI/gal}}$$

$$= 100 \text{ gal of herbicide concentrate needed}$$

Aquatic Weed Identification

Aquatic plants are commonly classified into several categories depending on the location in the water column they inhabit. Aquatic plants may be free floating, emersed, submersed, or shoreline. Free-floating plants are rarely, if ever, rooted into the soil and their leaves are located above the water. Emersed aquatic plants are rooted in the soil under water with their leaves on or above the water surface. Submersed aquatic plants are usually rooted in the soil with all or most of their leaves growing under water. Ditch bank plants are not true aquatic plants, but are often associated with the moist soils that are located in and around the canals and ditches within citrus groves. The following describes several common plants that may be in or around ditches and canals. The reader is directed to IFAS Publication SP168 and Circular 707 for further and updated information. Note: Herbicides listed with an asterisk (*) should not be applied directly to water.

Floating Plants

Common duckweed
(*Spirodela polyrhiza*)

Figure 37-8.
Common duckweed.
V. Ramey - CAIP/UF.

Description:
Small, footprint-shaped leaves, no more than 1/8 inch long having one root. Leaves are pale green and float flat on the water surface. Reproduction occurs by seeds and rapidly through budding.

Control:
Biological: grass carp
Herbicides: diquat, fluridone

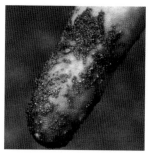

Common watermeal
(*Wolffia* species)

Figure 37-10.
Common watermeal.
ARM - CAIP/UF.

Description:
These tiny, floating, rootless plants are less than 1/32 inch long. The plant body is rounded and feels grainy when rolled between the fingertips. The plants are so small they appear to be merely green speckles. Often two to three are attached.

Control:
Biological: none
Herbicides: partial control with fluridone

Common salvinia
(*Salvinia minima*)

Figure 37-9.
Common salvinia.
David Sutton -
Circ. 707.

Description:
Circular leaves 1/4 to 1/2 inch in diameter with dense leaf hairs on the upper leaf surface. Leaves are brownish green and float flat on the surface. Salvinia is a fern and reproduces by spores and fragmentation.

Control:
Biological: partial control with grass carp
Herbicides: diquat, glyphosate

Water-lettuce
(*Pistia stratiotes*)

Figure 37-11.
Water-lettuce.
Vernon Vandiver, Jr. -
Circ. 707.

Description:
Resembles a head of lettuce. Grows in a rosette with spongy, dense hairy leaves 6 to 8 inches in diameter. Daughter plants are the major means of reproduction.

Control:
Biological: water-lettuce weevil
Herbicides: diquat, endothall

Water-hyacinth
(*Eichhornia crassipes*)

Figure 37-12.
Water-hyacinth.
Alison Fox - Circ. 707.

Description:
Plants several inches to 2 feet in height. Smooth leaves attached to spongy bulb-shaped stalks. Reproduction is primarily through the production of daughter plants.

Control:
Biological: water-hyacinth weevil
Herbicides: 2,4-D, diquat, glyphosate

Emersed Plants

Pickerelweed
(*Pontederia lanceolata*)

Figure 37-13.
Pickerelweed.
Vernon Vandiver - Circ. 707.

Description:
An erect plant with lance-shaped leaves up to 10 inches long. Each stem has violet-blue flowers at the top. Reproduction occurs by seed and creeping roots.

Control:
Biological: none
Herbicides: triclopy, partial control with 2,4-D and glyphosate

*Should not be applied directly to water.

Alligator weed
(*Altemanthera philoxeroides*)

Figure 37-14.
Alligator weed.
Ken Langeland - Circ. 707.

Description:
Hollow-stemmed perennial capable of forming dense mats. Leaves are opposite, between 2 and 4 inches long, and football shaped. Stems have a solitary white flower head at the tip. Reproduction by fragmentation.

Control:
Biological: alligator weed flea beetles and thrips
Herbicides: triclopyr*, partial control with 2,4-D and glyphosate

Pennywort
(*Hydrocotyle umbellata*)

Figure 37-15. Pennywort.
V. Ramey - CAIP/UF.

Description:
Dark green, shiny, rounded leaves that are centrally attached to a long stalk. Leaves may lie flat on the water surface or be erect. Pennywort reproduces by seed and creeping stems.

Control:
Herbicides: diquat, 2,4-D, glyphosate

Cattail
(*Typha* species)

Figure 37-16.
Cattail.
Vernon Vandiver -
Circ. 707.

Description:
Erect perennials (up to 9 feet) that can reproduce by seed or creeping roots. Grasslike leaves are flat and smooth to the touch. Flowers look like a "cat's tail" and can be found in a tightly packed spike usually 6 to 8 inches long.

Control:
Herbicides: glyphosate

Smartweed
(*Polygonum* species)

Figure 37-17.
Smartweed.
V. Ramey.
CAIP/UF.

Description:
Leaves are alternate, lance shaped, and attached to swollen joints on the stem. The flower stalk consists of many small pinkish-white flowers in a single spike. Smartweed spreads by seed, and may form large floating mats.

Control:
Herbicides: triclopyr*, partial control with glyphosate (species dependent) and 2,4-D

*Should not be applied directly to water.

White water-lily
(*Nymphaea odorata*)

Figure 37-18.
Water-lily.
Vernon Vandiver -
Circ. 707.

Description:
Leaves are flat, rounded, and attached at the center to the stalk. Leaves are often 10 inches in diameter and split to the center on one side. The flower is sweet scented, white, and showy. Reproduction is by seed and branching stems.

Control:
Herbicides: fluridone, 2,4-D liquid and granular, glyphosate

Spatterdock
(*Nuphar luteum*)

Figure 37-19.
Spatterdock.
A. Murray - CAIP/UF.

Description:
Large heart-shaped leaves arising from a stalk attached to a thick creeping root system. The flower is yellow and about one inch in diameter. Reproduction is by seed and new sprouts.

Control:
Herbicides: glyphosate, fluridone

Submersed Plants

Hydrilla
(*Hydrilla verticillata*)

Figure 37-20.
Hydrilla.
V. Ramey - CAIP/UF.

Description:
Long-stemmed, branching plant that is rooted to the bottom and often forms large surface mats. Leaves grow in a whorl with toothed margins that feel rough. Hydrilla can spread by plant fragments, underground stems, seed, leaf buds, or buds located on the underground stems.

Control:
Biological: grass carp (insect biocontrol under investigation)
Herbicides: copper, diquat, endothall (liquid and granular), fluridone

Coontail
(*Ceratophyllum demersum*)

Figure 37-21.
Coontail.
V. Ramey - CIAT/UF.

Description:
Leaves grow in a whorl, are finely dissected, and have teeth on one side of the leaf margin. Leaves are 1/2 to 1 inch in length and crowded toward the stem tip giving the appearance of a raccoon's tail. Coontail is rootless and floats near the surface in the warmer months. Reproduction is by seed and fragmentation.

Control:
Biological: grass carp
Herbicides: diquat, endothall liquid and granular, fluridone, 2,4-D granular

Southern naiad
(*Najas guadalupensis*)

Figure 37-22.
Southern naiad.
D. Sutton - IFAS

Description:
Bottom-rooted, slender-leaved, dark green to greenish-purple plant with branching stems. Leaves are less than 1 inch in length and narrow. Reproduction is by seed and fragmentation.

Control:
Biological: grass carp
Herbicides: diquat, endothall liquid and granular, fluridone, 2,4-D granular

Fanwort
(*Cabomba caroliniana*)

Figure 37-23.
Fanwort. Kerry Dressler.

Description:
Leaves of fanwort are finely dissected and fan-shaped. Leaves are opposite and generally no more than to 1 to 1/2 inches wide. The flower is white or cream colored, about 1/2 inch in dia meter, and blooms above the water surface. Reproduction is by seed and fragmentation.

Control:
Biological: grass carp
Herbicides: diquat, fluridone, 2,4-D granular

Pondweed
(*Potamogeton* species)

Figure 37-24.
Pondweed.
Alison Fox - Circ. 707.

Description:
Several species of pondweed are found in Florida; Illinois pondweed (*P. illinoensis*) is most frequently encountered. It has both floating and submersed leaf forms. The football-shaped floating leaves are not always present, but are easily distinguishable from the lance-shaped, submersed leaves. The flowers are clustered together on a spike 1 to 2 inches long located just above the water surface at the stem tip. Reproduction is by seed and from underground stems.

Control:
Biological: grass carp
Herbicides: diquat, endothall liquid

Grasses and Sedges

Torpedo grass
(*Panicum repens*)

Figure 37-25.
Torpedo grass.
A. Murray - CAIP/UF

Description:
Narrow-leaved (less than 1/4 inch wide), with stems often several feet in length. Torpedo grass creeps horizontally by underground stems and forms large floating mats. Reproduction is by seed and creeping stems.

Control:
Biological: partial control with grass carp
Herbicides: partial control with glyphosate, fluridone, imazpyr*

Maidencane
(*Panicum hemitomon*)

Figure 37-26.
Maidencane.
V. Ramey - CAIP/UF.

Description:
Maidencane leaves usually grow at 90° angles from the stem and generally 1/2 inch in width. An extensive creeping root system allows maidencane to form dense floating mats with stems often several feet in length. Reproduction is by seed and creeping roots.

Control:
Biological: partial control with grass carp
Herbicides: glyphosate

Proliferating spikerush
(roadgrass, hairgrass)
(*Eleocharis baldwinii*)

Figure 37-27.
Proliferating spikerush.
Ken Langland -
Circ. 707.

Description:
Proliferating spikerush has two growth forms. When it occurs on moist soils at the edge of ponds or lakes, it is erect and the leafless stems are 1 to 4 inches tall. When submersed, the stems become long and proliferate throughout the water column. Leaves occur only as bladeless sheaths at stem bases.

Control:
Biological: grass carp
Herbicides: fluridone (repeat applications)

*Should not be applied directly to water.

Sedge
(*Cyperus* species)
Figure 37-28.
Sedge. V. Ramey - CAIP/UF

Description:
Many sedges are found in Florida and are generally difficult to identify by species. In general, sedges can be identified by the triangular stem and leaf blades, which are generally rough to the touch. Flower stalks arise from the center, forming a compact group or headlike cluster of flower spikes. Reproduction is by seed.

Control:
Herbicides: partial control with glyphosate

Ditch Bank Brush

Wax Myrtle
(*Myrica cerifera*)

Figure 37-29.
Wax Myrtle, Mark Hoyer - Circ. 707.

Description:
Shrub or small tree usually 10 feet tall. Leaves are alternate, pale green, and lance shaped. When crushed, leaves emit a pleasant aroma. Close inspection of the leaves will reveal numerous small dark scales on top and bright orange scales below. Reproduction is by seed.

Control:
Herbicides: imazapyr*, triclopyr*

*Should not be applied directly to water.

Brazilian Pepper
(*Schinus terebinthifolius*)

Figure 37-30.
Brazilian pepper. Amy Ferriter - Circ. 707.

Description:
An extremely fast growing shrub found predominantly in disturbed areas of South Florida. This aggressive nonnative species produces large quantities of seeds contained in red fruit usually about 1/4 inch in diameter. Reproduction is by seed.

Control:
Herbicides: glyphosate, 2,4-D, imazapyr*, triclopyr*

Water-primrose
(*Ludwigia* species)
Figure 37-31.
Water-primrose. V. Ramey - CAIP/UF.

Description:
Small shrub attaining height of up to 6 feet with multiple branching stems. Leaves are lance shaped with small soft hairs on both sides. Flowers are yellow with four symmetrical petals. Reproduction is by seed and underground stems.

Control:
Herbicides: 2,4-D, imazapyr*

Willow
(*Salix* species)

Figure 37-32.
Willow.
V. Ramey - CAIP/UF.

Description:
Fast growing shrub that can become a tree in a short period of time. Leaves are alternate and lance shaped with finely toothed margins. The fruit capsule contains many small hairy seeds that drift in air currents.

Control:
Herbicides: 2,4-D, bromacil, glyphosate, imazapyr*, triclopyr*

*Should not be applied directly to water.

by Larry Parsons and Brian Boman

Millions of boxes of fruit and thousands of acres of citrus trees have been lost in freezes and frosts. More than nearly any other factor, freezes have caused some of the most dramatic changes in fruit supply, availability, and price. Thus, any method that provides some cold protection can be of major importance to citrus growers.

Many cold protection methods have been used over the years. These methods include heaters, wind machines, fog generators, high-volume over-tree irrigation, and low-volume, under-tree microsprinkler irrigation.

High fuel costs have made grove heating during freeze nights prohibitively expensive, except for high-value crops. Wind machines are effective, but they require maintenance and need a strong temperature inversion for optimum effectiveness. Fog can provide cold protection, but light winds can blow the fog away from the grove and obscure nearby roadways. In South Florida, where temperatures do not normally go far below freezing, high-volume, over-tree sprinkler irrigation has been used effectively on limes and avocados. In Central and North Florida, temperatures are usually colder, and over-tree sprinklers should not be used on large citrus trees because the weight of the ice formed can break off limbs and cause tree collapse. With overhead systems, all leaves are wetted and susceptible to damaging evaporative cooling during low humidity or windy freezes. Many trees were killed in the windy 1962 freeze because of evaporative cooling of wetted leaves when overhead sprinklers were used.

Low-volume, under-tree microsprinkler irrigation is an alternative method for partial frost protection and can be more affordable than other methods (Fig. 38-1). Microsprinklers have proven effective during several freeze nights in Central Florida tests. In addition to frost protection, microsprinklers can provide effective year-round irrigation. Microsprinklers, or spray jets, are small, low-volume irrigation sprinklers that discharge 5 to 50 gallons/hour. In citrus groves, the most commonly used spray jets discharge from 5 to 25 gallons/hour and cover a diameter of 5 to 21 feet. Usually 1 or 2 microsprinklers per tree are installed at the ground level or on short risers. Unlike overhead sprinklers, microsprinklers do not commonly wet leaves and branches above a height of about 3 feet and do not usually cause serious limb damage.

How Irrigation Works for Cold Protection
Various forms of irrigation have been used for frost and freeze protection for many years. When used

Figure 38-1. Ice on young citrus as a result of microsprinkler irrigation for freeze protection.

properly, water can provide partial or complete cold protection for a number of crops. On the other hand, improper use of water can increase cooling or ice loading and cause greater damage than if no water were used at all. Because water can provide protection in one situation and cause damage in another, it is important to know what principles are involved. To better understand what can happen when using water during a freeze, several commonly used terms need to be understood. With a knowledge of these terms, one can better evaluate the risks and benefits and successfully use irrigation for cold protection.

Heat of fusion

The heat that is released when liquid water freezes to solid ice is called the heat of fusion. The amount of heat generated when water freezes is 1200 BTUs/gallon or 80 calories/gram of water frozen. As long as enough water is continuously applied to a plant, the heat generated when water freezes can keep the plant at or near 32° F (0° C). This is the principle used by strawberry, fern, or citrus nursery growers when they apply high volumes of water by sprinkler irrigation to protect their plants. At least 0.25 inch/hour or more is required for cold protection. With very low temperatures, low humidity, or high winds, more water must be applied to get adequate protection. Many citrus nurserymen need to apply water at rates of 0.35 inch/hour or higher.

Heat of vaporization

The heat lost when water changes from a liquid to water vapor is called the heat of vaporization. At 32° F, the heat of vaporization is about 8950 BTUs/gallon or 596 calories/gram of water evaporated. Note that the heat of vaporization is about 7.5 times greater than the heat of fusion. This means that to maintain a stable situation when both freezing and evaporation occur, for every gallon of water that evaporates, 7.5 gallons of water need to be frozen to balance out the heat in the grove. Anything that promotes evaporation, such as low humidity and high windspeed, will promote overall cooling.

If the water application rate is high enough on the trunk of a young tree, it will be protected by the ice formation. However, on the edge of and outside of the iced zone, temperatures will not be maintained at 32° F, and those parts will probably be damaged or killed. Therefore, usually the tops of young trees or branches above the iced zone are more severely damaged after a freeze.

Dry bulb temperature (Tdb)

The dry bulb temperature is the temperature of the ambient air, which is the same thing as the normal air temperature read with a grove thermometer.

Wet bulb temperature (Twb)

The wet bulb temperature is defined as the lowest temperature to which air can be cooled solely by the addition of water. An example of wet bulb temperature is the temperature one feels when coming out of a swimming pool on a windy day. As long as a surface is wet while in the wind, its temperature will drop to the prevailing wet bulb temperature of the air.

The wet bulb temperature is between the dew point and dry bulb temperatures, and normally closer to the dry bulb than the dew point temperature. When the air is saturated with water vapor and the relative humidity is 100%, all three temperatures (dew point, wet bulb, and dry bulb) are equal (Fig. 38-2 and Appendix 11).

Humidity

Humidity refers to the amount of water vapor in the air. There are various ways to express humidity, but the most commonly used terms are relative humidity and dew point temperature.

Relative humidity (RH)

Relative humidity is the percentage (or ratio) of water vapor in the air in relation to the amount needed to saturate the air at the same temperature. Although commonly used, relative humidity is not the best measure of humidity because it depends on the air temperature. Warm air holds more water vapor than cool air.

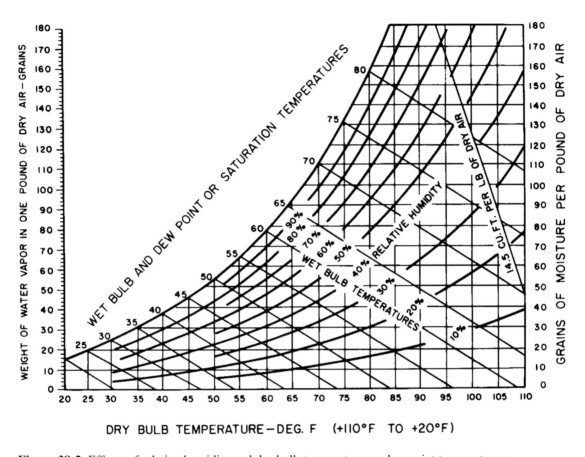

DRY BULB TEMPERATURE—DEG. F (+110°F TO +20°F)

Figure 38-2. Effects of relative humidity and dry bulb temperature on dew point temperatures.

Example:

Compare the amount of water vapor in air at 40° F and 70% RH to air at 90° F and 70% RH.

From Figure 38-2, enter the graph at 40° F on the horizontal axis and go straight up until midway between the 60% and 80% RH curves. Then read the weight of water vapor in 1 lb of dry air on the left vertical axis—about 25 grains.

Following the same procedure for 90° F and 70% RH, the weight of water vapor in 1 lb of dry air is about 150 grains, or 6 times as much.

Example:

For a dry bulb temperature of 45° F with a relative humidity of 50%, determine the dew point temperature.

From Appendix 11, select Figure A11-3 (for relationships between Tdb, RH, and Tdp). Enter the figure on the bottom axis at Tdb = 45° F and go straight up until the RH = 50% line is intersected. Read the Tdp from the vertical axis as Tdp ≈ 28° F.

Dew point temperature (Tdp)

The dew point temperature is the temperature at which dew begins to form or the temperature at which water vapor condenses to liquid water. It is also the temperature at which air reaches water vapor saturation. A common example of condensation is the water that forms on the outside of a glass of ice water. This happens because the temperature of the glass surface is lower than the dew point temperature of the ambient air in the room. Thus, some of the water vapor in the surrounding air condenses on the outside of the cold glass.

When referring to cold protection, dew point is one of the better ways to describe the humidity or amount of water vapor in the air. When the dew point is below 32° F, it is often called the frost point because frost can form when the temperature is below freezing. The dew point is important on freeze nights because water vapor in the air can slow the rate of temperature fall. With a relatively high dew point on a cool night, radiant heat losses from a grove are reduced, and the temperature may be expected to fall slowly. But if the dew point is quite low, the temperature can be expected to fall rapidly. Water vapor absorbs infrared radiation. Water droplets or fog are even more effective radiation absorbers than water vapor. Therefore, fog can reduce the rate of temperature drop on a frost night.

In addition to affecting the rate of radiation loss, the dew point is often a "basement" temperature, and the air temperature will not go much below it unless drier air moves in. The reason for this is that when dew condenses or ice forms, heat is given off. The amount of heat from condensation is the same as the heat of vaporization (about 8950 BTUs per gallon or 596 calories per gram of water) because vapor is changed to liquid water. This heat release during condensation slows the rate at which the air temperature drops. If dew forms, water vapor is condensed from the air, and the humidity or dew point of that air is lowered. This is how the evaporation coil in an air conditioner removes water vapor and dehumidifies the air.

Dew point temperatures are commonly higher on the coasts than they are inland. In the Central Florida citrus belt (e.g., near Lake Alfred), dew point temperatures on a moderate frost night can be in the vicinity of 20° to 30° F. On more severe freeze nights, dew point temperatures can be 10° F or lower. For example, in the damaging Christmas 1983 and January 1985 freezes, dew point temperatures in Lake Alfred approached 5° F, which is exceedingly low for Central Florida.

Example:
Determine the dew point for a dry bulb

temperature of 52° F and a wet bulb temperature of 42° F.

From Appendix 11, select Figure A11-2 (for a T_{db} in the range of 40°- 60° F). Enter the figure on the bottom axis at $T_{db} = 52°$ F and go straight up until the $T_{wb} = 42°$ F line is intersected. Read the T_{dp} from the vertical axis as $T_{dp} \approx 29°$ F.

Wind chill

Wind chill refers to the cooling effect of air moving over a warm body and is expressed in terms of the amount of heat lost per unit area per unit of time. Wind chill was developed to estimate heat loss rate from humans and other warm-blooded organisms. It does not apply to plants or vegetation. Even though wind chill does not apply to plants, wind can remove heat from a grove rapidly. In a windy freeze, the temperature of a dry leaf is usually fairly close to the air temperature. If the leaf is wet and water is not freezing on it, the leaf can theoretically cool to the wet bulb temperature. Thus, the length of time a grove will be at low temperatures can be longer on a windy night than on a calm one. Therefore, more damage can potentially occur during a windy freeze.

Psychrometer

A psychrometer is a device used to determine atmospheric humidity by reading the wet bulb and dry bulb temperatures. The wet bulb thermometer is kept wet by a moistened cotton wick or sleeve. With a psychrometer, one determines how much cooler the wet bulb is than the dry bulb and then calculates humidity by using appropriate graphs or tables. "Psychros" comes from the Greek word meaning "cold"; hence, a psychrometer measures humidity by determining how much colder the wet bulb thermometer is than the dry bulb thermometer.

For an accurate reading, the wet bulb thermometer must have air moving over it. With a sling psychrometer, air flow is created by rotating the two thermometers through the air by hand (Fig. 38-3). The sling psychrometer consists of two thermometers. One thermometer measures the air temperature while the other one measures the wet bulb temperature. After the wick is dipped in distilled water, the sling psychrometer is whirled around using the handle. Water evaporates from the wick on the wet bulb thermometer and cools the thermometer because of the latent heat of vaporization. The wet bulb thermometer is cooled to the lowest value possible in a few minutes. This value is known as the wet bulb temperature. Drier air results in more thermometer cooling and, therefore, a lower wet bulb temperature. With a fan-ventilated psychrometer, a fan blows air across the two thermometers. Fan-ventilated psychrometers cost more than sling psychrometers, but they are more convenient to operate on freeze nights.

Sling psychrometers work well at temperatures above freezing but are more difficult to operate at temperatures below freezing. The reason for this is that at temperatures much below freezing, the water on the wet bulb freezes, releases its heat of fusion, and raises the wet bulb temperature to around 32° F. Eventually, it is possible to get a "frost" wet bulb temperature if one rotates the sling psychrometer long enough. A battery-powered fan psychrometer avoids some of the problems of a sling psychrometer, but it may take 20 minutes or more to get a valid wet bulb temperature when the air is below freezing. A slightly different chart is used for humidity calculations when the wet bulb sleeve has ice on it.

Effectiveness of Microsprinklers

Microsprinkler irrigation is more effective for cold protection when high volumes of water are used. A system that delivers the maximum amount of water per acre and is practical or affordable is best for frost protection. Irrigation rates of 2000 gallons/ acre/hour or 33.3 gallons/acre/minute are recommended. This can be accomplished with one 20 gallon/hour jet or two 10 gallon/hour jets per tree in a grove with 100 trees/acre (Fig. 38-4). If there are 200 trees per acre, one 10 gallon/hour jet is adequate. Rates below this level will provide some protection but not as much as higher rates. Application rates of 3000 gallons/acre/hour or more are even more effective.

At high application rates, average warming with spray jets at a 4-foot height is only 2° to 3° F. Spot readings have occasionally shown temperature increases of 4° F or more, but 1° F or less is also common during freeze nights. At heights greater than 8 feet, warming is usually less than 1° F.

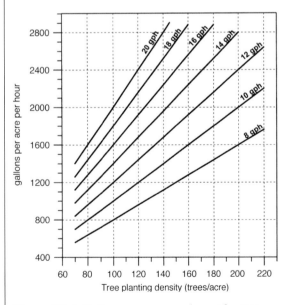

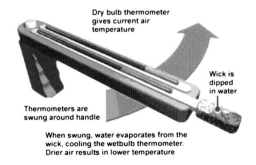

Figure 38-3. Operation of sling psychrometer to obtain wet and dry bulb temperature.

Figure 38-4. Gallons per acre per hour of water available for freeze protection based on microsprinkler discharge rate (ranging from 8 to 20 gph per tree) and tree planting density.

When compared to a nonirrigated area, lower-volume systems provide slight warming, but higher volume systems (2000 gallons/acre/hour) provide more warming. Low irrigation rates can provide some, although not much, protection on calm frost nights. While a small amount of water can provide a little protection, it is generally best not to go below 10 gallons/hour/tree. Emitters that put out less than 10 gallons/hour usually have small orifices that can plug easily. If a jet next to a tree is plugged, that tree will suffer more damage.

Example:

Determine the gallons per hour per acre available for freeze protection for a block that has 1 microsprinkler/tree that discharges 12 gph. Trees are planted at a 12 ft in-row and 24 ft across-row spacing.

Determine the number of trees/acre

$$\text{trees/ac} = \frac{43,560 \text{ ft}^2}{\text{ac}} \times \frac{1 \text{ tree}}{12 \text{ ft} \times 24 \text{ ft}} = 151 \text{ trees/ac}$$

Calculate gal/hr/ac

gal/hr/ac = 151 trees/ac x 12 gal/hr/tree = 1812 gal/hr/ac
or, use Fig. 28-4
Enter the figure on the bottom axis at a density of 151 trees/ac and go straight up until the 12 gph line is intersected. Read the per-acre application rate from the vertical axis as ≈1800 gal/hr/ac.

Microsprinklers can provide some protection to leaves and wood, particularly on the lower and inner part of the canopy. A dense canopy tends to retain heat from the soil and provide better protection than a thin canopy. Damage commonly will be seen on the outer and upper parts of the tree after severe freezes. Since fruit is more sensitive to cold temperatures than leaves or wood, microsprinklers generally do not protect the fruit. At higher volumes, spray jets will help protect fruit a little better than no irrigation, but generally microsprinkler irrigation is best for tree protection rather than fruit protection.

There is a limit to the effectiveness of microsprinklers. Factors such as tree health and cold acclimation affect tree survival. Depending on the volume of water applied, the lower limit of effectiveness for microsprinkler irrigation is around 17° F. The lower parts of young trees have been protected at even colder temperatures, but damage usually increases as it gets colder.

Young Tree Protection with Microsprinklers
Microsprinklers have been effective in protecting the bud union and lower portion of young trees. In young trees, the microsprinkler protects the lower trunk by the direct application of water. When water freezes, it releases heat. If the application rate is high enough, the freezing water will maintain the trunk at a temperature near 32° F. The spray jet must be close enough to the young tree so that water sprays directly on the trunk and lower part of the tree. Recommended distances between the trunk and the microsprinkler are 1 to 2.5 feet. If the microsprinkler is too far away from the young tree, wind can blow the water away. If the water freezes before it hits the tree, milky white ice can form on the tree. Protection under milky ice is usually not as good as under clear ice. During most Florida freeze nights, the wind comes from the north or northwest, making it best to put the microsprinkler on the north or northwest (upwind) side of the tree. In this position, the wind will carry the water into the tree and not blow it away from the tree.

It is common for protection to be seen only in the iced zone on young trees. Damage commonly occurs, particularly in severe freezes, above the iced zone. Figure 38-1 shows the kind of icing that can occur on a young tree during a freeze night. Figure 38-5 shows an example of microsprinklers that iced the lower part of young trees and protected the bud union and lower branches. The tree in the center received no freeze protection due to a plugged microsprinkler, and it subsequently died. Young trees are usually more tender and do not retain heat as well as mature trees. Therefore, protection down to 17° F cannot always be assured, even if the tree is in good health.

Insulating tree wraps placed around the trunks of young trees slow the rate of temperature fall. Tree wraps alone provide some trunk protection. Tree wraps in combination with microsprinkler irrigation provide even better cold protection insurance. If the irrigation system fails during the night, the tree wrap (particularly if it is a good insulator or has enclosed water pouches) can slow the temperature drop and protect the tree longer.

Operation of Microsprinklers on Freeze Nights
Compared to high-volume systems, microsprinklers operate at relatively low pressures of 20 to 25 psi. Because of this low pressure and the small openings in the emitters, spray jets can freeze up if they drop below 32° F before the water is turned on. The water should be turned on before the temperature reaches 32° F so the jets do not freeze. On frost nights, it is recommended that microsprinklers be turned on when the air temperature reaches 36° F. Be careful of thermometer placement because in low-lying cold pockets the ground surface can be below 32° F when it is 36° F at the thermometer height. When spray jets ice up, they are very difficult to thaw, and usually the emitters must be replaced in the field.

Figure 38-5. Example of freeze protection of young trees with microsprinkler irrigation. The dead tree is due to a plugged microsprinkler. Trees that survived had water continuously applied to the trunk and lower scaffold branches during the entire freeze period.

Once the microsprinklers are turned on, the system must keep running all night. If the system stops or fails when the temperature is below 32° F, it will be very difficult to restart the system because the emitters can rapidly freeze up. The situation becomes especially critical for young trees if the system fails when the temperature is below freezing. Shortly after the water stops spraying on a young tree, the trunk temperature can drop to the wet bulb temperature.

If the irrigation system stops, the trunk can rapidly drop below the actual air temperature by evaporative cooling. Thus, more tree damage can occur than if the jets were never turned on. Even if the system has only stopped for a short time, many of the emitters probably will have frozen, and it will be difficult to get the entire system fully operational again. Because of evaporative cooling, damage can be greater to trees in the area where a system stopped during a freeze night than where the system kept operating continuously. Do not turn off a system when the temperature is near the critical tree killing temperature. If the pumping system is unreliable, or if the pump is electrically driven in an area that commonly has electrical brownouts on freeze nights, it may be wise to convert to a more reliable system.

In the morning, when the temperature warms up, it is not necessary to wait until all the ice has melted before turning off the system. When the wet bulb temperature is above 34° F, the system can be safely turned off with no damage to the trees. At temperatures below freezing, it is possible for the wet bulb thermometer to erroneously read 32° F and actually appear to be warmer than the dry bulb thermometer. This occurs because the water in the wick freezes, releases the heat of fusion, and raises the wet bulb temperature to 32° F. Therefore, one should wait until the wet bulb temperature reaches 34° F before turning off the irrigation system.

Generally speaking, if the air temperature (dry bulb) has risen to 40° or 45° F, the irrigation system can be turned off safely. If the grove contains only mature hardened-off trees, the system can be turned off at 40° F. Under the most adverse conditions of low dew point and high wind, the grower may want to wait until the temperature is above 45° F. If it is a two-night freeze and the daytime temperature never gets above 40° F, then the system should be run continuously throughout the day and into the second night. If drainage or water conservation is of major concern, the system possibly can be turned off slightly before 40° F under less severe nonwindy conditions, but that may increase the freeze damage risk.

Why Microsprinklers Provide Cold Protection

Several factors contribute to the cold protection effectiveness of microsprinkler systems. Most well water in Florida is around 68° to 70° F. This warm water contributes a small amount of sensible heat to the grove as it drops from the initial temperature to 32° F. When temperatures drop below freezing, the latent heat of fusion is released when the water freezes. Depending on the amount of ice that forms, the heat released can raise temperatures in the lower part of the canopy. Water has a high heat capacity and can store a fair amount of heat. Therefore, a moist soil can hold more heat than a dry soil.

Microsprinklers can also raise the dew point or frost point temperature in the grove. When the temperature drops to the frost point, heat is released as the water vapor is converted to ice crystals. When the grove air temperature reaches the dew point temperature, the rate of cooling slows down because heat is released as the water vapor in the air condenses. It has been suggested that microsprinklers can provide some protection above the spray zone because moist air rises and condenses higher up in the canopy. The heat of condensation (8950 BTUs/gallon) may help warm the upper canopy and protect more of the tree.

Depending on the dew point temperature, microsprinklers can sometimes create fog on cold nights. Fog is beneficial for frost protection, and if the fog is dense enough and the droplets are of the proper size, the rate of cooling can be slowed because fog can act like a blanket and reduce the rate of radiation loss. Similarly, clouds, which consist of water droplets, can act like a blanket and slow radiation loss.

Operation on Windy Nights

Like other methods used for citrus cold protection, microsprinklers are less effective during windy or advective freeze nights. They provide little or no protection to mature trees. There is a risk when using water during windy or low humidity freezes. When dew point temperatures are low and winds are high, high evaporation rates can occur and cool the wetted part of a tree below the air temperature. This happened in the windy 1962 freeze. Where overhead sprinklers were used, evaporative cooling occurred, and trees were killed. The heat lost by evaporative cooling is approximately 7.5 times greater than the heat gained by ice formation.

The irrigation application rate on the wetted area influences the level of protection. A higher application rate can protect trees at lower temperatures. One way to increase the application rate is to reduce the spray pattern size. This can be done by changing a 360° full-circle spray pattern to a half-circle (180°) or quarter-circle (90°) pattern. This essentially doubles or quadruples the application rate by concentrating the amount of water on a smaller area. With a higher application rate, the protection level is better. Because of changing winds, a half-circle cap may do a better job of directing water into the young tree than a quarter-circle cap.

Summary

Under-tree microsprinkler irrigation is an affordable alternative to other forms of cold protection. It does not provide complete protection and generally will not protect fruit. Weak trees will receive little or no protection. On calm nights, microsprinklers have given partial protection to healthy and well-hardened trees down to 17° F. On windy severe freeze nights, little, if any, protection will be provided for mature trees, and only higher volume systems will provide protection for the lower portion of young trees.

by Mark Wade and Brian Boman

An economic analysis of alternatives is essential if maximum profits are to be achieved from a citrus irrigation system in Florida. A complete economic analysis includes an estimate of the initial investment required along with annual costs and returns, including financing costs. Decisions on what system to use, how to modify existing systems, and when to replace components of an existing system should all be evaluated on a technical as well as an economic basis. The profitability of an irrigation investment calculation greatly depends upon engineering estimates of the life expectancy of the equipment, energy usage, and the maintenance and repairs required in operating the equipment. Labor requirements should also be determined and evaluated relative to current and future availabilities.

Contingencies can also have an important influence upon the profitability of an irrigation decision. For example, it is necessary to consider the level of management required and to evaluate the implications of failure to maintain the assumed level of management. Similar considerations apply to possible equipment breakdowns and the availability of parts and service. System costs may change dramatically depending on the flexibility of the system. For example, a system that is designed for freeze protection will have a considerably higher per-acre cost than a zoned system.

Annual Ownership Costs

Annual irrigation costs include annual ownership costs and annual operating costs. Annual ownership costs include all costs that are associated with ownership and generally do not depend upon the level of use. These costs include the decrease in value of the initial investment due to age and obsolescence. Ownership costs also include an opportunity cost to reflect the returns that could be earned from the funds invested elsewhere. Other ownership costs include taxes and insurance.

The most accurate procedure for estimating average annual costs is to estimate the cash flows (out-of-pocket costs) for each year and determine the average annual equivalent by first discounting (finding the present value of) each cash cost, and then using an amortization (cost recovery) factor to determine the equivalent average annual cost. To be accurate, however, this method requires an estimate of all cash costs each year, including repairs, and many of these estimates are not readily available on an annual basis. As a result, the average annual costs are typically estimated from simple averages rather than derived from discounted annual cash flows.

Depreciation

Depreciation provides for the recovery of the initial investment over the investment period. Average annual depreciation is calculated by estimating the amount an asset will decline in value during its period of use and dividing by the years of use. The formula for average annual depreciation is:

$$Davg = \frac{Cost_1 - Salvage}{n} \qquad \textbf{Eq. 39-1}$$

where,

Davg = average annual depreciation
$Cost_1$ = purchase cost
Salvage = salvage value after n years
n = number of years the system is used

The salvage value is the value of the asset at the end of its use whether traded in for replacement equipment or scrapped. The trade-in or scrap value of a piece of equipment can be determined by both the annual level of use and the number of years of use. The number of years of use depends upon replacement decisions. The value of the asset at the end of its useful life can be positive, zero, or even negative if additional expenditure would be required to inactivate the system.

Table 39-1 lists typical useful life and annual maintenance costs for various microirrigation system components that have been tabulated from various sources. Considerable variability can occur for many of these components because of different physical conditions, repair levels, operation and maintenance practices, and the amount of time the system is used each year. Lower expected lifetimes are generally used for smaller units and are based on normal operation and maintenance practices that have generally occurred with their use. The upper ranges of life expectancy are suggested as guidelines for well-engineered, carefully constructed and maintained systems.

Careful judgement should be used when applying depreciation and life expectancy values during the economic analysis. If the depreciation period is based on an average usage of 1000 hours per year, factors such as higher or lower hours of operation and level of maintenance will affect the life of a particular component in the irrigation system, and hence, the rate of depreciation.

> Example:
> Determine the average annual depreciation for a coated steel screen filter with a new cost of $8,000, a salvage value of $500, and an estimated useful life of 10 years.
>
> Using Eq. 39-1,
>
> $$Davg = \frac{\$8,000 - \$500}{10} = \$750 \text{ per year}$$

The estimated useful life for each of the system components assumes considerable annual use of the equipment. The pump and power unit would have a longer life expectancy in systems designed to water all trees at once for freeze protection than in systems where the pump station serves several zones that are run independently.

There are many factors to be considered in determining the depreciation period and salvage value. Operating conditions, care, and maintenance, as well as wet-season operation, are contributing factors to the life of equipment and should be considered when determining the rate of depreciation and salvage value.

Productive life expectancies and salvage values of irrigation system components are also influenced by replacement policy and the rate of development of new technology. Where frequent replacement is practiced to avoid breakdowns, the years of use will be shortened and salvage values increased. Where new technology becomes available, equipment may be replaced more often, but salvage values may fall as new technologically enhanced components render older components obsolete, even ones that are well maintained.

Opportunity Costs

Accurate costing of resources used includes accounting for the value of funds invested. The economic cost, or opportunity cost, of any resource in producing a product is its value or worth in its best alternative use. A useful procedure to follow in calculating opportunity costs is to determine the returns that could be earned from the funds if invested elsewhere. For investments that have an indefinite useful life, the annual opportunity cost is estimated by multiplying the primary interest rate by the purchase price. The average cost of funds invested (Eq. 39-2) in depreciable items can be estimated by multiplying the average annual investment times the annual interest rate. The average annual investment (Eq. 39-3) is a simple average of the initial value of the investment (the purchase cost) and the value of the investment at the beginning of the last year of use (salvage value + average annual depreciation).

$$ACFI = AAI \times i \qquad \text{Eq. 39-2}$$

where,

ACFI = Average Cost of Funds Invested (Average Opportunity Cost)

i = Primary Interest Rate

Table 39-1. Typical useful life and annual maintenance costs (expressed as a percentage of initial cost) for microirrigation system components.

Component		Useful life		Annual maintenance and repairs
		(years)	(hours)	(%)
Pump house		15-25		0.5-1.5
Reservoirs		50		1-2
Land grading and bed formation		10-20		1-3
Well and casing		15-25		0.5-1.5
Ditches (with annual maintenance)		15-25		1-2
Micro-irrigation system	Lateral tubing	8-12		1-3
	PVC piping, underground	40		0.5-1
	PVC pipe, surface	10-15		1-3
	Aluminum components	10-20		1-3
	Valves	15		2-5
	Filters, coated steel	8-15		6-10
	Filters, galvanized	10-20		5-9
	Filters, stainless steel	15-25		4-8
	Emitter assemblies	5-10		5-8
Fertilization	Injection pump	3-5		4-8
	Solution tank	5-10		1-3
Electrical-mechanical components		4-8		5-10
Power units	Diesel engine	14-22	28,000	5-8
	Electric motor	25-35	50,000-70,000	1-3
	Gasoline engine, air-cooled	8-12	8,000	6-9
	Gasoline engine, water-cooled	10-16	18,000	5-8
	Propane engine	14-22	28,000	4-7
Pumps	Centrifugal pump	15-25	32,000-50,000	3-5
	Vertical turbine pump bowls	8-16	16,000-20,000	5-7
	Turbine pump column	15-20	32,000-40,000	3-5
Power transmission	Flat belt, fabric	6-10	10,000	5-7
	Flat belt, leather	8-12	20,000	5-7
	Flat belt, rubber	6-10	10,000	5-7
	Gear head	15-20	30,000	5-7
	V-Belt	4-8	6,000	5-7

$$AAI = \frac{Cost_1 + (Salvage + D_{avg})}{n}$$ **Eq. 39-3**

where,

AAI = Average Annual Investment

> **Example:**
> Determine the average opportunity cost of funds invested from the previous example.
>
> *Using Eq. 39-3,*
>
> $$AAI = \frac{\$8,000 + (\$500 + \$750)}{10}$$
> = \$925 per year
>
> *Using Eq. 39-2,*
>
> ACFI = \$925 x 0.06 = \$55.50 per year

According to this example, the average annual cost of forgoing the opportunity to invest the financial resources and gain a 6% annual interest return, instead of purchasing the screen filter, is \$55.50 per year.

Taxes and Insurance Costs

Insurance costs depend upon coverage levels selected, and can increase or decline over time based upon the type of coverage (replacement versus present value) and sales value of the asset. Property taxes and insurance costs can be approximated by multiplying the average annual investment by the annual tax and insurance rate. Taxes are typically calculated based upon an assessment rate that is multiplied by the full cash value to arrive at an assessed value. The assessed value is then multiplied times the tax rate per \$1.00 assessed value. The annual tax rate to be applied to the average annual investment is, therefore, the assessment rate times the tax rate per \$1.00 assessed value. However, the property that will be taxed, assessment procedures, and tax rates will all depend upon local tax provisions. The combined cost for taxes and insurance normally runs in the range of 1.5% to 2.5% of the initial value (purchase price) of the irrigation facilities.

Present Value

To carry out an annual cost calculation when individual items in the system are fully depreciated in less time than the period of analysis requires some means of accounting for component replacement. This is accomplished by first determining the time at which the replacement would occur, and then calculating the present value of the replacement cost. The present value is calculated by applying the interest rate being used for the analysis to the replacement cost of the item. The present value factor (PVF) and the present value (PV) are calculated by:

$$PVF = (1 + i)^n$$ **Eq. 39-4**

$$PV = S \times (1 + i)^{-n}$$ **Eq. 39-5**

where,

S = replacement cost
i = interest rate
n = number of years in the future the replacement purchase will be made.

> **Example:**
> Determine the present value of a new self-priming centrifugal pump costing \$1,152, to be replaced in 2 years. The current investment rate is 8%.
>
> *Using Eq. 39-4,*
>
> $$PVF = (1 + .08)^2$$
>
> *Using Eq. 39-5,*
>
> $$PV = \$1,152 \times (1 + .08)^{-2}$$
> = \$1,152 (.8573) = \$987.61

In this example, the cost of the pump is adjusted by a discount rate equal to 8% interest compounded annually (PVF) and represents the amount of money that should be invested today to purchase the pump at a future date.

Amortization

The annual cost of capital invested in the irrigation system can be calculated as the present value of the investment plus the interest incurred during the period of analysis. A commonly used approach for determining annual costs is to calculate a uniform series of annual values for depreciation and interest over the analysis period that is equivalent to the single present value. The value of this uniform series of annual costs is determined by application of an amortization factor, generally referred to as the capital recovery factor (CRF). This factor and the annual amortization value (AV) are calculated by:

$$CRF = \frac{i \times (1 + i)^n}{(1 + i)^n - 1}$$ **Eq. 39-6**

$$AV = PV \times CRF$$ **Eq. 39-7**

Example:
Compare two alternative filters using an interest rate of 8%. Filter 1 is an epoxy-coated steel filter with an initial cost of $8,000 and an expected life of 10 years. Filter 2 is a stainless steel filter that has an initial cost of $12,000 and a life expectancy of 20 years.

Epoxy-Coated Steel Filter:
Using Eq. 39-5 and the Compound Interest Rate Table (Table 39-2),

Using Eq. 39-5,

$$PV = \$8,000 \times (1.08^{-10}) = (\$8,000)(.4632)$$
$$= \$3,705.60$$

Using Eq. 39-6,

$$CRF = \frac{.08 \times (1 + .08)^{10}}{1.08^{10} - 1}$$

$$= \frac{.08 \times 2.1589}{2.1589 - 1}$$

$$= \frac{.08 \times 2.1589}{1.1589}$$

$$= 0.1490$$

Using Eq. 39-7,

$$AV = PV \times CRF = \$3,705.60 \times .1490$$
$$= \$552.13$$

Stainless Steel Filter:
Using Eq. 39-5,

$$PV = \$12,000 \times (1.08)^{-20} = (\$12,000)(.2145)$$
$$= \$2,574$$

Using Eq. 39-6,

$$CRF = \frac{.08 \times (1 + .08)^{20}}{(1 + .08)^{20} - 1}$$

$$= \frac{.08 \times 4.6610}{4.6610 - 1}$$

$$= \frac{.08 \times 4.6610}{3.6610}$$

$$= \frac{.08 \times 1.2731}{0.1019}$$

Using Eq. 39-7,

$$AV = PV \times CRF = \$2,574 \times .1019 = \$262.29$$

In this example, the stainless steel filter required less than half the annual capital investment ($262.29) as the epoxy-coated filter ($552.13) when amortized over the life of the filter.

Annual Operating Costs

The annual operation and maintenance (O&M) cost for an irrigation system includes the costs incurred for water, energy, lubrication, repairs, and labor. Improper design or operation of the system may increase overall O&M costs. Proper system design takes into consideration all economic factors when selecting each of the components of the system. A reduction in initial costs may result in an increase in the total annual per-acre cost of the system. For example, removing labor-saving features, such as valves, may increase labor costs enough to more than offset any savings earned by purchasing the lower cost equipment. Reduction in pipe sizes may increase fuel expenditures enough to more than offset equipment savings and may increase

Table 39-2. Compound interest (i) rate table.

	i = 6%		i = 8%		i = 10%	
	Future value of present sum	Present value of future sum	Future value of present sum	Present value of future sum	Future value of present sum	Present value of future sum
n	$(1+i)^n$	$(1+i)^{-n}$	$(1+i)^n$	$(1+i)^{-n}$	$(1+i)^n$	$(1+i)^{-n}$
1	1.0600	0.9434	1.080	0.9259	1.100	0.9091
2	1.1236	0.8900	1.166	0.8573	1.210	0.8264
3	1.1910	0.8396	1.260	0.7938	1.331	0.7513
4	1.2625	0.7921	1.360	0.7350	1.464	0.6830
5	1.3382	0.7473	1.469	0.6806	1.611	0.6209
6	1.4185	0.7050	1.587	0.6302	1.772	0.5645
7	1.5036	0.6651	1.714	0.5835	1.949	0.5132
8	1.5938	0.6274	1.851	0.5403	2.144	0.4665
9	1.6895	0.5919	1.999	0.5002	2.358	0.4241
10	1.7908	0.5584	2.159	0.4632	2.594	0.3855
11	1.8983	0.5268	2.332	0.4289	2.853	0.3505
12	2.0122	0.4970	2.518	0.3971	3.138	0.3186
13	2.1329	0.4688	2.720	0.3677	3.452	0.2897
14	2.2609	0.4423	2.937	0.3405	3.797	0.2633
15	2.3965	0.4173	3.172	0.3152	4.177	0.2394
16	2.5404	0.3936	3.426	0.2919	4.595	0.2176
17	2.6928	0.3714	3.700	0.2703	5.054	0.1978
18	2.8543	0.3503	3.996	0.2502	5.560	0.1799
19	3.0256	0.3305	4.316	0.2317	6.116	0.1635
20	3.2071	0.3118	4.661	0.2145	6.727	0.1486
21	3.3996	0.2942	5.034	0.1987	7.400	0.1351
22	3.6035	0.2775	5.437	0.1839	8.140	0.1228
23	3.8198	0.2618	5.871	0.1703	8.954	0.1117
24	4.0489	0.2470	6.341	0.1577	9.850	0.1015
25	4.2919	0.2330	6.848	0.1460	10.835	0.0923

total annual cost. Therefore, it is extremely important that the design engineer be thoroughly acquainted with all costs involved so the system can be designed to operate most economically, thus contributing more to overall operation profits.

Annual Energy Costs

Annual costs for energy can be estimated by using observed average costs. Also, many engine manufacturers give average values for fuel consumption in terms of gallons or pounds of fuel per brake horsepower. Fuel consumption will vary depending on the condition of the engine and the manner in which it is maintained. The load imposed on the engine can be an important factor, if it is operated at throttle settings beyond manufacturers' recommendations, or if the system planner imposes an overloading condition on the engine. The annual energy costs will depend on the type of power unit used, cost of fuel or energy, and the overall efficiency of the pumping plant.

Except for electrical installations, power costs will vary directly with the horsepower delivered and the number of operating hours during the season. Internal combustion power use can be obtained from fuel consumption curves for the specific engine/pump combination used. If these curves are not available, average consumption data (per BHP-hour) rates can be used (Table 39-3). Power rates for electric motors can be obtained from local power utility supplies. Electric power schedules are frequently based on a fixed standby charge for the hp rating of the motor and a schedule of rates that decreases with the energy actually consumed. Off-peak use rates apply in some areas.

Table 39-3. Average fuel consumption for internal combustion engines.

Engine type	Fuel Consumption
Gasoline, air cooled	1/8 gal/BHP-hr
Gasoline, water cooled	1/10 gal/BHP-hr
Diesel	1/12 gal/BHP-hr
Propane	1/7 gal/BHP-hr

Fixed and Variable Costs

A distinction between fixed costs and variable costs is vital to any capital investment and production decision. Fixed costs are those costs that do not vary with changes in output. For example, the cost of a well pump assembly and power unit is the same whether the pump is operating or not. Fixed assets, such as underground irrigation pipe or wells, are fixed costs that are referred to as "sunken costs" because the expense is very difficult to recoup once it has been incurred. Variable costs are those costs that change with the level of output. Examples include fuel expense, labor, maintenance costs, and materials.

This distinction between fixed costs and variable costs is key to decisions revolving around the "produce or don't produce" question. In attempting to determine if output should be produced, in this case irrigation water, the level of fixed costs versus variable costs is paramount. If a grower decides it is not cost-effective to irrigate a grove, only the variable costs should be considered because the fixed costs will be incurred regardless of the amount of water being pumped. If the economic benefits of irrigating are greater than or equal to average variable costs, then it is profitable to irrigate in the short run as the operation will incur a loss equal to its fixed costs only. In the long run, all costs are variable because, if given enough time, any factor of production can be changed. Therefore, in the long run, all costs, fixed and variable, must be covered in order to maximize profits or minimize losses.

Replacement Decisions

Many irrigation decisions involve the replacement of one or more system components. When a component is no longer repairable, the least cost replacement can be determined. However, often a component or system is operating at less than peak efficiency because of wear or obsolescence but is still serviceable. The replacement decision can then be considered in the context of continuing for another year without replacement versus making the replacement. Projected years of use should be determined for the replacement that will result in minimum average annual cost. The minimum average annual cost can then be compared to the estimated cost of continuing with the existing system for another year. Replacement would be indicated if the average annual cost with replacement is less than the cost for the next year without replacement. This decision should also be considered each subsequent year if costs are to be minimized.

Where reduced efficiency of the system increases costs as well as reducing the amount of water that can be pumped, yield or quality of fruit may be reduced. Returns above variable costs should then be maximized rather than simply minimizing costs.

Example:
Compare the minimum average annual cost of replacing a diesel engine with that of maintaining an existing power system. A new engine costs $25,000 and has a 20-year useful life expectancy. Annual tax and insurance expenses total 2% of the purchase price for both engines. The existing 18-year-old system cost $20,000 new and has a salvage value of $3,000. Annual maintenance and repair costs total $1,000 for the new engine and $3,000 for the existing engine. The existing system has fully depreciated. Current interest rates are 8%.

New Engine:
Using Eq. 39-6,

CRF = .08 x (1 + .08)^{20} x [(1 + .08)^{20} -1]$^{-1}$
= .373 x .273 = .1019

Using Eq. 39-5,

PV = $25,000 x (1 + .08)^0 = $25,000

Using Eq. 39-7,

AV = PV x CRF = $2,547.15

Using Eq. 39-1,

$$D_{avg} = \frac{\$25,000 - \$3,000}{20} = \$1,100$$

Using Eq. 39-3,

$$AAI = \frac{\$25,000 + (\$25,000 + \$1,100)}{20} = \$2,555$$

Using Eq. 39-2,

ACFI = $2,555 x .08 = $204.40

Total Annual Expense Summary - New Engine

Amortization Value	$2,547.15
Maintenance	$1,000.00
Tax and Insurance	$500.00
Opportunity Cost of Investment	$204.40
Total	$4,251.55

Existing Engine:
Using Eq. 39-3,

$$AAI = \frac{\$20,000 + (\$3,000 + 0)}{18} = \$1,277.78$$

Using Eq. 39-2,

ACFI = AAI x i = $1,277.78 x .08 = $102.22

Total Annual Expense Summary - Existing Engine

Maintenance	$3,000.00
Tax and Insurance	$400.00
Opportunity Cost of Investment	$102.22
Total	$3,502.22

The total annual expenses of the existing engine ($3,502.22) are less than the total annual expenses of the new engine ($4,251.55). Therefore, it is most economical to continue using the existing engine.

Example:
The existing engine referenced in the previous example has reduced water volume to a 15-year-old, 50-acre grove by 20%, from 11 inches to 9 inches of water, reducing average grapefruit yield by 12%. Per-acre yield at 11 inches of water had averaged 450 boxes. The on-tree price per box is $1.96. Determine if the engine should be replaced.

Lost Production: Current average yield
= 450 x .12
= 54 boxes of lost fruit per acre
   54 boxes per acre lost x 50 acres
= 27,00 boxes lost
   27,00 boxes x $1.96 per box
= $5,292

Total Expense New Engine: $4,251.55

Total Expense Existing Engine:

Annual Expense	$3,502.22
Value of Lost Production	$5,292.00
Total	$8,794.22

The value of lost production substantially increases the cost of the existing engine. Replacing the existing engine will reduce the real costs by one half.

APPENDIX 1 - Conversion Factors

To Convert	Into	Multiply by
A		
acre	hectare	0.4047
	sq feet	43,560
	sq meters	4047
	sq miles	1.562×10^{-3}
	sq yards	4840
acre-inch	gallons	17,158
	cu feet	43,560
	gallons	3.259×10^{5}
atmospheres	cm of mercury	76.0
	ft of water (at 39° F)	33.90
	in. of mercury (at 32° F)	29.92
	kg/sq cm	1.0333
	kg/sq meter	10,332
	pounds/sq in. (psi)	14.70
B		
bars	atmospheres	0.9869
	kg/sq meter	1.020×10^{4}
	pounds/sq ft	2089
	pounds/sq in. (Psi)	14.50
Btu	horsepower-hrs	3.931×10^{-4}
	kilowatt-hrs	2.928×10^{-4}
Btu/hr	foot-pounds/sec	0.2162
	gram-cal/sec	0.0700
	horsepower-hrs	3.929×10^{-4}
	watts	0.2931
Btu/min	foot-lbs/sec	12.96
	horsepower	0.02356
	kilowatts	0.01757
	watts	17.57

Btu/sq ft/min	watts/sq in.	0.1221
bushels	cu ft	1.2445
	cu in.	2150.4
	cu meters	0.03524
	liters	35.24
	quarts (dry)	32.0

C

Centigrade	Fahrenheit	(°C x 9/5)+32
centimeters	feet	3.281×10^{-2}
	inches	0.3937
	kilometers	10^{-5}
	meters	0.01
	miles	6.214×10^{-6}
	millimeters	10.0
	mils	393.7
	yards	1.094×10^{-2}
centimeters of mercury	atmospheres	0.01316
	feet of water	0.4461
	kg/sq meter	136.0
	pounds/sq ft	27.85
	pounds/sq in. (psi)	0.1934
centimeters/sec	feet/min	1.1969
	feet/sec	0.03281
	kilometers/hr	0.036
	knots	0.1943
	meters/min	0.6
	miles/hr	0.02237
	miles/min	3.728×10^{-4}
centimeters/sec/sec	feet/sec/sec	0.03281
	km/hr/sec	0.036
	meters/sec/sec	0.01
	miles/hr/sec	0.02237

cubic centimeters	cubic feet	3.531×10^{-5}
	cubic inches	0.06102
	cubic meters	10^{-6}
	cubic yards	1.308×10^{-6}
	gallons (U.S. liq.)	2.642×10^{-4}
	liters	0.001
	pints (U.S. liq.)	2.113×10^{-3}
	quarts (U.S. liq.)	1.057×10^{-3}
cubic feet	acre-feet	0.00028
	bushels (dry)	0.8036
	cubic cm	28,320
	cubic inches	1728
	cubic meters	0.02832
	cubic yards	0.03704
	gallons (U.S. liq.)	7.48052
	liters	28.32
	pints (U.S. liq.)	59.84
	quarts (U.S. liq.)	29.92
cubic feet/min	cubic cm/sec	472
	gallons/sec	0.1247
	liters/sec	0.4720
	pounds of water/min	62.43
cubic feet/sec	million gals/day	0.646317
	gallons/min	448.831
cubic inches	cubic cm	16.39
	cubic feet	5.787×10^{-4}
	cubic meters	1.639×10^{-5}
	cubic yards	2.143×10^{-5}
	gallons	4.329×10^{-3}
	liters	0.016391
	pints (U.S. liq.)	0.03463
	quarts (U.S. liq.)	0.01732
cubic meters	bushels (dry)	28.38
	cubic cm	106
	cubic feet	35.31

	cubic inches	61,023
	cubic yards	1.308
	gallons (U.S. liq.)	264.2
	liters	1000
	pints (U.S. liq.)	2113
	quarts (U.S. liq.)	1057
cubic yards	cubic cm	7.646×10^5
	cubic feet	27
	cubic inches	46,656
	cubic meters	0.7646
	gallons (U.S. liq.)	202
	liters	764.6
	pints (U.S. liq.)	1615.9
	quarts (U.S. liq.)	807.9
cubic yards/min	cubic ft/sec	0.45
	gallons/sec	3.367
	liters/sec	12.74

D

days	seconds	86,400

F

feet	centimeters	30.48
	kilometers	3.048×10^{-4}
	meters	0.3048
	miles (naut.)	1.645×10^{-4}
	miles (stat.)	1.894×10^{-4}
	millimeters	304.8
	mils	1.2×10^4
feet of water	atmospheres	0.02950
	in. of mercury	0.8826
	kg/sq cm	0.03048
	kg/sq meter	304.8
	pounds/sq ft	62.43
	pounds/sq in. (psi)	0.4335

feet/min	cm/sec	0.5080
	feet/sec	0.01667
	km/hr	0.01829
	meters/min	0.3048
	miles/hr	0.01136
feet/sec	cm/sec	30.48
	km/hr	1.097
	knots	0.5921
	meters/min	18.29
	miles/hr	0.6818
	miles/min	0.01136
feet/100 feet	percent grade	1.0

G

gallons	cubic cm	3785
	cubic feet	0.1337
	cubic inches	231
	cubic meters	3.785×10^{-3}
	cubic yards	4.951×10^{-3}
	liters	3.785
gallons of water	pounds of water	8.3453
gallons/min	cubic ft/sec	2.228×10^{-3}
	liters/sec	0.06308
	cubic ft/hr	8.0208
grams	kilograms	0.001
	milligrams	1000
	pounds	2.205×10^{-3}
grams/cm	pounds/inch	5.600×10^{-3}
grams/cubic cm	pounds/cubic ft	62.43
	pounds/cubic in.	0.03613
grams/liter	grains/gal	58.417
	pounds/cubic ft	0.062427
	parts/million	1000
grams/sq cm	pounds/sq ft	2.0481
	kg/ha	4.75

H

hectares	sq feet	1.076×10^5
horsepower	Btu/min	42.44
	foot-lbs/min	33,000
	foot-lbs/sec	550
	kg-calories/min	10.68
	kilowatts	0.7457
	watts	745.7
horsepower-hrs	Btu	2547
	kg-meters	2.737×10^5
	kilowatt-hrs	0.7457
hours	days	4.167×10^{-2}
	weeks	5.952×10^{-3}

I

inches	centimeters	2.540
	meters	2.540×10^{-2}
	miles	1.578×10^{-5}
	millimeters	25.40
	mils	1000
	yards	2.778×10^{-2}
inches of mercury	atmospheres	0.03342
	feet of water	1.133
	kg/sq cm	0.03453
	kg/sq meter	345.3
	pounds/sq ft	70.73
	pounds/sq in. (psi)	0.4912
inches of water (at 40° F)	atmospheres	2.458×10^{-3}
	inches of mercury	0.07355
	kg/sq cm	2.540×10^{-3}
	ounces/sq in.	0.5781
	pounds/sq ft	5.204
	pound/sq in. (psi)	0.03613

K

kilograms	grams	1000
	pounds	2.205
	tons (long)	9.842×10^{-4}
	tons (short)	1.102×10^{-3}
kilograms/cubic meter	grams/cubic cm	0.001
	pounds/cubic ft	0.06243
	pounds/cubic in.	3.613×10^{-5}
kilograms/ha	pounds/acre	0.8922
kilograms/meter	pounds/ft	0.6720
kilograms/sq cm	atmospheres	0.9678
	feet of water	32.81
	inches of mercury	28.96
	pounds/sq ft	2048
	pounds/sq in. (psi)	14.22
kilograms/sq meter	atmospheres	9.678×10^{-5}
	bars	98.07×10^{-6}
	feet of water	3.281×10^{-3}
	inches of mercury	2.896×10^{-3}
	pounds/sq ft	0.2048
	pounds/sq in. (psi)	1.422×10^{-3}
kilograms/sq mm	kg/sq meter	10^{6}
kilogram meters	Btu	9.294×10^{-3}
	kilowatt-hrs	2.723×10^{-6}
kilometers	centimeters	10^{5}
	feet	3281
	inches	3.937×10^{4}
	meters	1000
	miles	0.6214
	millimeters	10^{6}
	yards	1094
kilometers/hr	cm/sec	27.78
	feet/min	54.68
	feet/sec	0.9113
	knots	0.5396

	meters/min	16.67
	miles/hr	0.6214
kilowatts	Btu/min	56.92
	foot-lbs/min	4.426×10^4
	foot-lbs/sec	737.6
	horsepower	1.341
	kg-calories/min	14.34
	watts	1000
kilowatt-hrs	Btu	3413
	foot-lbs	2.655×10^6
	horsepower-hrs	1.341
	kg-meters	3.671×10^5

L

liters	bushels (U.S. dry)	0.02838
	cubic cm	1000
	cubic feet	0.03531
	cubic inches	61.02
	cubic meters	0.001
	cubic yards	1.308×10^{-3}
	gallons (U.S. liq.)	0.2642
	pints (U.S. liq.)	2.113
	quarts (U.S. liq.)	1.057
liters/ha	gal/acre	0.107
liters/min	cubic ft/sec	5.886×10^{-4}
	gals/sec	$4.40.3 \times 10^{-3}$

M

meters	centimeters	100
	feet	3.281
	inches	39.37
	kilometers	0.001
	miles (naut.)	5.396×10^{-4}
	miles (stat.)	6.214×10^{-4}
	millimeters	1000

	yards	1.094
meters/min	cm/sec	1.667
	feet/min	3.281
	feet/sec	0.05468
	km/hr	0.06
	knots	0.03238
	miles/hr	0.03728
meters/sec	feet/min	196.8
	feet/sec	3.281
	kilometers/hr	3.6
	kilometers/min	0.06
	miles/hr	2.237
	miles/min	0.03728
metric tons/ha	tons/ac	0.4461
microns	meters	10^{-6}
miles (statute)	centimeters	1.609×10^5
	feet	5280
	inches	6.336×10^4
	kilometers	1.609
	meters	1609
	miles (nautical)	0.8684
	yards	1760
miles/hr	cm/sec	44.70
	feet/min	88
	feet/sec	1.467
	km/hr	1.609
	km/min	0.02682
	knots	0.8684
	meters/min	26.82
milligrams	grains	0.01543236
	grams	0.001
milligrams/liter	parts/million	1.0
milliliters	liters	0.001
millimeters	centimeter	0.1
	feet	3.281×10^{-3}

	inches	0.03937
	kilometers	10^{-6}
	meters	0.001
	miles	6.214×10^{-7}
	mils	39.37
	yards	1.094×10^{-3}
million gals/day	cubic ft/sec (cfs)	1.54723
mils	centimeters	2.540×10^{-3}
	feet	8.333×10^{-5}
	inches	0.001
	kilometers	2.540×10^{-8}
	yards	2.778×10^{-5}

O

ounces	grains	437.5
	grams	28.349527
	pounds	0.0625
	ounces (troy)	0.9115
	tons (long)	2.790×10^{-5}
	tons (metric)	2.835×10^{-5}
ounces (fluid)	cubic inches	1.805
	liters	0.02957

P

parts/million	grains/U.S. gal	0.0584
	pounds/million gal	8.345
pints (dry)	cubic inches	33.60
pints (liq.)	cubic cm	473.2
	cubic feet	0.01671
	cubic inches	28.87
	cubic meters	4.732×10^{-4}
	cubic yards	6.189×10^{-4}
	gallons	0.125
	liters	0.4732
	quarts (liq)	0.5

pounds	grains	7000
	grams	453.6
	kilograms	0.4536
	ounces	16.0
	tons (short)	0.0005
pounds of water	cubic feet	0.01602
	cubic inches	27.68
	gallons	0.1198
pounds of water/min	cubic ft/sec	2.670×10^{-4}
pounds/acre	kg/ha	1.121
pounds/cubic ft	grams/cubic cm	0.01602
	kg/cubic meter	16.02
	pounds/cubic in.	5.787×10^{-4}
	gm/cubic cm	27.68
	kg/cubic meter	2.768×10^{4}
	pounds/cubic ft	1728
pounds/ft	kg/meter	1.488
pounds/in.	gm/cm	178.6
pounds/sq ft	atmospheres	4.725×10^{-4}
	feet of water	0.01602
	inches of mercury	0.01414
	kg/sq meter	4.882
	pounds/sq in.	6.944×10^{-3}
pounds/sq in.	atmospheres	0.06804
	feet of water	2.307
	inches of mercury	2.036
	kilopascals (kPa)	6.895
	kg/sq meter	703.1
	pounds/sq ft	144

Q

quarts (dry)	cubic inches	67.20
quarts (liq)	cubic cm	946.4
	cubic feet	0.03342
	cubic inches	57.75

	cubic meters	9.464×10^{-4}
	cubic yards	1.238×10^{-3}
	gallons	0.25
	liters	0.9463

R

revolutions/min	degrees/sec	6.0
	revs/sec	0.01667
revolutions/sec	degrees/sec	360
	revs/min	60
rods (surveyors' meas.)	meters	5.029
	yards	5.5
	feet	16.5

S

square centimeters	circular mils	1.973×10^{5}
	sq feet	1.076×10^{-3}
	sq inches	0.1550
	sq meters	0.0001
	sq miles	3.861×10^{-11}
	sq millimeters	100
	sq yards	1.196×10^{-4}
square feet	acres	2.296×10^{-5}
	circular mils	1.833×10^{8}
	sq cm	929
	sq inches	144
	sq meters	0.09290
	sq miles	3.587×10^{-8}
	sq millimeters	9.290×10^{4}
	sq yards	0.1111
square inches	circular mils	1.273×10^{6}
	sq cm	6.452
	sq feet	6.944×10^{-3}
	sq millimeters	645.2
	sq mils	10^{6}

	sq yards	7.716×10^{-4}
square kilometers	acres	247.1
	sq cm	1010
	sq ft	10.76×10^{6}
	sq inches	1.550×10^{9}
	sq meters	10^{6}
	sq miles	0.3861
	sq yards	1.196×10^{6}
square meters	acres	2.471×10^{-4}
	sq cm	10^{4}
	sq feet	10.76
	sq inches	1550
	sq miles	3.861×10^{-7}
	sq millimeters	10
	sq yards	1.196
square miles	acres	640
	sq feet	27.88×10^{6}
	sq km	2.590
	sq meters	2.590×10^{6}
	sq yards	3.098×10^{6}
square millimeters	circular mils	1973
	sq cm	0.01
	sq feet	1.076×10^{-5}
	sq inches	1.550×10^{-3}
square yards	acres	2.066×10^{-4}
	sq cm	8361
	sq feet	9.0
	sq inches	1296
	sq meters	0.8361
	sq miles	3.228×10^{-7}
	sq millimeters	8.361×10^{5}

T

temperature ($^\circ$ C)	absolute temperature ($^\circ$ C)	$^\circ$ C + 273
temperature ($^\circ$ C)	temperature ($^\circ$ F)	$1.8 \times {}^\circ$ C + 32

temperature (° F)	absolute temperature (° F)	° F + 460
temperature (° F)	temperature (° C)	5/9 x (° F - 32)
tons (long)	kilograms	1016
	pounds	2240
	tons (short)	1.120
tons (metric)	kilograms	1000
	pounds	2205
tons (short)	kilograms	907.2
	ounces	32,000
	pounds	2000
	tons (long)	0.89287
	tons (metric)	0.9078
tons/acre	metric tons/ha	2.2416
tons water/24 hrs	pounds of water/hr	83.333
	gallons/min	0.16643
	cubic ft/hr	1.3349

W

watts	Btu/hr	3.4129
	Btu/min	0.05688
	foot-lbs/min	44.27
	foot-lbs/sec	0.7378
	horsepower	1.341×10^{-3}
	horsepower (metric)	1.360×10^{-3}
	kg-calories/min	0.01433
	kilowatts	0.001
watt-hours	Btu	3.413
	foot-pounds	2656
	gram-calories	859.85
	horsepower-hrs	1.341×10^{-1}
	kilogram-calories	0.8605
	kilogram-meters	367.2
	kilowatt-hrs	0.001

Y

yards		
	centimeters	91.44
	kilometers	9.144×10^{-4}
	meters	0.9144
	miles (naut.)	4.934×10^{-4}
	miles (stat.)	5.682×10^{-4}
	millimeters	914.4

APPENDIX 2 - Pressure and Head Conversion Factors

Unit	Pound per in² (psi)	Feet of water (head)	Meters of water (head)	Inches of Mercury	Atmos.	kg per cm²
Psi	1	2.31	0.704	2.04	0.0681	0.0703
feet of water (32° F to 62° F)	0.433	1	0.305	0.882	0.0295	0.0305
meters of water (32° F to 62° F)	1.421	3.28	1	2.89	0.0967	1
inch Hg (32° F to 62° F)	0.491	1.134	0.3456	1	0.0334	0.0345
atmosphere (at sea level)	14.70	33.93	10.34	29.92	1	1.033
kg/cm²	14.22	32.8	10	28.96	0.968	1

APPENDIX 3 - Flow Conversion Factors

Unit	m³/sec	m³/day	L/sec	ft³/sec	ft³/day	ac-ft/day	gal/min	gal/day	mgd
m³/sec	1	8.64×10^4	10^3	35.31	3.051×10^6	70.05	1.58×10^4	2.282×10^7	22.82
m³/day	1.157×10^{-5}	1	0.0116	4.09×10^{-4}	35.31	8.1×10^{-4}	0.1835	264.2	2.64×10^{-4}
L/sec	0.001	86.4	1	0.0353	3051	0.070	15.85	2.28×10^4	2.28×10^{-2}
ft³/sec	0.0283	2446	28.32	1	8.64×10^4	1.984	448.8	6.46×10^5	0.646
ft³/day	3.28×10^{-7}	0.02832	3.28×10^{-4}	1.16×10^{-5}	1	2.3×10^{-5}	5.19×10^{-3}	7.48	7.48×10^{-6}
ac-ft/day	0.0143	1234	14.28	0.5042	43,560	1	226.3	3.259×10^5	0.3258
gal/min	6.3×10^{-5}	5.45	0.0631	2.33×10^{-3}	192.5	4.42×10^{-3}	1	1440	1.44×10^{-3}
gal/day	4.3×10^{-8}	3.79×10^{-3}	4.382×10^{-5}	1.55×10^{-6}	0.1337	3.07×10^{-6}	6.94×10^{-4}	1	10^{-6}
mgd	4.38×10^{-2}	3785	43.82	1.55	1.337×10^5	3.07	694	10^6	1

APPENDIX 4 - Common Liquid Fertilizer Solutions

Analysis (%)			lb/gal	Materials & formulation	Additional Elements (%)
N	P	K			
Component solutions					
21	0	0	10.8	Ammonium nitrate solution	
9	0	0	10.4	Ammonium sulfate solution	10 S
10	34	0	11.8	Ammonium polyphosphate	
12	0	0	11.0	Ammonium thiosulfate	26 S
9	0	0	12.2	Calcium nitrate solution	11 Ca
32	0	0	11.1	Urea ammonium nitrate solution	
0	54	0	14.5	Phosphoric acid – merchant grade	
3	0	11	9.7	Potassium nitrate solution	
0	0	62	16.5	Muriate of potash solution	
7	0	0	11.3	Magnesium nitrate	6 Mg
7	0	0	13.4	Manganese nitrate	15 Mn
7	0	0	13.3	Zinc nitrate	
					17 Zn
Fertilizer solutions					
5	0	10	10.0	NH_4NO_3, KCl	
5	0	10	10.5	NH_4NO_3, KCl, $Mg(NO_3)_2$ with micronutrients	
8	0	8	9.8	NH_4NO_3, KNO_3	
8	0	8	9.7	NH_4NO_3, KCl	
8	0	8	11.6	$CA(NO_3)_2$, KNO_3	
8	0	8	12.0	$CA(NO_3)_2$, KCl	
8	0	8	10.2	NH_4NO_3, KCl, $Mg NO_3$	1 Mg
8	2	8	10.3	NH_4NO_3, KNO_3, Phosphoric acid	
8	2	8	10.0	NH_4NO_3, KCl, Phosphoric acid	
8	4	8	10.3	NH_4NO_3, KNO_3, Phosphoric acid	
8	4	8	10.0	NH_4NO_3, KCl, Phosphoric acid	
9	0	9	10.2	NH_4NO_3, KNO_3	
9	0	9	10.1	NH_4NO_3, KCl	
9	2	9	10.7	NH_4NO_3, KNO_3, Phosphoric acid	
9	2	9	10.6	NH_4NO_3, KCl, Phosphoric acid	
9	4	9	10.7	NH_4NO_3, KNO_3, Phosphoric acid	
9	4	9	10.6	NH_4NO_3, KCl, Phosphoric acid	
10	0	10	10.4	NH_4NO_3, KNO_3	
10	0	10	10.3		

APPENDIX 5 - Salt Index of Fertilizer Sources

Material and Analysis	Salt Index (Sodium Nitrate = 100)	
	Per equal weight of materials basis	Per unit (lb) of plant nutrients
Nitrogen		
Ammonium nitrate, 35% N	105	3.0
Ammonium nitrate, 20.5% N	61	3.0
Ammonium sulfate, 21.2% N	69	3.3
Calcium nitrate, commercial grade, 15.5% N	65	4.2
Sodium nitrate, 16.5% N	100	6.1
Urea, 46.6% N	75	1.6
Nitrate of Soda Potash, 15% N, 14% K	92	3.2
Natural organic, 5% N	4	0.7
Phosphate		
Normal Superphosphate, 20% P_2O_5	8	0.4
Concentrated Superphosphate, 45% P_2O_5	10	0.2
Concentrated Superphosphate, 48% P_2O_5	10	0.2
Monoammonium phosphate, 12.2% N, 61.7% P_2O_5	30	0.4
Diammonium phosphate, 18% N, 46% P_2O_5	34	0.5
Potash		
Potassium chloride, 60% K_2O	116	1.9
Potassium chloride, 63.2% K_2O	114	1.8
Potassium nitrate, 13.8% N, 46.6% K_2O	74	1.2
Potassium sulfate, 46% K_2O	46	0.9
Monopotassium Phosphate, 52.2% P_2O_5, 34.6% K_2O	8	0.1
Sulfate of potash-magnesia, 21.9% K_2O	43	2.0

APPENDIX 6 - Backflow Prevention Requirements for Florida Irrigation Systems

Specific Authority: 570.07(23) FS.; Laws of Florida, Ch. 84-338, Sec. 17 (Sec. 487.055 (3) FS.)
Law Implemented: Laws of Florida, Ch. 84-338, Sec. 17 (Sec. 487.055 FS.)

Definitions

Antisyphon device: a safety device used to prevent backflow of a mixture of water and chemicals into the water supply.

Chemical: any substance that is intentionally added to water for agricultural purposes.

Emergency exemption: an exemption as authorized in Section 18 of the Federal Insecticide, Fungicide and Rodenticide Act.

Irrigation system: any device or combination of devices having a hose, pipe, or other conduit which connects directly to any source of ground or surface water, through which water or a mixture of water and chemicals is drawn and applied for agricultural purposes. The term does not include any handheld hose sprayer or other similar device which is constructed so that an interruption in water flow auto matically prevents any backflow to the water source.

Toxic Chemical: any pesticide whose label bears the signal word "Danger" or "Poison."

Section 487.055, Florida Statutes:

487.055 Antisyphon requirements for irrigation systems.
  (1) Any irrigation system used for the application of chemicals shall be equipped with an antisyphon device adequate to protect against contamination of the water supply, provided that an irrigation system installed prior to the effective date of this act shall be equipped with such a device within 18 months from the effective date of this act.

  (2) It shall be unlawful for any person to apply chemicals through an irrigation system that is not equipped with an antisyphon device as required by this section.

  (3) The department shall establish specific requirements for antisyphon devices by rule. The department shall adopt such rules on or before November 1, 1984.

  (4) Any governmental agency requiring antisyphon devices on irrigation systems used for the application of chemicals shall use the specific antisyphon device requirements adopted by the department.

Antisyphon Requirements for Irrigation Systems

Any irrigation system designed or used for the application of chemicals shall be equipped with the following components:

  a) Functional check valve on the irrigation pipe. This valve shall be located in the irrigation supply line between the irrigation pump and the point of injection of chemicals. This valve, when installed, shall be on a horizontal plane and level. A deviation of not more than 10 degrees from the horizontal shall be set.

(b) Low-pressure drain. Such drain shall have an orifice size of at least 3/4-inch diameter. It shall be located on the bottom of the horizontal pipe between the functional check valve and the irrigation pump. It must be level and must not extend beyond the inside surface of the bottom of the pipe as shown (Fig. A6-1). The outside opening of the drain shall be at least two (2) inches above grade.

(c) Vacuum breaker. A vacuum breaker shall be installed on the top of the horizontal pipe between the functional check valve and the irrigation pump, and opposite to the low-pressure drain. The vacuum breaker shall have an orifice size of at least 3/4-inch diameter.

(d) Functional check valves on the chemical injection line. A check valve shall be installed on the chemical injection line. If injector pumps are used, they shall be installed so that when water flow ceases, the injector pumps will not operate. A method shall be provided for positive shutoff of the chemical supply when the injection system is not in use.

A single antisyphon device assembly (Fig. A6-1) shall be used for those systems where nontoxic chemicals such as fertilizers will be injected. A double antisyphon device assembly as shown (Fig. A6-2) shall be used for those systems where toxic chemicals will be injected. The functioning of each device in the double assembly system must be capable of being checked independently of the others to ensure effectiveness of the system.

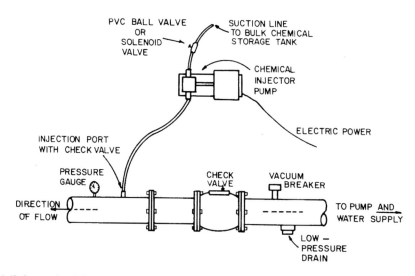

Figure A6-1. Minimum backflow requirements for injection of nontoxic chemicals into irrigation systems.

Chemical Storage Tanks.
Tanks shall be constructed and maintained in a manner to ensure containment of the chemical and to prevent contamination.

Variances.
(a) None of the antisyphon device components shall be altered in any manner that would render the antisyphon system inoperative or ineffective.

(b) An irrigation system where only fertilizer is injected into the irrigation pipes, where surface water is the only water source, and where both a check valve on the output side of the pump and a foot valve at water intake are present, will be approved as a variance to the rule.

(c) Specific variances of equipment not covered by this rule but which may be in compliance with this rule shall be considered on a case-by-case basis by the department.

Maintenance.

All check valves, low-pressure drains and vacuum breakers shall be maintained free of corrosion or other buildup and operative at all times during operation of the system. Cleaning agents used exclusively to maintain or clean an irrigation system shall not be subject to the regulations provided for herein.

Penalty.

Any person who shall use any irrigation system for the application of chemicals, without the required antisyphon device installed or without the antisyphon device in operating condition, shall be subject to an administrative fine not to exceed $1,000 for each violation.

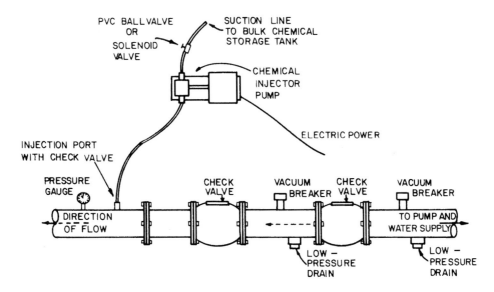

Figure A6-2. Minimum backflow requirements for injection of toxic chemicals into irrigation systems.

APPENDIX 7 - Approximate Gallons Contained in Partially Filled Horizontal Cylindrical Tanks (Flat Ends)

Table A7-1. Approximate gallons contained in partially filled horizontal cylindrical tanks (flat ends) where: TD = filled percentage of total tank diameter, and TC = percent of total tank capacity (see Figure A7-1).

TD	TC	TD	TC	TD	TC	TD	TC	TD	TC
0	0.0000	20	14.24	40	37.36	60	62.65	80	85.76
1	0.1692	21	15.27	41	38.60	61	63.89	81	86.77
2	0.4773	22	16.31	42	39.86	62	65.13	82	87.76
3	0.8742	23	17.38	43	41.12	63	66.39	83	88.73
4	1.342	24	18.46	44	42.38	64	67.59	84	89.67
5	1.869	25	19.55	45	43.64	65	68.81	85	90.59
6	2.450	26	20.66	46	44.91	66	70.02	86	91.49
7	3.077	27	21.79	47	46.19	67	71.22	87	92.36
8	3.748	28	22.92	48	47.46	68	72.41	88	93.20
9	4.458	29	24.07	49	48.73	69	73.59	89	94.02
10	5.204	30	25.23	50	50.00	70	74.77	90	94.80
11	5.985	31	26.41	51	51.27	71	75.93	91	95.54
12	6.797	32	27.59	52	52.54	72	77.08	92	96.25
13	7.639	33	28.78	53	53.81	73	78.22	93	96.92
14	8.509	34	29.98	54	55.09	74	79.34	94	97.55
15	9.406	35	31.19	55	56.36	75	80.45	95	98.13
16	10.33	36	32.41	56	57.62	76	81.54	96	98.66
17	11.27	37	33.64	57	58.88	77	82.62	97	99.13
18	12.24	38	34.87	58	60.14	78	83.69	98	99.52
19	13.23	39	36.11	59	61.40	79	84.73	99	99.83

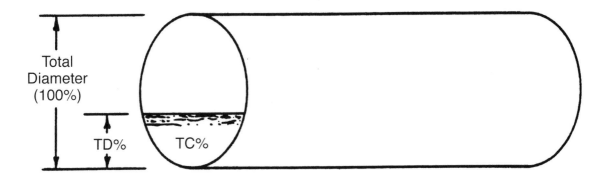

Figure A7-1. Relationship between fluid depth (TD%), tank diameter (TD), and percent of total tank capacity (TC%) for partially filled horizontal tank (see Table A7-1).

Example: Tank diameter = 48 inches with a capacity of 500 gallons when full.

Fluid level is 32 inches from bottom of tank.

$$TD\% = \frac{\text{fluid depth}}{\text{tank depth}} \times 100 = \frac{32}{48} \times 100 = 67\%$$

From Table A7-1, TC% = 71.22

Therefore, volume in tank is 500 gallons x 71.22% = 356 gallons.

APPENDIX 8 - Equations

Percent of Area Wetted by Emitter

Equation: $A(\%) = \dfrac{D^2 \times 0.785 \times 100}{S_1 \times S_2}$

where,

D = Wetted diameter (ft)
S_1 = Spacing between laterals (ft)
S_2 = Within-row spacing of emitters (ft)
A = Area in %
100 = %
0.785 = constant ($\pi/4$)

Data:

Flow rate = 12.5 gph
Lateral spacing = 24 ft
Emitter spacing = 12 ft
Wetted diameter = 15 ft

Solution: $\dfrac{15^2 \times 0.785}{24 \times 12} \times 100 = 61\%$

Working Pressure of Nonregulated Microsprinkler

Equation: $P = P_{nom} \times \left(\dfrac{q}{q_{nom}}\right)^2$

where,

P_{nom} = Nominal pressure (psi)
q = Actual flow rate (gph)
q_{nom} = Nominal flow rate (gph)
P = Working pressure

Data:

Nominal pressure = 20 psi
Nominal flow rate = 15.0 gph
Actual flow rate = 13.5 gph

Solution: $20 \times \left(\dfrac{13.5}{15.0}\right)^2 = 16$ psi

Working Flow Rate of Microsprinkler

Equation: $q = q_{nom} \times \left(\dfrac{P}{P_{nom}} \right)^{0.5}$

where,

q_{nom} = Nominal flow rate (gph)
P = Working pressure (psi)
P_{nom} = Nominal pressure (psi)
q = Actual flow rate (gph)

Data:

Nominal pressure = 20 psi
Nominal flow rate = 15.0 gph
Working pressure = 15 psi

Solution: $15.0 \times \left(\dfrac{15}{20} \right)^{0.5} = 13.0$ gph

Required Irrigation Time For Microsprinkler (100% efficiency)

Equation: $T = \dfrac{ET \times I}{q}$

where,

ET = Daily tree evapotranspiration (gal/tree)
q = Emitter flow rate (gph)
I = Irrigation interval (days)
T = Required irrigation time (hr)

Data:

Daily ET = 40 gallons/tree
Emitter flow rate = 12.5 gph
Irrigation interval = 2 days

Solution: $\dfrac{40 \times 2}{12.5} = 6.4$ hr

Flow Velocity - ft/sec (for gpm)

Equation: $V = \dfrac{Q \times 0.4085}{D^2}$

where,

  V = Flow velocity (fps)
  Q = Discharge (gpm)
  0.4085 = Constant
  D = Pipe I.D. (inches)

Data:

  Pipe I.D. = 4.154 inches. (Class 160 PVC)
  Discharge = 200 gpm

Solution: $\dfrac{200 \times 0.4085}{4.154^2} = 4.7$ fps

Pipe Diameter Selection - inches (for gpm)

Equation: $D = \sqrt{\dfrac{Q \times 0.4085}{V}}$

where,

  Q = Discharge (gpm)
  0.4085 = Constant
  V = Flow velocity (fps)
  D = Pipe I.D. diameter (inches)

Data:

  Discharge = 50 gpm
  Flow velocity = 5.0 fps

Solution: $D = \sqrt{\left(\dfrac{50 \times 0.4085}{5.0}\right)} = 2.0$ inches

(2-inch class 160 PVC pipe
has an I.D. of 2.193 inches)

Pipe Discharge - gpm

Equation: $Q = \dfrac{V \times D^2}{0.4085}$

where,

    V = Flow velocity (fps)
    D = Pipe I.D. diameter (inches)
    0. 4085 = Constant
    Q = Discharge (gpm)

Data:

    Pipe diameter = 6.115 in. (6 inch class 160 PVC)
    Flow velocity = 5.0 fps

Solution: $\dfrac{5.0 \times 6.115^2}{0.4085} = 458$ gpm

Friction Loss - Hazen-Williams % (for gpm)

Equation: $H_F = 1045 \times \left(\dfrac{Q}{C}\right)^{1.852} \times D^{-4.871}$

where,

    Q = Discharge (gpm)
    C = Friction factor
    D = Pipe I.D. diameter (in.)
    H_F = Friction loss (%)

Data:

    Discharge = 15.0 gpm
    Friction factor = 145 (Hazen-Williams)
    Pipe diameter = 1.049 in. (1-inch polyethylene tubing)

Solution: $1045 \times \left(\dfrac{15.0}{145}\right)^{1.852} \times 1.049^{-4.871} = 12.4\%$

Head Loss in PE or PVC Pipes - feet

Equation: $H_F = 0.001038 \times \left(\dfrac{Q^{1.852}}{D^{4.871}}\right) \times L$

where,

Q = Discharge (gpm)
D = Pipe diameter (in.)
L = Pipe length (ft)
H_F = Head loss (ft)

Data:

Discharge = 750 gpm
Pipe ID diameter = 7.961 (8-inch Class 160 PVC)
Pipe length = 400 ft

Solution: $0.001038 \times \left(\dfrac{200^{1.852}}{4.154^{4.871}}\right) \times 350 = 6.4$ ft

Motor Pump Plant Power Consumption - HP (for gpm)

Equation: $hp = \dfrac{Q \times H}{39.6 \times E}$

where,

Q = Pump discharge (gpm)
H = Head (ft)
E = Pump efficiency (%)
39.6 = Conversion factor
hp = Horsepower

Data:

Pump discharge = 480 gpm
Head = 165 ft
Pump efficiency = 70%

Solution: $\dfrac{480 \times 165}{39.6 \times 70} = 29$ hp

Pump Efficiency - % (for gpm)

Equation: $E = \dfrac{Q \times H}{53 \times kw}$

where,

Q = Pump discharge (gpm)
H = Head (ft)
kw = Power consumption
E = Efficiency (%)
53 = Constant (39.6/0.7457)

Data:

Pump discharge = 480 gpm
Head = 165 ft
Power consumption = 21.3 kw

Solution: $\dfrac{480 \times 165}{53 \times 21.3} = 70\%$

Pump Plant Energy Consumption - kw (for gpm)

Equation: $kw = \dfrac{Q \times H}{53 \times E}$

where,

Q = Discharge (gpm)
H = Head (ft)
E = Pumping efficiency (%)
53 = Constant

Data:

Discharge = 480 gpm
Pressure head = 165 ft
Pumping efficiency = 70%

Solution: $\dfrac{480 \times 165}{53 \times 70} = 21.3$ kw

Table A9-1. Equivalent number of feet of straight pipe for PVC fittings.

Fitting size (inches)	1/2	3/4	1	1 1/4	1 1/2	2	2 1/2	3	4	5	6	8	10
90° Ell	1.5	2.0	2.7	3.5	4.3	5.5	6.5	8.0	10	14	15	20	25
45° Ell	0.8	1.0	1.3	1.7	2.0	2.5	3.0	3.8	5.0	6.3	7.1	9.4	12
Long Sweep Ell	1.0	1.4	1.7	2.3	2.7	3.5	4.2	5.2	7.0	9.0	11	14	
Close Return Run	3.6	5.0	6.0	8.3	10	13	15	18	24	31	37	39	
Tee - Straight Run	1	2	2	3	3	4	5						
Tee - Side inlet	3.3	4.5	5.7	7.6	9.0	12	14	17	22	27	31	40	
Globe Valve	17.0	22	27	36	43	55	67	82	110	140	160	220	
Angle Valve	8.4	12	15	18	22	28	33	42	58	70	83	110	
Gate Valve	0.4	0.5	0.6	0.8	1.0	1.2	1.4	1.7	2.3	2.9	3.5	4.5	
Swing Check Valve	4	5	7	9	11	13	16	20	26	33	39	52	65
Spring Check Valve	4	6	8	12	14	19	23	32	43	58			

Example:
Determine equivalent length of 8-inch pipe for: 500 feet of 8-inch pipe with an 8-inch gate valve, and two 90° elbows.

Solution:

500 feet of 8-inch pipe	=	500 ft of straight pipe
8-inch gate valve	=	4.5 ft of straight pipe
Two 90° elbows	=	2 x 20 = 40 ft of straight pipe
Total	=	544.5 ft of straight pipe

APPENDIX 10 - Flow Velocity (V) and Friction Loss (F) for PVC Pipe Calculated by the Hazen-Williams Equation (in feet per 100 feet)

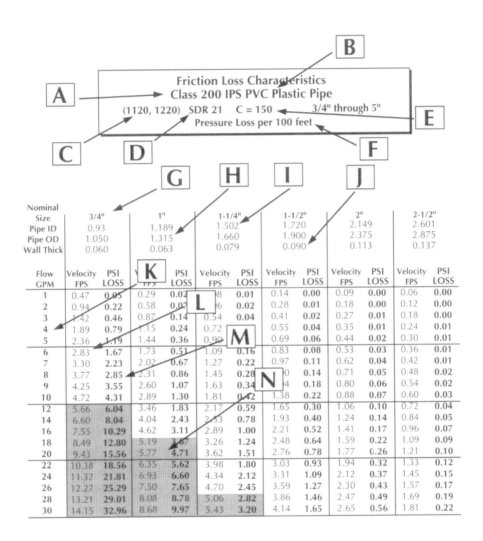

B

Friction Loss Characteristics
Class 200 IPS PVC Plastic Pipe
(1120, 1220) SDR 21   C = 150   3/4" through 5"
Pressure Loss per 100 feet

A **C** **D** **E** **F**

G **H** **I** **J**

Nominal Size	3/4"		1"		1-1/4"		1-1/2"		2"		2-1/2"	
Pipe ID	0.93		1.189		1.502		1.720		2.149		2.601	
Pipe OD	1.050		1.315		1.660		1.900		2.375		2.875	
Wall Thick	0.060		0.063		0.079		0.090		0.113		0.137	
Flow GPM	Velocity FPS	PSI LOSS	FPS	PSI LOSS	Velocity FPS	PSI LOSS	Velocity FPS	PSI LOSS	Velocity FPS	PSI LOSS	Velocity FPS	PSI LOSS
1	0.47	0.05	0.29	0.02	8	0.01	0.14	0.00	0.09	0.00	0.06	0.00
2	0.94	0.22	0.58	0.07	6	0.02	0.28	0.01	0.18	0.00	0.12	0.00
3	1.42	0.46	0.87	0.14	0.54	0.04	0.41	0.02	0.27	0.01	0.18	0.00
4	1.89	0.79	1.15	0.24	0.72		0.55	0.04	0.35	0.01	0.24	0.01
5	2.36	1.19	1.44	0.36	0.90		0.69	0.06	0.44	0.02	0.30	0.01
6	2.83	1.67	1.73	0.51	1.09	0.16	0.83	0.08	0.53	0.03	0.36	0.01
7	3.30	2.23	2.02	0.67	1.27	0.22	0.97	0.11	0.62	0.04	0.42	0.01
8	3.77	2.85	2.31	0.86	1.45	0.28	0	0.14	0.71	0.05	0.48	0.02
9	4.25	3.55	2.60	1.07	1.63	0.34	4	0.18	0.80	0.06	0.54	0.02
10	4.72	4.31	2.89	1.30	1.81	0.42	1.38	0.22	0.88	0.07	0.60	0.03
12	5.66	6.04	3.46	1.83	2.17	0.59	1.65	0.30	1.06	0.10	0.72	0.04
14	6.60	8.04	4.04	2.43	2.53	0.78	1.93	0.40	1.24	0.14	0.84	0.05
16	7.55	10.29	4.62	3.11	2.89	1.00	2.21	0.52	1.41	0.17	0.96	0.07
18	8.49	12.80	5.19	3.67	3.26	1.24	2.48	0.64	1.59	0.22	1.09	0.09
20	9.43	15.56	5.77	4.71	3.62	1.51	2.76	0.78	1.77	0.26	1.21	0.10
22	10.38	18.56	6.35	5.62	3.98	1.80	3.03	0.93	1.94	0.32	1.33	0.12
24	11.32	21.81	6.93	6.60	4.34	2.12	3.31	1.09	2.12	0.37	1.45	0.15
26	12.27	25.29	7.50	7.65	4.70	2.45	3.59	1.27	2.30	0.43	1.57	0.17
28	13.21	29.01	8.08	8.78	5.06	2.82	3.86	1.46	2.47	0.49	1.69	0.19
30	14.15	32.96	8.68	9.97	5.43	3.20	4.14	1.65	2.65	0.56	1.81	0.22

Explanation of a pressure loss chart

A. Type of pipe represented in the chart.

B. IPS (Iron Pipe Size): Indicates that the pipe's outside diameter dimensions correspond to that of iron pipe. All IPS PVC pipe of the same nominal size will have the same outside diameter. For example, all 1/2-inch PVC irrigation pipe will have an outside diameter of 0.840 inch; therefore, all 1/2 inch slip fittings will fit on the outside of all types of 1/2-inch PVC pipe.

C. (1120, 1220): Represents a designation for the specifications of the plastic pipe.

D. SDR (Standard Dimension Ratio): Indicates the pipe's wall thickness as a ratio of the outside diameter. Outside diameter of 1-inch pipe is 1.315 inch. If you divide 1.135 by the SDR, 21, it will give you a minimum wall thickness[1]. Minimum wall thickness for 1-inch Class 200 PVC pipe 1.315 ÷ 21 = 0.063 inch. Class-rated pipes (SDR pipes) maintain a uniform maximum operating pressure across all pipe sizes. This is not true of schedule-rated pipes such as Schedule 40 PVC. In schedule-rated pipes, the maximum operating pressure decreases as pipe size increases.

E. C = 150: Indicates the value of the C factor, which is a measure of the roughness of the inside of the pipe. The lower the number, the rougher the inside of the pipe and the greater the pressure loss. For PVC, C = 150; for galvanized pipe, C = 100.

F. Designated pressure losses shown in the chart are per 100 ft of pipe.

G. Size indicates the nominal pipe size; however, none of the actual pipe dimensions are exactly that size. For example, in the 3/4-inch pipe, none of the dimensions are actually 3/4-inch.

H. OD: Outside pipe diameter in inches.

I. ID: Inside pipe diameter in inches.

J. Wall Thick: Wall thickness in inches.

K. Flow (gpm): Flow rate in gallons per minutes

L. Velocity (fps): Speed of water in feet per second at the corresponding flow rate.

M. PSI Loss: Pressure loss per 100 ft of pipe in pounds per square inch at the corresponding flow rate.

N. The shaded area on the chart designates those flow rates that exceed 5 fps. It is recommended that caution be used with flow rates above 5 fps in mainlines where water hammer will be a concern.

---

[1]There may be exceptions to this rule in the smallest pipe size of each class of pipe, as the minimum wall thickness allowed is 0.060 inch. In these cases, either the wall thickness is rounded up to 0.060 inch (1-inch Class 160 PVC wall thickness is rounded up to 0.060 inch) or that pipe classification is not made in the smaller sizes (Class 200 PVC is not made in 1/2-inch size).

Table A10-1. Friction loss characteristics for Bowsmith, Inc. polyethylene tubing, C = 140 (pressure loss in psi per 100 ft of tubing; shaded area represents velocities over 5 fps that should generally be avoided).

Nominal size	1/2" (700P50)		3/4" (3/4"-50")		1" (1"-50")		1-1/4"	
Pipe ID (in.)	0.590		0.818		1.057		1.360	
Pipe OD (in.)	0.700		0.935		1.195		1.530	
Wall width (in.)	0.055		0.059		0.069		0.085	
Flow (gpm)	V (fps)	Loss (psi)	V (fps)	Loss (psi)	V (fps)	Loss (psi)	V (fps)	Loss (psi)
1	1.17	0.63	0.61	0.13	0.37	0.04	0.22	0.01
2	2.34	2.28	1.22	0.46	0.73	0.13	0.44	0.04
3	3.52	4.82	1.83	0.98	1.10	0.28	0.66	0.08
4	4.69	8.21	2.44	1.68	1.46	0.48	0.88	0.14
5	5.86	12.42	3.05	2.53	1.83	0.73	1.10	0.21
6	7.03	17.41	3.66	3.55	2.19	1.02	1.32	0.30
7	8.20	23.16	4.27	4.72	2.56	1.36	1.54	0.40
8	9.38	29.66	4.88	6.05	2.92	1.74	1.76	0.51
9	10.55	36.88	5.49	7.52	3.29	2.16	1.99	0.63
10	11.72	44.83	6.10	9.14	3.65	2.63	2.21	0.77
12	14.06	62.84	7.32	12.82	4.38	3.68	2.65	1.08
14			8.54	17.05	5.11	4.90	3.09	1.44
16			9.73	21.83	5.84	6.27	3.53	1.84
18			10.98	27.16	6.57	7.80	3.97	2.29
20			12.20	33.01	7.30	9.48	4.41	2.78
22			13.41	39.38	8.03	11.31	4.85	3.32
24			14.63	46.26	8.76	13.29	5.29	3.90
26			15.85	53.66	9.49	15.41	5.74	4.52
28			17.07	61.55	10.23	17.68	6.18	5.19
30			18.29	69.94	10.96	20.09	6.62	5.89
32			19.51	78.82	11.69	22.64	7.06	6.64
34					12.42	25.33	7.50	7.43
36					13.15	28.16	7.94	8.26
38					13.88	31.13	8.38	9.13
40					14.61	34.23	8.82	10.04
42					15.34	37.47	9.26	10.99
44					16.07	40.84	9.71	11.98
46					16.80	44.34	10.15	13.01
48					17.53	47.98	10.59	14.07
50					18.26	51.75	11.03	15.18
55							12.13	18.11
60							13.24	21.28
65							14.34	24.68
70							15.44	28.31
75							16.54	32.16
80							17.65	36.25
85							18.75	40.55
90							19.85	45.08

519

Table A10-2. Friction loss characteristics for PVC Class 100 pipe (1120, 1220), SDR 26, C = 150 (pressure loss in psi per 100 ft of pipe; shaded area represents velocities over 5 fps that should be avoided where water hammer is a concern).

Nominal size	3.5"		4"		5"		6"		8"		10"		12"	
Pipe ID (in.)	4.000		4.500		5.563		6.625		8.625		10.750		12.750	
Pipe OD (in.)	3.804		4.280		5.291		6.301		8.205		10.266		12.128	
Wall width (in.)	0.098		0.110		0.136		0.162		0.210		0.262		0.311	
Flow (gpm)	V (fps)	Loss (psi)	V (fps)	Loss (psi)	V (fps)	Loss (psi)	V (fps)	Loss (psi)	V (fps)	Loss (psi)	V (fps)	Loss (psi)	V (fps)	Loss (psi)
80	2.26	0.21	1.78	0.12	1.17	0.04	0.82	0.02	0.48	0.01				
90	2.54	0.27	2.00	0.15	1.31	0.05	0.92	0.02	0.55	0.01				
100	2.82	0.32	2.23	0.18	1.46	0.06	1.03	0.03	0.61	0.01				
110	3.10	0.39	2.45	0.22	1.60	0.08	1.13	0.03	0.67	0.01				
120	3.38	0.45	2.67	0.26	1.75	0.09	1.23	0.04	0.73	0.01				
130	3.67	0.53	2.90	0.30	1.89	0.11	1.34	0.05	0.79	0.01				
140	3.95	0.60	3.12	0.34	2.04	0.12	1.44	0.05	0.85	0.01	0.54	0.00		
150	4.23	0.69	3.34	0.39	2.19	0.14	1.54	0.06	0.91	0.02	0.58	0.01		
160	4.51	0.77	3.56	0.43	2.33	0.16	1.64	0.07	0.97	0.02	0.62	0.01		
170	4.79	0.86	3.79	0.49	2.48	0.17	1.75	0.07	1.03	0.02	0.66	0.01		
180	5.08	0.96	4.01	0.54	2.62	0.19	1.85	0.08	1.09	0.02	0.70	0.01		
190	5.36	1.06	4.23	0.60	2.77	0.23	1.95	0.09	1.15	0.02	0.74	0.01		
200	5.64	1.17	4.45	0.66	2.91	0.23	2.06	0.10	1.21	0.02	0.77	0.01	0.55	0.00
250	7.05	1.76	5.57	0.99	3.64	0.35	2.57	0.15	1.52	0.03	0.97	0.01	0.69	0.01
300	8.46	2.47	6.68	1.39	4.37	0.50	3.08	0.21	1.82	0.04	1.16	0.02	0.83	0.01
350	9.87	3.29	7.80	1.85	5.10	0.66	3.60	0.28	2.12	0.06	1.35	0.03	0.97	0.01
400	11.28	4.21	8.91	2.37	5.83	0.85	4.11	0.36	2.42	0.08	1.55	0.03	1.11	0.01
450	12.69	5.24	10.02	2.95	6.56	1.05	4.62	0.45	2.73	0.10	1.74	0.04	1.25	0.02
500	14.10	6.37	11.14	3.59	7.29	1.28	5.14	0.55	3.03	0.12	1.94	0.05	1.39	0.02
550	15.51	7.60	12.25	4.28	8.02	1.53	5.65	0.65	3.33	0.15	2.13	0.06	1.53	0.03
600	16.92	8.93	13.36	5.03	8.74	1.79	6.17	0.77	3.64	0.18	2.32	0.07	1.66	0.03
650	18.33	10.35	14.48	5.83	9.47	2.08	6.68	0.89	3.94	0.21	2.52	0.08	1.80	0.04
700	19.74	11.88	15.59	6.69	10.20	2.38	7.19	1.02	4.24	0.25	2.71	0.09	1.94	0.04
750			16.70	7.60	10.93	2.71	7.71	1.16	4.55	0.28	2.90	0.11	2.08	0.05
800			17.82	8.57	11.66	3.05	8.22	1.31	4.85	0.32	3.10	0.12	2.22	0.05
850			18.93	9.59	12.39	3.42	8.73	1.46	5.15	0.36	3.29	0.14	2.36	0.06
900					13.12	3.80	9.25	1.62	5.45	0.40	3.48	0.15	2.50	0.07
950					13.85	4.20	9.76	1.79	5.76	0.50	3.68	0.17	2.64	0.07
1000					14.57	4.62	10.28	1.97	6.06	0.55	3.87	0.18	2.77	0.08
1100					16.03	5.51	11.30	2.35	6.67	0.65	4.26	0.22	3.05	0.10
1200					17.49	6.47	12.33	2.77	7.27	0.77	4.65	0.26	3.33	0.11
1300					18.95	7.50	13.36	3.21	7.88	0.89	5.03	0.30	3.61	0.13
1400							14.39	3.68	8.48	1.02	5.42	0.34	3.88	0.15
1500							15.41	4.18	9.09	1.16	5.81	0.39	4.16	0.17
1600							16.44	4.71	9.70	1.30	6.19	0.44	4.44	0.19
1700							17.47	5.27	10.30	1.46	6.58	0.49	4.72	0.22
1800							18.50	5.86	10.91	1.62	6.97	0.54	4.99	0.24
1900							19.53	6.48	11.51	1.79	7.36	0.60	5.27	0.27
2000									12.12	1.97	7.74	0.66	5.55	0.29
2100									12.73	2.16	8.13	0.72	5.83	0.32
2200									13.33	2.35	8.52	0.79	6.10	0.35
2300									13.94	2.55	8.90	0.83	6.38	0.38
2400									14.55	2.76	9.29	0.93	6.66	0.41
2500									15.15	2.98	9.68	1.00	6.93	0.44
2600									15.76	3.20	10.07	1.08	7.21	0.48

Table A10-3. Friction loss characteristics for PVC Class 125 pipe (1120, 1220), SDR 26, C = 150 (pressure loss in psi per 100 ft of pipe; shaded area represents velocities over 5 fps that should be avoided where water hammer is a concern).

Nominal size	1"		1.25"		1.5"		2"		2.5"		3"		4"	
Pipe ID (in.)	1.315		1.660		1.900		2.375		2.875		3.500		4.500	
Pipe OD (in.)	1.211		1.548		1.784		2.229		2.699		3.284		4.224	
Wall width (in.)	0.052		0.056		0.058		0.073		0.088		0.108		0.138	
Flow (gpm)	V (fps)	Loss (psi)	V (fps)	Loss (psi)	V (fps)	Loss (psi)	V (fps)	Loss (psi)	V (fps)	Loss (psi)	V (fps)	Loss (psi)	V (fps)	Loss (psi)
1	0.27	0.02	0.17	0.01	0.12	0.00								
2	0.55	0.06	0.34	0.02	0.25	0.01	0.16	0.00						
3	0.83	0.13	0.51	0.04	0.38	0.02	0.24	0.01						
4	1.11	0.22	0.68	0.07	0.51	0.03	0.32	0.01	0.22	0.00				
5	1.39	0.33	0.85	0.10	0.64	0.05	0.41	0.02	0.28	0.01				
6	1.66	0.46	1.02	0.14	0.76	0.07	0.49	0.02	0.33	0.01				
7	1.94	0.62	1.19	019	0.89	0.09	0.57	0.03	0.39	0.01	0.26	0.00		
8	2.22	0.79	1.36	0.24	1.02	0.12	0.65	0.04	0.44	0.02	0.30	0.01		
9	2.50	0.98	1.53	0.30	1.15	0.15	0.73	0.05	0.50	0.02	0.34	0.01		
10	2.78	1.19	1.70	0.36	1.28	0.18	0.82	0.06	0.56	0.02	0.37	0.01		
11	3.06	1.42	1.87	0.43	1.41	0.22	0.90	0.07	0.61	0.03	0.41	0.01		
12	3.33	1.67	2.04	0.51	1.53	0.25	0.98	0.09	0.67	0.03	0.45	0.01	0.27	0.00
14	3.89	2.22	2.38	0.67	1.79	0.34	1.14	0.11	0.78	0.05	0.52	0.02	0.32	0.01
16	4.45	2.85	2.72	0.86	2.05	0.43	1.31	0.15	0.89	0.06	0.60	0.02	0.36	0.01
18	5.00	3.54	3.06	1.07	2.30	0.54	1.47	0.18	1.00	0.07	0.68	0.03	0.41	0.01
20	5.56	4.31	3.40	1.30	2.56	0.65	1.64	0.22	1.12	0.09	0.75	0.03	0.45	0.01
22	6.12	5.14	3.74	1.56	2.82	0.78	1.80	0.26	1.23	0.10	0.83	0.04	0.50	0.01
24	6.67	6.04	4.08	1.83	3.07	0.92	1.97	0.31	1.34	0.12	0.90	0.05	0.54	0.01
26	7.23	7.00	4.42	2.12	3.33	1.06	2.13	0.36	1.45	0.14	0.98	0.05	0.59	0.02
28	7.78	8.03	4.76	2.43	3.58	1.22	2.29	0.41	1.56	0.16	1.05	0.06	0.64	0.02
30	8.34	9.13	5.10	2.76	3.84	1.39	2.46	0.47	1.68	0.18	1.13	0.07	0.68	0.02
35	9.73	12.14	5.95	3.68	4.48	1.84	2.87	0.62	1.96	0.25	1.32	0.09	0.80	0.03
40	11.12	15.55	6.81	4.71	5.12	2.36	3.28	0.80	2.24	0.31	1.51	0.12	0.91	0.04
45	12.51	19.34	7.66	5.86	5.76	2.94	3.69	0.99	2.52	0.39	1.70	0.15	1.02	0.04
50	13.91	23.50	8.51	7.12	6.40	3.57	4.10	1.21	2.80	0.48	1.89	0.18	1.14	0.05
55	15.30	28.04	9.36	8.49	7.05	4.26	4.51	1.44	3.08	0.57	2.08	0.22	1.25	0.06
60	16.69	32.94	10.21	9.98	7.69	5.00	4.92	1.69	3.36	0.67	2.26	0.26	1.37	0.08
65	18.08	38.21	11.06	11.57	8.33	5.80	5.33	1.96	3.64	0.77	2.45	0.30	1.48	0.09
70	19.47	43.83	11.91	13.27	8.97	6.65	5.74	2.25	3.92	0.89	2.64	0.34	1.60	0.10
75			12.76	15.08	9.61	7.56	6.15	2.56	4.20	1.01	2.83	0.39	1.71	0.11
80			13.62	17.00	10.25	8.52	5.56	2.88	4.48	1.14	3.02	0.44	1.82	0.13
90			14.47	19.02	10.89	9.53	6.98	3.23	4.76	1.27	3.21	0.49	1.94	0.14
95			15.32	21.14	11.53	10.60	7.39	3.59	5.04	1.41	3.40	0.54	2.05	0.16
100			16.17	23.37	12.17	11.71	7.80	3.96	5.32	1.56	3.59	0.60	2.17	0.18
110			17.02	25.69	12.81	12.88	8.21	4.36	5.60	1.78	3.78	0.66	2.28	0.19
120			18.72	30.65	14.10	15.37	9.03	5.20	6.16	2.05	4.16	0.79	2.51	0.23
130					15.38	18.06	9.85	6.11	6.72	2.41	4.53	0.93	2.74	0.27
140					16.66	20.94	10.67	7.09	7.28	2.79	4.91	1.08	2.97	0.32
150					17.94	24.02	11.49	8.13	7.84	3.20	5.29	1.23	3.20	0.36
160					19.22	27.30	12.31	9.24	8.40	3.64	5.67	1.40	3.43	0.41
170							13.13	10.41	8.96	4.10	6.05	1.58	3.65	0.46
180							13.96	11.65	9.52	4.59	6.43	1.77	3.88	0.52
190							14.78	12.95	10.08	5.10	6.80	1.96	4.11	0.58
200							15.60	14.31	10.64	5.64	7.18	2.17	4.34	0.64
210							16.42	15.74	11.20	6.20	7.56	2.39	4.57	0.70

Table A10-3. Friction loss characteristics for PVC Class 125 pipe (1120, 1220), SDR 26, C = 150 (pressure loss in psi per 100 ft of pipe; shaded area represents velocities over 5 fps that should be avoided where water hammer is a concern).

Nominal size	2.5"		3"		4"		6"		8"		10"		12"	
Pipe ID (in.)	2.875		3.500		4.500		6.625		8.625		10.750		12.750	
Pipe OD (in.)	2.699		3.284		4.224		6.217		8.095		10.088		11.966	
Wall width (in.)	0.088		0.108		0.138		0.204		0.265		0.331		0.392	
Flow (gpm)	V (fps)	Loss (psi)	V (fps)	Loss (psi)	V (fps)	Loss (psi)	V (fps)	Loss (psi)	V (fps)	Loss (psi)	V (fps)	Loss (psi)	V (fps)	Loss (psi)
225	12.60	7.72	8.51	2.97	5.14	0.87	2.37	0.13	1.40	0.04	0.90	0.01	0.64	0.01
250	14.00	9.38	9.45	3.61	5.71	1.06	2.63	0.16	1.55	0.04	1.00	0.02	0.71	0.01
275	15.40	11.19	10.40	4.31	6.28	1.27	2.90	0.19	1.71	0.05	1.10	0.02	0.78	0.01
300	16.80	13.15	11.34	5.06	6.86	1.49	3.16	0.23	1.86	0.06	1.20	0.02	0.85	0.01
325	18.20	15.25	12.29	5.87	7.43	1.72	3.43	0.26	2.02	0.07	1.30	0.02	0.92	0.01
350	19.60	17.49	13.24	6.73	8.00	1.98	3.69	0.30	2.17	0.08	1.40	0.03	0.99	0.01
375			14.18	7.65	8.57	2.25	3.95	0.34	2.33	0.09	1.50	0.03	1.06	0.01
400			15.13	8.62	9.14	2.53	4.22	0.39	2.49	0.11	1.60	0.04	1.13	0.02
425			16.07	9.65	9.71	2.83	4.48	0.43	2.64	0.12	1.70	0.04	1.21	0.02
450			17.02	10.75	10.29	3.15	4.75	0.48	2.80	0.13	1.80	0.05	1.28	0.02
475			17.96	11.85	10.86	3.48	5.01	0.53	2.95	0.15	1.90	0.05	1.35	0.02
500			18.91	13.03	11.43	3.83	5.27	0.58	3.11	0.16	2.00	0.06	1.42	0.02
550					12.57	4.57	5.80	0.70	3.42	0.19	2.20	0.07	1.56	0.03
600					13.72	5.37	6.33	0.82	3.73	0.23	2.40	0.08	1.70	0.03
650					14.86	6.23	6.86	0.95	4.04	0.26	2.60	0.09	1.85	0.04
700					16.00	7.14	7.38	1.09	4.35	0.30	2.80	0.10	1.99	0.05
750					17.15	8.12	7.91	1.24	4.66	0.34	3.00	0.12	2.13	0.05
800					18.29	9.15	8.44	1.39	4.98	0.39	3.20	0.13	2.27	0.06
850					19.43	10.23	8.97	1.56	5.29	0.43	3.40	0.15	2.42	0.06
900							9.50	1.73	5.60	0.48	3.60	0.16	2.56	0.07
950							10.02	1.92	5.91	0.53	3.80	0.18	2.70	0.08
1000							10.55	2.11	6.22	0.58	4.00	0.20	2.84	0.09
1050							11.08	2.31	6.53	0.64	4.20	0.22	2.99	0.10
1100							11.61	2.52	6.84	0.70	4.41	0.24	3.13	0.10
1150							12.13	2.73	7.16	0.76	4.61	0.26	3.27	0.11
1200							12.66	2.96	7.41	0.82	4.88	0.28	3.41	0.12
1250							13.19	3.19	7.78	0.88	5.01	0.30	3.56	0.13
1300							13.72	3.43	8.09	0.95	5.21	0.33	3.70	0.14
1350							14.25	3.68	8.40	1.02	5.41	0.35	3.84	0.15
1400							14.77	3.93	8.71	1.09	5.61	0.37	3.98	0.16
1450							15.30	4.20	9.02	1.16	5.81	0.40	4.13	0.17
1500							15.83	4.47	9.33	1.24	6.02	0.42	4.27	0.18
1550							16.36	4.75	9.65	1.31	6.21	0.45	4.41	0.20
1600							16.88	5.04	9.96	1.39	6.41	0.48	4.55	0.21
1650							17.41	5.33	10.27	1.48	6.61	0.51	4.70	0.22
1700							17.94	5.63	10.58	1.56	6.81	0.53	4.84	0.23
1750							18.47	5.94	10.89	1.65	7.01	0.56	4.98	0.25
1800							19.00	6.26	11.20	1.73	7.21	0.59	5.12	0.26
1850							19.52	6.59	11.51	1.82	7.41	0.63	5.27	0.27

Table A10-4. Friction loss characteristics for PVC Class 160 pipe (1120, 1220), SDR 26, C = 150 (pressure loss in psi per 100 ft of pipe; shaded area represents velocities over 5 fps that should be avoided where water hammer is a concern).

Nominal size	1"		1.25"		1.5"		2"		2.5"		3"		4"	
Pipe ID (in.)	1.315		1.660		1.900		2.375		2.875		3.500		4.500	
Pipe OD (in.)	1.195		1.532		1.754		2.193		2.655		3.230		4.154	
Wall width (in.)	0.060		0.064		0.073		0.091		0.110		0.135		0.173	
Flow (gpm)	V (fps)	Loss (psi)	V (fps)	Loss (psi)	V (fps)	Loss (psi)	V (fps)	Loss (psi)	V (fps)	Loss (psi)	V (fps)	Loss (psi)	V (fps)	Loss (psi)
1	0.28	0.02	0.17	0.01	0.13	0.00								
2	0.57	0.06	0.34	0.02	0..26	0.01	0.16	0.00						
3	0.85	0.14	0.52	0.04	0.39	0.02	0.25	0.01						
4	1.14	0.23	0.69	0.07	0.53	0.04	0.33	0.01	0.23	0.00				
5	1.42	0.35	0.86	0.11	0.66	0.05	0.42	0.02	0.28	0.01				
6	1.71	0.49	1.04	0.15	0.79	0.08	0.50	0.03	0.34	0.01	0.23	0.00		
7	1.99	0.66	1.21	0.20	0.92	0.10	0.59	0.03	0.40	0.01	0.27	0.01		
8	2.28	0.84	1.39	0.25	1.06	0.13	0.67	0.04	0.46	0.02	0.31	0.01		
9	2.57	1.05	1.58	0.31	1.19	0.16	0.76	0.05	0.52	0.02	0.35	0.01		
10	2.85	1.27	1.73	0.38	1.32	0.20	0.84	0.07	0.57	0.03	0.39	0.01		
11	3.14	1.52	1.91	0.45	1.45	0.23	0.93	0.08	0.63	0.03	0.43	0.01		
12	3.42	1.78	2.08	0.53	1.59	0.28	1.01	0.09	0.69	0.04	0.46	0.01	0.28	0.00
14	3.99	2.37	2.43	0.71	1.85	0.37	1.18	0.12	0.81	0.05	0.54	0.02	0.33	0.01
16	4.57	3.04	2.78	0.91	2.12	0.47	1.35	0.16	0.92	0.06	0.62	0.02	0.37	0.01
18	5.14	3.78	3.12	1.13	2.38	0.58	1.52	0.20	1.04	0.08	0.70	0.03	0.42	0.01
20	5.71	4.59	3.47	1.37	2.65	0.71	1.69	0.24	1.15	0.09	0.78	0.04	0.47	0.01
22	6.28	5.48	3.82	1.64	2.91	0.85	1.86	0.29	1.27	0.11	0.86	0.04	0.52	0.01
24	6.85	6.44	4.17	1.92	3.18	1.00	2.03	0.34	1.38	0.13	0.93	0.05	0.56	0.02
26	7.42	7.47	4.51	2.23	3.44	1.15	2.20	0.39	1.50	0.15	1.01	0.06	0.61	0.02
28	7.99	8.57	4.86	2.56	3.71	1.32	2.37	0.45	1.62	0.18	1.09	0.07	0.66	0.02
30	8.57	9.74	5.21	2.91	3.97	1.50	2.54	0.51	1.73	0.20	1.17	0.08	0.70	0.02
35	9.99	12.95	6.08	3.87	4.64	2.00	2.96	0.68	2.02	0.27	1.36	0.10	0.82	0.03
40	11.42	16.59	6.95	4.95	5.30	2.56	3.39	0.86	2.31	0.34	1.56	0.13	0.94	0.04
45	12.85	20.63	7.82	6.16	5.96	3.19	3.81	1.08	2.60	0.42	1.75	0.16	1.06	0.05
50	14.28	25.07	8.69	7.49	6.63	3.88	4.24	1.31	2.89	0.52	1.95	0.20	1.18	0.06
55	15.71	29.91	9.56	8.93	7.29	4.62	4.66	1.56	3.18	0.62	2.15	0.24	1.30	0.07
60	17.14	35.14	10.43	10.49	7.95	5.43	5.09	1.83	3.47	0.72	2.34	0.28	1.41	0.08
65	18.57	40.76	11.29	12.17	8.62	6.30	5.51	2.12	3.76	0.84	2.54	0.32	1.53	0.09
70	19.99	46.76	12.16	13.96	9.28	7.23	5.93	2.44	4.05	0.96	2.73	0.37	1.65	0.11
75			13.03	15.86	9.94	8.21	6.36	2.77	4.34	1.09	2.93	0.42	1.77	0.12
80			13.90	17.88	10.60	9.25	6.78	3.12	4.63	1.23	3.12	0.47	1.89	0.14
85			14.77	20.00	11.27	10.35	7.21	3.49	4.91	1.38	3.32	0.53	2.00	0.16
90			15.64	22.23	11.93	11.51	7.63	3.88	5.20	1.53	3.51	0.59	2.12	0.17
95			16.51	24.58	12.59	12.72	8.05	4.29	5.49	1.69	3.71	0.65	2.24	0.19
100			17.38	27.03	13.26	13.99	8.48	4.72	5.78	1.86	3.91	0.72	2.36	0.21
110			19.12	32.24	14.58	16.69	9.33	5.63	6.36	2.22	4.30	0.86	2.60	0.25
120					15.91	19.61	10.18	6.61	6.94	2.61	4.69	1.01	2.83	0.30
130					17.24	22.74	11.02	7.6	7.52	3.03	5.08	1.17	3.07	0.34
140					18.56	26.09	11.87	8.80	8.10	3.47	5.47	1.34	3.31	0.39
150					19.89	29.64	12.72	10.00	8.68	3.94	5.86	1.52	3.54	0.45
160							13.57	11.27	9.26	4.45	6.25	1.71	3.78	0.50
170							14.42	12.61	9.83	4.97	6.64	1.92	4.01	0.56
180							15.27	14.02	10.41	5.53	7.03	2.13	4.25	0.63
190							16.11	15.49	10.99	6.11	7.43	2.35	4.49	0.69
200							16.96	17.03	11.57	6.72	7.82	2.59	4.72	0.76

Table A10-4. Friction loss characteristics for PVC Class 160 pipe (1120, 1220), SDR 26, C = 150 (pressure loss in psi per 100 ft of pipe; shaded area represents velocities over 5 fps that should be avoided where water hammer is a concern).

Nominal size	2.5"		3"		4"		6"		8"		10"		12"	
Pipe ID (in.)	2.875		3.500		4.500		6.625		8.625		10.750		12.750	
Pipe OD (in.)	2.655		3.230		4.154		5.993		7.805		9.728		11.538	
Wall width (in.)	0.110		0.135		0.173		0.316		0.410		0.511		0.606	
Flow (gpm)	V (fps)	Loss (psi)	V (fps)	Loss (psi)	V (fps)	Loss (psi)	V (fps)	Loss (psi)	V (fps)	Loss (psi)	V (fps)	Loss (psi)	V (fps)	Loss (psi)
225	13.02	8.36	8.79	3.22	5.31	0.95	2.45	0.14	1.44	0.04	0.93	0.01	0.66	0.01
250	14.47	10.16	9.77	3.91	5.91	1.15	2.72	0.18	1.60	0.05	1.03	0.02	0.73	0.01
275	15.91	12.12	10.75	4.67	6.50	1.37	3.00	0.21	1.77	0.06	1.13	0.02	0.80	0.01
300	17.36	14.24	11.73	5.49	7.09	1.61	3.27	0.25	1.93	0.07	1.24	0.02	0.88	0.01
325	18.81	16.51	12.70	6.36	7.68	1.87	3.54	0.29	2.09	0.08	1.34	0.03	0.95	0.01
350			13.68	7.30	8.27	2.15	3.81	0.33	2.25	0.09	1.44	0.03	1.03	0.01
375			14.66	8.29	8.86	2.44	4.09	0.37	2.41	0.10	1.55	0.04	1.10	0.02
400			15.64	9.35	9.45	2.75	4.36	0.42	2.57	0.12	1.65	0.04	1.17	0.02
425			16.62	10.46	10.04	3.07	4.63	0.47	2.73	0.13	1.76	0.04	1.25	0.02
450			17.59	11.62	10.63	3.42	4.90	0.52	2.89	0.14	1.86	0.05	1.32	0.02
475			18.57	12.85	11.23	3.78	5.18	0.58	3.05	0.16	1.96	0.05	1.39	0.02
500			19.55	14.13	11.82	4.15	5.45	0.63	3.21	0.18	2.07	0.06	1.47	0.03
550					13.00	4.96	6.00	0.76	3.54	0.21	2.27	0.07	1.61	0.03
600					14.18	5.82	6.54	0.89	3.86	0.25	2.48	0.08	1.76	0.04
650					15.36	6.75	7.09	1.03	4.18	0.29	2.69	0.10	1.91	0.04
700					16.55	7.75	7.63	1.18	4.50	0.33	2.89	0.11	2.06	0.05
750					17.73	8.80	8.18	1.34	4.82	0.37	3.10	0.13	2.20	0.05
800					18.91	9.92	8.72	1.51	5.15	0.42	3.31	0.14	2.35	0.06
850							9.27	1.69	5.47	0.47	3.52	0.16	2.50	0.07
900							9.81	1.88	5.79	0.52	3.72	0.18	2.65	0.08
950							10.36	2.08	6.11	0.58	3.93	0.20	2.79	0.09
1000							10.91	2.29	6.43	0.63	4.14	0.22	2.94	0.09
1050							11.45	2.50	6.75	0.69	4.34	0.24	3.09	0.10
1100							12.00	2.73	7.08	0.76	4.55	0.26	3.23	0.11
1150							12.54	2.96	7.40	0.82	4.76	0.28	3.38	0.12
1200							13.09	3.20	7.72	0.89	4.97	0.30	3.53	0.13
1250							13.63	3.45	8.04	0.96	5.17	0.33	3.68	0.14
1300							14.18	3.72	8.36	1.03	5.38	0.35	3.82	0.15
1350							14.72	3.98	7.69	1.10	5.59	0.38	3.97	0.16
1400							15.27	4.26	9.01	1.18	5.79	0.40	4.12	0.18
1500							16.36	4.84	9.65	1.34	6.21	0.46	4.41	0.20
1600							17.45	5.46	10.30	1.51	6.62	0.52	4.71	0.23
1700							18.54	6.11	10.97	1.69	7.04	0.58	5.00	0.25
1800							19.63	6.79	11.58	1.88	7.45	0.64	5.30	0.28
1900									12.23	2.08	7.87	0.71	5.59	0.31
2000									12.87	2.29	8.28	0.78	5.89	0.34
2300									14.80	2.96	9.52	1.01	6.77	0.44
2600									16.73	3.72	10.77	1.27	7.65	0.55
2900									18.66	4.55	12.01	1.56	8.54	0.68

Table A10-5. Friction loss characteristics for PVC Class 200 pipe (1120, 1220), SDR 26, C = 150 (pressure loss in psi per 100 ft of pipe; shaded area represents velocities over 5 fps that should be avoided where water hammer is a concern).

Nominal size	1"		1.25"		1.5"		2"		2.5"		3"		4"	
Pipe ID (in.)	1.185		1.502		1.720		2.145		2.601		3.166		4.072	
Pipe OD (in.)	1.315		1.660		1.900		2.375		2.875		3.500		4.500	
Wall width (in.)	0.063		0.079		0.090		0.113		0.137		0.167		0.214	
Flow (gpm)	V (fps)	Loss (psi)	V (fps)	Loss (psi)	V (fps)	Loss (psi)	V (fps)	Loss (psi)	V (fps)	Loss (psi)	V (fps)	Loss (psi)	V (fps)	Loss (psi)
5	1.44	0.36	0.90	0.12	0.69	0.06	0.44	0.02	0.30	0.01				
6	1.73	0.51	1.09	0.16	0.83	0.08	0.53	0.03	0.36	0.01	0.24	0.00		
7	2.02	0.67	1.27	0.22	0.97	0.11	0.62	0.04	0.42	0.01	0.28	0.01		
8	2.31	0.86	1.45	0.28	1.10	0.14	0.71	0.05	0.48	0.02	0.33	0.01		
9	2.60	1.07	1.63	0.34	1.24	0.18	0.80	0.06	0.54	0.02	0.37	0.01		
10	2.89	1.30	1.81	0.42	1.38	0.22	0.88	0.07	0.60	0.03	0.41	0.01		
11	3.46	1.83	2.17	0.59	1.65	0.30	1.06	0.10	0.72	0.04	0.49	0.02	0.30	0.00
12	4.04	2.43	2.53	0.78	1.93	0.40	1.24	0.14	0.84	0.05	0.57	0.02	0.34	0.01
14	4.62	3.11	2.89	1.00	2.21	0.52	1.41	0.17	0.96	0.07	0.65	0.03	0.39	0.01
16	5.19	3.87	3.26	1.24	2.48	0.64	1.59	0.22	1.09	0.09	0.73	0.03	0.44	0.01
18	5.77	4.71	3.62	1.51	2.76	0.78	1.77	0.26	1.21	0.10	0.81	0.04	0.49	0.01
20	6.35	5.62	3.98	1.80	3.03	0.93	1.94	0.32	1.33	0.12	0.90	0.05	0.54	0.01
22	6.93	6.60	4.34	2.12	3.31	1.09	2.12	0.37	1.45	0.15	0.98	0.06	0.59	0.02
24	7.50	7.65	4.70	2.45	3.59	1.27	2.30	0.43	1.57	0.17	1.06	0.07	0.64	0.02
26	8.08	8.78	5.06	2.82	3.86	1.46	2.47	0.49	1.69	0.19	1.14	0.07	0.69	0.02
28	8.66	9.97	5.43	3.20	4.14	1.65	2.65	0.56	1.81	0.22	1.22	0.08	0.74	0.02
30	9.24	11.24	5.79	3.61	4.41	1.86	3.09	0.74	2.11	0.29	1.42	0.11	0.86	0.03
35	10.10	13.27	6.33	4.26	4.83	2.20	3.53	0.95	2.41	0.38	1.63	0.14	0.98	0.04
40	11.54	16.99	7.23	5.45	5.52	2.82	3.98	1.19	2.71	0.47	1.80	0.18	1.11	0.05
45	12.99	21.13	8.14	6.78	6.21	3.51	4.42	1.44	3.02	0.57	2.04	0.22	1.23	0.06
50	14.43	25.69	9.04	8.24	6.90	4.26	4.86	1.72	3.32	0.68	2.24	0.26	1.35	0.08
55	15.87	30.65	9.95	9.83	7.59	5.08	5.30	2.02	3.62	0.80	2.44	0.31	1.48	0.09
60	17.32	36.00	10.85	11.55	8.27	5.97	5.74	2.34	3.92	0.93	2.65	0.36	1.60	0.10
65	18.76	41.76	11.76	13.39	8.96	6.93	6.18	2.69	4.22	1.06	2.85	0.41	1.72	0.12
70			12.66	15.36	9.65	7.95	6.63	3.05	4.52	1.21	3.05	0.46	1.85	0.14
75			13.56	17.46	10.34	9.03	7.07	3.44	4.82	1.36	3.26	0.52	1.97	0.15
80			14.47	19.67	10.38	10.17	7.51	3.85	5.13	1.52	3.46	0.58	2..09	0.17
90			15.37	22.01	11.72	11.38	7.95	4.28	5.43	1.69	3.66	0.65	2.21	0.19
95			16.28	24.47	12.41	12.65	8.39	4.73	5.73	1.87	3.87	0.72	2.34	0.21
100			17.18	27.05	13.10	13.99	8.83	5.20	6.03	2.06	4.07	0.79	2.46	0.23
110			18.09	29.74	13.79	15.38	9.72	6.21	6.63	2.45	4.48	0.94	2.71	0.28
120			19.89	35.48	15.17	18.35	10.60	7.30	7.24	2.88	4.88	1.11	2.95	0.33
130					16.55	21.56	11.48	8.46	7.84	3.34	5.29	1.28	3.20	0.38
140					17.93	25.00	12.37	9.71	8.44	3.83	5.70	1.47	3.44	0.43
150					19.31	28.68	13.25	11.03	9.05	4.36	6.11	1.67	3.69	0.49
175							15.46	14.67	10.55	5.80	7.12	2.23	4.31	0.65
200							17.67	18.79	12.06	7.42	8.14	2.85	4.92	0.84

Table A10-5. Friction loss characteristics for PVC Class 200 pipe (1120, 1220), SDR 26, C = 150 (pressure loss in psi per 100 ft of pipe; shaded area represents velocities over 5 fps that should be avoided where water hammer is a concern).

Nominal size	2.5"		3"		4"		6"		8"		10"		12"	
Pipe ID (in.)	2.601		3.166		4.072		5.993		7.805		9.728		11.538	
Pipe OD (in.)	2.875		3.500		4.500		6.625		8.625		10.750		12.750	
Wall width (in.)	0.137		0.167		0.214		0.316		0.410		0.511		0.606	
Flow (gpm)	V (fps)	Loss (psi)	V (fps)	Loss (psi)	V (fps)	Loss (psi)	V (fps)	Loss (psi)	V (fps)	Loss (psi)	V (fps)	Loss (psi)	V (fps)	Loss (psi)
200	12.06	7.42	8.14	2.85	4.92	0.84	2.27	0.13	1.34	0.04	0.86	0.01	0.61	0.01
225	13.57	9.23	9.16	3.55	5.54	1.04	2.56	0.16	1.51	0.04	0.97	0.02	0.69	0.01
250	15.08	11.22	10.18	4.31	6.15	1.27	2.84	0.19	1.67	0.05	1.08	0.02	0.77	0.01
275	16.58	13.39	11.19	5.14	6.77	1.51	3.12	0.23	1.84	0.06	1.19	0.02	0.84	0.01
300	18.09	15.73	12.21	6.04	7.38	1.78	3.41	0.27	2.01	0.07	1.29	0.03	0.92	0.01
325			13.23	7.04	7.99	2.07	3.69	0.31	2.18	0.09	1.40	0.03	1.00	0.01
350			14.25	8.04	8.61	2.36	3.98	0.36	2.34	0.10	1.51	0.03	1.07	0.01
375			15.26	9.21	9.22	2.69	4.26	0.41	2.51	0.11	1.62	0.04	1.15	0.02
400			16.28	10.29	9.84	3.03	4.54	0.46	2.68	0.13	1.72	0.04	1.23	0.02
425			17.30	11.54	10.25	3.39	4.83	0.52	2.85	0.14	1.83	0.05	1.30	0.02
450			18.32	12.80	11.07	3.76	5.11	0.57	3.01	0.16	1.94	0.05	1.38	0.02
475					11.68	4.16	5.40	0.63	3.18	0.18	2.05	0.06	1.46	0.03
500					12.30	4.57	5.68	0.70	3.35	0.19	2.16	0.07	1.53	0.03
550					13.53	5.46	6.25	0.83	3.68	0.23	2.37	0.08	1.69	0.03
600					14.76	6.41	6.82	0.98	4.02	0.27	2.59	0.09	1.84	0.04
650					15.98	7.47	7.38	1.13	4.35	0.31	2.80	0.11	1.99	0.05
700					17.22	8.53	7.95	1.30	4.69	0.36	3.02	0.12	2.15	0.05
750					18.45	8.72	8.52	1.48	5.02	0.41	3.23	0.14	2.30	0.06
800					19.68	10.92	9.09	1.67	5.36	0.46	3.45	0.16	2.45	0.07
850							9.66	1.86	5.69	0.52	3.66	0.18	2.61	0.08
900							10.22	2.07	6.03	0.57	3.88	0.20	2.76	0.09
950							10.79	2.29	6.36	0.63	4.10	0.22	2.91	0.09
1000							11.36	2.52	6.70	0.70	4.31	0.24	3.06	0.10
1100							12.50	3.00	7.37	0.83	4.74	0.28	3.37	0.12
1200							13.63	3.53	8.04	0.98	5.17	0.33	3.68	0.15
1300							14.77	4.09	7.71	1.13	5.60	0.39	3.98	0.17
1400							15.90	4.70	9.38	1.30	6.04	0.44	4.29	0.19
1500							17.04	5.34	10.05	1.48	6.47	0.51	4.60	0.22
1600							18.18	6.01	10.72	1.66	6.90	0.57	4.90	0.25
1700							19.31	6.73	11.39	1.86	7.33	0.64	5.21	0.28
1800									12.06	2.07	7.76	0.71	5.52	0.31
1900									12.73	2.29	8.19	0.78	5.82	0.34
2000									13.40	2.51	8.62	0.86	6.13	0.38
2100									14.06	2.75	9.05	0.94	6.44	0.41
2200									14.73	3.00	9.48	1.03	6.74	0.45
2300									14.40	3.26	9.92	1.11	7.05	0.49
2400									16.07	3.52	10.35	1.21	7.36	0.53
2500									16.74	3.80	10.78	1.30	7.66	0.57
2600									17.41	4.09	11.21	1.40	7.97	0.61
2700									18.08	4.38	11.64	1.50	8.27	0.65
2800									18.75	4.69	12.07	1.60	8.58	0.70
2900									19.42	5.00	12.50	1.71	8.89	0.75

Table A10-6. Friction loss characteristics for PVC Schedule 40 pipe (1120, 1220), SDR 26, C = 150 (pressure loss in psi per 100 ft of pipe; shaded area represents velocities over 5 fps that should be avoided where water hammer is a concern).

Nominal size	1"		1.25"		1.5"		2"		2.5"		3"		4"	
Pipe ID (in.)	1.049		1.380		1.610		2.067		2.469		3.068		4.026	
Pipe OD (in.)	1.315		1.660		1.900		2.375		2.875		3.500		4.500	
Wall width (in.)	0.133		0.140		0.145		0.154		0.203		0.216		0.237	
Flow (gpm)	V (fps)	Loss (psi)	V (fps)	Loss (psi)	V (fps)	Loss (psi)	V (fps)	Loss (psi)	V (fps)	Loss (psi)	V (fps)	Loss (psi)	V (fps)	Loss (psi)
1	0.37	0.03	0.21	0.01	0.16	0.00								
2	0.74	0.12	0.43	0.03	0.31	0.02	0.19	0.00						
3	1.11	0.26	0.64	0.07	0.47	0.03	0.29	0.01	0.20	0.00				
4	1.48	0.44	0.86	0.12	0.63	0.05	0.38	0.02	0.27	0.01				
5	1.85	0.66	1.07	0.17	0.79	0.08	0.48	0.02	0.33	0.01	0.22	0.00		
6	2.22	0.93	1.29	0.25	0.94	0.12	0.57	0.03	0.40	0.01	0.26	0.01		
7	2.60	1.24	1.50	0.33	1.10	0.15	0.67	0.05	0.47	0.02	0.30	0.01		
8	2.97	1.59	1.71	0.42	1.26	0.20	0.76	0.06	0.54	0.02	0.35	0.01		
9	3.34	1.97	1.93	0.52	1.42	0.25	0.86	0.07	0.60	0.03	0.39	0.01		
10	3.71	2.40	2.14	0.63	1.57	0.30	0.95	0.09	0.67	0.04	0.43	0.01		
11	4.08	2.86	2.36	0.75	1.73	0.36	1.05	0.11	0.74	0.04	0.48	0.02		
12	4.45	3.36	2.57	0.89	1.89	0.42	1.15	0.12	0.80	0.05	0.52	0.02		
14	5.19	4.47	3.00	1.18	2.20	0.56	1.34	0.16	0.94	0.07	0.61	0.02		
16	5.93	5.73	3.43	1.51	2.52	0.71	1.53	0.21	1.07	0.09	0.69	0.03		
18	6.67	7.12	3.86	1.88	2.83	0.89	1.72	0.26	1.20	0.11	0.78	0.04		
20	7.42	8.66	4.28	2.28	3.15	1.08	1.91	0.32	1.34	0.13	0.87	0.05	0.50	0.01
22	8.16	10.33	4.71	2.72	3.46	1.28	2.10	0.38	1.47	0.16	0.95	0.06	0.55	0.01
24	8.90	12.14	5.14	3.20	3.78	1.51	2.29	0.45	1.61	0.19	1.04	0.07	0.60	0.02
26	9.64	14.08	5.57	3.71	4.09	1.75	2.48	0.52	1.74	0.22	1.13	0.08	0.65	0.02
28	10.38	16.15	6.00	4.25	4.41	2.01	2.67	0.60	1.87	0.25	1.21	0.09	0.70	0.02
30	11.12	18.35	6.43	4.83	4.72	2.28	2.86	0.68	2.01	0.28	1.30	0.10	0.76	0.03
35	12.98	24.41	7.50	6.43	5.51	3.04	3.34	0.90	2.34	0.38	1.52	0.13	0.88	0.04
40	14.83	31.26	8.57	8.23	6.30	3.89	3.82	1.15	2.68	0.49	1.73	0.17	1.01	0.04
45	16.68	38.88	9.64	10.24	7.08	4.83	4.30	1.43	3.01	0.60	1.95	0.21	1.13	0.06
50	18.54	47.26	10.71	12.44	7.87	5.88	4.77	1.74	3.35	0.73	2.17	0.25	1.26	0.07
55			11.78	14.84	8.66	7.01	5.25	2.08	3.68	0.88	2.38	0.30	1.38	0.08
60			12.85	17.44	9.44	8.24	5.73	2.44	4.02	1.03	2.60	0.36	1.51	0.10
65			13.93	20.23	10.23	9.55	6.21	2.83	4.35	1.19	2.82	0.41	1.64	0.11
70			15.00	23.20	11.02	10.96	6.68	3.25	4.69	1.37	3.03	0.48	1.76	0.13
75			16.07	26.36	11.81	12.45	7.16	3.69	5.02	1.55	3.25	0.54	1.89	0.14
80			17.14	29.71	12.59	14.03	7.64	4.16	5.35	1.75	3.47	0.61	2.01	0.16
85			18.21	33.24	13.38	15.70	8.12	4.65	5.69	1.96	3.68	0.68	2.14	0.18
90			19.28	36.95	14.17	17.45	8.59	5.17	6.02	2.18	3.90	0.76	2.27	0.20
95					14.95	19.29	9.07	5.72	6.36	2.41	4.12	0.84	2.39	0.22
100					15.74	21.21	9.55	6.29	6.69	2.65	4.33	0.92	2.52	0.25
110					17.31	25.31	10.50	7.50	7.36	3.16	4.77	1.10	2.77	0.29
120					18.89	29.74	11.46	8.82	8.03	3.71	5.20	1.29	3.02	0.34
130							12.41	10.22	8.70	4.31	5.63	1.50	3.27	0.40
140							13.37	11.73	9.37	4.94	6.07	1.72	3.52	0.46
150							14.32	13.33	10.04	5.61	6.50	1.95	3.77	0.52
160							15.28	15.02	10.71	6.33	6.94	2.20	4.03	0.59
170							16.23	16.80	11.38	7.08	7.37	2.46	4.28	0.67
180							17.19	18.68	12.05	7.87	7.80	2.73	4.53	0.73
190							18.14	20.65	12.72	8.71	8.24	3.02	4.78	0.81
200							19.10	22.71	13.39	9.56	8.67	3.32	5.03	0.89

Table A10-6. Friction loss characteristics for PVC Schedule 40 pipe (1120, 1220), SDR 26, C = 150 (pressure loss in psi per 100 ft of pipe; shaded area represents velocities over 5 fps that should be avoided where water hammer is a concern).

Nominal size	2.5"		3"		4"		6"		8"		10"		12"	
Pipe ID (in.)	2.469		3.068		4.026		6.065		7.981		10.020		11.814	
Pipe OD (in.)	2.875		3.500		4.500		6.625		8.625		10.750		12.750	
Wall width (in.)	0.203		0.216		0.237		0.280		0.322		0.365		0.406	
Flow (gpm)	V (fps)	Loss (psi)	V (fps)	Loss (psi)	V (fps)	Loss (psi)	V (fps)	Loss (psi)	V (fps)	Loss (psi)	V (fps)	Loss (psi)	V (fps)	Loss (psi)
100	6.69	2.65	4.33	0.92	2.52	0.25	1.11	0.03	0.64	0.01				
120	8.03	3.71	5.20	1.29	3.02	0.34	1.33	0.05	0.77	0.01	0.49	0.00		
140	9.37	4.94	6.07	1.72	3.52	0.46	1.55	0.06	0.90	0.02	0.57	0.01		
160	10.73	6.33	6.74	2.20	4.03	0.59	1.77	0.08	1.02	0.02	0.65	0.01		
180	12.05	7.87	7.80	2.73	4.53	0.73	2.00	0.10	1.15	0.03	0.73	0.01		
200	13.39	9.56	8.67	3.32	5.03	0.89	2.22	0.12	1.28	0.04	0.81	0.01	0.58	0.00
250	16.73	14.46	10.84	5.02	6.29	1.34	2.77	0.18	1.60	0.05	1.02	0.02	0.73	0.01
300			13.00	7.04	7.55	1.88	3.33	0.26	1.92	0.06	1.22	0.02	0.88	0.01
350			15.17	9.37	8.81	2.50	3.88	0.34	2.24	0.08	1.42	0.03	1.02	0.01
400			17.34	12.00	10.07	3.20	4.44	0.44	2.56	0.11	1.63	0.04	1.17	0.02
450					11.33	3.98	4.99	0.54	2.88	0.14	1.83	0.05	1.32	0.02
500					12.59	4.83	5.55	0.66	3.20	0.17	2.03	0.06	1.46	0.03
600					15.10	6.77	6.66	0.92	3.84	0.24	2.44	0.08	1.75	0.04
700					17.62	9.01	7.76	1.23	4.48	0.32	2.84	0.11	2.05	0.05
800							8.87	1.57	5.12	0.41	3.25	0.14	2.34	0.06
900							9.98	1.95	5.76	0.51	3.66	0.17	2.63	0.08
1000							11.09	2.38	6.41	0.62	4.06	0.21	2.92	0.09
1200							13.31	3.33	7.69	0.88	4.88	0.29	3.51	0.13
1400							15.53	4.43	8.97	1.16	5.69	0.39	4.09	0.17
1600							17.75	5.67	10.25	1.49	6.50	0.49	4.68	0.22
1800									11.53	1.86	7.31	0.61	5.26	0.28
2000									12.81	2.26	8.13	0.75	5.85	0.33
2200									14.09	2.69	8.94	0.89	6.43	0.40
2400									15.37	3.16	9.75	1.04	7.02	0.47
2600									16.65	3.67	10.57	1.21	7.60	0.54
2800									17.94	4.21	11.38	1.39	8.19	0.62
3000											12.19	1.58	8.77	0.71
3200											13.00	1.78	9.35	0.80
3400											13.82	1.99	9.94	0.89
3600											14.63	2.21	10.52	0.99
3800											15.44	2.45	11.11	1.10
4000											16.25	2.69	11.69	1.21
4500											18.29	3.35	13.15	1.50
5000													14.62	1.83
5500													16.08	2.18
6000													17.54	2.56
6500													19.00	2.97

APPENDIX 11 - Wet Bulb, Dry Bulb, and Dew Point Temperature Relationships

When wet bulb and dry bulb temperatures are known, Figures A11-1 and A11-2 can be used. (Fig. A11-1 for $T_{wb} < 40°$ F and A11-2 for $T_{wb} = 40$-$60°$ F). Locate T_{db} on the bottom axis and go up to the curve with the measured T_{wb}. The dew point temperature (T_{dp}) is read from the vertical axis value for the intersection of the T_{db} and T_{wb} lines.

If T_{db} and the relative humidity (RH) are known, T_{dp} can be determined from Fig. A11-3. Locate T_{db} on the bottom axis and go up to the line that represents the measured relative humidity. T_{dp} is read from the vertical axis value for the intersection of the T_{db} and RH lines.

Example:
For the following temperatures, determine the dew point temperature.

$T_{db} = 60°$ F
$T_{wb} = 46°$ F

From Fig. A11-2, locate $60°$ F on the bottom axis and go up to the $T_{wb} = 46°$ F curve. The value of T_{dp} can be read from the vertical axis as $30°$ F.

Example:
For the following conditions, determine the dew point temperature.

$T_{db} = 50°$ F
RH = 70%

From Fig. A11-3, locate $50°$ F on the bottom axis and go up to the RH = 70% line. The value of T_{dp} can be read from the vertical axis as $37°$ F.

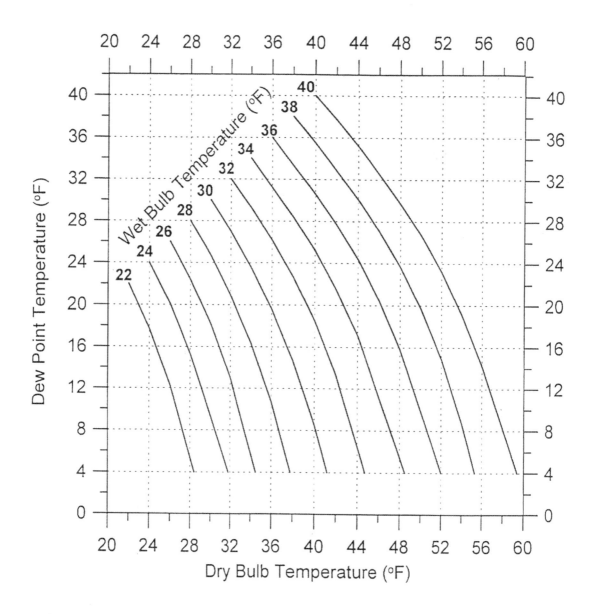

Figure A11-1. Wet bulb, dry bulb, and dew point temperature relationships for wet bulb temperatures less than 40°F.

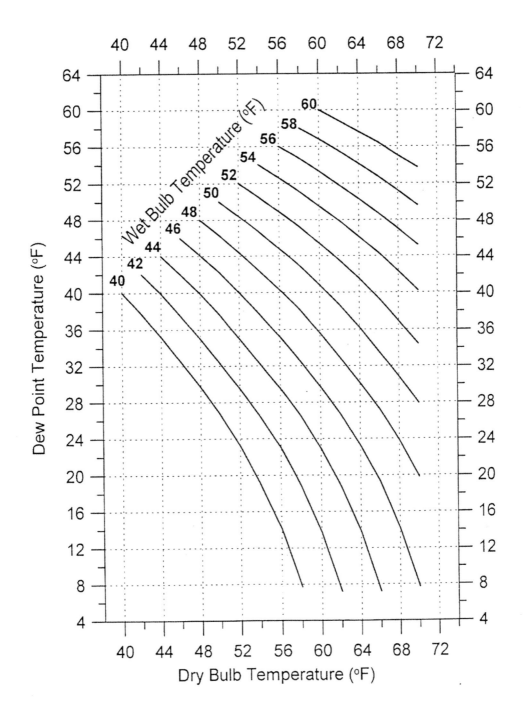

Figure A11-2. Wet bulb, dry bulb, and dew point temperature relationships for wet bulb temperatures from 40° to 60° F.

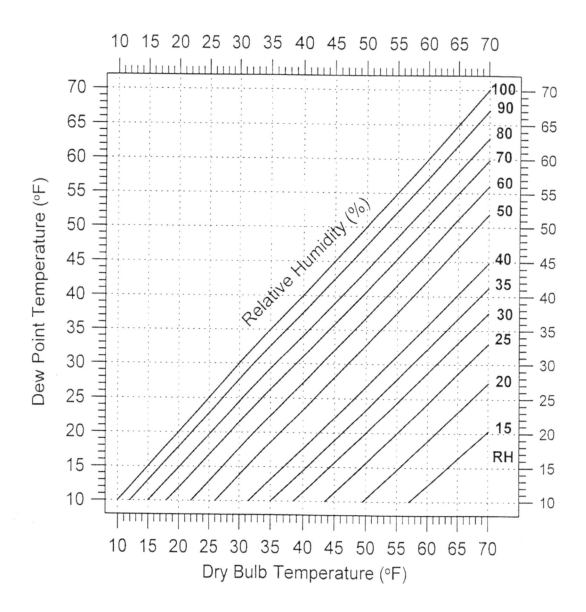

Figure A11-3. Relationships between dry bulb temperature, relative humidity, and dew point temperature.

APPENDIX 12 - EPA Maximum Contaminant Level Goals (MCLG) and Maximum Contaminant Levels (MCL) for Microorganisms, Organic, and Inorganic Constituents

Table A12-1. EPA maximum contaminant level goals (MCLG) and maximum contaminant levels (MCL) for microorganisms.

Microorganisms	MCLG	MCL	Potential Health Effects from Ingestion of Water	Sources of Contaminant in Drinking Water
Giardia lamblia	zero	TT[1]	Giardiasis, a gastroenteric disease	Human and animal fecal waste
Heterotrophic plate count	N/A	TT	Has no health effects, but can indicate how effective treatment is at controlling microorganisms	N/A
Legionella	zero	TT	Legionnaire's Disease, commonly known as pneumonia	Found naturally in water; multiplies in heating systems
Total Coliforms (including fecal coliform and E. Coli)	zero	5.00%	Used as an indicator that other potentially harmful bacteria may be present	Human and animal fecal waste
Turbidity	N/A	TT	Has no health effects but can interfere with disinfection and provide a medium for microbial growth; may indicate the presence of microbes	Soil runoff
Viruses (enteric)	zero	TT	Gastroenteric disease	Human and animal fecal waste

---

[1] When there is no reliable method that is economically and technically feasible to measure a contaminant at particularly low concentrations, a Treatment Technique (TT) is set rather than a MCL. A TT is an enforceable procedure or level of technology that a public water system must follow to ensure control of a contaminant.

Table A12-2. EPA maximum contaminant level goal (MCLG) and maximum contaminant level (MCL) for inorganic chemicals.

Inorganic Chemicals	MCLG	MCL	Potential Health Effects from Ingestion of Water	Sources of Contaminant in Drinking Water
Antimony	0.006	0.006	Increase in blood cholesterol; decrease in blood glucose	Discharge from petroleum refineries; fire retardants; ceramics; electronics; solder
Arsenic	none	0.05	Skin damage; circulatory system problems; increased risk of cancer	Discharge from semiconductor manufacturing; petroleum refining; wood preservatives; animal feed additives; herbicides; erosion of natural deposits
Asbestos	7 million fibers per Liter	7 MFL	Increased risk of developing benign intestinal polyps	Decay of asbestos cement in water mains; erosion of natural deposits
Barium	2	2	Increase in blood pressure	Discharge of drilling wastes; discharge from metal refineries; erosion of natural deposits
Beryllium	0.004	0.004	Intestinal damage	Discharge from metal refineries and coal-burning factories; discharge from electrical, aerospace, and defense industries

Table A12-2. EPA maximum contaminant level goal (MCLG) and maximum contaminant level (MCL) for inorganic chemicals.

Inorganic Chemicals	MCLG	MCL	Potential Health Effects from Ingestion of Water	Sources of Contaminant in Drinking Water
Cadmium	0.005	0.005	Kidney damage	Corrosion of galvanized pipes; erosion of natural deposits; discharge from metal refineries; runoff from waste batteries and paints
Chromium	0.1	0.1	Some people who use water containing chromium well in excess of the MCL over many years could experience allergic dermatitis.	Discharge from steel and pulp mills; erosion of natural deposits
Copper	1.3	Action Level = 1.3; TT	Short-term exposure: gastrointestinal distress. Long-term exposure: liver or kidney damage. Those with Wilson's Disease should consult their personal doctor if their water systems exceed the copper action level.	Corrosion of household plumbing systems; erosion of natural deposits; leaching from wood preservatives
Cyanide	0.2	0.2	Nerve damage or thyroid problems	Discharge from steel/metal factories; discharge from plastic and fertilizer factories
Flouride	4	4	Bone disease (pain and tenderness of the bones). Children may get mottled teeth.	Water additive that promotes strong teeth; erosion of natural deposits; discharge from fertilizer and aluminum factories
Lead Inorganic	zero	Action Level = 0.015; TT	Infants and children: delays in physical or mental development. Adults: kidney problems; high blood pressure	Corrosion of household plumbing systems; erosion of natural deposits

Table A12-2. EPA maximum contaminant level goal (MCLG) and maximum contaminant level (MCL) for inorganic chemicals.

Inorganic Chemicals	MCLG	MCL	Potential Health Effects from Ingestion of Water	Sources of Contaminant in Drinking Water
Inorganic Mercury	0.002	0.002	Kidney damage	Erosion of natural deposits; discharge from refineries and factories; runoff from landfills and cropland
Nitrate (measured as Nitrogen)	10	10	"Blue baby syndrome" in infants under six months - life threatening without immediate medical attention. Symptoms: infant looks blue and has shortness of breath	Runoff from fertilizer use; leaching from septic tanks, sewage; erosion of natural deposits
Nitrite (measured as Nitrogen)	1	1	"Blue baby syndrome" in infants under six months - life threatening without immediate medical attention. Symptoms: infant looks blue and has shortness of breath	Runoff from fertilizer use; leaching from septic tanks, sewage; erosion of natural deposits
Selenium	0.05	0.05	Hair or fingernail loss; numbness in fingers or toes; circulatory problems	Discharge from petroleum refineries; erosion of natural deposits; discharge from mines

Table A12-3. EPA maximum contaminant level goal (MCLG) and maximum contaminant level (MCL) for organic chemicals.

Organic Chemicals	MCLG	MCL	Potential Health Effects from Ingestion of Water	Sources of Contaminant in Drinking Water
Alachlor	zero	0.002	Eye, liver, kidney, or spleen problems; anemia; increased risk of cancer	Runoff from herbicide used on row crops
Atrazine	0.003	0.003	Cardiovascular system problems; reproductive difficulties	Runoff from herbicide used on row crops
Carbofuran	0.04	0.04	Problems with blood or nervous system; reproductive difficulties	Leaching of soil fumigant
Chlordane	zero	0.002	Liver or nervous system problems; increased risk of cancer	Residue of banned termiticide
Chlorobenzene	0.1	0.1	Liver or kidney problems	Discharges from chemical and agricultural chemical factories
2,4-D	0.07	0.07	Kidney, liver, or adrenal gland problems	Runoff from herbicide
Dalapon	0.2	0.2	Minor kidney changes	Runoff from herbicide used on rights of way
1,2-Dibromo-3-chloropropane (DBCP)	zero	0.0002	Reproductive difficulties; increased risk of cancer	Runoff/leaching from soil fumigant
Dinoseb	0.007	0.007	Reproductive difficulties	Leaching from PVC plumbing systems
Diquat	0.02	0.02	Cataracts	Runoff from herbicide used on soybeans and vegetables
Endothall	0.1	0.1	Stomach and intestinal problems	Emissions from waste incineration and other combustion

Table A12-3. EPA maximum contaminant level goal (MCLG) and maximum contaminant level (MCL) for organic chemicals.

Organic Chemicals	MCLG	MCL	Potential Health Effects from Ingestion of Water	Sources of Contaminant in Drinking Water
Endrin	0.002	0.002	Nervous system effects	Runoff from herbicide use
Epichlorohydrin	zero	TT	Stomach problems; reproductive difficulties; increased cancer risk	Runoff from herbicide use
Ethylbenzene	0.7	0.07	Liver or kidney problems	Residue of banned insecticide
Glyphosate	0.7	0.07	Kidney problems; reproductive difficulties	Runoff from herbicide use
Heptachlor	zero	0.0004	Liver damage; increased risk of cancer	Discharge from petroleum refineries
Heptachlor epoxide	zero	0.0002	Liver damage; increased risk of cancer	Runoff from herbicide use
Hexachloro benzene	zero	0.001	Liver or kidney problems; reproductive difficulties; increased risk of cancer	Residue of banned termiticide
Hexachlorocyclo-pentadiene	0.05	0.05	Kidney or stomach problems	Breakdown of hepatachlor
Lindane	0.0002	0.0002	Liver or kidney problems	Runoff/leaching from insecticide
Simazine	0.004	0.004	Problems with blood	Herbicide runoff
Total Trihalomethanes (TTHM's)	none	0.01	Liver, kidney, or central nervous system problems; increased risk of cancer	Byproduct of drinking water disinfection
2,4,5-TP (Silvex)	0.05	0.05	Liver problems	Residue of banned herbicide
Vinyl chloride	zero	0.002	Increased risk of cancer	Leaching from PVC pipes

APPENDIX 13 - Typical IPS PVC Pipe Dimensions

<table>
<tr><td rowspan="2">Class 63 psi
SDR 64</td><td>Nominal size
(inches)</td><td>OD
(inches)</td><td>ID
(inches)</td><td>Weight
(lb/ft)</td></tr>
<tr><td>

4
6
8
10
12

</td><td>

4.500
6.625
8.625
10.750
12.750

</td><td>

4.360
6.417
8.355
10.414
12.352

</td><td>

0.655
1.394
2.357
3.655
5.134

</td></tr>
</table>

<table>
<tr><td rowspan="2">Class 100 psi
SDR 41</td><td>Nominal size
(inches)</td><td>OD
(inches)</td><td>ID
(inches)</td><td>Weight
(lb/ft)</td></tr>
<tr><td>

4
5
6
8
10
12

</td><td>

4.500
5.563
6.625
8.625
10.750
12.750

</td><td>

4.280
5.291
6.301
8.205
10.226
12.128

</td><td>

0.993
1.518
2.153
3.633
5.650
7.954

</td></tr>
</table>

<table>
<tr><td rowspan="2">Class 125 psi
SDR 32.5</td><td>Nominal size
(inches)</td><td>OD
(inches)</td><td>ID
(inches)</td><td>Weight
(lb/ft)</td></tr>
<tr><td>

1.25
1.5
2
2.5
3
4
5
6
8
10
12

</td><td>

1.660
1.900
2.375
2.875
3.500
4.500
5.563
6.625
8.625
10.750
12.750

</td><td>

1.548
1.784
2.229
2.699
3.284
4.224
5.221
6.217
8.095
10.088
11.966

</td><td>

0.186
0.221
0.345
0.501
0.753
1.238
1.896
2.693
4.555
7.091
9.96

</td></tr>
</table>

	Nominal size (inches)	OD (inches)	ID (inches)	Weight (lb/ft)
Class 160 psi SDR 26	1.25	1.660	1.532	0.211
	1.5	1.900	1.754	0.276
	2	2.375	2.193	0.429
	2.5	2.875	2.655	0.628
	3	3.500	3.230	0.934
	4	4.500	4.154	1.539
	5	5.563	5.135	2.354
	6	6.625	6.115	3.340
	8	8.625	7.961	5.661
	10	10.750	9.924	8.778
	12	12.750	11.770	12.352

	Nominal size (inches)	OD (inches)	ID (inches)	Weight (lb/ft)
Class 200 psi SDR 21	0.75	1.050	0.930	0.123
	1	1.315	1.189	0.163
	1.25	1.660	1.502	0.258
	1.5	1.900	1.720	0.336
	2	2.375	2.149	0.528
	2.5	2.875	2.601	0.774
	3	3.500	3.166	1.144
	4	4.500	4.072	1.886
	5	5.563	5.033	2.887
	6	6.625	5.993	4.099
	8	8.625	7.803	6.925
	10	10.750	9.726	10.646

	Nominal size (inches)	OD (inches)	ID (inches)	Weight (lb/ft)	Pressure rating (psi)
Schedule 40	0.5	0.840	0.622	0.165	600
	0.75	1.050	0.824	0.219	480
	1	1.315	1.049	0.325	450
	1.25	1.660	1.380	0.434	370
	1.5	1.900	1.610	0.518	330
	2	2.375	2.067	0.696	280
	2.5	2.875	2.469	1.100	300
	3	3.500	3.068	1.458	260
	4	4.500	4.026	2.077	220
	6	6.625	6.065	3.580	180
	8	8.625	8.095	5.55	160
	10	10.75	10.02	7.65	140
	12	12.75	11.94	10.10	130

	Nominal size (inches)	OD (inches)	ID (inches)	Weight (lb/ft)	Pressure rating (psi)
Schedule 80	0.5	0.840	0.546	0.203	850
	0.75	1.050	0.742	0.276	690
	1	1.315	0.957	0.408	630
	1.25	1.660	1.278	0.562	520
	1.51	1.900	1.500	0.682	470
	2	2.375	1.939	0.944	400
	2.5	2.875	2.323	1.440	420
	3	3.500	2.900	1.930	370
	4	4.500	3.826	2.820	320
	6	6.625	5.761	5.390	280
	8	8.26	7.625	8.180	245
	10	10.75	9.564	12.100	230
	12	12.75	11.376	16.700	230

APPENDIX 14 - Agency Addresses

Department of Environmental Protection
3900 Commonwealth Blvd.
Tallahassee, FL 32399-3000
(850) 488-4805

> Division of Administrative and Technical Services
> 3900 Commonwealth Blvd.
> Tallahassee, FL 32399-3000
> (850) 488-2955
>
> Division of Air Resources Management
> 3900 Commonwealth Blvd.
> Tallahassee, FL 32399-3000
> (850) 488-0114
>
> Division of Environmental Resource Permitting
> 2600 Blairstone Road
> Tallahassee, FL 32399-2400
> (850) 488-3177
>
> Division of Waste Management
> 3900 Commonwealth Blvd.
> Tallahassee, FL 32399-3000
> (850) 487-3299
>
> Division of Water Facilities
> 3900 Commonwealth Blvd.
> Tallahassee, FL 32399-3000
> (850) 487-1855

Department of Agriculture & Consumer Services
Mayo Bldg. 2nd Floor
The Capitol
Tallahassee, FL 32399-0800
(850) 488-3022
(800) 435-7352

> Division of Standards
> 3125 Conner Boulevard
> Tallahassee, FL 32399-1650
> (850) 488-0645
>
> Division of Forestry
> 3125 Conner Boulevard
> Tallahassee, FL 32399-1650
> (850) 488-4274

<u>Division of Agriculture and Environmental Services</u>
3125 Conner Boulevard
Tallahassee, FL 32399-1650
(850) 488-3731

Department of Labor and Employment Security
2012 Capitol Circle SE
Hartman Bldg. Suite 303
Tallahassee, FL 32399-2152
(800) 342-9909
(850) 922-7021
    Call for information on Florida's Worker Right-To-Know Program.

EPA National Offices & Numbers
U. S. Environmental Protection Agency
Office of Water (4101)
1200 Pennsylvania Avenue, N. W.
Washington, DC 20460
(202) 382-5700
    Provides information on Clean Water Act and related water pollution regulations.

Fish and Wildlife Conservation Commission
Farris Bryant Building
620 South Meridian Street
101 Bryant Building
Tallahassee, FL 32399-1600
(850) 488-4746

National Pesticide Telecommunications Network
Provides information on pesticides and pesticide poisonings.
Operating 24 hours a day, 365 days a year.
Toll Free, 1-800-858-7378

Florida Administrator of EPA Pesticide Registration
Bureau of Pesticides/ Division of Inspection
Dept. of Agriculture and Consumer Services
3125 Conner Blvd., MD-2
Tallahassee, FL 32399-1650
(850) 487-2130

Water Management District Addresses

<u>Southwest Florida Water Management District</u>
2379 Broad Street
Brooksville, FL 34609-6899
(352) 796-7211

        Inverness Service Office
        2303 Highway 44 West
        Inverness, FL 34453
        (352) 637-1360

        Bartow Service Office
        170 Century Blvd.
        Bartow, FL 33830
        (941) 534-1448 or (800) 492-7862

        Venice Service Office
        115 Corporation Way
        Venice, FL 34292
        (941) 486-1212 or (800) 320-3503

        Tampa Service Office
        7601 U. S. Highway 301 North
        Tampa, FL 33637
        (813) 985-7481 or (800) 836-0797

<u>South Florida Water Management District</u>
3301 Gun Club Road
West Palm Beach, FL 33416-46680
        or
P. O. Box 24680
West Palm Beach, FL 33416
(800) 432-2045
(561) 686-8800

        Orlando Service Center Director
        (407) 858-6100 or (800) 250-4250

        Okeechobee Service Center
        (941) 462-5260 or (800) 250-4200

        Martin/St. Lucie Service Center Director
        (772) 223-2600 or (800) 250-4100

        Fort Myers Service Center
        (941) 338-2929 or (800) 248-1201

Big Cypress Basin/Naples Service Center
(941) 597-1505

Broward Service Center Director
(954) 713-3200

Miami/Dade Service Center
(305) 663-3521 or (800) 250-4300

Florida Keys Service Center
(305) 289-2739 or (800) 464-5067

St. Johns River Water Management District
Highway 100 West
Palatka, FL 32177
        or
P. O. Box 1429
Palatka, FL 32178-1429
(386) 329-4500

Jacksonville Service Center
(904) 730–6270

Melbourne Service Center
(321) 984–4940

Orlando Service Center
(407) 897–4300

Palatka Service Center
(386) 329–4240

University of Florida

<u>Gainesville</u>

Vice President for Agriculture and Natural Resources	352-392-1971
Office of the Dean for Extension	352-392-3583
Office of the Dean for Research	352-392-1784
Agricultural and Biological Engineering Department	352-392-1864
Food Resource and Economics	352-392-1826
Horticultural Sciences	352-392-1928
Plant Pathology	352-392-3631
Soil and Water Science	352-392-1803

<u>Research and Education Centers</u>

Citrus REC - Lake Alfred	941-956-1151
Indian River REC - Fort Pierce	772-468-3922
Southwest Florida REC - Immokalee	941-658-3400
Tropical REC - Homestead	305-246-7001

<u>County Extension Offices</u>

Brevard	321-633-1702	Manatee	941-722-4524
Broward	954-370-3725	Marion	352-620-3440
Charlotte	941-639-6255	Martin	772-288-5654
Citrus	352-726-2141	Okeechobee	863-763-6469
Collier	941-353-4244	Orange	407-836-7570
Dade	305-248-3311	Osceola	407-846-4181
DeSoto	863-993-4846	Palm Beach	772-233-1712
Glades	863-946-0244	Pasco	352-521-4288
Hardee	863-773-2164	Pinellas	727-582-2100
Hendry	863-674-4092	Putnam	904-329-0318
Hernando	352-754-4433	St. Lucie	772-462-1660
Highlands	863-402-6540	Sarasota	941-316-1000
Hillsborough	813-744-5519	Seminole	407-655-5551
Indian River	772-770-5030	Sumter	352-793-2728
Lake	352-343-4101	Volusia	904-822-5778
Lee	941-338-3232		

USDA-ARS Horticultural Research Laboratory, Ft. Pierce
772-462-5800

Glossary

BMP: Best Management Practice

CERCLA: Comprehensive Environmental Response, Compensation and Liability Act (or Superfund)

CWA: Clean Water Act

DACS: Department of Agriculture and Consumer Services

DEP: Department of Environmental Protection

EPA: Environmental Protection Agency

ESA: Endangered Species Act

FAC: Florida Administrative Code

FIFRA: Federal Insecticide, Fungicide, and Rodenticide Act

FWPCA: Federal Water Pollution Control Act

MSSW: Management and Storage of Surface Waters

NEPA: National Environmental Policy Act

NPDES: National Pollutant Discharge Elimination System

RCRA: Resource Conservation and Recovery Act

NRCS: Natural Resources Conservation Service

SDWA: Safe Drinking Water Act

SWCD: Soil and Water Conservation District

SWIM: Surface Water Improvement Management

TSCA: Toxic Substances Control Act

UIC: Underground Injection Control

WMD: Water Management District

WPS: Watershed Protection Section

Access Road: A road located and constructed to avoid soil erosion caused by haphazard traffic patterns, yet still provide needed access.

Acidification: Applying acid or acidifying compounds (e.g., acidifying fertilizers) to irrigation water to control the formation of chemical precipitates and clogging in a microirrigation system by lowering the pH of the water.

Acid Rain: Sulfates from fossil fuel combustion are washed out of the atmosphere by rain, acidifying lakes and streams.

Acre Foot: A unit of measure equal to 43,560 ft^3, 1 foot of water covering 1 acre of surface area; also acre-inch = 27,154 gallons.

Ad Valorem Tax: A tax imposed on the value of property.

Advective Energy Transfer: Process of energy transfer by means of the mass motion of the atmosphere.

Air Release Valve: An on-line valve to release air bubbles that may accumulate in pipelines or filters during the operation of the irrigation system. Accumulated air in the system can restrict water flow and may cause water hammer.

Air/Vacuum Valve: An on-line valve, sometimes referred to as an air vent. Usually used to allow air to enter pipelines upon system shutoff to avoid developing vacuum that may lead to collapse of pipes. It also allows air to escape from the system upon filling the lines with water when starting the system to avoid entrapping air in the lines during operation. Entrapped air restricts water flow and may cause water hammer that could result in serious damage to pipelines.

Albedo: Ratio of the amount of solar radiation reflected by a surface to the amount incident upon it.

Algal Bloom: Excessive growth of algae in lakes and other freshwater bodies, caused by excessive nutrient enrichment.

Allocation Coefficient: A multiplier used in calculating permit allocations which accounts for the irrigation system efficiency and the effects on the relevant water storage system (see Resource Efficiency).

Allowable Head Loss: The loss in pressure allowed by the irrigation system designer to maintain the required flow rate and the desired emission uniformity.

Anaerobic: A condition with the absence of oxygen.

Anion: Negatively charged ion such as chloride, sulfate, nitrate, and bicarbonate.

Annual Withdrawal: The quantity of water permitted to be withdrawn during any 12-month time period.

Aquatic Filter Ponds: Utilization of ponds, basins, or channels containing aquatic vegetation in order to filter or assimilate nutrients from drainage water.

Aquifer: A geological formation that is saturated with water and consists of highly permeable material that will yield significant volumes of water to a well or spring.

Aquifer Remediation: A use of water involving the withdrawal of groundwater for the authorized removal of contaminants in order to restore water quality.

Aquifer Storage and Recovery: Projects involving approved Class V injection wells for the injection and recovery of freshwater into a groundwater reservoir.

Artesian (Aquifer or Well): Water that is held under pressure in porous rock or soil confined by impermeable geologic formations.

Artesian Pressure: Pressure within a groundwater aquifer developed as a result of hydrostatic head.

Artificial Barriers: Fencing, rocks, earthen banks, and similar facilities that provide protection for highly erodible areas.

Autotrophs: Organisms that produce organic matter utilizing CO_2 usually obtained from solar radiation as the primary energy source.

Available Soil Moisture (or Available Water): The depth of water held in the soil between field capacity and wilting point, usually expressed in unit depth of water per unit length of soil, (e.g., inch of water per foot of soil). The availability of water in this range decreases with the decrease in soil moisture content.

Backwashing: Cleaning filters by reversing the direction of water flow or by forcing water through screens to wash away contaminants.

Base Flow: The groundwater contribution to stream flow that comes from springs or seepage into a stream channel.

Basin: An area drained by a river and its tributaries, also termed catchment or watershed.

Best Efficiency Point (BEP): The point on a pump's performance curve that corresponds to the highest efficiency.

Best Management Practices (BMPs): Management and cultural practices designed to efficiently use and conserve the land.

Biochemical Oxygen Demand (BOD): The quantity of dissolved oxygen required for the aerobic decomposition of organic matter in water.

Biological Control of Pests: Use of natural enemies as part of an integrated pest management (IPM) program which can reduce the use of pesticides.

Brackish: A description of water with a high content of soluble salts, often caused by the mixing of seawater with freshwater.

Brush Management: Management and manipulation of brush to improve or restore a quality plant cover in order to reduce soil erosion.

Canopy: A vegetative cover formed by the tops of plants.

Capillarity: The process by which the surface of a liquid in a very narrow tube rises against the pull of gravity (related to unsaturated hydraulic conductivity).

Capillary Fringe: The zone immediately above the water table that is nearly saturated.

Capillary Water: Water that remains in the soil pore spaces after gravity drainage has occurred.

Casing: The body of the pump which encloses the impeller (volute).

Catchment: The watershed, drainage basin, or area drained.

Cation: Positively charged ion such as sodium, calcium, magnesium, and potassium.

Cation Exchange Capacity: The sum of cations that a soil can absorb and exchange for other cations in the soil solution. It is usually expressed in milliequivalents per 100 grams of soil.

Cavitation: The sudden collapse of gas bubbles due to the sudden pressure increase. This can happen when the pressure is below the vapor pressure at the pump suction. The sudden increase in pressure as the fluid moves through the impeller causes the vapor bubbles to collapse. Those bubbles that are close to the impeller surface produce high localized stresses which damages the surface. Remember that the pressure at the eye of the impeller is even lower than the pressure at the suction flange. This is because the fluid is accelerated by the impeller movement as it moves through the eye.

Centrifugal Force: A force associated with a rotating body. In the case of a pump, the rotating impeller pushes fluid on the back of the impeller blade, imparting circular and radial motion. A body moving in a circular path will have a centrifugal force associated with it. The impeller pushes the fluid against a fixed pump casing, thereby pressurizing the fluid.

Centrifugal Pump: A pump with a fan-shaped impeller rotating in a circular housing, pushing water toward a discharge opening. Centrifugal pumps are usually of simple design, with the only wearing parts being the shaft seal and bearings. They are usually used where large flow of water at relatively low pressure (head, lift) is desired. Self-priming centrifugals have the same features as straight centrifugals, but they will self-prime without a foot valve after an initial filling of the pump casing. Centrifugals that are not self-priming work best with the water source higher than the pump (flooded suction/gravity feed system). As the discharge pressure (head) increases, flow and drive power requirements decrease. Maximum flow and motor loading usually occur at minimum head.

Channel: The flow path of a river or stream that meanders through an area, transporting, eroding, and depositing alluvial sediments.

Channel Capacity: Flow rate of a ditch, canal, or natural channel when flowing full.

Channelization: The digging of a channel through the center of a streambed to increase flow velocity or expedite the removal of floodwater.

Channel Stabilization: The prevention of erosion by the use of jetties, drop structures, vegetation, or other means.

Check Valve: An in-line valve, sometimes referred to as a one-way valve because it allows flow in one direction only. A check valve installed on the discharge side of the pump acts as an automatic shutoff valve when the pump stops, preventing reversed flow into the pump. Reversed flow could cause damage to pump components when they rotate in reverse. Usually, an air vent is installed upstream from the check valve to prevent vacum development and a pressure relief is installed downstream from it to protect the system from surge pressure. Otherwise, both of these problems could develop when the pump stops.

Chemical Oxygen Demand (COD): The amount of a strong chemical oxidant required to oxidize the organic matter in a sample; expressed as an equivalent amount of oxygen.

Chemical Precipitation: Separation of chemical from irrigation water in the form of deposits (which can clog emitters).

Chemigation: Applying chemicals by injecting them through an irrigation system.

Chlorination: Injecting chlorine gas or chlorine compounds into microirrigation systems to control microbial growth in these systems.

Chlorinator: A device mounted on the source of chlorine gas (cylinder or container) to regulate and meter the flow of the gas for safe injection in the irrigation system or similar water flow.

Chlorosis: Yellowing or bleaching of leaves, often induced by nutrient deficiency, specific ion toxicity, or diseases.

Clay: A soil particle less than 0.002 mm in equivalent diameter.

Closed System: System that exchanges neither matter nor energy with its surroundings.

Coagulation: The process of individual particles coming together to form clusters and aggregates.

Coefficient of Roughness: A factor in fluid flow formulas expressing the character of a channel surface and its frictional resistance to flow.

Common Law: Law determined by courts or custom, in contrast with statutory law or legislatively made law.

Cone of Depression: The conical shape taken by the potentiometric surface showing the variation of drawdown with distance due to pumping from a well or wellfield within its area of influence.

Confined Aquifer: An aquifer containing groundwater that is confined under pressure and bounded between significantly less permeable materials, such that water will rise in a fully penetrating well above the top of the aquifer. In cases where the hydraulic head is greater than the elevation of the overlying land surface, a fully penetrating well will naturally flow at the land surface without means of pumping or lifting.

Confining Bed: A related, less permeable zone adjacent to an aquifer.

Confining Unit: A body of significantly less permeable material than the aquifer or aquifers that it stratigraphically separates. The hydraulic conductivity (K) may range from nearly zero to some value significantly lower than that of the adjoining aquifers.

Conservation: The beneficial reduction of water use through voluntary or mandatory altering of water use practices, reduction of distribution losses or installation and maintenance of low-volume water use systems, fixtures, or devices.

Conservation Cropping System: Growing crops utilizing needed cultural and management measures to improve the soil and protect it during periods when erosion occurs. Includes cover cropping, which protects soil from wind and water erosion.

Constant Drawdown: In dewatering systems, the practice of pumping the source unit to a static level for a long duration. Also used in context with aquifer performance tests associated with flowing wells.

Consumptive Use: Any use of water which reduces the supply from which it is withdrawn or diverted.

Contaminant: Measurable amount of a foreign substance.

Continuous Ponding: The process of reclaiming saline soils by ponding water on the soil surface until enough salts have been leached from the crop root zone.

Convective Storms: A meteorological phenomena associated primarily with surface heating.

Conveyance Losses: Water lost from a canal or ditch by evaporation or downward seepage.

Correct Application of Pesticides: Spraying when conditions for drift are minimal. Mixing chemicals properly and according to label. Avoiding application when heavy rain is forecast.

Correct Pesticide Container Disposal: Following accepted methods for pesticide container disposal.

Critical Area Planting: Planting vegetation to stabilize the soil and reduce erosion and runoff.

<u>Crop Coefficient</u>: A factor that relates actual evapotranspiration to potential evapotranspiration, given crop, age, and cultural practice.

<u>Crop Residue Use</u>: Using plant residues to protect cultivated areas during critical erosion periods.

<u>Crop Water Use</u>: The amount of water used by a crop in a given period of time (also see evapotranspiration).

<u>Cultural Control of Pests</u>: Using cultural practices, such as elimination of host sites and adjustment of planting schedules, to partly substitute for pesticides.

<u>Darcy's Law</u>: A mathematical description of fluid flow through a permeable media (velocity of flow is equal to the hydraulic gradient multiplied by the flow characteristics of the media).

<u>Datum Plane</u>: A reference plane. A conveniently accessible known surface which all vertical measurements are based on.

<u>Debris Basin</u>: A barrier or berm constructed across a watercourse or at other suitable locations to form a silt or sediment basin.

<u>Deep Percolation</u>: Water which percolates below the root zone and cannot be used by the plant.

<u>Deep-Well Submersible Pump</u>: A centrifugal pump in which a number of impeller assemblies in a housing are mounted on a shaft directly coupled to a submersible motor. The entire assembly is located at the bottom of the well. Power is brought to the motor by a waterproof cable.

<u>Demand Management</u>: Reducing the demand for water through activities that alter water use practices, improve efficiency in water use, reduce losses of water, reduce waste of water, alter land management practices, and/or alter land uses.

<u>Denitrification</u>: The reduction of nitrates to nitrogen gas and oxides of nitrogen.

<u>Depression Storage</u>: Initial storage of rain in small surface puddles.

<u>Desalination</u>: The process of removing or reducing salts and other chemicals from seawater or other highly mineralized water sources.

<u>Detention</u>: The delay of storm water runoff prior to discharge into receiving waters.

<u>Detention Basin</u>: A constructed basin for the temporary storage of surface runoff and drainage water that will be released at a controlled rate.

<u>Diaphragm Pump</u>: Pump type with a flexible diaphragm that moves up and down in a chamber, creating suction and pressure. As the diaphragm is moved up, it creates vacuum, which opens the suction valve and draws fluid into the chamber. When the diaphragm is forced down, fluid is forced out through the discharge valve. Diaphragm pumps handle fluid mixtures with a greater percentage of solids (e.g., silt, mud, sludge, and waste) than other types of pumps.

<u>Differential Pressure Tank</u>: Tank used for injecting chemicals through the irrigation system. The tank inlet is connected to the irrigation system at a point where pressure is higher than that at the outlet connection, causing irrigation water to flow through the tank containing the chemical to be injected.

<u>Dissolved Oxygen</u>: Atmospheric oxygen that is held in solution within water.

<u>Drainage Basin</u>: The area from which a stream collects water and through which it travels.

Drawdown: The difference between static water level and water level in the well during the pumping process.

Dynamic Discharge Head: The sum of static discharge head, friction head, and velocity head.

Dynamic Suction Lift: Includes static suction lift, drawdown, friction head loss, and velocity head.

Ecosystem: The interconnected community of organisms and the natural environment in which they exist.

Effective Rainfall: The portion of rainfall that infiltrates into the soil and is stored for plant use within the root zone.

Effluent: Liquid discharge from a point source; sometimes refers to liquid that comes out of a treatment plant after completion of the treatment process.

Effluent Stream: A stream intersecting the water table and receiving groundwater flow.

Electrical Conductivity (EC): The property of conducting electric current (reciprocal of resistance). The electrical conductivity of water is proportional to the concentration of dissolved salts in it and is therefore used as an estimate of total dissolved salts in water or in an extract of a water-saturated soil paste. It is expressed in units of dS/m, mS/cm, or mmho/cm at 25° C.

Electrodialysis: The electrochemical process where water is desalinated by ions passing through a semipermeable membrane from a less concentrated to a saturated solution.

Elevation: The height in feet above mean sea level according to National Geodetic Vertical Datum (NGVD, 1929). May also be expressed in feet above mean sea level (MSL) as reference datum.

Elevation Head: Energy possessed by a fluid due to its position above some reference point.

Emission Uniformity (EU): A measure of uniformity of water application over a field, calculated as the minimum depth of applied water (usually the average discharge rate of the lowest 25% of all measured emitters), divided by the average discharge of all emitters measured, multiplied by 100. Sometimes referred to as distribution uniformity.

Emitter: A device through which water is discharged to the soil surface or to the root zone from the lateral lines in microirrigation systems.

Encrustation: The accumulation of deposits in the perforations of the well casing, in well screen openings, and in the voids of the gravel pack and water-bearing soil. Encrustation clogs the openings, decreases the open area in the well casing or screen, impedes water flow into the well, and reduces well efficiency.

Energy Gradient: The change in energy per unit length in the direction of flow or motion.

Erosion: The detachment and movement of soil from the land surface by wind, water, or ice.

Estuary: A partially enclosed body of brackish water with a connection to the sea.

Eutrophication: The condition of high concentration of nutrients within a water body that results in the consumption of available oxygen.

Evaporation: The change of liquid water to water vapor in the atmosphere.

Evaporative Cooling: Cooling or reduction in temperature caused by evaporation from a wetted area.

Evapotranspiration (ET): The total loss of water to the atmosphere by evaporation from land and water surfaces and by transpiration from plants.

Evapotranspiration Rate: The rate of water loss, during a specific period of time, through transpiration from vegetation plus evaporation from the soil. Estimates of crop evapotranspiration rate are either historical averages or real-time (current).

Existing Legal Use of Water: A water use that is authorized under a district water use permit or which is existing and exempt from permit requirements.

Fertigation: Applying fertilizers through an irrigation system.

Field Border: A border or strip of permanent vegetation established at field edges to control soil erosion and filter nutrients.

Field Capacity: The percentage of water remaining in the soil two or three days after having been saturated and after free drainage has practically ceased (typically 1 to 2 days in sandy soil).

Field Windbreak: A strip or belt of trees established to reduce wind erosion.

Filtration: The removal of suspended solids from irrigation water.

Flexible Impeller Pump: Pumps with a flexible vaned member (usually rubber) rotating in an eccentric housing. The volume of the spaces between the vanes changes as the pump rotates, creating pumping action.

Flocculation: The process of small particles coming together in water to form aggregates.

Flooded Suction: Liquid source is higher than the pump, and liquid flows to the pump by gravity. Flooded suction is preferable for centrifugal pump installations.

Floodplain: Area that is adjacent to a channel and that may be inundated during high water.

Floodwater Reduction Structure: A structure providing temporary storage of stormwater for its controlled release to reduce flooding, streambed erosion, and sedimentation.

Flow: The measure of the liquid volume capacity of a pump. Given in gallons per hour (gph) or gallons per minute (gpm). To convert gph to gpm, divide by 60.

Flowmeter: An instrument that is used for the accurate measurement of water flow through a closed pipe. Care should be taken to properly install and calibrate it.

Flume: A device for measuring the flow of water.

Foot Valve: A type of check valve with a built-in strainer. Used at point of water intake to retain water in the system and prevent loss of prime when the water source is lower than the pump.

Freshwater: An aqueous solution with a chloride concentration equal to or less than 250 milligrams per liter (mg/L).

Friction: The force produced as reaction to movement. All fluids produce friction when they are in motion. The higher the fluid viscosity, the higher the friction force for the same flow rate. Friction is produced internally as one layer of fluid moves with respect to another and also at the fluid/wall interface.

Friction Head: The pressure expressed in pounds per square inch or feet of liquid needed to overcome the resistance to flow in the pipe and fittings.

Friction Head Difference: The difference in head required to move a mass of fluid from one position to another at a certain flow rate within a piping system. It is also the specific energy required to overcome friction in the system.

Furrow: A small ditch for the conveyance of irrigation or drainage water.

Gaining Stream: A stream where flow increases due to inflow from groundwater.

Gear Pump: Pump type having two meshed gears in a housing. As gears rotate, fluid is carried in the space between the gears. Gear pumps will not handle abrasives because of close running tolerances of gears. Gear pumps are best suited for pumping higher viscous fluids at slower speeds.

Ghyben-Herzberg Principle: Principle is based on the fact that freshwater is 2.5 percent less dense than seawater, and consequently seawater will tend to intrude into coastal freshwater aquifers.

Grade: The degree of slope of a ground surface.

Grade Stabilization Structure: A structure to stabilize the streambed or to control erosion in natural or constructed channels.

Gradient: The change of elevation, velocity, pressure, energy, or temperature per unit length.

Grassed Waterway, or Outlet: A natural or constructed waterway or outlet maintained with vegetative cover in order to prevent soil erosion and filter nutrients.

Gravitational Water: Water that moves through the root zone under the influence of gravity forces.

Greenhouse Effect: A reduction in the net loss of energy from the earth's upper atmosphere resulting in an increase in the earth's mean temperature.

Groundwater: Water beneath the surface of the ground contained in a soil-saturated condition, as opposed to surface water and soil water.

Halophytes: Plants with a high tolerance for soluble salts.

Hardness: Relates to the encrusting potential of a water and is measured by the concentration of calcium and magnesium.

Hazardous Waste: Waste considered a threat to human health or the environment by EPA; does not include petroleum, although some petroleum products are hazardous wastes.

Head: Another measure of pressure, expressed in feet. Usually applies to centrifugal pumps. Indicates the height of a column of water being lifted by the pump, not including friction losses in piping. For water, divide head in feet by 2.31 to get pressure in pounds per sq inch (psi).

Headwaters: The upstream source of a river or other moving body of water.

Heat Stress Damage: Exposure to high temperature extremes such that the crop or plant is economically damaged.

Hydraulic Conductivity: The measure of the ease with which water can flow through a soil profile, expressed in units of length per time (velocity).

Hydroperiod: The pattern of inundation, saturation, and drying in a wetland, characterized in terms of both the range and the duration of water level fluctuation.

Impeller: The rotating element of a pump which imparts movement and pressure to a fluid.

Impoundment: Any lake, reservoir, or other containment of surface water occupying a depression or bed in the earth's surface and having a discernible shoreline.

Injection Well: Source of water or other liquids entering groundwater; can be a very deep well or several types of runoff.

Injunctive Relief: Court order to prohibit someone from doing some specified act or to command someone to undo some wrong or injury; an example is a restraining order.

Irrigation Return Flow: The flow of water under the influence of gravity to a watercourse, which occurs as surface water flow or shallow groundwater flow resulting from the application of water for supplemental irrigation purposes.

Irrigation System Efficiency: A measure of the effectiveness of an irrigation system in delivering water to a crop for irrigation and freeze protection purposes. It is expressed as the ratio of the volume of water used for supplemental crop evapotranspiration to the volume pumped or delivered for use.

Irrigation Water Conveyance: A pipeline or lined waterway constructed to prevent erosion and loss of water quality and quantity.

Irrigation Water Management: Determining and controlling the rate, amount, and timing of irrigation water application in order to minimize soil erosion, runoff, and fertilizer and pesticide movement.

Irrigation Water Use: A water use classification which incorporates all uses of water for supplemental irrigation purposes, including golf, nursery, agriculture, recreation, and landscape.

Jet Pump: A type of centrifugal pump utilizing water flow through a narrow opening or nozzle (jet or ejector) to bring water from a well. As water is forced through the nozzle, an area of low pressure is created and atmospheric pressure forces additional water from the well into the system. In shallow well systems (up to 25 ft lift), the jet is located at the pump. In deep-well systems, it is located at the bottom of the well.

Joint and Several Liability: The responsible offending parties have both individual and collective responsibility for the wrongs suffered by another party.

Kinetic Energy: A thermodynamic property. The energy associated with the mass and velocity of a body.

Lake Recharge: The withdrawal of water for the purpose of replacing a volume of water removed from a lake system or other water body utilized as a source of water supply or indirectly as a source of wellfield recharge. Lake recharge does not include artificial maintenance of the water level of a surface water body at a desired elevation for aesthetic purposes, but may include augmentation of the volume of water stored within a surface water body that is effecting recharge to an adjacent wellfield.

Laminar: A distinct flow regime that occurs at a low Reynolds number (Re < 2000). It is characterized by particles in successive layers moving past one another without mixing.

Land Absorption Areas and Use of Natural Wetland Systems: Providing an adequate land absorption area downstream from tilled or grazed areas so that soil and plants absorb nutrients and animal wastes.

Leaching: Process by which nutrient chemicals or contaminants are dissolved and carried away by water or are moved into a lower layer of soil.

Leaching Fraction (LF): The portion (fraction) of the irrigation water that is applied in excess of the soil moisture depletion and which should percolate below the root zone in order to prevent the build-up of salinity.

Leaching Requirement (LR): The leaching fraction needed to keep the root zone salinity level at or below the threshold tolerance by the crop. The leaching requirement is determined by the crop tolerance to salinity and by the salinity of the irrigation water.

Leakance: The vertical movement of water from one aquifer to another across a confining zone or zones due to differences in hydraulic head. Movement may be upward or downward depending on hydraulic head potential in the source aquifer and the receiving aquifer. This variable is typically expressed in units of gpd/ft^3.

Letter Modification: An administrative process that allows for the modification of an existing permit to account for minor changes that do not result in significant change to the terms and conditions of the permit.

Liability: The state of being bound or obliged in law to do, pay, or make good on something.

Lien: A claim or charge on property for payment of some debt, obligation, or duty.

Lift (Suction Lift): The suction achieved when the source of supply of the liquid is below the central line of the pump. Pumping action creates a partial vacuum, and atmospheric pressure forces liquid up to the pump. The theoretical limit of suction is 34 feet, but the practical limit is 25 feet or less, depending on pump type and elevation above sea level.

Lined Waterway or Outlet: A runoff water channel or outlet with an erosion-resistant lining to prevent erosion. Applicable to situations where unlined or grassed waterways would be inadequate.

Line Source: Drip tape or tubing with emission points at close regular intervals such that the discharge of the emission points develops a continuous wetting pattern in the soil.

Localized Irrigation: Irrigation by systems which wet, in particular, the area of soil at the base of the plant (e.g. trickle, drip, micro).

Mainlines: The water delivery pipelines that supply water to the submain lines.

Manufacturer Coefficient of Variability: A measure of the variability in discharge rates of the same model of emitter that results from the manufacturing process.

Maximum Daily Allocation: The maximum quantity permitted to be withdrawn in any single 24-hour period.

Maximum Monthly Allocation: The maximum quantity of water assigned to the permit to be withdrawn during the month in the growing season when the largest supplemental crop requirement is needed by the specific crop for which the allocation is permitted.

Media Filter: A filter consisting of a tank filled with filtering media such as silica sand or crushed granite that removes suspended solids from irrigation water by passing the water through the filtering media.

Microirrigation: The application of small quantities of water on or below the soil surface as drops or tiny streams of spray through emitters or applicators placed along a water delivery line. Microirrigation includes a number of methods or concepts such as drip or microsprinkler.

Minimum Water Levels: Level of water below which withdrawals would be harmful to the ecosystem or water resources; set by water management districts.

Mitigation: To make less severe; to reduce a penalty or punishment imposed by law.

Moisture Release Curve: A graph plotting soil water potential against soil water content for a particular soil.

Muck: An organic soil.

Mulching: Applying plant residues or other suitable materials to the soil surface in order to reduce water runoff and soil erosion.

National Geodetic Vertical Datum (NGVD): A geodetic datum derived from a network of information collected in the United States and Canada. It was formerly called the "Sea Level Datum of 1929" or "mean sea level." Although the datum was derived from the average sea level over a period of many years at 26 tide stations along the Atlantic, Gulf of Mexico, and Pacific Coasts, it does not necessarily represent local mean sea level at any particular place.

Negative Pressure: Pressure that is less than the pressure in the local environment (vacuum).

Negligence: Failure to use such care as a reasonably prudent and careful person would use under similar circumstances.

Negligence Per Se: A form of negligence that results from violation of a statute. Running a red light is negligence per se.

Net Positive Suction Head (NPSH.): The head in water column as measured or calculated at the pump suction flange minus the vapor pressure of the water converted to a water column depth.

Nonpoint Source Pollution: Pollution that is discharged over a wide area and enters into receiving waters at generally irregular intervals as a consequence of storm runoff.

Operating Point: The point on the system curve corresponding to the flow and head required to meet the irrigation system demands.

Osmosis: The tendency of a fluid to pass through a semipermeable membrane into a solution where its concentration is lower, thus equalizing the conditions on either side of the membrane.

Osmotic Effect: The additional energy that the plant must exert to extract and absorb water from a salty soil or solution.

Osmotic Potential: The force or negative pressure created by osmosis.

Overland Flow: Water flowing over the ground surface and over a wide area.

Oxbows: Permanently standing bodies of water that result from the cutoff of river meanders.

Perched Groundwater: A locally saturated zone above an impervious layer of limited extent.

Percolation: Downward movement of water in the soil due to gravity.

Performance Curve: A curve of total head vs. flow for a specific pump model and impeller diameter.

Permeability: The physical structure and texture of the soil that allows water to move through it.

Pesticide: Any substance used to regulate, prevent, repel, or destroy any pest or plant.

Pesticide Selection: Selecting pesticides which are less toxic, persistent, soluble, and volatile, whenever feasible.

pH: A measure of acidity or alkalinity of a media (free hydrogen ion concentration within a solution). A pH of 7.0 is neutral; a pH less than 7.0 is acidic; and a pH greater than 7.0 is alkaline.

Phloem: The principal food-conducting tissue of vascular plants, basically composed of sieve elements, parenchyma cells, and fibers.

Phreatic Water: Another term for groundwater; water occurring in a zone of saturation.

Phreatophytes: Plants that derive a majority of their water from groundwater as opposed to soil moisture (e.g. cypress).

Piezometric Level: The level to which water would rise in an unconfined tube, equivalent to the static head.

Piston Pump: A pump where fluid is drawn in and forced out by pistons moving within cylinders.

Plume: A body of contaminated groundwater originating from a specific source and influenced by such factors as the local groundwater flow pattern, density of contaminant, and character of the aquifer.

Point Source Pollution: Pollution being discharged into the environment from a specific location, i.e. pipes, sewers, channels.

Pollutant: Presence of contaminants in water, soil, or air to such a degree that the use of the resource is impaired; includes gasoline or oil, any pesticide, or any ammonia or chlorine compound or derivative.

Porosity: Ratio of the pore volume to the total volume of a material.

Portable Guns: Large sprinklers, including truck- or tractor-mounted units, which discharge high volumes of water at high pressures through the air and are moved from location to location irrigating in a circular spray pattern.   .

Positive Displacement Pump: Generally self-priming pump with pumping action created by moving chambers or pistons. The flow rate of this pump is almost the same at any pressure level. They should never be operated dry because of internal wearing of rubber parts. As discharge flow is restricted (higher pressure or head), the drive horsepower requirement increases. A relief device should be provided on the discharge to prevent overpressure and damage to the pump or motor if the discharge line is closed off or severely restricted.

Potable Water: Water that is suitable for drinking, culinary, or domestic purposes.

Potential Evapotranspiration (ETp): Evapotranspiration that would occur from a well-vegetated surface if water were not limited.

Potentiometric Surface: A surface which represents the hydraulic head in an aquifer. It is defined by the level to which water will rise in a cased well that penetrates the aquifer.

Prescribed Burning: Using fire, under conditions where the intensity of the fire is controlled, to improve plant cover so that runoff and erosion are reduced.

Pressure: The force exerted on the walls of a container (tank, pipe, etc.) by the liquid. Measured in pounds per square inch (psi).

Pressure Head: The specific energy of water pressure, normally stated as feet of head. One foot of head (= 0.433 psi) is equal to the pressure at the bottom of a 1-foot high column of water.

Pressure Relief Valve: An on-line valve, usually installed at low points in the irrigation system or where pressure buildup is expected. It opens when system pressure exceeds a preset limit. It should have enough water-releasing capacity to avoid any excessive pressure buildup.

Pressure-Sustaining Valve: An in-line valve to maintain a constant preset upstream pressure regardless of changing upstream flow rate.

Prime: A charge of liquid required to begin pumping action of centrifugal pumps when the liquid source is lower than the pump. Water may be held in the pump by a foot valve on the intake line or a chamber within the pump.

Propeller Flowmeter: A device consisting of a propeller linked by a cable or shafts and gears to a flow indicator for measuring flow rate and total applied fluid in pipelines.

Public Water Supply: Water that is withdrawn, treated, transmitted, and distributed as potable or reclaimed water.

Pump Capacity: The discharge rate or flow rate of the pump in gallons per minute (or liter per hour).

Pump Discharge Head: Pump discharge pressure in pounds per square inch (psi) multiplied by 2.31 = pump discharge head in feet of water.

Pump Discharge Pressure: Pressure required by the sprinklers or emitters in addition to the pressure needed to overcome the elevation difference and pressure losses caused by friction (the reading of the pressure gauge on the discharge side of the pump).

Pump Performance Curve: Measurements supplied by the pump manufacturer showing the relationship between the total head developed by the pump and flow rate.

Pumping Lift: The distance from the discharge pipe at the pump head to the water level in the pumped well while the pump is running and when the depth of water table is stabilized.

Pumping Plant Efficiency: The combined efficiency of the pump and the motor or engine, usually expressed as the pump discharge capacity, multiplied by the total head and divided by the input horsepower.

Pumps in Parallel: Several pumps discharging into a common pipeline to increase the flow rate.

Pumps in Series: Pumps installed so that one pump discharges into the intake of another pump. The head developed by the second pump (often called the booster pump) is added to the head of the first pump, thereby increasing the total discharge head.

Rainfall Duration: The length of time during which rain fell.

Rainfall Intensity: The amount (depth) of rainfall per unit of time.

Recharge: Inflow of water into an aquifer.

Recharge Area: An area that is connected to the aquifer by a highly porous media and has an ample source of surface water.

Reclaimed Water: Water that has received at least secondary treatment and is reused after flowing out of a wastewater treatment facility.

Reclamation: Process of increasing mined land or other used resource to a higher value by physically changing the land, e.g., wetland reclamation.

Reduced Threshold Area (RTA): An area established by a water management district for which the threshold separating a General Permit from an Individual Permit has been lowered from a maximum limit of 100,000 gpd to 20,000 gpd. These areas are typically resource-depleted areas where there has been an established history of substandard water quality, saline water movement into ground or surface water bodies, or lack of water availability to meet projected needs of a region.

Regulated Runoff Impoundment: Retention, or detention with filtration prior to discharge, to reduce runoff quantity and nutrient and pesticide discharge.

Relative Humidity: The ratio of the water content of the air to the maximum water-holding capacity of the air at a given temperature.

Relief Valve: Usually used at the discharge of a positive displacement pump. An adjustable, spring-loaded valve opens or relieves when a preset pressure is reached to prevent excessive pressure and pump or motor damage if the discharge line is closed off.

Reservoir: An open area where water is stored.

Residence Time: The average time period water remains in a wetlands; reciprocal of the turnover rate.

Resource Efficiency: The efficient use of water as measured in terms of the net impact on the relevant water storage system.

Restricted Allocation Area: Areas designated within the water management districts for which specific sources of water are under allocation restrictions because the water level is insufficient to meet the projected needs of the region.

Retention: The prevention of stormwater runoff from direct discharge into receiving waters; included as examples are systems which discharge through percolation, exfiltration, filtered bleed-down, and evaporation processes.

Retrofit: The replacement or changing cut of an irrigation system with a different irrigation system such as a conversion from an overhead sprinkler system to a microirrigation system.

Return Period: The reciprocal of the probability of an event (e.g., a 5-year return period for a storm of a certain magnitude means that, on the average, such a storm is not expected to occur more often than once in 5 years).

Reuse: The deliberate application of reclaimed water, in compliance with FDEP and water management district rules, for a beneficial purpose.

Reverse Osmosis: A method of desalinization and filtration of water.

Riparian: Pertaining to the land adjacent to a body of water.

Roller or Vane Pump: Pump that has a roller or vanes on a rotor that rotates in an eccentric housing. The volume of the spaces between the vanes or rollers changes as the pump rotates, creating pumping action.

Root Zone: The depth of the soil from which the roots of the crop extract water and nutrients.

Rotary Screw Pump: Pump with a screw-shaped rotor that turns within a flexible stator, usually made of rubber. Progressing cavities between the screw and the stator carry the fluid. Rotary screw pumps can handle abrasive mixtures or slurries at slower speeds.

Runoff: That component of rainfall which is not absorbed by soil, intercepted and stored by surface water bodies, evaporated to the atmosphere, transpired and stored by plants, or infiltrated to groundwater, but which flows to a watercourse as surface water flow.

Runoff Hydrograph: A curve showing the time distribution of runoff rates.

Run-on: The component of a water budget consisting of runoff from an adjacent field.

Saline Soil: A nonsodic soil containing sufficient soluble salts to impair its productivity. As a general guideline, soils with salinity of saturated paste extract of 4 ds/m (or 4 mmho/cm) or more are considered saline.

Saline Water: An aqueous solution with a chloride concentration greater than 250 mg/L and less than that of seawater.

Saline Water Interface: Hypothetical surface of chloride concentration between freshwater and saline water where the chloride concentration is 250 mg/L at each point on the surface.

Salinity: Soil condition in which the salt concentration in the crop root zone is too high and can adversely affect plant growth and crop production.

Salinization: The process of accumulation of salts in the soil.

Salt Index: An index of the extent to which a given amount of fertilizer increases the osmotic pressure (related to increasing the osmotic effect) of soil solution.

Saltwater Intrusion: The phenomenon occurring when saltwater moves laterally from the ocean into the groundwater to displace freshwater.

Sand Separator: A device used to separate dense (denser than water) sand-sized particles from water.

Scour Velocity: Flow velocity that needs to be maintained in the microirrigation system to scour deposits and flush them out during the lateral flushing process. Required scour flow velocity is about one foot per second, which is equivalent to a flow rate of about one gallon per minute in a 1/2-inch lateral at the downstream end of the system.

Seal: A device mounted in the pump housing and/or on the pump shaft to prevent leakage of liquid from the pump. Mechanical seals have a rotating part and a stationary part with highly polished touching surfaces. They have excellent sealing capability, but can be damaged by dirt or grit in the liquid. Lip seals have a flexible ring (usually rubber or similar material), with the inner edge held closely against the rotating shaft by a spring.

Seal-less Magnetic Drive Pump: Pump that transfers motor horsepower to the pump impeller by magnetic force, through a wall that completely separates the motor from the impeller; no seal is used.

Seasonal High Water Level: The elevation to which the groundwater or surface water can be expected to rise due to a normal wet season.

Seawater: An aqueous solution with a chloride concentration equal to or greater than 19,000 mg/L.

Sedimentation: The act or process of accumulation of sediments in layers.

Seepage Irrigation System: A means to artificially supply water for plant growth which relies primarily on gravity to move the water over and through the soil, and does not rely on emitters, sprinklers or any other type of device to deliver water to the vicinity of expected plant use.

Semiconfined Aquifer: A completely saturated aquifer that is bounded above by a semipervious layer, which has a low, though measurable permeability, and below by a layer that is either impervious or semipervious.

Sensible Heat: Energy absorbed by a substance which results in a change in temperature rather than a change in state.

Shutoff Head: The total head corresponding to zero flow on the pump performance curve.

Sinkhole: In Florida, an area where the surface of the land has subsided or collapsed as a result of the underlying limestone being dissolved.

Slope: Degree of deviation of a surface from the horizontal.

Sloughs: Areas of standing water.

Slow Release Fertilizer: Applying slow release fertilizers to minimize nitrogen losses from soils prone to leaching.

Soil Aeration: Movement of air into a soil as water is drained out.

Soil Aggregate: A granule of soil consisting of many single soil particles held together.

Soil Amendment: A substance which when applied to the soil can improve its properties.

Soil Testing and Plant Analysis: Testing to avoid overfertilization and subsequent losses of nutrients in runoff water.

Soil Water: The amount of water held in the soil from rain, irrigation, or seepage (expressed in unit length per unit depth of the soil, e.g., inches of water per foot of soil).

Soil Water Deficit: The amount of water required to raise the moisture content of a soil to field capacity.

Soil Water Potential: The work required to move a unit mass of water in a soil from an arbitrary datum to the point in question.

Solid Waste: Trash, sludge, semiliquid, and gaseous wastes; does not include domestic sewage, irrigation return flow, or pollutants included in NPDES permits.

Specific Capacity: The ratio of discharge rate to drawdown in a pumping well, usually expressed in units of gpm/ft or similar units.

Specific Gravity: The ratio of the density of a fluid to that of water at standard conditions. Pumping heavier liquids (specific gravity greater than 1.0) will require more drive horsepower.

Specific Ion Toxicity: Any adverse effect caused by a salt constituent, rather than the total salinity, to the plant growth. Sodium, chloride, and boron can be toxic to citrus if their concentrations exceed tolerance levels.

Specific Yield: The amount of water that a unit volume of an aquifer will yield when pumped.

Stage: The elevation of a water surface in a stream or reservoir in reference to an established datum.

Staged Drawdown: In dewatering systems, the practice of pumping the source unit to discrete, incremental levels.

Standby Facility: The minimal operation of a withdrawal facility to maintain the mechanical integrity of the pumping apparatus as recommended by the manufacturer or for a limited time period each month.

Static Discharge Head: The vertical distance from the center line of the pump to the point of free discharge.

Static Suction Head: The vertical distance from the center line of the pump up to the free level of the liquid source.

Static Suction Lift: The vertical distance from the center line of the pump down to the free level of the liquid source.

Static Water Level: The stabilized level of the groundwater in a well before pumping commences.

Streambank Protection: Stabilizing and protecting banks of streams, lakes, estuaries, or excavated channels against scour and erosion with vegetative or structural means.

Strict Liability: Liability without fault, when one is responsible for all consequences regardless of one's fault.

Sublimation: The direct change of state of ice or snow to water vapor.

Submain Lines: The water delivery pipelines that supply water to the manifolds and lateral lines.

Subsurface Drain: A conduit, such as tile, installed beneath the ground surface to control the water level for increased production.

Suction Head: Head that exists when the source of supply is above the centerline of the pump.

Suction Lift: Distance from the water surface to the pump intake when the pump is located above the surface.

Suction Static Head: The difference in elevation between the liquid level of the fluid source and the centerline of the pump. This head also includes any additional pressure head that may be present at the suction tank fluid surface.

Supercooling: The process in which a liquid goes below its freezing point without forming a solid. An example would be water which goes below 32° F, remains liquid, and does not form ice.

Supplemental Crop Requirement (SCR): The volume of water, usually expressed in acre-inches, representing the difference between the estimated evapotranspiration of a given crop and the effective rainfall available in a specific geographic area over some prescribed time period and climatic event.

Surface Runoff: Water flowing off the lower end of the field, usually indicating that the irrigation water was applied at a rate that exceeded the soil intake rate.

Surface Water: Water on the surface of the ground, including water in man-made boundaries.

Suspended Sediment: Sediment carried by stream flow; the amount is a function of the stream velocity, particle size, and density.

Swamp: Wetlands dominated by trees and shrubs.

Tensiometer: A device consisting of a ceramic cup, column of water, and vacuum gauge that is used to measure soil-moisture tension (or suction) as an indication of water availability in the root zone. Moisture tension is displayed in centibars (1/100 of a bar) on the tensiometer gauge. Higher gauge readings are associated with drier soil, which means the plant must exert greater energy to extract water from the soil.

Threatened/Endangered Species: Species of plants or animals that are threatened with extinction or are in danger of extinction.

Timing and Placement of Fertilizers: Timing and placement of fertilizers for maximum utilization by plants and minimum leaching or movement by surface runoff.

Total Dynamic Head: The suction head necessary to lift water to the soil surface, plus the pressure head needed to supply water throughout the irrigation system and to overcome pressure losses caused by friction and elevation differences in the system.

Total Static Head: The difference between the discharge and suction static head, including the difference between the surface pressure of the discharge and suction tanks. It is also the specific potential energy of the system if we do not consider the pressure head in the discharge or suction tank.

Toxic Substances: Substances carrying a risk to produce birth defects, heart disease, emphysema, or other health effects in humans or carrying a risk to injure the environment.

Transmissivity: The rate at which water moves through an aquifer at a specific energy gradient.

Transpiration: The process by which plants lose water vapor.

Traveling Guns: Large sprinklers that discharge high volumes of water through the air above the level of the plant being irrigated at high pressures. These sprinklers are self-propelled and move slowly across the area being irrigated, such as lateral move or linear irrigation systems.

Treatment Facility: Any plant or other works used for the purpose of treating, stabilizing, or holding wastewater.

Tributary: A branch of a river, stream, or channel that contributes flow to the main channel.

Turbulent: A type of flow regime characterized by the rapid movement of fluid particles in many directions as well as the general direction of the overall fluid flow.

Turgor: The normal distension or rigidity of plant cells caused by the pressure of the cell contents exerted against the cell walls. Some turgor is necessary for cell expansion and growth. Loss of turgor pressure can cause wilting.

Turnover Rate: The ratio of flow-through volume to average volume within a wetlands or a reservoir.

<u>Unconfined Aquifer</u>: A permeable geologic unit or units only partly filled with water and overlying a relatively impervious layer. Its upper boundary is formed by a free water table or phreatic surface under atmospheric pressure. Also referred to as a water table aquifer.

<u>Upconing</u>: Upward migration of mineralized or saline water as a result of pressure variation caused by withdrawals.

<u>Vadose Zone</u>: The portion of the soil that is above the water table and unsaturated.

<u>Vane (Roller) Pump</u>: Pump that has a roller or vanes on a rotor that rotates in an eccentric housing. The volume of the spaces between the vanes or rollers changes as the pump rotates, creating pumping action.

<u>Vapor Pressure</u>: The pressure at which a liquid boils at a specified temperature.

<u>Vapor Pressure Deficit</u>: Difference between the existing vapor pressure and that of a saturated atmosphere at the same temperature.

<u>Velocity Head (Hv)</u>: The head needed to accelerate the water in the system. Knowing the velocity of the water, the velocity head can be calculated by a simple formula: $Hv = V^2/2g$, in which g is acceleration due to gravity (32.16 feet/second). Although the velocity head loss is a factor in figuring dynamic head, the value is usually small and in most cases negligible.

<u>Venturi Injector</u>: A device for injecting chemicals into pressurized irrigation systems. The device is installed in parallel with the water line like a bypass. It consists of a constriction in the middle of the bypass that results in a negative pressure (or suction) at the constriction. The inlet to the constriction is connected to the reservoir or container of liquid chemicals. The developed suction allows chemicals to flow through the constriction of the venturi and into the water flow line.

<u>Vicarious Liability</u>: Liability of employer based on actions of employee that occur within the scope of his/her employment.

<u>Viscosity</u>: The cohesive force existing between particles of a fluid which causes the fluid to offer resistance to flow. Temperature must be stated when specifying viscosity, since most liquids flow more easily as they get warmer. The more viscous the liquid, the slower the pump speed required.

<u>Waste Utilization</u>: Using wastes for fertilizer in a manner which improves the soil and protects water resources. This may also include recycling of waste solids for animal feed supplement.

<u>Wastewater</u>: The combination of liquid and water-carried pollutants from residences, commercial buildings, industrial plants, and institutions, together with any groundwater, surface runoff, or leachate that may be present.

<u>Water Amendment</u>: A chemical added to the water in order to improve certain soil water properties such as infiltration rate, which may be increased by modifying the chemical composition of the soil water complex.

<u>Water Application Efficiency</u>: The efficiency of a single irrigation event of a given field, calculated as the water stored in the crop root zone divided by the total water applied, multiplied by 100.

<u>Water Budget</u>: An account of additions and subtractions of water from a particular area or region.

Water Potential: The chemical potential of water. An indication of the driving force of water through a plant or soil. It is one measure of the amount of water stress a plant experiences and is expressed in units of pressure such as atmospheres, bars, or Pascals.

Water Table: The surface of a body of unconfined groundwater at which the pressure is equal to that of the atmosphere; defined by the level where water within an unconfined aquifer stands in a well.

Water Table Management: Control of the water table at the highest level consistent with the crop's needs. It reduces oxidation of organic soils and the release of nutrients to drainage water.

Water Use: Any use of water which reduces the supply from which it is withdrawn or diverted.

Water Well: Any excavation that is drilled, cored, bored, washed, driven, dug, jetted, or otherwise constructed when the intended use of such excavation is for the location, acquisition, development, or artificial recharge of groundwater.

Well Yield: Same as specific capacity.

Wetlands: Lands supporting vegetation suited to a wetland environment and/or covered periodically with water.

Wilting Point (or Permanent Wilting Point): The moisture content of the soil, on an oven-dry basis, at which plants wilt and fail to recover their turgidity when placed in a dark humid atmosphere.

Wind Stress Damage: Exposure to high wind such that the crop or plant is economically damaged.

Work: The energy required to drive the fluid through the system.

Xeriscape: A landscaping method that maximizes the conservation of water by the use of site-appropriate plants and an efficient watering system.

Xerophyte: A plant adapted to a limited supply of water.

Xylem: The principal water-conducting tissue in vascular plants. The xylem can also be a supporting tissue, especially the secondary xylem (wood).

Zone of Discharge: Predefined three-dimensional area underground around a source of water injected into an aquifer. The water quality standards are usually more relaxed in this zone than in a groundwater supply zone.

References

CHAPTERS

1. History of Florida Citrus Irrigation.

Tucker, D.H.P. 1985. Citrus irrigation management. Univ. of Florida, IFAS Coop. Ext. Serv. Circ. 444.

2. Overview of Grove Design and Development

Haeussler, E.F., Jr. and R. Paul. 1993. Introductory mathematical analysis. 7th ed. Prentice-Hall, Englewood Cliffs, N.J.

Randall, A. 1987. Resource economics. 2nd ed. Wiley, N.Y.

3. Environmental Acts and Regulatory Agencies.

Olexa, M.T. 1991. Handbook of Florida water regulation. Univ. of Florida, IFAS Coop. Ext. Serv. Circ. 1026.

4. Water Management Districts

Olexa, M.T. 1991. Handbook of Florida water regulation. Univ. of Florida, IFAS Coop. Ext. Serv. Circ. 1026.

SFWMD. 1996. Guidance for preparing an application for a water use permit. South Florida Water Management District, West Palm Beach, Fla.

SWFWMD. 1996. Tips about getting water management permits for agricultural operations. Southwest Florida Water Management District, Brooksville, Fla.

5. Conservation Practices and BMPs

Boman, B.J., P.C. Wilson, and J.W. Hebb (eds.). 2000. Water quality/quantity BMPs for Indian River area citrus groves. Fla. Dept. Environmental Prot., Tallahassee, Fla.

Boman, B.J. 1993. Nitrogen leaching from young Florida citrus, p. 651-658. In: R. G. Allen and M. U. Neale (eds.). Management of irrigation and drainage systems - Integrated perspectives. Proc. ASCE Irrig. and Drain. Spec. Conf. ASCE, New York.

Boman, B.J., C. Crandall, and H. Zhong. 1998. Groundwater nitrate in coastal Florida citrus groves. Proc. Ground Water Management Symposium, 1998 International Water Resources Engineering Conference. ASCE, Reston, Va. 1: 211-216.

Bottcher, D. and D. Rhue. 1984. Fertilizer management – Key to a sound water quality program. Univ. of Florida, IFAS Ext. Serv. Publ. SP-28.

Brown, R.B., A.G. Hornsby, and G.W. Hurt. Rev. 1996. Soil ratings for selecting pesticides for water quality goals. Univ. of Florida, IFAS Coop. Ext. Serv. Circ. 959.

Syvertsen, J.P., M.L. Smith, and B.J. Boman. 1993. Tree growth, mineral nutrition, and leaching losses from soil of salinized citrus. Agr. Ecosystems Environ. 45:319-334.

Tucker D.P.H, A.K. Alva, L.K. Jackson, and T.A. Wheaton (eds.). 1995. Nutrition of Florida citrus trees. Univ. of Florida, IFAS Coop. Ext. Serv. Publ. SP-169.

6. Environmental Concerns - Pesticides

Buttler, T.M., A.G. Hornsby, D.P. Tucker, J.L. Knapp, and J.W. Noling. Rev. 1995. Citrus: Managing pesticides for crop production and water quality protection. Univ. of Florida, IFAS Coop. Ext. Serv. Circ. 974.

Buttler, T., W. Martinkovic, and O.N. Nesheim. 1993. Factors influencing pesticide movement to ground water. Univ. of Florida, IFAS Coop. Ext. Serv. Pesticide Information Sheet RF-WQ-107.

Nesheim, O.N. Rev. 1998. Proper disposal of pesticide wastes. Univ. of Florida, IFAS Coop. Ext. Serv. Publ. PI-18.

Nesheim, O.N. 1993. Best management practices to protect groundwater from agricultural pesticides. Univ. of Florida, IFAS Coop. Ext. Serv. Fact Sheet RF-WQ108.

Rao, P.S.C. and A.G. Hornsby. Rev. 1999. Behavior of pesticides in soils and water. Univ. of Florida, IFAS Coop. Ext. Serv. Fact Sheet SL-40.

7. Water Resources of Florida

Boman, B.J. and D.R. Justice. 1992. Tapping shallow groundwater with horizontal wells, p. 45-50. In: E. T. Engman (ed.). Irrigation and drainage: Saving a threatened resource - in search of solutions. Proc. ASCE Water Forum '92. ASCE, New York.

Clark, G.A., A.G. Smajstrla, F.S. Zazueta, F.T. Izuno, B.J. Boman, D.J. Pitts, and D.Z. Haman. 1993. Uses of water in Florida crop production systems. Univ. of Florida, IFAS Coop. Ext Serv. Circ. 940.

Clark, G.A., F.S. Davies and M.A. Maurer. 1993. Reclaimed wastewater for irrigation of citrus in Florida. sHortTechnology 3:163-167.

Fitzpatrick, E.A. 1974. An introduction to soil science. Longman Singapore, Singapore.

Graham, W.D. 1991. Florida's groundwater resource: Vast quantity, good quality? Univ. of Florida, IFAS Coop. Ext. Serv. Circ. 944.

Maurer, M.A. and F.S. Davies. 1993. Microsprinkler irrigation of young 'Redblush' grapefruit trees using reclaimed wastewater. HortScience 28:1157-1161.

8. Well Design and Construction

Driscoll, F.G. 1986. Groundwater and wells. Johnson Division, St. Paul, Minn.

Frazee, J.M., Jr. 1985. Water well construction for agricultural needs. The Citrus Industry 66:50-59.

Haman, D.Z., A.G. Smajstrla, and G.A. Clark. 1988. Water wells for Florida irrigation systems. Univ. of Florida., IFAS Coop. Ext. Serv. Circ. 803.

9. Water Quality Parameters

Hem, J.D. 1970. Study and interpretation of the characteristics of natural waters. USGS Water Supply Paper 1473. US GPO, Washington D.C.

10. Water Quality Monitoring

Taylor, L.A., F.T. Izuno, and A.D. Bottcher. 1982. Water quality sampling and analysis instruments and procedures. Univ. of Florida, IFAS Coop. Ext. Serv. Circ. 1040.

11. Physiological Response to Irrigation And Water Stress

Boman, B.J. 1993. Effects of soil moisture depletion levels on Navel orange on microirrigated flatwoods. Proc. Fla. State Hort. Soc.105: 66-70.

Davenport, T.L.1990. Citrus flowering, p. 349-408. In: J. Janick (ed.). Horticultural reviews. Vol. 12. Timber Press, Portland, Ore.

Davies, F.S. and L.G. Albrigo. 1994. Citrus. CAB International Oxon, UK.

Graser, E.A. and L.H. Allen Jr. 1988. Water management for citrus production in the Florida flatwoods. Proc. Fla. State Hort. Soc. 100:126-136.

Simone, G.W. 1998. Post-harvest disease control in citrus (Citrus spp.). IFAS Coop.Ext. Serv. Publ. PDMG-V3-13.

Zekri, M. and L.R. Parsons. 1988. Water relations of grapefruit trees in response to drip, microsprinkler, and overhead sprinkler irrigation. J. American Soc. Hort. Sci. 113 (6): 819-823.

12. Post Harvest Effects of Water and Nutrients

Arpaia, M. L. 1994. Preharvest factors influencing postharvest quality of tropical and subtropical fruit. HortScience 29:982-985.

Boman, B.J. and J.W. Hebb. 1998. Post bloom and summer foliar K effects on grapefruit size. Proc. Fla. State Hort. Soc. 111:128-135.

Embleton, T.W., W.W. Jones, C.K. Labanauskas, and W. Reuther. 1973. Leaf analysis as a diagnostic tool and guide to fertilization, p. 183-210. In: W. Reuther, (ed.). The Citrus Industry, Rev. ed. Vol. 3. Univ. of California. Div. Agr. Sci., Berkeley, Calif.

Embleton, T.W., W.W. Jones, and R.G. Platt. 1975. Plant nutrition and citrus fruit crop quality and yield. HortScience 10:48-50.

Koo, R.C.J. and R.L. Reese. 1977. Influence of nitrogen, potassium and irrigation on citrus fruit quality. Proc. Int. Soc. Citriculture 1:34-38.

Koo, R.C.J. 1988. Fertilization and irrigation effects on fruit quality, p. 35-42. In: J. J. Ferguson and W.F. Wardowski (eds.). Factors affecting fruit quality. Univ. of Florida, IFAS Coop. Ext. Serv.

Monselise, S.P. and R. Goren. 1987. Preharvest growing conditions and postharvest behavior of subtropical and temperate-zone fruits. HortScience 22:1185-1189.

Reitz, H.J. and T.W. Embleton. 1986. Production practices that influence fresh fruit quality, p. 49-77. In: W. F. Wardowski, S. Nagy, and W. Grierson (eds.). Fresh citrus fruits. AVI Publ. Co., Westport, Conn.

Tucker, D.P. H., A.K. Alva, L.K. Jackson, and T.A. Wheaton. 1995. Nutrition of Florida citrus trees. Univ. of Florida, IFAS Coop. Ext. Serv. Publ. SP 169.

13. Salinity Problems

Boman, B.J. 1993. First-year response of 'Ruby Red' grapefruit on four rootstocks to fertilization and salinity. Proc. Fla. State Hort. Soc. 106:12-18.

Boman, B.J., J.P. Syvertsen, and D.P.H. Tucker. 1990. Salinity considerations in Florida citrus production. The Citrus Industry 71(5):66-68, 70.

Maas, E.V. 1992. Salinity and citriculture. Proc.VII Intl. Soc. Citricult. 3: 1290-1301.

Wadleigh, C.H. and J. Ayers. 1945. Plant physiology. Am. Soc. Plant Physiol., Rockville, Md.

Wander, I.W. and H.J. Reitz. 1951. The chemical composition of irrigation water used in citrus groves. Univ. of Florida, IFAS Coop. Ext.Serv. Bul. 480.

14. Soil and Water Relationships

Bouma, J., R.B. Brown, and P.S.C. Rao. 1982. Basics of soil-water relationships - Part I. Soil as a porous medium. Univ. of Florida, IFAS Soil Science Fact Sheet SL-37.

Bouma, J., R.B. Brown, and P.S.C. Rao. 1982. Basics of soil-water relationships - Part II. Retention of water. Univ. of Florida, IFAS Soil Science Fact Sheet SL-38.

Bouma, J., P.S.C. Rao, and R.B. Brown. 1982. Basics of soil-water relationships - Part III. Movement of water. Univ. of Florida, IFAS Soil Science Fact Sheet SL-39.

Brady, N.C. 1974. The nature and properties of soils. Macmillan, New York.

Fitzpatrick, E.A. 1974. An introduction to soil science. Longman Singapore, Singapore.

Kramer, P.J. and J.S. Boyer. 1995. Water relations of plants and soils. Academic Press, New York.

Obreza, T.A. and B.J. Boman. 1992. Simulated citrus water use from shallow groundwater, p.177-182. In: E. T. Engman (ed.). Irrigation and drainage: Saving a threatened resource - in search of solutions. Proc. ASCE Water Forum '92. ASCE, New York.

15. Soil Water Measuring Devices

Baker, J.M. and R.R. Allmaras. 1990. System for automating and multiplexing soil moisture measurement by time-domain reflectometry. J. Soil Sci. Soc. Am. 54(1):1-6.

Bouyoucos, G.J. and A.H. Mick. 1948. A comparison of electric resistance units for making a continuous measurement of soil moisture under field conditions. Plant Physiology 23: 532-543.

Dean, T.J., J.P. Bell and A.J.B. Baty. 1987. Soil moisture measurement by an improved capacitance technique, Part I. Sensor design and performance. J. of Hydrology. 93:67-78.

Erbach, D.C. 1983. Measurement of soil moisture and bulk density. ASAE Paper No. 83-1553.

Gardner, W.H. and D. Kirkham. 1952. Determination of soil moisture by neutron scattering. Soil Sci. 73: 391-401.

Smajstrla, A.G., D.S. Harrison, and F.X. Duran. 1985. Tensiometers for soil moisture measurement and irrigation scheduling. Univ. of Florida, IFAS Coop. Ext. Serv. Circ. 487.

Smajstrla, A.G. and D.S. Harrison. 1984. Measurement of soil water for irrigation management. Univ. of Florida., IFAS Coop. Ext. Serv. Circ. 532.

Smajstrla, A.G. and D.J. Pitts. 1997. Tensiometer service, testing and calibration. Univ. of Florida, IFAS Coop. Ext. Serv. Bul. 319.

16. Evapotranspiration

Clark, G.A. , A.G. Smajstrla, and F.S. Zazueta. 1989. Atmospheric parameters which affect evapotranspiration. Univ. of Florida, IFAS Coop. Ext. Serv. Circ. 822.

Jensen, M.E., R.D. Burman, and R.G. Allen (eds.). 1990. Evapotranspiration and irrigation water requirements. Amer. Soc. Civil Engineers, New York.

Jones, J.W., L.H. Allen, S.F. Shih, J.S. Rogers, L.C. Hammond, A.G. Smajstrla, and J.D. Martsolf. 1984. Estimated and measured evapotranspiration for Florida climate, crops and soils. Univ. of Florida, IFAS Coop. Ext. Serv. Tech. Bul. 840.

Reitz, H.J., L.G. Albrigo, A.H. Allen, Jr., J.F. Bartholic, D.V. Calvert, H.W. Ford, J.F. Gerber, L. C. Hammond, D.S. Harrison, R.C.J. Koo, R.S. Mansell, J.M, Myers, J.S. Rogers, and D.P. H. Tucker. 1981. Water requirements for citrus. Univ. of Florida, IFAS Coop. Ext. Serv. Publ. WRC-4.

Smajstrla, A.G. and F.S. Zazueta. 1995. Estimating crop irrigation requirements for irrigation system design and consumptive use permitting. Univ. of Florida, IFAS Coop. Ext. Serv. Fact Sheet AE-257.

USDA. 1970. Irrigation water requirements. Engineering Division, Soil Conservation Service, Tech. Release No. 21.

17. Citrus Water Use and Irrigation Scheduling

Boman, B.J. 1994. Evapotranspiration from young Florida flatwoods citrus trees. J. Irrig. and Drain. Eng. 120(1):81-88.

Boman, B.J. 1994. Microirrigation management considerations for flatwoods citrus. Proc. Soil Crop Sci. Soc. Florida 53:3-9.

Boman, B.J. 1997. Lysimeter ET measurements for developing 'Valencia' orange trees. IRREC Research Rept. 97-03. Indian River Research and Education Center, Ft. Pierce, Fla.

Boman, B.J. 1997. Effects of microirrigation frequency on Florida grapefruit. Proc. VIII Intl. Soc. Citricult. 2: 678-382.

Boman, B.J., Y. Levy, and L. Parsons. 1999. Water management, p.72-81. In: L.W. Timmer and L. W. Duncan (eds.). Citrus health management. APS Press, St. Paul, Minn.

Cooney, J.J. and J.E. Peterson. 1955. Avocado irrigation. California Agr. Exp. Sta. Leaflet 50.

Graser, E.A. and L.H. Allen Jr. 1987. Water management for citrus production in the Flatwoods. Proc. Fla. State Hort. Soc. 100:126-136.

Graser, E.A. and L.H. Allen Jr. 1988. Water relations of 7-year-old containerized citrus trees under drought and flooding stress. Proc. Soil Crop Sci. Fla. 47:165-174.

Koo, R.C.J. 1963. Effects of frequency of irrigations on yield of orange and grapefruit. Proc. Fla. State Hort. Soc. 76:1-5.

Koo, R.C.J. 1978. Response of densely planted Hamlin orange on two rootstocks to low volume irrigation. Proc. Fla. State Hort. Soc. 91:8-10.

Koo, R.C.J. 1985. Response of 'Marsh' grapefruit to drip, under tree spray and sprinkler irrigation. Proc. Fla. State Hort. Soc. 98:29-32.

Parsons, L.R. and R.C.J. Koo. 1993. Irrigation scheduling tables for Florida citrus. Univ. of Florida, IFAS Coop. Ext. Serv. Publ. HS166.

Parsons, L.R. 1989. Management of microirrigation systems for Florida citrus. Univ. of Florida, IFAS Coop. Ext. Serv. Fact Sheet FC-81.

Smajstrla, A.G., D.S. Harrison, F.S. Zazueta, L. R. Parsons, and K.G. Stone. 1987. Trickle irrigation scheduling for Florida. Univ. of Florida, IFAS Coop. Ext. Serv. Bul. 208.

Smajstrla, A.G., B.J. Boman, G.A. Clark, D.Z. Haman, F. T. Izuno, and F.S. Zazueta. 1988. Basic irrigation scheduling in Florida. Univ. of Florida, IFAS Coop. Ext. Serv. Bul. 249.

Smajstrla, A.G., D.S. Harrison, and G.A. Clark. 1986. Trickle irrigation scheduling 1: Durations of water applications. Univ.of Florida, IFAS Coop. Ext. Serv. Bul. 204.

Smajstrla, A.G., B.J. Boman, G.A. Clark, D.Z. Haman, D. S. Harrison, F.T. Izuno, D.J. Pitts, and F.S. Zazueta. 1991. Efficiencies of Florida irrigation systems. Univ. of Florida, IFAS Coop. Ext. Serv. Bul. 247.

Smajstrla, A.G. and R.C.J. Koo. 1984. Effects of trickle irrigation methods and amounts of water applied on citrus yield. Proc. Fla. State Hort. Soc. 97:3-7.

Smajstrla, A.G., L.R. Parsons, K. Aribi, and G.Velledis. 1985. Response of young citrus trees to irrigation. Proc. Fla. State Hort. Soc. 98: 25-28.

Tucker, D.H.P. 1985. Citrus irrigation management. Univ. of Florida, IFAS Coop. Ext. Serv. Circ. 444.

Zekri, M. and L.R. Parsons. 1988. Water relations of grapefruit trees in response to drip, micro-sprinkler, and overhead sprinkler irrigation. J. Amer. Soc. Hort. Sci. 113(6):819-823.

Zekri, M. and L.R. Parsons. 1989. Grapefruit leaf and fruit growth in response to drip, micro-sprinkler, and overhead sprinkler irrigation. J. Amer. Soc. Hort. Sci. 114(1):25-29.

18. *Weather Data for Irrigation Management*

Barfield, B.J. and J.F. Gerber. 1979. Modification of the aerial environment of plants. ASAE Monograph No. 2. St. Joseph, Mich.

Butson, K.D. and G.M. Prine. 1968. Weekly rainfall probabilities in Florida. Univ. of Florida, IFAS Coop Ext. Circ. S-187.

Parsons, L.R. and J. Jackson. 1999. Freeze protection—the Florida automated weather network and minimum temperature estimation for frost protection. Citrus Industry 80(11):12-14.

Smajstrla, A.G., G.A. Clark, S.F. Shih, F.S. Zazueta, and D.S. Harrison. 1984. Potential evapotranspiration probabilities and distributions in Florida. Univ. of Florida, Coop. Ext. Serv. Bul. 205.

19. *Drainage*

Boman, B.J., L.R. Parsons, D.P. H. Tucker, and S. H. Futch. 1995. Assessing damage from flooding in citrus groves. The Citrus Industry 76(12):28-30.

Ford, H.W. 1979. The complex nature of ochre. Kulturtechnik and Flurbereinigung 20: 226-232.

Ford, H.W. 1979. Characteristics of slimes and ochre in drainage and irrigation systems. Trans. ASAE 22(5):1093-1096.

Ford, H.W. 1982. Biological clogging of drain envelopes. Proc. 2nd Int. Drainage Workshop 1:215-220.

Ford, H.W. 1982. Estimating the potential for ochre clogging before installing drains. Trans. ASAE 25(6):1597-1600.

Ford, H.W., C.B. Beville, and V. W. Carlisle. 1985. A guide for plastic tile drainage in Florida citrus groves. Univ. of Florida, IFAS Coop. Ext. Serv. Circ. 661.

Graser, E.A. and L.H. Allen Jr. 1988. Water relations of 7-year-old containerized citrus trees under drought and flooding stress. Proc. Soil Crop Sci. Fla. 47:165-174.

Kuntze, H. 1982. Iron clogging in soils and pipes—analysis and treatment. Pitman Publishing, Marshfield, Mass.

Reitz, H.J. and W.T. Long. 1955. Water table fluctuations and depth of rooting of citrus trees in the Indian River area. Proc. Fla. State Hort. Soc. 68: 24-29.

20. *Water Table Measurement and Monitoring*

Boman, B.J. 1990. A low-cost water table indicator for shallow water table observation. Applied Eng. Agr. 6(3):305-307.

Boman, B.J. 1987. Effects of soil series on shallow water table fluctuations in bedded citrus. Proc. Fla. State Hort. Soc. 100:137-141.

Izuno, F.T., D.Z. Haman and G.A. Clark. 1988. Water table monitoring. Univ. of Florida, IFAS Coop. Ext. Serv. Bul. 251.

Izuno, F.T., G.A. Clark, D.Z. Haman, A.G. Smajstrla and D. J. Pitts. 1989. Manual monitoring of farm water tables. Univ. of Florida, IFAS Coop Ext. Serv. Circ. 731.

Obreza, T.A. 1988. Water table behavior under multi-row citrus beds. Proc. Fla. State Hort. Soc. 101:53-58.

Obreza, T.A. and K.E. Admire. 1985. Shallow water table fluctuation in response to rainfall, irrigation, and evapotranspiration in Flatwoods citrus. Proc. Fla. State Hort. Soc. 98: 32-37.

21. *Hydraulics*

Boswell, M.J. 1990. Micro-irrigation design manual. James Hardie Irrigation, Inc., El Cajon, Calif.

Clark, G.A., A.G. Smajstrla, and D.Z. Haman. 1994. Water hammer in irrigation systems. Univ. of Florida, IFAS Coop. Ext. Serv. Circ. 828.

Howell, T.A. , D.S. Stevenson, K.A. Aljibury, H.M. Gitlin, A.W. Warrick, and P. A.C. Raats. 1980. Design and operation of trickle (drip) systems, p. 663-717. In: M.E. Jensen (ed.). Design and operation of farm irrigation systems. ASAE Monograph No. 3. Amer. Soc. of Agr. Eng., St. Joseph, Mich. Hunter Irrigation. 1997. Irrigation hydraulics student manual. Hunter Industries, Inc. San Marcos, Calif.

22. *Water Measurement*

Pair, C.H., W.H. Hinz, K.R. Frost, R.E. Sneed, and T.J. Schiltz. 1983. Irrigation. The Irrigation Assn. Silver Spring, Md.

Smajstrla, A.G. 1998. Shunt flow meters for irrigation water measurement. Univ. of Florida, IFAS Coop. Ext. Serv. Fact Sheet AE-155.

Smajstrla, A.G. 1998. Measuring irrigation water. Univ. of Florida, IFAS Coop. Ext. Serv. Fact Sheet AE-156.

Smajstrla, A.G. and D.S. Harrison. 1982. Selection and use of impeller meters for irrigation water measurement. Univ. of Florida, IFAS Coop. Ext. Serv. Fact Sheet AE-22.

23. Pumps

Haman, D.Z., F.T. Izuno, and A.G. Smajstrla. 1994. Pumps for Florida irrigation and drainage systems. Univ. of Florida, IFAS Coop. Ext. Serv. Circ. 832.

Harrison D.S. and R.E.Choate. 1968. Selection of pumps and power units for irrigation systems in Florida. Univ. of Florida, IFAS Coop. Ext. Serv. Circ. 330.

Holland, F.A. and F.S. Chapman. 1966. Pumping of liquids. Reinhold Publishing, New York.

Hydraulic handbook. 1979. Colt Industries, Fairbanks Morse Pump Division, Kansas City, Kans.

Karassik, I.J., W.C. Krutzsch, W.H. Fraser and J.P. Messina. 1976. Pump handbook. McGraw-Hill, New York.

Longenbaugh, R.A. and H.R. Duke. 1980. Farm pumps, p. 347-391. In: M.E. Jensen (ed.). Design and operation of farm irrigation systems. ASAE Monograph No. 3. Amer. Soc. of Agr. Eng., St. Joseph, Mich.

Pair, C.H., W.H. Hinz, K.R. Frost, R.E. Sneed and T.J. Schiltz. 1983. Irrigation. The Irrigation Assn., Silver Spring, Md.

Smajstrla A.G. and D.S. Harrison. 1982. Net positive suction head (NPSH) required and pump installation. Univ. of Florida, IFAS Coop. Ext. Serv. Publ. AE-82-7.

Walker, R. 1975. Pump selection - a consulting engineer's manual. Ann Arbor Science Publishers, Ann Arbor, Mich.

24. Power Units

Boman, B.J. and M.L. Parsons. 1995. Intake screen plugging effects on microirrigation pump station performance, p. 544-549. In: F. R. Lamm (ed.). Microirrigation for a changing world. Proc. 5th Intl. Microirrigation Cong. ASAE, St. Joseph, Mich.

Smajstrla, A.G. and F.S. Zazueta. 1994. Loading effects on irrigation power unit performance. Univ. of Florida, IFAS Coop. Ext. Serv. Publ. AE-242.

25. Pump and Power Unit Evaluation

Harrison D.S. and R.E. Choate. 1968. Selection of pumps and power units for irrigation systems in Florida. Univ. of Florida, IFAS Coop. Ext. Serv. Circ. 330.

Pair, C.H., W.H. Hinz, K.R. Frost, R.E. Sneed, and T.J. Schiltz. 1983. Irrigation. The Irrigation Assn. Silver Spring, Md.

26. System, Pump, and Engine Maintenance

Cornell Pump Company. 1992. Installation and care of Cornell pumps. Cornell Pump Company, Portland, Ore.

27. Filtration

Dvir, Y. 1997. Flow control devices. Control Appliances Books, Lehavot Habashan, Israel.

Haman, D.Z., A.G. Smajstrla, and F.S. Zazueta. 1989. Screen filters in trickle irrigation systems. Univ. of Florida, IFAS Coop. Ext.Serv. Publ. AE-61.

Haman, D.Z., A.G. Smajstrla, and F.S. Zazueta. 1994. Media filters for trickle irrigation in Florida. Univ. of Florida, IFAS Coop. Ext. Serv. Publ. AE-57.

28. On-Off and Check Valves

Dvir, Y. 1997. Flow control devices. Control Appliances Books, Lehavot Habashan, Israel.

Haman, D.Z., T.T. Izono, and F.S. Zazueta. 1989. Valves in irrigation systems. Univ. of Florida, IFAS Coop. Ext. Serv. Circ. 824.

29. Control, Regulating, and Air Valves

Dvir, Y. 1997. Flow control devices. Control Appliances Books, Lehavot Habashan, Israel.

Haman, D.Z., T.T. Izuno, and F.S. Zazueta. 1989. Valves in irrigation systems. Univ. of Florida, IFAS Coop. Ext. Serv. Circ. 824.

30. Control and Automation

Zazueta, F.S., A.G. Smajstrla, and G.A. Clark. 1993. Irrigation system controllers. Univ. of Florida, IFAS Coop. Ext. Serv. Publ. AGE-32.

31. Materials and Installation

Bliesner, R.D. 1987. Designing, operating, and maintaining piping systems using PVC fittings. Irrigation Association, Arlington, Va.

Microirrigation Committee, Soil and Water Division, ASAE. 1988. EP-405. Design, installation and performance of trickle irrigation systems. In: R.H. Hahn and E.E Rosentreter (eds.). Standards 1988. 35th ed., ASAE, St. Joseph, Mich.

Haman, D.Z. and G.A. Clark. 1995. Fittings and connections for flexible polyethylene pipe used in microirrigation systems. Univ. of Florida, IFAS Coop. Ext. Serv. Fact Sheet AE-69.

Jensen, M.E. 1980. Design and operation of farm irrigation systems. ASAE, St. Joseph, Mich.

Keller, J. and D. Karmeli. 1975. Trickle irrigation design. Rain Bird Sprinkler Manufacturing Corporation, Glenora, Calif.

Nakayama, F.S. and D.A. Bucks. 1986. Trickle irrigation for crop production - Design, operation and management. Elesvier, Amsterdam, Netherlands.

32. Emitter Selection Considerations

Boman, B.J. 1989. Distribution patterns of microirrigation spinner and spray emitters. Applied Eng. Agr. 5(1):50-56.

Boman, B.J. 1989. Emitter and spaghetti tubing effects on microsprinkler flow uniformity. Trans. ASAE 32(1):168-172.

Boman, B.J. 1990. Clogging characteristics of various microsprinkler designs in a mature citrus grove. Proc. Fla. State Hort. Soc. 103:327-330.

Boman, B.J. 1991. Micro-tubing effects on microsprinkler discharge rates. Trans. ASAE. 34(1):106-112.

Boman, B.J. 1995. Effects of orifice size on microsprnkler clogging rates. Applied Eng. Agr. 11(6):839-843.

Boman, B.J. 1996. Effects of microsprinkler pattern area on Florida grapefruit. 1996 Proc. Intl. Soc. Citricult. Dynamic Ad, Nelspruit, South Africa. 2:673-677.

Boman, B.J. and E.O. Ontermaa. 1994. Citrus microsprinkler clogging: Costs, causes, and cures. Proc. Fla. State Hort. Soc. 107:39-47.

Boman, B.J. and M.L. Parsons. 1993. Changes in microsprinkler discharges resulting from long-term use. Applied Eng. Agr. 9(3):281-284.

Boman, B.J. and R.C. Bullock. 1994. Damage to microsprinkler riser assemblies from Selenisa sueroides caterpillars. Applied Eng. Agr. 10(2):221-223.

Boman, B.J., R.C. Bullock, and M.L. Parsons, 1995. Ant damage to microsprinkler pulsator assemblies. Appl. Eng. Agr. 11(6):835-837.

Boman, B.J. and L.R. Parsons. 1999. Microsprinkler experiences in Florida citrus. Applied Eng. in Agr. 15(5): 465-475.

33. Chemigation Equipment and Techniques

Burt, C. and S. Styles. 1994. Drip and microirrigation for trees, vines, and row crops. Irrigation Training and Research Center, California Polytechnic State Univ., San Luis Obispo, Calif.

Clark, G.A., D.Z. Haman, and F.S. Zazueta. 1998. Injection of chemicals into irrigation systems: rates, volumes, and injection periods. Univ. of Florida, IFAS Coop. Ext. Serv. Bul. 250.

Haman, D.Z., F.T. Izuno, and A.G. Smajstrla.1989. Positive displacement pumps for agricultural applications. Univ. of Florida, IFAS Coop. Ext. Serv. Circ. 826.

Haman, D.Z., A.G. Smajstrla, and F.S. Zazueta. 1994. Chemical injection methods for irrigation. Univ. of Florida, IFAS Coop. Ext. Serv. Circ. 864.

Nakayama, F.S. and D.A. Bucks. 1986. Trickle irrigation for crop production: Design, operation, and management. Elsevier Science Publishers, Amsterdam, Netherlands.

Smajstrla, A.G., D.Z. Haman, and F.S. Zazueta. 1992. Calibration of fertilizer injectors for agricultural irrigation systems. University of Florida, IFAS Coop. Ext. Serv. Circ. 1033.

Smajstrla, A.G., D.S. Harrison, W.J. Becker, F.S. Zazueta, and D.Z. Haman. 1985. Backflow prevention requirements for Florida irrigation systems. Univ. of Florida, IFAS Coop. Ext. Serv. Bul. 217.

Smajstrla, A.G., D.Z. Haman, and F.S. Zazueta. 1986. Chemical injection (chemigation): Methods and calibration. Univ. of Florida, IFAS Coop. Ext. Serv. Publ. AE 85-22 (revised).

34. Efficiency, Uniformity, and System Evaluation

Bralts,V.F. and C.D. Kesner. 1982. Drip irrigation field uniformity estimation. ASAE Paper No. 82-2062. ASAE, St. Joseph, Mich.

Burt, C. and S. Styles. 1994. Drip and microirrigation for trees, vines, and row crops. Irrigation Training and Research Center, California Polytechnic State Univ., San Luis Obispo, Calif.

Fischback, P.E. and M.A. Schroeder (eds.). 1982. Irrigation pumping plant performance handbook. 4th ed. University of Nebraska-Lincoln, Nebr.

Smajstrla, A.G., B.J. Boman, G.A. Clark, D.Z. Haman, D.J. Pitts, and F.S. Zazueta. 1990. Field evaluation of microirrigation application uniformity. Univ. of Florida, IFAS Coop. Ext.Serv. Bul. 265.

Smajstrla, A.G., D.S. Harrison, and J.C. Good. 1985. Performance of irrigation pumping systems in Florida. Univ. of Florida, IFAS Coop. Ext. Serv. Circ. 653.

Smajstrla, A.G., D.S. Harrison, and J.M. Stanley. 1984. Evaluating irrigation pumping systems. Univ. of Florida, IFAS Coop. Ext. Serv. Publ. AE-24.

35. Fertigation and Nutrient Management

Burt, C., K. O'Conner, and T. Ruehr. 1995. Fertigation. Irrigation Training and Research Center, California Polytechnic State Univ., San Luis Obispo, Calif.

California Fertilizer Association. 1980. Western fertilizer handbook, 7th ed. The Interstate Printers and Publishers, Ill.

Tucker, D.P. H, A.K. Alva, L.K. Jackson, and T.A. Wheaton (eds.). 1995. Nutrition of Florida citrus trees. Univ. of Florida, IFAS Coop. Ext. Serv. Publ. SP-169.

36. Prevention of Emitter Clogging

Boman, B.J. 1990. Clogging characteristics of various microsprinkler designs in a mature citrus grove. Proc. Fla. State Hort. Soc. 103:327-330.

Boman, B.J. and E.O. Ontermaa. 1994. Citrus microsprinkler clogging: Costs, causes, and cures. Proc. Fla. State Hort. Soc. 107:39-47.

Clark, G.A. and A.G. Smajstrla. 1992. Treating irrigation systems with chlorine. Univ. of Florida, IFAS Coop. Ext. Serv. Circ.1039.

Cowan, J.C. and D.J. Weintritt. 1976. Water-formed scale deposits. Gulf Publishing Co., Houston, Texas.

Ford, H.W. and D.P.H. Tucker. 1975. Blockage of drip irrigation filters and emitters by iron-sulfur-bacterial products. HortScience 10 (1): 62-64.

Ford, H.W. 1979. The present status of research on iron deposits in low pressure irrigation systems. Univ. of Florida, IFAS Coop. Ext. Serv. Publ. FC 79-3.

Ford, H.W. 1979. The use of chlorine in low pressure systems where bacterial slimes are a problem. Univ. of Florida, IFAS Coop. Ext. Serv. Publ. FC 79-5.

Ford, H.W. 1979. A key for determining the use of sodium hypochlorite (liquid chlorine) to inhibit iron and slime clogging of low pressure irrigation systems in Florida. Univ. of Florida, IFAS, Lake Alfred CREC Research Report CS 79-3.

Ford, H.W. 1977. Controlling certain types of slime clogging in drip/trickle irrigation systems. Proc. 7th Intl Agr. Plastics Congr. 1: 118-124.

Ford, H.W. 1987. Iron ochre and related sludge deposits in subsurface drain lines. Univ. of Florida, IFAS Coop. Ext. Serv. Circ. 671.

Ford, H.W. 1979. Using a DPD test kit for measuring free chlorine residual. Univ. of Florida, IFAS Coop. Ext. Serv. Publ. FC79-4.

Gilbert, R.G and H.W. Ford. 1986. Operational principles, p. 159-178. In: F.S. Nakayama and D.A. Bucks (eds.). Trickle irrigation for crop production. Elsevier Science Publishers, Amsterdam, Netherlands.

Haman, D.Z., A.G. Smajstrla, and F.S. Zazueta. 1987. Water quality problems affecting micro-irrigation in Florida. Univ. of Florida, IFAS Coop. Ext. Serv. Publ. AE 87-2.

Nakayama, F.S. and D.A. Bucks. 1986. Trickle irrigation for crop production. Elesvier, Amsterdam, Netherlands.

Pitts, D.J., D.Z. Haman, and A.G. Smajstrla. 1990. Causes and prevention of emitter plugging in microirrigation systems. Univ. of Florida, IFAS Coop. Ext. Serv. Bul. 258.

Pitts, D.J., J.A. Ferguson, and J.T. Gilmour. 1985. Plugging characteristics of drip-irrigation emitters using backwash from a water-treatment plant. Univ. of Arkansas, Coop. Ext. Serv. Bul. 880.

Pitts, D.J. and P.L. Tacker. 1986. Trickle irrigation: Causes and prevention of emitter plugging. Univ. of Arkansas, Coop. Ext. Serv. Publ. MP 271

Pitts, D.J. and J. Capece. 1997. Iron plugging of micro irrigation emitters. Citrus and Vegetable Magazine 61(7): 6-9.

Smajstrla, A.G., D.S. Harrison, W.J. Becker, F.S. Zazueta, and D.Z. Haman. 1988. Backflow prevention requirements for Florida irrigation systems. Univ. of Florida, IFAS Coop. Ext. Serv. Bul. 217.

Tyson, A.W. and K.A. Harrison. 1985. Chlorination of drip irrigation systems to prevent emitter clogging. Serv. Univ. of Georgia. Coop. Ext. Serv. Publ. MP 183.

Smajstrla, A.G. and B.J. Boman. 1998. Flushing procedures for microirrigation systems. Univ. of Florida, IFAS Coop. Ext. Serv. Publ. AE-98-6.

Zazueta, F.S. and A.G. Smajstrla. 1985. Diagnosis and treatment of iron and slime clogging problems. Univ. of Florida, IFAS Coop. Ext. Serv. Circ. 585.

37. *Aquatic Weed Control*

Thayer, D.D., K.A. Langeland, W.T. Haller, and J.C. Joyce. Weed control in aquaculture and farm ponds. Univ. of Florida, IFAS Ext. Serv. Circ.707.

Langland, K. 1991. Aquatic pest control training manual (for aquatic category exam). Univ. of Florida, IFAS Ext. Serv. Publ. SM 3.

Vandiver, V.V., Jr. 1998. Florida citrus aquatic weed control guide. Univ. of Florida, IFAS Coop. Ext. Serv. Publ. SP-168.

38. *Microsprinkler Irrigation for Cold Protection*

Buchanan, D.W., F.S. Davies, and D. S. Harrison. 1982. High and low volume under-tree irrigation for citrus cold protection. Proc. Fla. State Hort. Soc. 95:23-26.

Parsons, L.R., T.A. Wheaton, N.D. Faryna, and J.L. Jackson. 1991. Elevated microsprinklers improve protection of citrus trees in an advective freeze. HortScience 26(9): 1149-1151.

Parsons, L.R. and T.A. Wheaton. 1987. Microsprinkler irrigation for freeze protection: Evaporative cooling and extent of protection in an advective freeze. J. Amer. Soc. Hort. Sci. 112:897-902.

Parsons, L.R., T.A. Wheaton, D.P.H. Tucker, and J.D. Whitney. 1982. Low volume microsprinkler irrigation for citrus cold protection. Proc. Fla. State Hort. Soc. 95:20-23.

39. *Economic Considerations*

Haeussler E.F., Jr. and R. Paul. 1993. Introductory mathematical analysis, 7th ed., Prentice-Hall, Englewood Cliffs, N.J.

McConnell, C.R. and S.L. Brue. 1996. Microeconomics, 13th ed., McGraw- Hill, New York.

Index

relationship to dry bulb and dew point temperatures, 472, 529-532
Wetlands, 13-14, 27-29, 567
    Clean Water Act on, 20, 27-28
    dredge and fill activities in, 20, 27-28, 36
    endangered species in, 23
    Environmental Resource Permit for activities affecting, 36, 37, 38-39
    Everglades Forever Act on, 28-29
    federal regulations on, 39
    soil and plant characteristics identifying, 28
    surface water management in, 13-14
Wetting patterns of emitters. *See* Emitters, wetting patterns of
    Wildlife
    dissolved oxygen levels affecting, 45, 94, 97, 98, 456
    Fish and Wildlife Conservation Commission responsibilities for, 25, 49, 453
    nitrogen levels in water affecting, 45, 94, 107
    pesticide use affecting, 56, 58-64
        in herbicide use for aquatic weeds, 456-459
    pH range of water affecting, 90-91
    sentinel species in, 60
    threatened or endangered species in, 11, 23-24, 565
    in wetland areas, 14, 23
Willows, 454, 468
Wilt
    in drought-induced bloom, 116
    in flooding damage, 198
    midday, 112
    permanent, 112
Wilting point in soil water, 112, 137, 139, 177, 567
Wind chill, 472
Wind velocity, 472
    and application efficiency, 402, 403
    and cold protection with microsprinklers, 476
    and evapotranspiration rate, 170
    measurement of, 170, 171-172
        sensors in, 346, 348
Winder soil, 147

Y
Y-valves, 311, 312, 321

Z
Zinc, 89, 106
    in fertigation, 421, 423